Methods in Cell Biology

VOLUME 50
Methods in Plant Cell Biology, Part B

ASCB

Series Editors

Leslie Wilson
Department of Biological Sciences
University of California, Santa Barbara
Santa Barbara, California

Paul Matsudaira
Whitehead Institute for Biomedical Research and
Department of Biology
Massachusetts Institute of Technology
Cambridge, Massachusetts

Methods in Cell Biology

Prepared under the Auspices of the American Society for Cell Biology

VOLUME 50
Methods in Plant Cell Biology, Part B

Edited by

David W. Galbraith
Department of Plant Sciences
University of Arizona
Tucson, Arizona

Don P. Bourque
Department of Biochemistry
University of Arizona
Tucson, Arizona

Hans J. Bohnert
Department of Biochemistry
University of Arizona
Tucson, Arizona

ACADEMIC PRESS

San Diego   New York   Boston   London   Sydney   Tokyo   Toronto

Cover photograph (paperback edition only): From Chapter 20 by Carrington. For details see the legend to Fig. 2.

This book is printed on acid-free paper. ∞

Academic Press, Inc.
A Division of Harcourt Brace & Company
525 B Street, Suite 1900, San Diego, California 92101-4495

United Kingdom Edition published by
Academic Press Limited
24-28 Oval Road, London NW1 7DX

International Standard Serial Number: 0091-679X

International Standard Book Number: 0-12-564152-4 (hardcover)

International Standard Book Number: 0-12-273872-1 (comb)

PRINTED IN THE UNITED STATES OF AMERICA
95 96 97 98 99 00 EB 9 8 7 6 5 4 3 2 1

CONTENTS

PART I   Subcellular Fractionation and Analysis of Function

1. Flow Cytometric Analysis of Transgene Expression in Higher Plants: Green Fluorescent Protein

David W. Galbraith, Robert J. Grebenok, Georgina M. Lambert, and Jen Sheen

2. Functional Effects of Structural Changes in Photosystem II as Measured by Chlorophyll Fluorescence Kinetics

Wim F. J. Vermaas

3. Determination of Protein Isoprenylation *in Vitro* and *in Vivo*

Jian-Kang Zhu, Ray A. Bressan, and Paul M. Hasegawa

4. Extraction and Assay of Protein from Single Plant Cells

William H. Outlaw, Jr.

5. Photoaffinity Labeling and Strategies for Plasma Membrane
Protein Purification

Joachim Feldwich, Andreas Vente, Narciso Campos, Rolf Zettl, and Klaus Palme

6. Cell Cycle Synchronization, Chromosome Isolation, and Flow-Sorting
in Plants

Sergio Lucretti and Jaroslav Doležel

7. Principles and Applications of Recombinant Antibody Phage Display
Technology to Plant Biology

William L. Crosby and Peter Schorr

PART II Molecular Methods for Analysis of Cell Function *in Vivo*

26. Electroporation of Plant Protoplalsts and Tissues

George W. Bates

27. Particle Bombardment

Paul Christou

28. Preparation and Transformation of Monocot Protoplasts

C. Maas, C. Reichel, J. Schell, and H.-H. Steinbiß

29. Tobacco Protoplast Transformation and Use for Functional Analysis of
Newly Isolated Genes and Gene Constructs

Regina Fischer and Rüdiger Hain

CONTRIBUTORS

Numbers in parentheses indicate the pages on which the authors' contributions begin.

Shunnosuke Abe (209, 223), Laboratory of Molecular Cell Biology, College of Agriculture, Ehime University, Matsuyama 790, Japan

Yoram Altschuler[1] (497), Department of Plant Genetics, The Weizmann Institute of Science, Rehovot 76100, Israel

George W. Bates (363), Department of Biological Science, Florida State University, Tallahassee, Florida 32306

Rebecca S. Boston (309), Department of Botany, North Carolina State University, Raleigh, North Carolina 27695

Barbara Brady (177), Department of Molecular and Cell Biology, University of California, Berkeley, Berkeley, California 94720

Ray A. Bressan (31), Center for Plant Environmental Stress Physiology, Purdue University, Lafayette, Indiana 47907

Narciso Campos (51), Max-Planck Institute für Züchtungsforschung, D-50829 Köln, Germany

W. Zacheus Cande (177), Department of Molecular and Cell Biology, University of California, Berkeley, Berkeley, California 94720

James C. Carrington (283), Department of Biology, Texas A&M University, College Station, Texas 77843

Aldo Ceriotti (295, 497), Istituto Biosintesi Vegetali, Consiglio Nazionale delle Ricerche, I-20133 Milano, Italy

Paul Christou (375), Laboratory for Transgenic Technology and Metabolic Pathway Engineering, John Innes Center, Norwich, United Kingdom

William L. Crosby (85), Molecular Genetics Group, Plant Biotechnology Institute, National Research Council of Canada, Saskatoon, Saskatchewan, Canada S7N 0W9

John C. Cushman (113), Department of Biochemistry and Molecular Biology, Oklahoma State University, Stillwater, Oklahoma 74078

Eric Davies (209, 223), School of Biological Sciences, University of Nebraska, Lincoln, Nebraska 68588

Marcella de Silvestris (295), Dipartimento di Farmacologia, Università di Milan, I-20129 Milano, Italy

Jürgen Denecke[2] (335), Department of Molecular Genetics, Uppsala Genetic Center, Swedish University of Agricultural Sciences, S-75007 Uppsala, Sweden

[1] Present Address: Department of Anatomy and Cardiovascular Research Institute, University of California School of Medicine, San Francisco, California 94143

[2] Present Address: The Plant Laboratory, Department of Biology, University of York, Heslington, York Y01 5DD, United Kingdom

Natalie D. DeWitt[3] (129), Department of Horticulture, University of Wisconsin, Madison, Wisconsin 53706

Jaroslav Doležel (61), Norman Borlang Centre for Plant Science, Institute of Experimental Botany, De Montfort University, CZ-77200 Olomouc, Czech Republic

Bernard L. Epel (237), Botany Department, George S. Wise Faculty of Life Sciences, Tel Aviv University, Tel Aviv 69978, Israel

Michael Erlanger (237), Botany Department, George S. Wise Faculty of Life Sciences, Tel Aviv University, Tel Aviv 69978, Israel

Siegfried Feldmar (461), Institute of Genetics, University of Cologne, D-50931 Cologne, Germany

Joachim Feldwisch (51), Max-Planck Institute für Züchtungsforschung, D-50829 Köln, Germany

Regina Fischer (401), PF-F/Biotechnologie, Bayer AG, D-51368 Leverkusen, Germany; and Institute of Plant Physiology, University of Hohenheim, D-70593 Stuttgart, Germany

Elizabeth P. B. Fontes (309), BIOAGRO-Sector de Biologia Molecular de Plantas, Universidade Federal de Viçosa, Viçosa MG 36570.000, Brasil

Heidi Fusswinkel (461), Institute of Genetics, University of Cologne, D-50931 Cologne, Germany

David W. Galbraith (3), Department of Plant Sciences, University of Arizona, Tucson, Arizona 85721

Gad Galili (497), Department of Plant Genetics, The Weizmann Institute of Science, Rehovot 76100, Israel

Chritiane Gatz (411), Institute für Genetik, Universität Bielefeld, D-33501 Bielefeld, Germany

Jeffrey W. Gillikin (309), Department of Botany, North Carolina State University, Raleigh, North Carolina 27695

Elzbieta Glaser (269), Department of Biochemistry, Arrhenius Laboratories for Natural Science, Stockholm University, S-106 91 Stockholm, Sweden

Robert J. Grebenok (3), Department of Plant Sciences, University of Arizona, Tucson, Arizona 85721

Jean Michel Grienenberger (161), Institut de Biologie Moléculaire des Plantes du CNRS, Université Louis Pasteur, F-67084 Strasbourg, France

Jose Manuel Gualberto (161), Institut de Biologie Moléculaire des Plantes, Université Louis Pasteur, F-67084 Strasbourg, France

Tom J. Guilfoyle (101), Department of Biochemistry, University of Missouri, Columbia, Missouri 65211

Noureddine Hadjab (189), Waksman Institute, Rutgers University, Piscataway, New Jersey 08855

Rüdiger Hain (401), PF-F/Biotechnologie, Bayer AG, D-51368 Leverkusen, Germany

Hirokazu Handa[4] (161), Institut de Biologie Moléculaire des Plantes, Université Louis Pasteur, F-67084 Strasbourg, France

[3] Present Address: Department of Biology, University of York, Heslington, York Y01 5DD, United Kingdom

[4] Present Address: Department of Cell Biology, National Institute of Agrobiological Resources, Tsukuba Science City 305, Japan

Paul M. Hasegawa (31), Center for Plant Environmental Stress Physiology, Purdue University, Lafayette, Indiana 47907

Marie Hugosson (269), Department of Biochemistry, Arrhenius Laboratories for Natural Science, Stockholm University, S-106 91 Stockholm, Sweden

Manfred Klemm (425), Max-Planck-Institut für Züchrungsforschung, D-50829 Köln, Germany

Carina Knorpp (269), Department of Biochemistry, Arrhenius Laboratories for Natural Science, Stockholm University, S-106 91 Stockholm, Sweden

Guy Kotlizky (237), Botany Department, George S. Wise Faculty of Life Sciences, Tel Aviv University, Tel Aviv 69978, Israel

Bella Kuchuck (237), Botany Department, George S. Wise Faculty of Life Sciences, Tel Aviv University, Tel Aviv 69978, Israel

Reinhard Kunze[5] (461), Institute of Genetics, University of Cologne, D-50931 Cologne, Germany

Georgina M. Lambert (3), Department of Plant Sciences, University of Arizona, Tucson, Arizona 85721

Garrett J. Lee (325), Department of Biochemistry, University of Arizona, Tucson, Arizona 85721

Sergio Lucretti (61), ENEA Research Centre Casaccia, Biotechnology and Agriculture Sector, 00060 S. M. di Galeria (Rome), Italy

C. Maas (383), Abteilung Genetische Grundlagen der Pflanzenzüchtung, Max-Planck-Institut für Züchtungsforschung, D-50829 Köln, Germany

June B. Nasrallah (439), Section of Plant Biology, Division of Biological Sciences, Cornell University, Ithaca, New York 14853

Lee A. Newman (189), Waksman Institute, Rutgers University, Piscataway, New Jersey 08855

William H. Outlaw, Jr. (41), Department of Biological Science, Florida State University, Tallahassee, Florida 32306

Klaus Palme (51), Max-Planck Institute für Züchtungsforschung, D-50829 Köln, Germany

Emanuela Pedrazzini (295), Istituto Biosintesi Vegetali, Consiglio Nazionale delle Ricerche, I-20133 Milano, Italy

C. A. Price (189), Waksman Institute, Rutgers University, Piscataway, New Jersey 08855

Ellen M. Reardon (189), Waksman Institute, Rutgers University, Piscataway, New Jersey 08855

C. Reichel (383), Abteilung Genetische Grundlagen der Pflanzenzüchtung, Max-Planck-Institut für Züchtungsforschung, D-50829 Köln, Germany

Bernd Reiss (425), Max-Planck-Institut für Züchtungsforschung, D-50829 Köln, Germany

G. Eric Schaller (129), Department of Botany, University of Wisconsin, Madison, Wisconsin 53706

[5] Present Address: Institute of Genetics and Microbiology, University of Munich, 80638 Munich, Germany

J. Schell (383), Abteilung Genetische Grundlagen der Pflanzenzüchtung, Max-Planck-Institut für Züchtungsforschung, D-50829 Köln, Germany

Peter Schorr (85), Molecular Genetics Group, Plant Biotechnology Institute, National Research Council of Canada, Saskatoon, Saskachewan, Canada S7N 0W9

Peter H. Schreier (449), Department of Genetic Princples of Plant Breeding, Max-Planck-Institut für Züchtungsforschung, D-50829 Köln, Germany; and Bayer AG, PF-F Institut für Biotechnologie, D-51368 Leverkusen, Germany

Julian I. Schroeder (519), Department of Biology and Center for Molecular Genetics, University of California, San Diego, La Jolla, California 92093

Ramón Serrano (481), Departamento de Biotechnología, Esc. Tec. Sup. Ing. Agronomos, Universidad Politécnica, 46022 Valencia, Spain

Jen Sheen (3), Department of Genetics, Harvard Medical School, and Department of Molecular Biology, Massachusetts General Hospital, Boston, Massachusetts 02114

Shomrat Shurtz (237), Botany Department, George S. Wise Faculty of Life Sciences, Tel Aviv University, Tel Aviv 69978, Israel

Jürgen Soll (255), Botanisches Institut der Universität, 24098 Kiel, Germany

H.-H. Steinbisse (383), Abteilung Genetische Grundlagen der Pflanzenzüchtung, Max-Planck-Institut für Züchtungsforschung, D-50829 Köln, Germany

Peter Steinecke[6] (449), Department of Genetic Principles of Plant Breeding, Max-Planck-Institut für Züchtungsforschung, D-50829 Köln, Germany

Clemens Suter-Crazzolara[7] (425), Max-Planck-Institut für Züchrungsforschung, D-50829 Köln, Germany

Heven Sze (149), Department of Plant Biology, University of Maryland, College Park, Maryland 20742

Mary K. Thorsness (439), Department of Molecular Biology, University of Wyoming, Laramie, Wyoming 82071

Michael E. Vayda (349), Department of Biochemistry, Microbiology, and Molecular Biology, University of Maine, Orono, Maine 04469

Andreas Vente (51), Max-Planck Institute für Züchtungsforschung, D-50829 Köln, Germany

Wim F. J. Vermaas (15), Department of Botany, and Center for the Study of Early Events in Photosynthesis, Arizona State University, Tempe, Arizona 85287

José-Manuel Villalba (481), Departamento de Biología Celular, Facultad de Ciencias, Universidad de Córdoba, 14004 Córdoba, Spain

Alessandro Vitale (295, 335), Istituto Biosintesi Vegetali, Consoglio Nazionale delle Ricerche, I-20133 Milano, Italy

Erik von Stedingk (269), Department of Biochemistry, Arrhenius Laboratories for Natural Science, Stockholm University, S-106 91 Stockholm, Sweden

Karin Waegemann (255), Botanisches Institut der Universität, 24098 Kiel, Germany

John M. Ward (149), Department of Biology, University of California, San Diego, La Jolla, California 92093

[6] Present Address: BIOPLANT GmbH, 29574 Ebstorf, Germany
[7] Present Address: Department of Anatomy and Cell Biology III, 69120 Heidelberg, Germany

Harrison Wein (177), Department of Molecular and Cell Biology, University of California, Berkeley, Berkeley, California 94720

Avital Yahalom (237), Botany Department, George S. Wise Faculty of Life Sciences, Tel Aviv University, Tel Aviv 69978, Israel

Rolf Zettl (51), Max-Planck Institute für Züchtungsforschung, D-50829 Köln, Germany

Jian-Kang Zhu (31), Center for Plant Environmental Stress Physiology, Purdue University, Lafayette, Indiana 47907

PREFACE

It is taken for granted that research scientists should employ the most powerful, most appropriate, and most up-to-date methods to investigate specific questions in biology. Yet biologists, as a group, have an ambivalent relationship with the methods they employ. This situation may reflect the observation that, in many cases, those methods providing the most powerful insights arise not from biology but from other scientific disciplines, notably physics, chemistry, and the mathematical sciences. Intelligent use of these techniques tends to require reeducation of the biologist in these disciplines, a process that often encounters a certain degree of inertia. It is also true that the emergence and widespread adoption of a particularly useful experimental method or instrument are not necessarily accompanied by a recognition of the difficulties inherent in the development and reduction to practice of this method. Consequently, the process of methods development within the biological sciences does not enjoy the general degree of support from the research community that is deserved in terms of the benefits that it provides. The elements of creativity, innovation, and sophistication in methods development are seldom appreciated and indeed are frequently the subject of active criticism.

Yet, if traced from the historical record, the biological sciences are replete with illustrations of the profound influence of novel methods and techniques on biological understanding. Few would question the impact of radioisotopes on studies of metabolism, or that of simple fixation solutions and embedding resins on revolutionizing techniques of visualizing cellular structure by transmission electron microscopy. Now readily accessible methods and equipment for computer-linked capture, storage, and manipulation of enormous amounts of data have further revolutionized microscopy, delivering a quantum leap in the provision of tools, again derived from other disciplines. Electrophysiological techniques have allowed increasingly greater insight into mechanisms of metabolite and ion transport in individual cells or cell membrane patches. Techniques for manipulation of DNA have fundamentally altered the ways in which biological questions are addressed. The recent development of methods for the amplification of specific gene fragments by iterative cycles of DNA annealing and synthesis has emerged as perhaps the single most powerful, yet elegantly simple, technique of modern biology, and these methods are well on their way to making an indelible mark on culture and society. It is not too difficult to predict that the continued development of computer and communications technologies will lead to increased use of computer-based modeling for analysis of biological questions. These tools should enhance the use and importance of theoretical, predictive approaches within the laboratory setting, in which experimental observation is used less to chart out and categorize unknown components and more to confirm or reject the predictions of these models.

The purpose of Volumes 49 and 50 in the series *Methods in Cell Biology* is threefold. First, we have brought together a comprehensive collection of different methods applicable to plant cell biology. This compilation should permit researchers ready access to those methods that are most appropriate for their work. Second, we have attempted to demystify the individual methods by requesting from the authors a chapter format that clearly and succinctly explains the principles behind the individual methods, as well as providing a step-by-step "cookbook" approach to implementing these methods. Finally, we would like the reader to come away with an appreciation of the breadth and depth of sophisticated thought that went into the development of the individual methods. The diligence with which our colleagues have responded to our charge in dealing with the topics of their expertise is what makes these two volumes particularly important compilations of current technical knowledge. The authors have gone beyond the requisite provision of useful hints and tricks by lucidly explaining the conceptual framework of the techniques. Many chapters outline paths to future improvements. However, the essential theme reiterated within each chapter is that cutting-edge research in plant cell biology requires complex, multifaceted technical expertise and that interdisciplinary scientific efforts are integral to ensuring continued advancements in the field.

The Coeditors thank the Academic Press staff, past and present, for their expert help during the preparation of these volumes. We gratefully acknowledge the contributions of the authors, who promptly provided manuscripts requiring minimal editorial changes, and we look forward eagerly to the discoveries that will be made possible through the use of the methods within these volumes.

David W. Galbraith
Don P. Bourque
Hans J. Bohnert

PART I

Subcellular Fractionation and Analysis of Function

CHAPTER 1

Flow Cytometric Analysis of Transgene Expression in Higher Plants: Green–Fluorescent Protein

David W. Galbraith, * **Georgina M. Lambert,** *
Robert J. Grebenok, * **and Jen Sheen** [†]

* Department of Plant Sciences
University of Arizona
Tucson, Arizona 85721
[†] Department of Genetics
Harvard Medical School
and Department of Molecular Biology
Massachusetts General Hospital
Boston, Massachusetts 02114

I. Introduction

The growth and development of higher plants is governed to a large degree by the regulated expression of genes. The availability of robust techniques for the analysis of cell- and tissue-specific gene expression is therefore critical for progress in this area. One approach has been to develop transgenic markers, employing the coding regions of heterologous protein whose expression is placed under the control of various plant-derived regulatory DNA sequences. Biochemical and histological examination of patterns of expression of these markers can then provide information about the way in which gene expression patterns are regulated. Most heterologous protein markers are enzymes, since the amplification step inherent to biochemical assays provides great sensitivity (see Chapter 31, this volume). Probably the most widely-used of these to date in higher plants has been the β-glucuronidase (GUS) activity encoded by the *uidA* locus of *Escherichia coli* (Jefferson, 1987). Particular advantages of the GUS system include a lack of background activities in plants, the fact that the protein is readily synthesized in the cells of different genera of higher plants transformed or transfected with appropriate DNA constructs, the relative stability of the GUS enzyme, and the availability of substrates and antisera for biochemical, histochemical, and immunological assays. The main disadvantage of the GUS system centers around its assay *in vivo*. Thus, although qualitative assays for GUS expression have been devised for a wide variety of living plant tissues, fine-resolution histochemical localization utilizing chromogenic substrates requires tissue dissection and fixation, in order for the substrate to gain even access to the cytoplasmic contents of the different cells within the plant. It has not been possible to devise conditions under which fluorogenic substrates provide quantitative measurement *in vivo* of the distributions and amounts of GUS activities within different cells. This is due to factors including problems of substrate access, a lack of differential retention between fluorogenic substrate and fluorescent product, alterations to product fluorescence levels associated with pH microenvironments, and differential accumulation of product within different subcellular compartments. The ability to be able to quantitatively assay promoter activity *in vivo* on a cell-by-cell basis using marker gene technologies takes on particular importance in view of the availability of flow cytometric techniques for quantitative analysis of fluorescence, accompanied by the subsequent use of cell sorting for the selective purification of minor subpopulations of cells that express particular genes (Darzynkiewicz *et al.*, 1994).

A candidate gene appropriate for the quantitative *in vivo* analysis of gene expression based on fluorescence and flow cytometry has recently been described (Prasher *et al.*, 1992; Chalfie *et al.*, 1994). This gene encodes the green-fluorescent protein (GFP) of *Aequorea victoria*. GFP is typical of a class of photoproteins found in bioluminescent coelenterates. Its primary sequence comprises 238 amino acids, and these undergo a series of post-translational intramolecular reactions, involving cyclization and autoxidation of amino acids 65–67 (Ser–Tyr–Gly) to generate a *p*-hydroxybenzylidine–imidazolidinone fluorescent chromophore (Cody *et al.*, 1993; Heim *et al.*, 1994). The fluorescence spectra of GFP exhibit excitation peaks at 395 and 475 nm and an emission peak at 509 nm with a shoulder at 540 nm. GFP can therefore be readily detected via fluorimetry and fluorescence microscopy, using conventional filter sets. Expression of GFP in prokaryotes results in emission of green fluorescence (Chalfie *et al.*, 1994), and its use as a marker for examining developmentally regulated patterns of gene expression has already been reported for eukaryotic organisms including *Caenorhabditis elegans* and *Drosophila melanogaster* (Chalfie *et al.*, 1994; Wang and Hazelrigg, 1994).

In this chapter, we describe methods for the expression of GFP in higher plant cells, for its detection using flow cytometry, and for the sorting of these cells under conditions that allow recovery of viable cells.

II. Materials

A. Plant Materials

Maize hybrid lines FR9^{cms} X FR37 or FR992 X FR637 (Illinois Foundation Seed, Champaign, IL) were used for all experiments. Seeds were germinated (25 seeds/pot, 6 in. in diameter) and plants grown either in sterile potting soil (Hyponex, Marysville, IL) or a 2:1 vermiculite:peatilite mix (Grace Sierra, Milpitas, CA) within an environmental chamber (Revco Model 122LTP) in darkness at 25°C. Etiolated leaves were harvested 11–12 days after germination. No additions of fertilizer were made, only water, during plantlet growth.

B. Chemicals

Cellulase RS and macerozyme were obtained either from the Yakult Honsha Co. (Tokyo, Japan) or from Karplan Co. (Torrance, CA). Restriction enzymes were obtained from BRL (Grand Island, NY). Geneclean kits were from BIO 101 (La Jolla, CA). All other chemicals were obtained from the Sigma Chemical Company (St. Louis, MO).

C. Plasticware

Sterile plastic 50-ml centrifuge tubes (Falcon 2098) and culture dishes (Falcon 3047) were used for protoplast preparation and transfection. Other disposables included 1.5-ml microcentrifuge tubes, pipet tips, and 12×75-mm test tubes. These were obtained from either VWR Scientific (Phoenix, AZ) or Fisher Scientific (Tustin, CA). Plastic electroporation cuvettes were obtained from BTX (San Diego, CA).

D. Equipment

An IEC Centra 7R refrigerated centrifuge (Damon Labs, Needham Hts., MA) equipped with a swinging-bucket rotor (Cat. No. 269; rotating radius, 7.8 in.) was used during protoplast preparation. A Sorvall RC5-B (Dupont, Wilmington, DE) preparative centrifuge and a Beckman TL-100 ultracentrifuge (Beckman Instruments, Fullerton, CA) were used during DNA preparation. Protoplast electroporation was done using a BTX electroporation system (Model 600; San Diego, CA). Other equipment used during protoplast preparation included a gyratory shaker (Model G2; New Brunswick Instruments, Edison, NJ), and a Boekel aspirator pump (Model 177000; Aldrich Chemical Company, Milwaukee, WI).

PCR was done using a Perkin–Elmer DNA cycler (Perkin–Elmer, Norwalk, CT). For microscopy, we employed a Zeiss standard microscope equipped with epifluorescence (Ploem) illumination and Neofluar 10/0.30, 25/0.060, and 40/0.075 objectives. Filters appropriate for GFP observation were provided by filter set XF-13 (Omega Optical, Brattleboro, VT). This set comprises excitation filter 405DF40, which transmits light with a half-bandwidth of 40 nm centered at 405 nm, dichroic beam-splitter 450DRLPO2, which is centered at 450 nm, and emission filter 450EFLP, which transmits light of wavelengths between 450 and about 670 nm. If necessary, a BG24 blue glass filter was employed to eliminate red light from chlorophyll autofluorescence. Flow cytometry and cell sorting was done using an EPICS Elite (Coulter Electronics, Miami Lakes, FL) equipped with a 20-mW argon-ion laser and a 100-μm-diameter sense-in-quartz flow tip.

III. Methods

A. Recombinant Constructions

(1) 35S-C4ppdk-GFP. Plasmid D66 (constructed by D. C. Prasher) was first linearized by cutting 5' to the GFP coding sequence with *Kpn*I; this site was then rendered blunt-ended by treatment with T4 DNA polymerase. The GFP coding sequence was then excised by restriction with *Pst*I. Plasmid 35SC4pppdk1-CAT (Sheen, 1993), containing the 35SC4ppdk promoter and *nos* terminator, was linearized using *Bam*HI; the site was blunt-ended with T4 DNA polymerase and the plasmid digested with *Pst*I. The GFP fragment was then ligated into this vector.

(2) CAMV35S-GFP. This recombinant plasmid (pRJG2; 4.4 Kb) was formed by excising the coding sequence for GFP from D66, using *Hin*dIII and *Eco*RI.

The 0.75-kb fragment was GeneCleaned and ligated into *Hin*dIII/*Eco*RI-cut pJIT 117 (Guerineau *et al.,* 1988). This places the GFP coding sequence under the transcriptional regulation of a doubled CaMV 35S promoter and the transcriptional terminator and polyadenylation addition signals from the same gene.

Large-scale plasmid preparations were done according to Sambrook *et al.* (1989), and the DNA was stored at −20°C.

B. Protoplast Preparation

Second leaves from 11- to 12-day-old maize plants were excised. Leaves (20–25) were stacked in a sterile plastic petri dish (diameter, 9 cm) and trimmed, removing 4 cm from the top. The remaining portion was cut into 0.5-mm strips using fresh razor blades sterilized with 95% ethanol. The strips were placed in a 250-ml sterile side-arm filtration flask containing 20 ml of the following enzyme solution: 1% cellulase RS, 0.25% macerozyme R10, 0.6 M mannitol, 10 mM 2-[N-morpholino]ethanesulfonic acid (Mes), 1 mM CaCl$_2$, 5 mM β-mercaptoethanol, and 0.1% BSA. Prior to use, the solution was filter-sterilized using a Millipore Millex-GS filter unit (0.22-μm filter). After placing the leaf strips in the enzyme solution, vacuum was applied for 15 min with constant swirling of the flask. The flask was then placed on a rotary shaker at 40 rpm for 2 h. The shaker speed was then increased to 100 rpm for 10 min, and the suspension was filtered through a funnel covered with 60-μm mesh (Tetko, Elmsford, NY) into a 50-ml sterile plastic centrifuge tube (Falcon 2098) and pelleted at 100g for 5 min. The supernatant was removed and the pellet resuspended in electroporation buffer comprising 0.6 M mannitol, 20 mM KCl, 4 mM Mes, and 5 mM EGTA, pH 5.7. The protoplasts were pelleted by centrifugation at 100g for 10 min and then resuspended to a concentration of 1–2 × 10^6/ml.

C. Protoplast Transfection

Protoplasts (1.5 × 10^5) were resuspended in 0.3 ml of electroporation buffer and transferred into a plastic electroporation cuvette (gap width, 4 mm; BTX). Plasmid DNA was added (25–50 μg in 30–60 μl TE). Electroporation conditions were as follows: 10 -ms pulse length, 400–500 V/cm, 200 μF capacitance, and three total pulses. The electroporated protoplasts were placed in plastic six-well culture plates (Falcon 3046) prerinsed with 5% calf serum in electroporation buffer. The protoplasts were kept on ice for 10 min after electroporation and then were diluted by addition of electroporation solution (0.7 ml/well). The tissue culture plates were covered with foil and were incubated at room temperature for 15–18 h.

D. Microscopy

Protoplasts were observed using a Zeiss standard microscope following transfer onto a microscope slide having a "raised coverslip" configuration. This involved cementing two parallel strips (4 × 22 mm) of coverslip glass onto a standard (76 × 25 mm) glass microscope slide. The protoplasts were placed in a droplet (ca. 10–20 μl) between the two coverslip strips, and a second coverslip was placed over them. Photographs were taken using Kodak Ektachrome 160T color film.

E. Flow Cytometry

The Elite flow cytometer was operated with a laser output of 20 mW at 488 nm. Signals were collected for forward-angle light scatter, 90° light scatter (PMT1), and fluorescence (PMTs 2 (green; 505–545 nm) and 4 (red; 670–680 nm)). High-voltage and amplification settings were as follows: FS: 260/10, PMT1 350/7.5, PMT2 850/10, PMT4 950/5. The FALS discriminator was set to 50 and all other discriminators were switched off. A sort-sense flow cell was used having an orifice of 100 μm. The cytometer was aligned using calibration fluorospheres (DNA Check; Coulter Electronics, Miami Lakes, FL). Biparametric histograms of log 90° light-scatter versus log green fluorescence and log red fluorescence (chlorophyll) versus log green fluorescence were collected to a total count of 100,000. The sample flow rate was 100–200/s. The sheath fluid comprised 0.47 M mannitol, 50 mM KCl, 10 mM CaCl$_2$, and 4 mM Mes, pH 5.7, and was filtered through a 0.22-μm filter prior to use.

F. Cell Sorting

The Elite was operated at a sheath pressure of 8.0 psi, a sample pressure of 7.3 psi, a drive frequency of 15 kHz, and a drive amplitude of 11.5%. These settings usually produced a sort stream with six drops free above the deflection plate. Delay settings were optimized by batch sorting groups of 25 particles on a slide and counting the number of particles recovered. Calibration beads (10 μm, DNA Check) and *Lycopodium* spores (28 μm; Polysciences, Warrington, PA) were used for this purpose. Delay settings between 17 and 19 gave stable sort streams at a deflection amplitude of 70%. Samples were sorted at a rate of 25–50 per second into 12 × 75-mm plastic tubes.

IV. Critical Aspects of the Procedures

A. Plant Growth

Strict control of the environmental conditions under which the maize seedlings are grown ensures a reproducible protoplast yield and survival rate after electroporation. Growth media (soil or vermiculite/peatlite) and planting density should be evaluated for seed type to obtain seedlings which demonstrate vigorous growth. Germination rates should be determined for each batch of seeds to ensure adequate numbers of plants available for use. The two maize lines are recommended as particularly suitable for the procedures.

B. Protoplast Transfection

The electroporation procedure must be optimized for each instrument and particular application. This may involve empirical determination of protoplast viability and transfection efficiency for various voltage settings and pulse lengths. The concentration of protoplasts used for electroporation should be evaluated. We found that protoplast concentrations greater than 5×10^5 reduced the survival rate after electroporation. The condition of the starting material is also critical to successful transfection. The second leaves should be well hydrated, uniformly yellow, and absent of any signs of disease. Attention to aseptic technique will reduce problems with contamination introduced during the protoplast preparation. Alternatively, since sterile technique can be time-consuming, bacterial growth can be controlled simply by addition of ampicillin ($100 \mu g/ml$). The vessel into which the protoplasts are placed after electroporation should be shallow enough to allow adequate oxgenation. Tissue culture plates are ideal for this purpose. Prerinsing the culture plates with 5% calf serum reduced problems of adhesion of the protoplasts to the surfaces of the plastic dishes.

C. Flow Cytometry

The first consideration in flow cytometric analysis is verification of proper alignment and instrument performance. This is usually done by checking the data obtained from running calibration beads or other standard cells. Once instrument performance has been verified, nontransfected protoplasts should be analyzed and photomultiplier voltages adjusted to obtain histograms with a clearly identifiable protoplast population. Since in this case the protoplasts are etiolated, it is not practical to trigger the flow cytometer based on red fluorescence (this will not be the case for mature leaf protoplasts). Triggering is thus done based on FALS; however, careful adjustment of the discriminator is needed to exclude contributions from the large numbers of small particles present in the protoplast preparations derived from subcellular debris. If small particles are not excluded from detection by adjustment upward of the discriminator setting, their contribution overwhelms that of the less-numerous protoplasts, and it becomes difficult if not impossible to define their positions within the two-dimensional displays. Empirically, we have found that FALS data rates of 1000–2000/s still permit a full definition of the positions of the protoplast populations. Filtration of the sheath, while time-consuming, will reduce background light scattering and lessen this problem. Finally, protoplasts will settle to the bottom of the sample tube readily; therefore, gentle periodic resuspension is typically necessary.

D. Cell Sorting

In addition to the considerations mentioned above, several aspects are critical when sorting. The sheath and sample pressures must be reduced to permit formation of large droplets (Harkins and Galbraith, 1987). Careful optimization of the delay settings as described under Methods is necessary. It is essential to use calibration particles near to the size of the particle of interest for this process as delay settings will differ for different sized particles (Harkins and Galbraith, 1987). Periodic reanalysis of sorted samples and visual observation under the microscope during the sort procedure ensures detection of any drift in the sort parameters.

V. Results and Discussion

Protoplasts prepared from etiolated maize seedlings contain etioplasts and therefore appear as translucent spheroids, within which can be seen the nucleus, a large vacuole, and cytoplasmic strands containing numerous granules (Color Plate 1A). The protoplasts approximate 25 μm in diameter. When subjected to biparametric analysis using flow cytometry, control (nontransfected) protoplasts characteristically were located as a single cluster having a very low level of green autofluorescence and a somewhat higher level of red autofluorescence (Fig. 1A). This autofluorescence is a consequence of the presence of pigments other than GFP within the cells. Since the degree of chloroplast development is sensitive to the amount of light experienced during seedling growth and during protoplast preparation, we assume that part of the contribution to the autofluorescence comes from photosynthetic pigments within the developing chloroplasts. In contrast, protoplasts examined 10–18 h after transfection with either GFP construct exhibited obvious green fluorescence, which had a cytoplasmic location (Color Plate 1B). Correspondingly, flow cytometric analysis revealed the appearance of a second cluster of protoplasts, exhibiting greatly enhanced green fluorescence but unaltered red fluorescence (Fig. 1B). This is consistent with the supposition that the upper cluster contains protoplasts that are expressing GFP. Integration of the two protoplast clusters indicated that in this experiment approximately 14% fell within the upper cluster and 86% in the lower. In a second series of experiments, we explored the use of other flow cytometric parameters for detection of protoplast transfection. Flow cytometric analysis of nontransfected protoplasts according to surface granularity (90° light scatter) and green fluorescence revealed a single cluster (Fig. 1C), whereas protoplasts analyzed 18 h posttransfection were distributed between two clearly separated clusters (Fig. 1D). The upper cluster in this case contained 34% of the protoplasts and the lower 66%. The fact that 90° light scatter can be employed to detect intact protoplasts and, biparametrically, to monitor GFP expression, may prove useful in applications involving multicolor fluorescent protein expression, as such expression systems come on-line.

We noted a degree of variation between experiments, with protoplasts expressing GFP ranging from 4 to 34% of the total. However, it should be noted that the lower cluster will include nonviable as well as viable, nontransfected

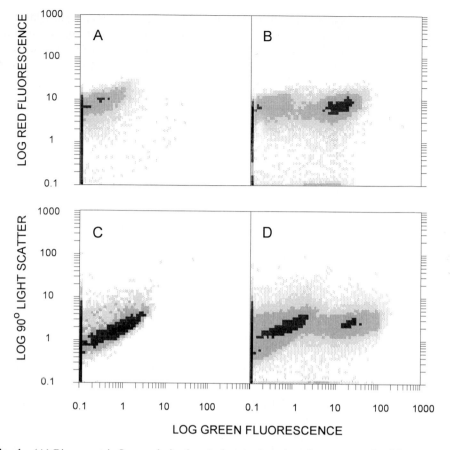

Fig. 1 (A) Biparametric flow analysis of control protoplasts, based on green and red fluorescence emission. (B) Biparametric flow analysis, based on green and red fluorescence emission, of transfected maize protoplasts. (C) Biparametric flow anaylsis of control protoplasts, based on green fluorescence and 90° light scatter. (D) Biparametric flow analysis of transfected protoplasts, based on green fluorescence and 90° light scatter.

protoplasts. The average fluorescence emission intensity from GFP-expressing protoplasts was about 80-fold that of nontransfected cells. Similar results were obtained in transfecting tobacco mesophyll protoplasts by electroporation or using polyethylene glycol (data not shown); the percentage of GFP-expressing protoplasts ranged from 2 to 10%.

Conditions for the sorting of intact maize protoplasts according to GFP expression were based on previously described methods (Galbraith, 1991, 1992, 1994; Galbraith and Lambert, 1995). Sort parameters were optimized using standard microspheres and *Lycopodium* spores (which approximate the size of the maize protoplasts). An amorphous bit-mapped sort window was placed around the upper cluster of protoplasts (Color Plate 2), and the sorting circuitry was enabled. The sorted protoplasts remained intact and exhibited green fluorescence (Color Plate 2, right). Upon reanalysis using flow cytometry, the sorted protoplasts fell within the bit-mapped region that had been employed for defining the sort window (data not shown).

VI. Conclusions and Prospects

Our data indicate that transient expression of GFP in higher plants can be readily achieved by applying conventional transfection methods to protoplasts. Expression of GFP can be detected both by immunofluorescence microscopy and by flow cytometry, and it is subsequently possible to sort protoplasts based on GFP fluorescence. We and others have previously established that protoplasts can be employed as reporters of signal transduction pathways, and that protoplast-derived results can be representative of the tissues from which the protoplasts are prepared (Harkins *et al.,* 1990; Sheen, 1993). This opens up the possibility of employing flow cytometry and cell sorting, coupled to GFP expression following protoplast transfection with expression libraries, as a powerful means for the uncovering of novel trans-activating factors involved in pathways of signal transduction.

Expression of GFP in permanently transformed plants would also provide an entrée into the use of flow cytometry and cell sorting for isolation of specific cells within plants, assuming GFP expression is placed under the control of developmentally interesting promoter sequences, and this could provide a powerful means for the rapid characterization of networks of coordinated gene expression. At present, it is becoming clear that work remains to be done to define conditions for the recovery of transgenic, GFP-expressing plants. In some species (for example, *Arabidopsis thaliana*) it appears that efficient splicing of a cryptic intron prevents expression from the native GFP coding sequence (J. Haseloff, personal communication). Systematic replacement of the consensus splice junction sequences, coupled to an elevation of GC content within the cryptic intron, eliminates splicing and can allow recovery of plants expressing sufficient levels of GFP to be detectable using confocal microscopy. For other species, similar approaches might prove necessary.

Further technical modifications to the GFP system are underway in several laboratories. Reports have already emerged concerning improvements to excitation efficiencies and altered fluorescence emission profiles (Delagrave *et al.,* 1995; Heim *et al.,* 1994, 1995). This would permit multicolor flow cytometric and sorting applications and would be of considerable importance in either of the applications to plant biology outlined above. Finally, exploration of modifications to the half-lives of biosynthesis and destruction of GFP (Heim *et al.,* 1994) should lead to versions of the molecule that more accurately reflect the kinetics of gene expression.

Application of GFP as a fluorescent marker for intracellular movement targeting also offers considerable promise. Through construction of translational fusions, GFP chimeric proteins should allow tracking of cytoplasmic proteins, as well as targeting to intracellular organelles (although the small size of GFP will probably not permit its use, alone, as a marker for nuclear targeting). GFP chimeric proteins that functionally complement mutations within specific genes should provide a uniquely powerful means for the analysis of the subcellular functions of gene products.

In summary, the future appears particularly bright for GFP and for the other representatives, cloned and uncloned, of the cnidarian class of fluorescent proteins in the study of many important and unsolved problems in plant cell and molecular biology.

Acknowledgements

This work was supported by grants to D.W.G. from the N.S.F. Instrumentation and Instrument Development Program (BIR 91-16067) and the N.S.F.-D.O.E.-U.S.D.A. Joint Program of Collaborative Research in Plant Biology (92-20332), and to J.S. of a N.S.F. Career Advancement Award (MCB 94-07834) and by Hoechst AG.

References

Chalfie, M., Tu, Y., Euskirchen, G., Ward, W. W., and Prasher, D. C. (1994). Green-fluorescent protein as a marker for gene expression. *Science* **263**, 802–805.

Cody, C. W., Prasher, D. C., Westler, W. M., Prendergast, F. G., and Ward, W. W. (1993). Chemical structure of the hexapeptide chromophore of the *Aequorea* green-fluorescent protein. *Biochemistry* **32**, 1212–1218.

Darzynkiewicz, Z., Robinson, J. P., and Crissman, H. A. (1994). "Flow Cytometry," 2nd ed., Parts A and B, this series, Volumes 41 and 42.

Delagrave, S., Hawtin, R. E., Silva, C. M., Yang, M. M., and Youvan, D. C. (1995). Red-shifted excitation mutants of the green fluorescent protein. *Bio/Technology* **13**, 151–154.

Galbraith, D. W. (1991). Fluorescence-activated cell sorting of protoplasts and somatic hybrids. *In* "Plant Tissue Culture Manual: Fundamentals and Applications" (J. M. Brevis, ed.), pp. 1–19. Kluwer Academic, Dordrecht/Norwell, MA.

Galbraith, D. W. (1992). Large particle sorting. *In* "Flow Cytometry and Cell Sorting" (A. Radbruch, ed.), pp. 189–204. Springer-Verlag, Berlin.

Galbraith, D. W. (1994). Flow cytometry and sorting of plant protoplasts and cells. *In* "Methods in Cell Biology," Vol. 42, "Methods in Flow Cytometry (Z. Darzynkiewicz, H. Crissman, and J. P. Robinson, eds.), pp. 539–561. Academic Press, San Diego.

Galbraith, D. W., and Lambert, G. M. (1995). Advances in the flow cytometric characterization of plant cells and tissues. *In* "Flow Cytometric Applications in Cell Culture" (M. Al-Rubeai and A. N. Emery, eds.). Marcel Dekker, New York. In press.

Guerineau, F., Woolston, S., Brooks, L., and Mullineaux, P. (1988). An expression cassette for targeting foreign proteins into chloroplasts. *Nucleic Acids Res.* **16**, 11380.

Harkins, K. R., and Galbraith, D. W. (1987). Factors governing the flow cytometric analysis and sorting of large biological particles. *Cytometry* **8**, 60–71.

Harkins, K. R., Jefferson, R. A., Kavanagh, T. A., Bevan, M. W., and Galbraith, D. W. (1990). Expression of photosynthesis-related gene fusions is restricted by cell-type in transgenic plants and in transfected protoplasts. *Proc. Natl. Acad. Sci. U.S.A.* **87**, 816–820.

Heim, R., Prasher, D. C., and Tsien, R. Y. (1994). Wavelength mutations and post-translational autoxidation of green fluorescent protein. *Proc. Natl. Acad. Sci. U.S.A.* **91,** 12501–12504.

Heim, R., Cubitt, A. B., and Tsien, R. Y. (1995). Protein marker for development. *Nature* **373,** 663–664.

Jefferson, R. A. (1987). Assaying chimeric genes in plants: The GUS gene fusion system. *Plant Mol. Biol. Rep.* **5,** 387–405.

Prasher, D. C., Eckenrode, V. K., Ward, W. W., Prendergast, F. G., and Cormier, M. J. (1992). Primary structure of the *Aequorea victoria* green-fluorescent protein. *Gene* **111,** 229–233.

Sambrook, J., Fritsch, E. F., and Maniatis, T. (1989). "Molecular Cloning, a Laboratory Manual," 2nd ed. Cold Spring Harbor Laboratory Press, Cold Spring Harbor, NY.

Sheen, J. (1993). Protein phosphatase activity is required for light-inducible gene expression in maize. *EMBO J.* **12,** 3497–3505.

Wang, S., and Hazelrigg, T. (1994). Implications for bcd mRNA localization from spatial distribution of exu protein in Drosophila oogenesis. *Nature* **369,** 400–403.

CHAPTER 2

Functional Effects of Structural Changes in Photosystem II as Measured by Chlorophyll Fluorescence Kinetics

Wim F. J. Vermaas

Department of Botany and
Center for the Study of Early Events in Photosynthesis
Arizona State University
Tempe, Arizona 85287-1601

I. Introduction

Thylakoid membranes of plants and cyanobacteria contain a number of integral membrane protein complexes, including photosystems I and II (PS I and PS II), the cytochrome $b_6 f$ complex, and ATP-synthase. These four complexes together are involved in the catalysis of the light-induced production of ATP and NADPH, which are utilized for carbon fixation. Membrane–protein complexes generally are poorly understood in terms of relationships between structure and function.

This is due in part to difficulties in purifying membrane–protein complex to homogeneity without a significant loss of properties and activity, and to pleiotropic effects usually accompanying mutations in one of the subunits of a membrane–protein complex. However, with the advent of molecular-genetic techniques to introduce targeted mutations or deletions, a detailed study of the effects of protein subunits and of individual domains and residues in a protein subunit now has become possible [for example, see reviews by Bryant (1992), Nixon *et al.* (1992a), and Vermaas (1993)]. The membrane–protein complex that has been analyzed in most detail by a combination of molecular–genetic and biochemical and biophysical techniques is PS II, the pigment–protein complex that is responsible for the first steps in the photosynthetic electron transport chain, which lead to the light-induced oxidation of water (resulting in oxygen) and reduction of plastoquinone.

PS II consists of about 20 polypeptides (of which at least 13 are integral membrane proteins), 30–50 chlorophyll *a* molecules, and several carotenoids. The total molecular mass of the PS II complex exceeds 300,000. Over the years, a large amount of information regarding the structure and function of the PS II complex has accumulated, mostly because of the fact that a number of sensitive techniques have been developed for in-depth studies of PS II function. These analysis techniques take advantage of the fact that PS II redox reactions are induced by single photons of light, and thus are easily triggered by single light flashes (of ps–μs duration) or by continuous illumination. Examples of such techniques include absorption difference spectroscopy, electron paramagnetic resonance spectroscopy, fluorescence spectroscopy, and flash-induced or steady-state oxygen evolution measurements. The opportunities for detailed analysis have contributed to having PS II become arguably the best-characterized membrane–protein complex of its size. In this chapter, the commonly used (and misused) technique of measuring chlorophyll fluorescence will be described as a sensitive and relatively simple tool to study the function of PS II. When used properly, fluorescence measurements can provide detailed insight into the effects of mutation-induced structural changes on functional parameters of cofactors and prosthetic groups associated with PS II.

To properly understand parameters influencing fluorescence, first a short introduction regarding PS II structure and function is provided. Upon excitation of one of the pigments associated with the PS II complex by light, the exciton (essentially the light energy) travels through the pigment bed (the "antenna") until it is trapped by P680, which is chlorophyll *a* in a specific protein environment and associated with the reaction center proteins D1 and D2 (see Fig. 1). P680 in excited state can transfer an electron to a nearby pheophytin *a* (Pheo), which in turn can transfer an electron to the plastoquinone Q_A, which is bound to the D2 protein. Q_A^- reduces another plastoquinone, Q_B, which is in a homologous position at the D1 protein (the D1 and D2 proteins form a heterodimeric complex,

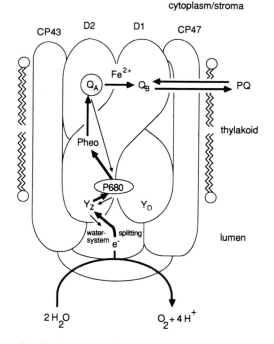

Fig. 1 Simplified model of photosystem II, showing electron transport pathways and the major photosystem II proteins, including the reaction center proteins D1 and D2, and the chlorophyll-binding antenna proteins CP47 and CP43. The two main cofactors whose redox state affects chlorophyll fluorescence yield, Q_A and P680, have been circled. The physiological electron transport pathway has been indicated by fat arrows. The pathway of Q_A^- reoxidation by donor-side components if electron transfer to Q_B has been blocked has been indicated by thin arrows.

with a twofold symmetry axis essentially perpendicular to the thylakoid membrane; Fig. 1). A non-heme iron is located between Q_A and Q_B.

Even though Q_A and Q_B are chemically identical, their properties are different in at least two respects. First, in contrast to Q_A, Q_B can be reduced to the quinol form in two subsequent electron-transfer events. Q_A is reduced only to the semiquinone form. Another fundamental difference between Q_A and Q_B is the fact that Q_A remains tightly bound to its binding site on the D2 protein, whereas Q_B in its fully oxidized and fully reduced form exchanges readily with plastoquinone that is free in the thylakoid (Q_B^- is tightly bound to the D1 protein). This exchange of Q_B with plastoquinone is functionally relevant, as by this mechanism electrons are transported from the PS II complex to the next membrane complex in the photosynthetic electron transport chain, the cytochrome b_6f complex.

When P680 donates an electron to Pheo, Q_A, and Q_B, then the oxidized form of P680, P680$^+$, is formed. This is reduced by an electron coming eventually from water via the water-splitting system and a redox-active Tyr residue in the D1 protein (Y_Z), the immediate physiological donor to P680$^+$. After four successive redox turnovers of P680, and thus four electron-transport events to fill electron holes on P680$^+$, sufficient oxidizing equivalents have accumulated in the water-oxidizing system of PS II to oxidize two water molecules to give rise to one oxygen molecule (see Fig. 1).

Chlorophyll a in the PS II complex is associated with the reaction center proteins D1 and D2, with the chlorophyll-binding proteins CP47 and CP43 (the "core antenna proteins"), and in plants with the peripheral antenna, the light-harvesting complex II (LHC II). Upon excitation of chlorophyll, the excited pigment rapidly relaxes to its first excited state (S_1), and then can transfer its excitation energy to another pigment, and eventually to P680, where the energy can be used for photochemistry. Alternatively, chlorophyll in excited state can lose its excitation via internal conversion (heat) or via fluorescence. Thus, upon excitation of chlorophyll, some fluorescence (peaking at 680–690 nm at room temperature in intact systems) is emitted.

The chlorophyll fluorescence yield observed is a function of the redox state of Q_A (Duysens and Sweers, 1963). When Q_A is oxidized, the chlorophyll fluorescence yield is low, whereas upon formation of Q_A^- the chlorophyll fluorescence yield increases up to fivefold. The redox state of Q_B has little effect on the chlorophyll fluorescence yield. The reasons for the large effect of Q_A reduction on chlorophyll fluorescence is that in the presence of Q_A^- charge separation between P680 and Pheo is relatively slow, and even if charge separation occurs, then Pheo$^-$ cannot be oxidized by Q_A^-, and recombination to the P680*·Pheo state occurs (see Holzwarth, 1992, for a review). Consequently, the exciton traveling through the antenna chlorophylls cannot be trapped by the reaction center if Q_A^- is present. Under such conditions, the exciton will continue to travel through the antenna complex, until the excited state fluoresces or is dissipated by internal conversion (heat). Thus, in the absence of net photochemistry in the PS II reaction center, the chlorophyll fluorescence intensity will be high.

At least on time scales of illumination of 10 μs to 1 s, this influence of the redox state of Q_A on the chlorophyll fluorescence yield is a main factor causing the phenomenon of variable fluorescence. When dark-adapted thylakoids, cells, or whole plants are illuminated, the chlorophyll fluorescence yield rises reflecting the reduction of Q_A. This phenomenon of the rise of chlorophyll fluorescence upon illumination is named fluorescence induction [reviewed, for example, by Papageorgiou (1976) and Krause and Weis (1991)]. However, factors other than the redox state of Q_A govern chlorophyll fluorescence yields as well. P680$^+$ is a fluorescence quencher (Butler et al., 1973), but the lifetime of P680$^+$ is so short (20 ns–1 μs) under physiological conditions that it does not affect fluorescence yields at time scales >10 μs provided that the electron donor side of PS II is intact. A number of factors (summarized in part in Section IV) causes slow (>2 s) changes in the chlorophyll fluorescence yield; however, these slower changes are difficult to interpret unequivocally because of the many different processes influencing the fluorescence yield. The difficulties in understanding the factors influencing fluorescence yield and the indiscriminate use of using fluorescence induction by some groups led to the recent publication of an opinion by Holzwarth (1993), in which he argued that variable fluorescence should not be used until factors influencing this parameter are better established. Even though this opinion probably reflects an overreaction to overinterpretation of fluorescence results, it represents a justified concern that fluorescence measurements need to be evaluated objectively, and often in correlation with results obtained by other techniques. However, when applied wisely and conservatively, fluorescence properties are good indicators regarding the redox state and function of PS II, and also regarding other factors, such as the vicinity of PS I. In the following sections, use of variable fluorescence to study functional properties in structurally altered PS II will be highlighted.

II. Materials

One of the reasons that fluorescence measurements are popular with respect to investigations of PS II properties is that chlorophyll fluorescence is relatively simple to measure. Currently, several instruments that are highly suitable for fluorescence induction measurements are commercially available. Suitable instruments are marketed by, among others, Hansatech Ltd (England), Richard Brancker (Ottawa, Canada), and Heinz Walz GmbH (Effeltrich, Germany). The instrument marketed by Walz (Schreiber et al., 1986), named the PAM fluorometer, currently is the most popular commercial fluorescence monitoring device for plant research in the laboratory. One of the most attractive features of this instrument is that fluorescence is measured by weak, short pulses of red light emitted by a light-emitting diode (LED) at a 1.6- to 100-kHz frequency. Fluorescence elicited by these pulses can be separated from that elicited by continuous, actinic illumination by subtraction of the fluorescence level when the pulse is off vs when the pulse is on. Thus, fluorescence (reflecting, in part, the redox state of Q_A) can be measured independent of actinic illumination. This allows an accurate measurement of the F_0 level (the fluorescence level in dark-adapted state) as well as of fluorescence yield changes upon turning off actinic illumination (reflecting mostly Q_A^- oxidation). In addition, with this apparatus Q_A^- oxidation after single flashes can be measured, providing information regarding the rate of electron transport between Q_A^- and $Q_B^{(-)}$, or of charge recombination between Q_A^- and the PS II donor side if electron transport between Q_A and Q_B is blocked by an inhibitor (such as diuron).

The Walz fluorometer easily interfaces with a computer for data acquisition. Walz markets a software and interface package (DA100), but recently another package has been developed (Tyystjärvi and Karunen, 1990) that for many applications may be preferable to the DA100 combination. This package is marketed through Q_A Data (Turku, Finland).

Fluorescence induction can be measured in many forms of plant material, ranging from intact plants to PS II particles. To measure properties of the PS

II complex, it often is preferable to utilize isolated, but reasonably intact systems (for example, thylakoids). Suspensions of intact cells can also be adequate. However, in intact systems, proper control experiments are needed to determine the origin and cause of chlorophyll fluorescence yield changes. Interpretation of the origin of various quenching mechanisms observed in intact systems often is equivocal (see Krause and Weis, 1991).

III. Methods

A. Preparation of Biological Material

For many applications, fluorescence measurements in single cells or in isolated thylakoids are preferable. If one works with cells from algae or cyanobacteria, it generally suffices to centrifuge a suspension of cells in logarithmic phase to harvest the cells, and then to resuspend cells in fresh growth medium. The resulting suspension then can be dark-adapted. In most intact systems a 1-minute dark adaptation period is adequate for the redox state of Q_A to reach a steady-state (mostly oxidized) level. These dark-adapted cells then can be used for fluorescence induction measurements, or for measurements of Q_A^- oxidation after single flashes (see Section V).

If one is to work with thylakoids (which often is advantageous in that physiological changes during the course of the experiment are less significant in the isolated system), then thylakoids need to be isolated from cells or tissue. For most experimental systems, thylakoid preparation procedures have been worked out. For higher plants, this generally involves cutting off 10–200 g of leaves (remove the midrib and other large veins); after washing the leaf tissue in ice-cold water, leaves are put into a blender (a household blender generally will work well) with 25–250 ml isotonic, buffered medium [for example, 25 mM tricine/NaOH (pH 7.6), 25 mM NaCl, 5 mM MgCl$_2$, and 0.4 M sorbitol]. The blender and medium should be cold, and all subsequent thylakoid isolation steps should be carried out at 0–4°C. After cell disruption, the slurry is passed through two-to-four layers of cheesecloth to remove large leaf fragments, and subsequently through 10 layers of cheesecloth to remove smaller unbroken parts of the leaf. The homogenate is centrifuged (1000 × g, 5 min) to pellet chloroplasts and other cell organelles, and after removal of the supernatant and resuspension of the pellet with a paint brush in remaining liquid, chloroplasts are broken by addition of a hypotonic buffer, such as the thylakoid isolation buffer but without the sorbitol. Thylakoids and other fragments are spun down (1000 × g, 5 min) and resuspended with a paint brush in a small volume of an isotonic solution (such as the thylakoid isolation buffer including the sorbitol). The chlorophyll concentration of the thylakoid suspension can be determined by chlorophyll extraction in 80% acetone, followed by centrifugation (to remove insoluble components) and measurement of absorbance at 645 and 663 nm; for organisms containing only chlorophyll a and no chlorophyll b, a 663-nm reading suffices. From the absorbance values, the chlorophyll concentration can be determined (Mackinney, 1941; Lichtenthaler, 1987).

To isolate thylakoids from algae and cyanobacteria, a somewhat different approach needs to be taken. A practical and relatively inexpensive method to break algae and cyanobacteria is to mix cells (washed and resuspended in a isotonic or hypertonic medium) with small glass beads (0.1 mm for cyanobacteria, a little larger for algae), and to shake vigorously (at several hundred rpm). The choice of the medium appears to be more critical than is the case for thylakoid isolation from plants. For example, for many cyanobacteria a suitable thylakoid isolation buffer consists of 25 mM Hepes/NaOH, pH 7, 5 mM MgCl$_2$, 20 mM CaCl$_2$, 10 mM NaCl, and 10% (v/v) glycerol. In the absence of CaCl$_2$, PS II electron transport may be easily inhibited. Commercial devices for breaking cells include the Mini BeadBeater (BioSpec Products, Bartlesville, OK) and the larger-scale Braun homogenizer (Braun, Melsungen, Germany). In both cases, one volume of a concentrated cell suspension is mixed with one volume of glass beads. The container needs to be filled to about 90% of its capacity, and after vigorous shaking for 3×20 s (with cooling on ice between bursts) most cells are broken. An economical large-scale device has been developed by BioSpec Products, and involves blending in the presence of glass beads. If in doubt whether effective cell breakage has occurred, one can add an aliquot of the cell suspension before and after breakage to 80% acetone in water. Chlorophyll in intact cells from many species will not be extracted in this way, whereas chlorophyll from thylakoids is extracted effectively. After mixing briefly and centrifugation, a comparison of the color of the pellet and supernatant will yield information on the breaking efficiency. After cells have been broken, thylakoids are separated from glass beads by incubation on ice for a few minutes to have glass beads settle. The thylakoid fraction can be carefully removed, and glass beads can then be washed with thylakoid buffer to remove remaining thylakoids. If the high viscosity of the highly concentrated broken cells presents a problem in the settling of glass beads, dilution of the sample or addition of some DNase will help. The chlorophyll concentration of the thylakoid sample can be determined as described above.

B. Fluorescence Induction

The measurement of fluorescence induction (chlorophyll fluorescence as a function of time upon the start of actinic illumination) is a useful first step in analysis of PS II properties. When actinic light is turned on, and as Q$_A$ is reduced by PS II activity, the fluorescence yield increases. Examples of fluorescence induction curves will be presented in Sections IV and V. To measure a fluorescence induction curve, one may utilize a commercial fluorometer (such as the PAM fluorometer), but it is also possible to monitor reliably fluorescence induction in a cuvette that is illuminated from one side with blue light [for example, with light from a tungsten lamp filtered through a LS-450 filter (Corion, Holliston

MA) transmitting blue light only]. An electronic shutter (Uniblitz, Vincent Associates, Rochester NY), with an opening time of 1–3 ms, regulates the start of actinic illumination. A photodiode, protected from scattered actinic light by a red cutoff filter [LG-650 (Corion) or equivalent], can detect fluorescence emitted by the sample. Any type of photodiode can be utilized, as long as it is sensitive to red (670–700 nm) light. Fluorescence detection at a 90° angle compared to the path of actinic illumination is preferred, to minimize artifacts caused by direct detection of actinic light. The photodiode signal may be amplified and recorded by a storage oscilloscope or by a computer equipped with an analog-to-digital converter. The sample concentration for fluorescence induction measurements ranges from 2 to 20 μg/ml chlorophyll, depending on the amount of PS II in the sample and on the geometry of the sample chamber.

If fluorescence induction is measured in the presence of diuron or another inhibitor of electron transport between Q_A and Q_B, and if a commercial fluorometer measuring fluorescence from weak modulated light pulses is used, it is important to make sure that the measuring light pulses (which usually stay on even when the actinic light is off) do not have an actinic effect. As Q_A^- is rather stable under such conditions, any Q_A^- formed by excitation by the measuring light will remain around for a relatively long time (0.5–20 s), and this would lead to an artificial increase in the "dark" fluorescence level. This can be easily checked by measuring the ratio of variable fluorescence and F_0 at different intensities of modulated light pulses.

C. Fluorescence Decay

Fluorescence decay measurements provide information regarding Q_A^- oxidation kinetics after illumination by a single-turnover flash, or by continuous light. Both assays are easily performed with the Walz PAM fluorometer. Measuring Q_A^- oxidation after a single flash provides information on the rate of electron transport to $Q_B^{(-)}$, as well as on the location of the semiquinone equilibrium $Q_A^- \cdot Q_B \leftrightarrows Q_A \cdot Q_B^-$. Kinetics of Q_A^- oxidation (electron transport to $Q_B^{(-)}$) show a major component of 300–600 μs, which can be measured upon data acquisition at 25 μs/point [available in the standard version of the PAM fluorometer; the acquisition time constant (damping) should be set to an appropriately small value]. Alternatively, the amount of Q_A^- remaining can be measured by giving weak probing flashes at different intervals after the actinic flash. Essentially identical information can be obtained in this way (Bowes et al., 1980).

In the presence of diuron, Q_A^- is reoxidized in part through a back-reaction with an oxidized component on the PS II donor side (Robinson and Crofts, 1983). This reaction generally is much slower (500 ms–20 s), unless the water-splitting apparatus has been inactivated, and Q_A^- is oxidized by Y_Z^{ox} ($t_{1/2}$ of 10–20 ms) (see Fig. 1). Fluorescence decay (Q_A^- oxidation) thus can be used as a probe to monitor intactness of the water-splitting apparatus, as well as to investigate Q_B properties. For fluorescence decay measurements, essentially the same chlorophyll concentrations can be used as for fluorescence induction measurements.

IV. Parameters Influencing Chlorophyll Fluorescence

The redox state of Q_A to a large extent determines the fluorescence yield shortly after turning on the actinic light. However, a number of factors influence the degree of variable fluorescence. For example, thylakoid stacking has a profound effect on the level of the variable fluorescence observed. In stacked thylakoids, the variable fluorescence is much higher than that in unstacked ones (Murata, 1969). The reason for this phenomenon most likely is the physical separation of PS II and PS I in stacked systems, whereas the two photosystems exist in the same thylakoid domain in unstacked thylakoids. If PS II and PS I are close to each other, energy transfer can occur from closed PS II centers to PS I, with which no variable fluorescence is associated [see Krause and Weis (1991) for a discussion]. Therefore, a large variable fluorescence yield is observed only if PS II and PS I are in separate domains (such as in stacked thylakoids of plants), and if energy transfer from PS II to PS I is unlikely. An interesting case in this respect is offered by cyanobacteria, where stacking does not occur, and where PS II and PS I are not segregated into different domains of the thylakoids. Cyanobacteria in general have a low variable fluorescence yield, whereas upon genetic deletion of the PS I core complex (Shen *et al.,* 1993) the ratio of variable fluorescence and F_0 is increased by a factor of 6–8, and becomes comparable to the ratio seen in stacked thylakoids from higher plants (Fig. 2; Vermaas *et al.,* 1994a).

Another physiological phenomenon influencing variable fluorescence involves the so-called state transitions. In plants, energy can be funneled preferentially toward PS II (state 1) or toward PS I (state 2), depending on the redox state of the photosynthetic system. If the electron acceptor pool between the two photosystems is reduced, this means that PS II (generating electrons for the pool) has been more active than PS I (which takes away from the pool in linear electron flow); the plant corrects this situation by funneling more excitation energy to PS I (state 2). In contrast, an oxidized pool between the photosystems indicates a relative overoxidation of PS I, and energy then is funneled preferentially toward PS II.

A third factor influencing fluorescence induction in plants and several algal groups is the so-called xanthophyll cycle (reviewed by Demmig-Adams and Adams, 1992). In this cycle, zeaxanthin is accumulated under high light conditions and is converted to violaxanthin at lower light intensity. Zeaxanthin, and not violaxanthin, appears to be effective in thermal dissipation of excitation energy, thus quenching fluorescence. Accumulation of zeaxanthin thus leads to photoprotection under high light conditions, even though the molecular mechanisms of this protection are not yet known in any detail (Demmig-Adams and Adams, 1992). The xanthophyll cycle operates on a time scale much slower than that of

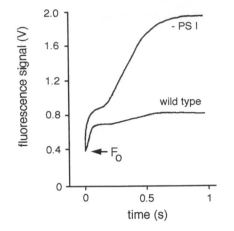

Fig. 2 Fluorescence induction curves in intact cells of the wild-type cyanobacterium *Synechocystis* sp. PCC 6803, and of a mutant in which photosystem I has been removed by deletion of the *psaAB* operon, which codes for the photosystem I reaction center core complex. Actinic illumination was started at time 0. The F_0 level (the chlorophyll fluorescence yield in the absence of actinic illumination) has been indicated. The chlorophyll concentration was 10 μg/ml for the wild type, and 2 μg/ml for the photosystem I-less mutant; with this ratio of concentrations, the PS II concentration in the samples is virtually identical. The ratio of variable and F_0 fluorescence yields is 0.9 for the wild type, and 3.7 for the PS I-less mutant. These data have been obtained with a commercial PAM fluorometer.

fluorescence induction, and thus usually is not an important factor for fluorescence induction measurements in laboratory-grown material.

On very short time scales (ns–μs) or under conditions where the donor side of PS II has been inhibited, one also needs to consider $P680^+$ as a fluorescence quencher (Butler *et al.,* 1973). It should be kept in mind, however, that under conditions where the PS II donor side has been inhibited, Q_A^- will not accumulate unless an inhibitor of electron transport between Q_A and Q_B has been added. Thus, upon donor-side inhibition, generally variable fluorescence is expected to be low, even if P680 is reduced.

V. Fluorescence as a Probe for Photosystem II

A. Relationship between Variable Fluorescence Yield and Q_A^- Concentration

The variable fluorescence yield generally is not directly proportional to the amount of Q_A^- (Joliot and Joliot, 1964). A simple explanation of this nonlinearity is best provided in qualitative terms: If a fraction of the PS II reaction centers is closed (contains Q_A^-), and if one exciton can visit more than one PS II reaction center complex during its lifetime (which is 500 ps–2 ns if PS II is closed), then the exciton can travel to another PS II center, and if it is open, it can lead to

PS II charge separation there. Hence, this exciton is not available for fluorescence anymore. Thus, if a fraction of the PS II centers is closed, and if the PS II centers arc in close contact with each other, then the amount of variable fluorescence is less than this same fraction of the maximal variable fluorescence. The more connectivity exists between PS II reaction centers, the more deviation from linearity between variable fluorescence and relative Q_A^- concentration occurs. This nonlinearity can be treated quantitatively by introduction of a "connectivity parameter," p (Joliot and Joliot, 1964). This parameter is most significant for stacked thylakoids, and may be lower or close to 0 for systems where the excitonic connectivity between PS II centers is less (for example, in cyanobacteria).

B. Variable Fluorescence to Probe PS II Acceptor Side Parameters

Variable fluorescence can be used most effectively to monitor redox kinetics involving Q_A^-. This pertains both to the quantum yield of Q_A^- formation and to the rate of Q_A^- oxidation by $Q_B^{(-)}$ and the semiquinone equilibrium between $Q_A^- \cdot Q_B$ and $Q_A \cdot Q_B^-$. With regard to the latter, this was first illustrated by experiments involving thylakoids from triazine-resistant biotypes of various weeds. In such biotypes, a single-base change has occurred in codon 264 of the gene coding for the D1 protein, leading to a change of Ser264 to Ala (Hirschberg and McIntosh, 1983). When measuring the decay of variable fluorescence after a single-turnover flash, the variable fluorescence decayed within several milliseconds to a steady-state level that was significantly above the F_0 level (Bowes *et al.,* 1980). This is an indication that a sizeable steady-state concentration of Q_A^- remains present in this system after a flash, whereas in the wild-type within the same time scale the variable fluorescence decays essentially to 0. These data are best compatible with a change in the equilibrium constant of the $Q_A^- \cdot Q_B \leftrightarrows Q_A \cdot Q_B^-$ reactions, which suggests a change in the Q_B/Q_B^- midpoint redox potential; a change in the midpoint potential of Q_A/Q_A^- is less likely, as the mutation in the D1 protein is far removed from Q_A, but is close to the binding site for Q_B.

A change in the midpoint redox potential of the Q_B/Q_B^- couple to more negative values (without necessitating a change in the Q_A^- oxidation rate by Q_B) is also indicated by the results of fluorescence induction. In thylakoids from a triazine-resistant biotype of *Amaranthus hybridus,* fluorescence induction is indicative of a higher initial Q_A^- level upon turning on the light, followed by a slower filling of the plastoquinone pool (Fig. 3; Vermaas and Arntzen, 1983). This phenomenon is independent of the light intensity used, indicating that a kinetic limitation of the $Q_A^- \rightarrow Q_B$ electron transfer reaction can be ruled out. Experimentation monitoring the pattern of oxygen evolution upon excitation by single flashes after a period of dark adaptation confirmed the notion that in this triazine-resistant biotype the Q_B/Q_B^- redox couple had been affected in its midpoint potential. The origin of the apparent decrease in the midpoint redox potential may be decreased binding of Q_B to its binding site, and/or a destabiliza-

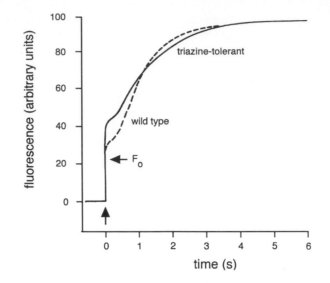

Fig. 3 Fluorescence induction kinetics of dark-adapated thylakoids from a triazine-resistant biotype (solid line) and the wild type (broken line, where different from the solid line) of *Amaranthus hybridus*. Illumination was started at time 0 (indicated by an arrow). The F_0 level has been indicated. These results were obtained using a classical fluorometer, in which actinic illumination with blue light was started by opening of an electronic shutter. The chlorophyll concentration in both samples was 10 μg/ml.

tion of Q_B^-. Since the work on triazine-resistant weeds, a large number of mutants in the D1 protein has been generated, many with modified fluorescence characteristics, and most of these with a fluorescence phenotype similar to that of the triazine-resistant *Amaranthus hybridus* biotype [reviewed in Vermaas (1989)].

Fluorescence induction also can provide information regarding the quantum efficiency of Q_A reduction. In wild-type systems, this quantum efficiency is close to one. However, mutations may affect the quantum yield of Q_A reduction considerably. For example, mutation of His268 of the D2 protein to Gln reduces the apparent quantum yield of Q_A reduction to about 2%. This results in very slow fluorescence induction kinetics (Fig. 4; Vermaas *et al.*, 1994b). Also, the kinetics of Q_A^- oxidation upon turning off the light were very slow (half-time of about 5 s), implying that Q_A^- has become isolated from other prosthetic groups. On the basis of homology with the reaction center from purple bacteria, His268 of the D2 protein is predicted to serve as a ligand to the non-heme iron (see Fig. 1). This suggests that a functional absence of the non-heme iron has significant implications for electron transfer pathways at the acceptor side of PS II. This is in apparent contrast with the situation in purple bacteria, where the non-heme iron can be removed with much less, if any, dramatic consequences (Dutton *et al.*, 1978; Debus *et al.*, 1986; Kirmaier *et al.*, 1986).

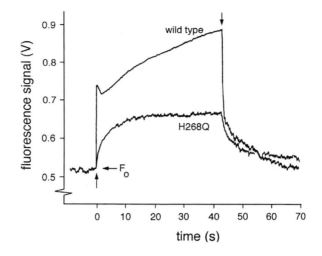

Fig. 4 Fluorescence induction and decay kinetics in whole cells of wild-type *Synechocystis* sp. PCC 6803, and of a mutant (H268Q), in which His268 of the D2 protein, which is thought to provide a ligand to the nonheme iron of PS II, has been mutated to Gln. Note the long time scale of fluorescence induction. The slow rise in the fluorescence yield in wild type most likely is not related to Q_A reduction; only the fast rise (to about 0.75 V) is. In wild type, upon turning off the light (arrow down) Q_A^- oxidation is essentially instantaneous on the time scale of the figure (fluorescence decay to about 0.65 V). The slow decay represents mostly phenomena unrelated to redox states in PS II. Fluorescence induction as well as Q_A^- oxidation in H268Q is very slow. For clarity, the fluorescence decay curve of H268Q has been interrupted at the intersection point with the curve of the wild type. The chlorophyll concentration for wild type was 10 μg/ml; for H268Q, 20 μg/ml.

A further application of fluorescence techniques with respect to the H268Q mutant showed that addition of small concentrations of benzoquinone derivatives have a large effect on the quenching of variable fluorescence. In contrast, similar quinone concentrations do not affect variable fluorescence significantly in wild type. These results were interpreted to indicate that the quinone in oxidized form could replace the native Q_A in the H268Q mutant, but not in wild type (Vermaas *et al.,* 1994b). Also, once Q_A had been reduced, exchange between the native plastoquinone and the artificial quinone does not seem to occur. Thus, after functional removal of the non-heme iron from PS II, Q_A may have functional similarities to Q_B, which also can be exchanged with quinones in its oxidized, but not in its semireduced, form. These results indicate the usefulness of variable fluorescence to study mutants in which the acceptor side of PS II has been modified by mutations. These results help in establishing structure/function relationships in the PS II complex, in that functional modifications of PS II as a result of structural changes can be easily screened.

C. Variable Fluorescence to Monitor Donor-Side Modifications in PS II

In addition to acceptor-side events, also changes in donor-side activity can be monitored by fluorescence measurements. The reason for this is that, as indicated earlier, the presence of Q_A as well as of $P680^+$ leads to quenched fluorescence. Thus, if $P680^+$ is not reduced efficiently, flash-induced fluorescence will be low: $P680^+$ will be formed in a flash, but will not be able to obtain an electron from the donor side. This property has been taken advantage of in the investigation of mutants impaired in water splitting (Nixon *et al.,* 1992b). In such mutants, after dark adaptation followed by one single-turnover flash, $P680^+$ that has been generated will be reduced quickly by Y_Z, a redox-active Tyr residue of the D1 protein that is the physological donor to $P680^+$. However, Y_Z^{ox}, which usually is reduced by the water-splitting apparatus, remains in the oxidized form in mutants in which the oxygen-evolving apparatus cannot donate an electron. Therefore, in subsequent flashes in such mutants generation of a high-fluorescent state by the flash is impaired: $P680^+$ is not rapidly reduced, and fluorescence is quenched (Nixon *et al.,* 1992b).

A complementary assay to probe the functionality of the donor side of PS II by fluorescence is to add an inhibitor of electron transfer between Q_A and Q_B (so that Q_A^- will be oxidized mainly by donor side components via $P680^+$), and to then measure reoxidation kinetics of Q_A^-. If the water-splitting system is intact, a fluorescence decay with a half-time of 300–600 ms is expected. However, if Y_Z^{ox} is not reduced rapidly by the water-splitting system, a charge recombination between Q_A^- and Y_Z^{ox} can occur with a half-time of about 10–20 ms. Thus, from the kinetics of Q_A^- oxidation in the presence of an inhibitor of electron transport to Q_B the intactness of the PS II donor side can be probed (Nixon *et al.,* 1992b). However, often clearly biphasic Q_A^- oxidation kinetics are seen in mutants with an impaired oxygen evolving system under such conditions; the reason for this most likely is that in a fraction of the PS II centers Y_Z^{ox} is rapidly reduced by exogenous components, or that the water-splitting system, with altered thermodynamic properties, can donate an electron in part of the centers.

The kinetics of reduction of Y_Z^{ox} or $P680^+$ by components other than the water-splitting system can be monitored in more detail by illumination (for several seconds) in the presence of an inhibitor of electron transfer between Q_A and Q_B. If an intact water-splitting apparatus is present, the net result of the illumination is that an electron will be transferred from the water-splitting complex to Q_A. This situation will remain during the illumination, and after the illumination stops, recombination of charges can occur, and Q_A^- will reoxidize with a half-time of 300–500 ms. However, if the water-splitting system is not active, and a Y_Z^{ox}/Q_A^- state is generated by illumination, Q_A^- will be relatively stable after the illumination has stopped: Y_Z^{ox} is easily reduced by other reductants during the illumination, and the electron "hole" on the PS II donor side thus can disappear easily (Chu *et al.,* 1994). Thus, by following the kinetics of the decay of variable fluorescence in mutants with impaired donor side, the extent of damage to the water-splitting system and the accessibility of Y_Z^{ox} can be easily probed.

==== **VI. Conclusions and Perspectives**

Chlorophyll fluorescence provides a simple and sensitive probe toward detailed studies of PS II. A major advantage is that fluorescence measurements can be done using *in vivo* systems, most easily using single-cell organisms. Now that genetic modification of a membrane–protein complex such as PS II is becoming a relatively routine procedure, a simple and informative method to determine the nature and extent of functional alterations in the PS II complex as a consequence of introduced mutations is highly valuable. The concurrent development of commercial, relatively sophisticated fluorescence detection equipment with good time resolution (up to 25 μs) and relatively moderate price brings detailed fluorescence studies into the reach of most laboratories involved in PS II analysis. It is hoped that this chapter will be helpful toward a meaningful and correct interpretation of results obtained, and toward dispelling unfounded distrust in the application of fluorescence toward the measurement of PS II reaction kinetics.

Acknowledgments

Recent research from my laboratory in this chapter is supported by the National Science Foundation (MCB-9316857). This is Publication 179 from the Arizona State University Center for the Study of Early Events in Photosynthesis.

References

Bowes, J. M., Crofts, A. R., and Arntzen, C. J. (1980). Redox reactions on the reducing side of photosystem II in chloroplasts with altered herbicide-binding proteins. *Arch. Biochem. Biophys.* **200,** 303–308.

Bryant, D. A. (1992). Molecular biology of photosystem I. *In* "Topics in Photosynthesis," (J. Barber, ed.), Vol. 11, pp. 501–549. Amsterdam: Elsevier.

Butler, W. L., Visser, J. W. M., and Simons, H. L. (1973). The kinetics of light-induced changes of C-550, cytochrome b_{559} and fluorescence yield in chloroplasts at low temperature. *Biochim. Biophys. Acta.* **292,** 140–151.

Chu, H.-A., Nguyen, A. P., and Debus, R. J. (1994). Site-directed photosystem II mutants with perturbed oxygen-evolving properties. I. Instability or inefficient assembly of the manganese cluster *in vivo. Biochemistry* **33,** 6137–6149.

Debus, R. J., Feher, G., and Okamura, M. Y. (1986). Iron-depleted reaction centers from *Rhodopseudomonas sphaeroides* R-26.1: Characterization and reconstitution with Fe^{2+}, Mn^{2+}, Co^{2+}, Ni^{2+}, Cu^{2+}, and Zn^{2+}. *Biochemistry* **25,** 2276–2287.

Demmig-Adams, B., and Adams, W. W., III (1992). Photoprotection and other responses of plants to high light stress. *Annu. Rev. Plant Physiol. Plant Mol. Biol.* **43,** 599–626.

Dutton, P. L., Prince, R. C., and Tiede, D. M. (1978). The reaction center of photosynthetic bacteria. *Photochem. Photobiol.* **28,** 939–949.

Duysens, L. N. M., and Sweers, H. E. (1963). Mechanism of the two photochemical reactions in algae as studied by means of fluorescence. *In* "Studies on Microalgae and Photosynthetic Bacteria" (J. Ashida, ed.), pp. 353–372. Tokyo: Univ. of Tokyo Press.

Hirschberg, J., and McIntosh, L. (1983). Molecular basis of herbicide resistance in *Amaranthus hybridus. Science* **222,** 1346–1349.

Holzwarth, A. R. (1992). Excited-state kinetics in chlorophyll systems and its relationship to the functional organization of the photosystems. *In* "Chlorophylls" (H. Scheer, ed.), pp. 1125–1151. Boca Raton, FL: CRC Press.

Holzwarth, A. R. (1993). Is it time to throw away your apparatus for chlorophyll fluorescence induction? *Biophys. J.* **64,** 1280–1281.

Joliot, A., and Joliot, P. (1964). Étude cinétique de la réaction photochimique libérant l'oxygène au cours de la photosynthèse. *C. R. Acad. Sci. Paris* **258,** 4622–4625.

Kirmaier, C., Holten, D., Debus, R. J., Feher, G., and Okamura, M. Y. (1986). Primary photochemistry of iron-depleted and zinc-reconstituted reaction centers from *Rhodopseudomonas sphaeroides. Proc. Natl. Acad. Sci. U.S.A.* **83,** 6407–6411.

Krause, G. H., and Weis, E. (1991). Chlorophyll fluorescence and photosynthesis: The basics. *Annu. Rev. Plant Physiol. Plant Mol. Biol.* **42,** 313–349.

Lichtenthaler, H. K. (1987). Chlorophylls and carotenoids: Pigments of photosynthetic biomembranes. *Methods Enzymol.* **148,** 350–382.

Mackinney, G. (1941). Absorption of light by chlorophyll solutions. *J. Biol. Chem.* **140,** 315–322.

Murata, N. (1969). Control of excitation transfer in photosynthesis: II. Magnesium ion-dependent distribution of excitation energy between two pigment systems in spinach chloroplasts. *Biochim. Biophys. Acta* **189,** 171–181.

Nixon, P. J., Chisholm, D. A., and Diner, B. A. (1992a). Isolation and functional analysis of random and site-directed mutants of photosystem II. *In* "Plant Protein Engineering" (P. Shewry and S. Gutteridge, eds.), pp. 93–141. Cambridge: Cambridge Univ. Press.

Nixon, P. J., Trost, J. T., and Diner, B. A. (1992b). Role of the carboxy terminus of polypeptide D1 in the assembly of a functional water-oxidizing manganese cluster in photosystem II of the cyanobacterium *Synechocystis* sp. PCC 6803: Assembly requires a free carboxyl group at the C-terminal position 344. *Biochemistry* **31,** 10859–10871.

Papageorgiou, G. (1976). Chlorophyll fluorescence: An intrinsic probe of photosynthesis. *In* "Bioenergetics of Photosynthesis" (Govindjee, ed.), pp. 319–371. New York: Academic Press.

Robinson, H. H., and Crofts, A. R. (1983). Kinetics of the oxidation-reduction reactions of the photosystem II quinone acceptor complex, and the pathway for deactivation. *FEBS Lett.* **153,** 221–226.

Schreiber, U., Schliwa, U., and Bilger, W. (1986). Continuous recording of photochemical and non-photochemical chlorophyll quenching with a new type of modulation fluorometer. *Photosynth. Res.* **10,** 51–62.

Shen, G., Boussiba, S., and Vermaas, W. F. J. (1993). *Synechocystis* sp. PCC 6803 strains lacking photosystem I and phycobilisome function. *Plant Cell* **5,** 1853–1863.

Tyystjärvi, E., and Karunen, J. (1990). A microcomputer program and fast analogue to digital converter card for the analysis of fluorescence induction transients. *Photosynth. Res.* **26,** 126–132.

Vermaas, W. F. J. (1989). The structure and function of photosystem II. *In* "Techniques and New Developments in Photosynthesis Research" (J. Barber and R. Malkin, eds.), NATO ASI Series A168, pp. 35–59. New York: Plenum.

Vermaas, W. (1993). Molecular-biological approaches to analyze photosystem II structure and function. *Annu. Rev. Plant Physiol. Plant Mol. Biol.* **44,** 457–481.

Vermaas, W. F. J., and Arntzen, C. J. (1983). Synthetic quinones influencing herbicide binding and photosystem II electron transport: The effect of triazine resistance on quinone binding properties in thylakoid membranes. *Biochim. Biophys. Acta* **725,** 483–491.

Vermaas, W. F. J., Shen, G., and Styring, S. (1994a). Electrons generated by photosystem II are utilized by an oxidase in the absence of photosystem I in the cyanobacterium *Synechocystis* sp. PCC 6803. *FEBS Lett.* **337,** 103–108.

Vermaas, W. F. J., Vass, I., Eggers, B., and Styring, S. (1994b). Mutation of a putative ligand to the non-heme iron in photosystem II: Implications for Q_A reactivity, electron transfer, and herbicide binding. *Biochim. Biophys. Acta* **1184,** 263–272.

NOTE ADDED IN PROOF: Since this chapter was written, several pertinent studies have appeared. Time constraints did not allow their inclusion here.

CHAPTER 3

Determination of Protein Isoprenylation *in Vitro* and *in Vivo*

Jian-Kang Zhu, Ray A. Bressan, and Paul M. Hasegawa

Center for Plant Environmental Stress Physiology
Purdue University
W. Lafayette, Indiana 47907-1165

I. Introduction

Protein isoprenylation is a fundamentally important post-translational modification of proteins that involves the transfer of a polyisoprene moiety to a conserved cysteine residue near the carboxyl-terminus through the formation of a thioether linkage. The attached polyisprenoid is either farnesyl or geranygeranyl that are derived from pyrophosphate conjugates via the mevalonic acid pathway. Protein isoprenylation has been determined to function in cell cycle control and growth and signaling reactions in eukaryotic cells. Proteins known to be isoprenylated include the oncoprotein Ras and many Ras-related small GTP-binding proteins, the γ subunit of the heterotrimeric GTP-binding proteins, nuclear lamins, cGMP phosphodiesterase, fungal mating pheromones, rhodopsin kinase, hepatitis δ virus large antigen, and certain molecular chaperones (Clarke, 1992; Zhu et al., 1993a). For every isoprenylated protein examined, the post-translational modification has been determined to be a requisite for function. It has been postulated that the hydrophobic isoprene moiety targets isoprenylated proteins to and facilitates association with cellular membranes (Rine and Kim, 1990). Another possible function is that isoprenylation affects protein–protein interactions.

Most isoprenylated proteins share a common C-terminal CaaX (C, cysteine; a, aliphatic amino acid; X, any amino acid) motif (Clarke, 1992). The terminal X is known to confer isoprenylation specificity. Polypeptides containing a carboxyl-terminal serine, alanine, methionine, or glutamine are farnesylated, whereas those containing a leucine residue are, in most instances, geranylgeranylated. Following protein isoprenyltransferase-catalyzed conjugation of the isoprenoid to the cysteine residue, a protease cleaves the -aaX and then the new α-carboxyl group is methylated. Some of the proteins containing a carboxyl-terminal CC or CXC motif are also geranylgeranylated (Clarke, 1992).

Present understanding about protein isoprenylation has been obtained largely from research on mammals and yeast, A survey of sequences in the GENBANK database indicates that numerous plant proteins may be isoprenylated, based on the occurrence of a C-terminal Caax (Table I), CC, or CXC motif. In this chapter, we describe the principles and protocols for assaying plant protein isoprenylation using the *Atriplex nummularia* molecular chaperone ANJ1 as a prototype.

Table I
Examples of Higher Plant Proteins Containing a CaaX Isoprenylation Motif

Protein	DNA-encoded C-terminal sequence	Comments	Reference[a]
ANJ1	CAQQ	Molecular chaperone, farnesylated	1
Apetala1	CFAA	Arabidopsis floral homeotic gene product	2
Opaque2	CRRR	Maize regulatory protein	3
barley nitrate reductase	CLVF		4
Potato pyruvate kinase	CVVK	Cytosolic enzyme	5
Potato type H phosphorylase	CRVP		6
Tomato proline-rich protein	CPST		7
Ras-related small G-proteins ara, ara2, ara3, ara4, ara5 (from arabidopsis)	CSST, CCSN, CCGT, CCSR, CCST		8
yptm1, yptm2 (from maize)	CCST, CCSS		9
rgp1 (from rice)	CCMS		10
Rho1Ps	CSIL	Geranylgeranylated	11

[a] 1. Zhu et al., 1993b; 2. Mandel et al., 1992; 3. Hartings et al., 1989; 4. Miyazaki et al., 1991; 5. Blakeley et al., 1990; 6. Mori et al., 1991; 7. Salts et al., 1991; 8. Anai et al., 1991; 9. Palme et al., 1992; 10. Sano and Youssefian, 1991; 11. Yang and Watson, 1993.

II. Analysis of Protein Isoprenylation *in Vitro*

Although proteins with a C-terminal CaaX, CC, or CXC motifs are candidates for isoprenylation, not all of these proteins can be modified by this post-translational process. The *Arabidopsis* metallothionins MT1 and MT2 contain a C-terminal CNC and CTCK, respectively (J. Zhou and P. Goldsbrough, unpublished data), but neither could be isoprenylated (J.-K. Zhu, J. Zhou, P. Goldsbrough, P. M. Hasegawa, unpublished data) in an *in vitro* assay that has been demonstrated to be functional with plant, yeast, and animal proteins (Zhu *et al.,* 1993a; Caplan *et al.,* 1992; Adamson *et al.,* 1992). The isoprenylation assay is usually accomplished with recombinant proteins obtained from an *Escherichia coli* expression vector. Recombinant proteins are not modified in *E. coli*, because bacteria do not have protein isoprenyltransferases. The radioactive isoprenyl donor can be purchased as farnesylpyrrophosphate (F-PP) or geranylgeranyl pyrophosphate (GG-PP). Cell-free extract of the plant species from which the gene has been isolated can be used to catalyze the isoprenyl modification.

A. Recombinant Protein Production

Recombinant ANJ1 is made using one of the pET vectors (pET9c). It is usually desirable to make another recombinant protein that contains a serine for cysteine substitution in the CaaX box. This mutant protein should not be isoprenylated and thus serves as a negative control.

The entire open reading frame of ANJ1 cDNA is isolated from clone pANJ1 (Zhu *et al.,* 1993) by polymerase chain reaction (PCR) with a sense primer (5′ ATGGCCGGATCCCATATGTTTGGAAGAGCACCAA 3′) and an anti-sense primer (5′ TCCACGGGATCCAGATCACTGTTGAGCACA 3′). The amplified DNA is digested with *Nde*I and *Bam*HI and inserted between the *Nde*I and *Bam*HI sites of the bacterial expression vector pET9c (Studier *et al.,* 1990) to generate construct pETANJ1. Depending on the cloning sites available, in the expression vector of choice, other appropriate restriction sites can be included on the 5′ end of the PCR primers for convenient cloning. Constructs pETANJ1S and pETANJ1L are similarly generated by PCR using the same sense primer and antisense primers 5′ TCCACGGGATCCAGATCACTGTTGAGCAGA 3′ and 5′ CCGGAATTCGGATCCAGATCACAGTTGAGCACA 3′, respectively. The recombinant proteins produced by the three constructs are identical except for differences in the carboxyl-terminal four amino acids: CAQQ for ANJ1, SAQQ for ANJ1S and CAQL for ANJ1L. The plasmids are transformed into the *E. coli* strain BL21(DE3) for protein expression. Aliquots of 0.5 ml of BL21(DE3) overnight cultures containing either pET9c, pETANJ1C, pETANJ1S, and pETANJ1L are subcultured into 5 ml of fresh LB medium plus kanamycin (250 μg/ml) and allowed to grow for 1.5 h at 37°C. The cultures are then induced for 2.5 h by addition of IPTG to 0.4 mM.

Aliquots of the cultures are harvested and bacterial cells pelleted by centrifugation and resuspended in a sample buffer (2% SDS, 5% 2-mercaptoethanol, 0.002% bromophenol blue, 10% glycerol, and 62.5 mM Tris–HCl, pH 6.8) for SDS––PAGE analysis. The induced cultures are also harvested and washed once with extraction buffer (50 mM Tris–HCl, pH 8, 200 mM NaCl, 0.1 mM EDTA, and 5% glycerol). The cells are resuspended in 150 μl of extraction buffer plus protease inhibitors (1 mM phenylmethylsulfonyl fluoride and 1 μg/ml each of leupeptin, pepstatin, aprotinin, and chymostatin) and sonicated with 5 × 6-s bursts using a microprobe (cooling on ice for 45 s between each burst). The extracts are centrifuged at 13,000 × g for 10 min at 4°C. The supernatants are stored at −80°C for later isoprenylation assays.

B. Preparation of Cell-Free Extract

The cell-free extract that is the source of protein isoprenyltransferases is prepared by homogenizing actively growing *A. nummularia* cells (or other source of cells) with a glass–glass homogenizer in extraction buffer plus the proteinase inhibitors indicated above (cell:buffer = 6:1 v/v). The homogenate is centrifuged at 13,000 × g for 10 min at 4°C and the supernatant is used for isoprenylation assays. It is important that only a small amount of homogenization buffer be used during extraction in order to obtain a concentrated cell-free extract. The cell-free extract should contain 5 to 10 μg protein/μl.

C. Assay Conditions

To an Eppendorf tube, add 10 μl of 5× assay buffer (250 mM Tris–HCl, pH 8, 100 mM KCl, 25 mM MgCl$_2$, 50 μM ZnCl$_2$, 50 mM dithiothreitol), 20 μg (protein) of *E. coli* extract containing recombinant ANJl, 50–75 μg (protein) of cell-free extract from *A. nummularia* and 1 × 10^{-4} μmol [^{3}H]farnesyl pyrophosphate (triammonium salt, [1 − ^{3}H]; 60 Ci/mmol; American Radiolabelled Chemicals, Inc., St. Louis, MO) or [^{3}H]geranylgeranyl pyrophosphate (triammonium salt, [1 − ^{3}H]; 12 Ci/mmol; American Radiolabelled Chemicals, Inc.), with a final volume of 50 μl. The mixture is incubated at 37°C for 1 h. Zn^{2+} is present in the assay buffer because it is a cofactor for protein isoprenyltransferases.

D. Detection

The reaction is stopped by the addition of 50 μl of 2× SDS–PAGE sample buffer. A 10-μl aliquot of the mixture is separated by SDS–PAGE (10% gel). The gel is first fixed for 30 min in a solution of 50% (v/v) isopropanol plus 10% acetic acid, then soaked for 30 min in a solution of fluorographic reagent from Amersham (Amplify; Amersham International plc, Amersham, UK), dried onto filter paper, and exposed to X-ray film (Fig. 1). Exposure time is normally several hours to several days.

E. Quantitation of Isoprenyl Incorporation

If ^{3}H incorporation is to be quantified, the isoprenylation assay reaction is terminated by the addition of 500 μl of 4% SDS and proteins are precipitated with 500 μl of 30% trichloroacetic acid for 1 h at room temperature. The samples are filtered through glass fiber filters and washed sequentially with 60 ml of 2% SDS plus 6% trichloroacetic acid, 40 ml of 6% trichloroacetic acid, and 20 ml of 95% ice-cold ethanol. The filters are then dried by lyophilization and the amount of [^{3}H]farnesyl or [^{3}H]geranylgeranyl incorporated into protein is determined by liquid scintillation spectroscopy (Table II).

III. Analysis of Protein Isoprenylation *in Vivo*

A. Mobility Shift

When a protein becomes isoprenylated, its mobility in SDS–PAGE gels increases slightly. Therefore, if a protein is modified by isoprenoids *in vivo* in plant cells, it should migrate faster during SDS–PAGE compared to the recombinant version expressed in *E. coli*. Since the mobility difference is usually less than 1 kDa apparent molecular mass, it is necessary that the separation of proteins be sufficient to distinguish this difference. For ANJ1, with a molecular mass of

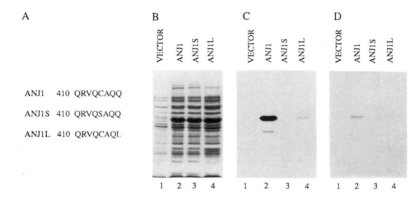

Fig. 1 Isoprenylation of ANJ1 *in vitro* with *A. nummularia* cell-free extracts. (A) C-terminal amino acid sequences of ANJ1, ANJ1S, and ANJ1L; (B) Coomassie blue-stained gel of soluble extracts from IPTG-induced *E. coli* containing pET9c (vector, lane 1), pETANJ1 (ANJ1, lane 2), pETANJ1S (ANJ1S, lane 3), and pETANJ1L (ANJ1L, lane 4); (C) Fluorogram (1-week exposure) of farnesylation reaction products. (D) Fluorogram (1-week exposure) of geranylgeranylation reaction products. Reproduced from Zhu *et al.* (1993a) with permission.

Table II
Quantitation of [³H]Farnesyl and [³H]Geranylgeranyl Incorporated into Proteins[a]

C-terminal sequence	[³H]Farnesyl (pmol/h)	[³H]Geranylgeranyl (pmol/h)
Vector control	0	0
CAQQ	0.65	0.15
SAQQ	0	0
CAQL	0.06	0

[a] *In vitro* farnesylation or geranylgeranylation assays were carried out as described in Section II,C. The substrates were soluble extracts (20 μg of protein) from IPTG-induced *E. coli* containing pET9c (vector control), pETANJ1 (CAQQ), pETANJ1S (SAQQ), and pETANJ1L (CAQL), respectively. After 1 h, the reactions were stopped by the addition of SDS, proteins were precipitated with trichloroacetic acid and filtered through glass fiber filters, and the amount of [³H]farnesyl or [³H]geranylgeranyl incorporated was determined by liquid scintillation spectroscopy.

approximately 51 kDa, the mobility difference between the plant form and the recombinant version from *E. coli* was readily visible when the samples were separated using a large-format (15 cm long) gel, which was run until a 49.5-kDa prestained marker was nearly eluted from the bottom of the gel (Fig. 2).

B. Mevalonic Acid Labeling

The polyisoprenoids conjugated to proteins, i.e., farnesyl and geranylgeranyl, are derived from mevalonic acid via the isoprenoid pathway. [³H]Mevalonic acid

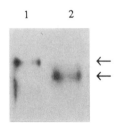

Fig. 2 Determination of ANJ1 isoprenylation *in vivo* by mobility shift analysis. Electrophoretic mobilities of recombinant ANJ1 expressed in *E. coli* and native ANJ1 from plant cells were compared. Cell extracts from IPTG-induced *E. coli* strain BL21(DE3) containing pETANJ1 (lane 1) or *A. nummularia* cells (lane 2) were separated by SDS–PAGE, electroblotted onto nitrocellulose, and probed with anti-ANJ1. The arrows indicate the two forms of ANJ1. Reproduced from Zhu *et al.* (1993a) with permission.

added to culture medium is taken up by cells and converted to farnesyl and geranylgeranyl moieties and some is incorporated into proteins that can be detected as radiolabeled bands on SDS–PAGE gels (Fig. 3). To increase the efficiency of mevalonic acid labeling, cells need to be pretreated with the hydroxymethylglutaryl–coenzyme A (HMG-CoA) reductase inhibitor lovastatin (mevinolin) to deplete the endogenous pool of mevalonic acid.

1. Preparation of Lovastatin

Lovastatin is supplied by Merck & Co. (Rahway, NJ) as an inactive lactone. To prepare a stock solution (4 mg/ml), 40 mg of lovastatin is dissolved in 1 ml of ethanol, and then 1.5 ml of 0.1 N NaOH is added. After the solution is heated at 50°C for 2 h, it is neutralized with 0.1 N HCl to pH 7.2 and brought to a volume of 10 ml with distilled H_2O. The stock solution is separated into several aliquots and stored at -20°C.

2. Lovastatin Pretreatment

To 100 ml of *A. nummularia* cell suspension culture (early log phase), add 200 μl of the lovastatin stock solution. The final lovastatin concentration is approximately 20 μM.

3. Mevalonic Acid Labeling

Sixteen hours after the addition of lovastatin, 5 ml of the pretreated cell suspension culture is transferred to a 50-ml flask for labeling. Mevalonolactone

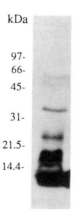

Fig. 3 Detection of isoprenylated proteins from plant cells after *in vivo* [³H]mevalonic acid labeling. Shown is a fluorogram of total proteins from metabolically labeled *A. nummularia* cells that were separated by SDS–PAGE.

(250 μCi) (RS [5 $-$ ^{3}H(N)]; 60 Ci/mmol; 1 mCi/ml; American Radiolabelled Chemicals, Inc.) is added to 5 ml of culture. The labeling is allowed to proceed for up to 24 h. We found that a 24-h incubation resulted in substantially more labeling of proteins than a 6-h incubation.

To detect labeled proteins, cells are collected by filtration and ground in liquid N_2. The cell powder is then incubated overnight with acetone at $-$ 20°C, collected on filter paper, and lyophilized. Five milligrams of the dry powder is mixed with 400 μl of SDS–PAGE sample buffer. Twenty microliters (approximately 20 μg) of the protein extract is then separated by SDS–PAGE, and the labeled protein detected by fluorography (Fig. 3). Typical exposure times are 1 week to 1 month.

The labeled cells may also be homogenized, and a specific labeled protein isolated from the homogenate by immunoprecipitation. The polyisoprenoid can be cleaved from the immunoprecipitated labeled protein using methyl iodide and analyzed by HPLC to determine the chain length of the isoprenoid moiety (i.e., farnesyl or geranylgeranyl). For a detailed procedure on the immunoprecipitation, methyl iodide cleavage, and HPLC analysis, readers are referred to Masterson and Magee (1992).

Acknowledgments

This work was supported in part by a McKnight Foundation Fellowship and by USDA/NRICP Grant 92-37100-7628.

References

Adamson, P., Marshall, C. J., Hall, A., and Tilbrook, P. A. (1992). Post-translational modification of p21^{rho} proteins. *J. Biol. Chem.* **267,** 20033–20038.

Anai, T., Hasegawa, K., Watanabe, Y., Uchimiya, H., Ishizaki, R., and Matsui, M. (1991). Isolation and analysis of cDNAs encoding small GTP-binding proteins of Arabidopsis thaliana. *Gene* **108,** 259–264.

Blakeley, S. D., Plaxton, W. C., and Dennis, D. T. (1990). Cloning and characterization of a cDNA for the cytosolic isozyme of plant pyruvate kinase: The relationship between the plant and non-plant enzyme. *Plant Mol. Biol.* **15,** 665–669.

Caplan, A. J., Tsai, J., Casey, P. J., and Douglas, M. G. (1992). Farnesylation of YDJ1p is required for function at elevated growth temperatures in Saccharomyces cerevisiae. *J. Biol. Chem.* **267,** 18890–18895.

Clarke, S. (1992). Protein isoprenylation and methylation at carboxyl-terminal cysteine residues. *Annu. Rev. Biochem.* **61,** 355–386.

Hartings, H., Maddaloni, M., Lazzaroni, N., Foιιzo, N. D., Motto, M., Salamini, F., and Thompson, R. (1989). The O2 gene which regulates zein deposition in maize endosperm encodes a protein with structural homologies to transcriptional activators. *EMBO J.* **8,** 2795–2801.

Mandel, M. A., Gustafson-Brown, C., Savidge, B., and Yanofsky, M. F. (1992). Molecular characterization of the *Arabidopsis* floral homeotic gene *APETALA1. Nature* **360,** 273–277.

Masterson, W. J., and Magee, A. I. (1992). Lipid modification involved in protein targeting. *In* "Protein Targeting: A Practical Approach" (A. I. Magee and T. Wileman, eds.), pp. 233–259. New York: IRL Press.

Miyazaki, J., Juricek, M., Angelis, K., Schnorr, K. M., Kleinhofs, A., and Warner, R. L. (1991). Characterization and sequence of a novel nitrate reductase from barley. *Mol. Gen. Genet.* **228,** 329–334.

Mori, H., Tanizawa, K., and Fukui, T. (1991). Potato tuber type H phosphorylase isozyme. *J. Biol. Chem.* **226**, 18446–18453.

Palme, K., Diefenthal, T., Vingron, M., Sander, C., and Schell, J. (1992). Molecular cloning and structural analysis of genes from *Zea mays* (L.) coding for members of the *ras*-related *rgp1* gene family. *Proc. Natl. Acad. Sci. U.S.A.* **89**, 787–791.

Salts, Y., Wachs, R., Gruissem, W., and Barg, R. (1991). Sequence coding for a novel proline-rich protein preferentially expressed in young tomato fruit. *Plant Mol. Biol.* **17**, 149–150.

Sano, H., and Youssefian, S. (1991). A novel ras-related *rgp1* gene encoding a GTP-binding protein has reduced expression in 5-azacytidine-induced dwarf rice. *Mol. Gen. Genet.* **228**, 227–232.

Studier, F. W., Rosenberg, A. H., Dunn, J. J., and Dubendorff, J. W. (1990). Use of T7 RNA polymerase to direct expression of cloned genes. *Methods Enzymol.* **185**, 60–89.

Yang, Z., and Watson, J. C. (1993). Molecular cloning and characterization of rho, a ras-related small GTP-binding protein from the garden pea. *Proc. Natl. Acad. Sci. U.S.A.* **90**, 8732–8736.

Zhu, J.-K., Bressan, R. A., and Hasegawa, P. M. (1993a). Isoprenylation of the plant molecular chaperone ANJ1 facilitates membrane association and function at high temperature. *Proc. Natl. Acad. Sci. U.S.A.* **90**, 8557–8561.

Zhu, J.-K., Shi, J., Bressan, R. A., and Hasegawa, P. M. (1993b). Expression of an *Atriplex nummularia* gene encoding a protein homologous to the bacterial molecular chaperone DnaJ. *Plant Cell* **5**, 341–349.

CHAPTER 4

Extraction and Assay of Protein from Single Plant Cells

William H. Outlaw, Jr.

Department of Biological Science
Florida State University
Tallahassee, Florida 32306-3050

I. Introduction
II. Preparation of the Sample
III. Extraction and Hydrolysis of the Sample
IV. Analysis of Glutamate in the Protein Hydrolysate
V. Advantages and Limitations
   References

I. Introduction

The protein complement of an organism, an organ, or a cell defines its function. Thus, selected knowledge of this complement is a usual goal. For quantitative interpretations, an attribute of some element, say, activity of a particular enzyme, is compared to an appropriate reference such as the correlate activity observed in a different physiological or developmental context. These cross-comparisons require a basis for expression, which is often protein mass. Thus, development and improvement of methods for assaying protein mass are old and ongoing endeavors. This chapter will focus on extraction and quantification of protein from single plant cells. One method will be described; a brief comparison with general methods in use and selected highly sensitive methods will be made. No attempt will be made to touch on the vast literature on protein methods; similarly, other special implementations (such as postcolumn detection, e.g., Chang *et al.,* 1989; Lee and Yeung, 1992) will not be addressed. For a bibliography of detection methods for protein in separation matrices, the reader is referred to Hames and

Rickwood (1990; also see Wallace and Saluz, 1992). For protocols on measuring protein in solution, refer to Stoscheck (1990).

The Lowry method (Lowry *et al.,* 1951) is the usual standard by which other methods are judged because it is simple, sensitive, and precise (Peterson, 1979). Color development is effected primarily by interaction of the chromogen with tyrosine and tryptophan residues and peptide bonds. An important and often unacknowledged advantage of the Lowry is that different pure proteins are measured accurately. Kjeldahl N varies by a factor of about 1.05 (the ratio, assay response (protein X) $\cdot$ mg (protein X)$^{-1}$: assay response (protein y) $\cdot$ mg (protein y)$^{-1}$), whereas the Lowry usually varies by a factor of 1.2 or less (references in Peterson, 1979). Notwithstanding interferences, the precision of the Lowry should compare with $A_{200-220}$ methods (which measure the peptide bond), the 10-fold less sensitive A_{280} methods (which measure aromatic amino acids), and the ninhydrin method (which measures α amines). Reported miniaturized Lowrys are sensitive to the 100-ng level (references in Peterson, 1979), but it should be possible to minaturize the Lowry further. An alternative to the Lowry is the bicinchoninic acid method (Smith *et al.,* 1985), which appears to be similar in accuracy and sensitivity but somewhat simpler. Various dye-binding methods, prototypically that of Bradford (1976), find widespread current use and are touted as being easier and more sensitive than the Lowry. As we (Tarczynski and Outlaw, 1987) have implemented the dye-binding method in a polyacrylamide gel matrix, one easily can measure 1 ng protein or less (which, for perspective, is about 40% of the mass of protein in a single palisade parenchyma cell of *Vicia faba* (Outlaw *et al.,* 1981). Although dye-binding methods are superior to silver methods in accuracy, they are inferior to the Lowry in this respect. Example results (Stoscheck, 1987) with three proteins and using BSA as a standard showed that the Lowry factor (see above) ranged from 0.7 to 1, whereas the Bradford factor ranged from <0.02 to 0.5.

Several methods offer the possibility of high sensitivity. It is well known that hydrophobic polymeric surfaces bind protein, which is the basis for a competitive protein micromethod (50ng $\cdot$ ml^{-1}, Sandwick and Schray, 1985). Staining with colloidal gold on nitrocellulose (Hunter and Hunter, 1987) or in solution (Stoscheck, 1987; Ciesiolka and Gabius, 1988) permits measurements to the 1-ng level. As mentioned, silver binding is another sensitive method (15 ng in solution, Krystal *et al.,* 1985; 0.01 ng in matrix, Merril, 1990). *In vitro* labeling of protein with ^{35}S (Kalinich and McClain, 1992) permits measurement of protein to 1 ng. In summary—because of the wide range of responses to various pure proteins—the methods referenced in this paragraph are most useful for a protein that has been calibrated by accurate methods. On the other hand, in homogenates or other complex mixtures, the differences among individual proteins may "average out" so that with caution these methods may be usefully applied to them.

Fluorometry is more sensitive than spectrophotometry by one to several orders of magnitude (Seitz, 1980). In particular, two reagents, fluorescamine (Udenfriend *et al.,* 1972) and *o*-phthalaldehyde (Roth, 1971), are α-amine-reacting fluorogenic reagents that have been used for high-sensitivity measurements, especially after protein hydrolysis (fluorescamine: Stowell *et al.,* 1978; *o*-phthalaldehyde: Butcher and Lowry, 1976). The *o*-phthalaldehyde assay is simpler to implement than is the fluorescamine assay (Viets *et al.,* 1978), and, if used on hydrolyzed samples, the *o*-phthalaldehyde method is highly accurate (Hernández *et al.,* 1990). By reduction in assay volume and subsequent measurement of fluorescence by quantitative microfluorometry, subnanogram quantities of protein can be measured (Zawieja *et al.,* 1984; Hammer and Nagel, 1986; see Advantages and Limitations).

In this chapter, I describe a method (Butcher and Lowry, 1976) based on hydrolysis of the protein, which is followed by determination of the liberated glutamate by enzymatic cycling methods. The method is sensitive to less than 1 ng and has been tested on a variety of extracts and pure proteins.

II. Preparation of the Sample

This section describes stabilization of the biological sample and collection of the cell.

Step 1: The sample (e.g., a leaf) that contains the cell of interest is stabilized by submersion in liquid nitrogen; portions (1–3 mm) fractured from the original frozen sample are freeze-dried.

Comment: Whereas good laboratory practice (use of slender forceps, avoidance of tissue around the forceps, use of thin and small samples) is recommended, extraordinary measures to speed freezing or to control freeze-drying rigorously, as implemented for measurement of metabolites, are not required.

Step 2: Individual cells are manually dissected from the dried tissue.

Comment: A small knife (Passonneau and Lowry, 1993) constructed of a razor-blade fragment connected to a wooden dowel is sufficient for dissection at high morphological precision (Hampp and Outlaw, 1987). Micromanipulators or other special tools are neither required nor recommended.

III. Extraction and Hydrolysis of the Sample

This section describes conversion of the cell protein into a low-volume glutamate extract. Unconventional methods are required to minimize contact with surfaces and avoid evaporation. Mineral oil, dispensed into the wells of a Teflon "oil-well rack," serves as the reagent "vessel."

Step 1: Fifteen nanoliters of 12 N H_2SO_4 is individually delivered to each required well in an oil-well rack. Next, 15 nl of H_2O (for blank and sample wells) *or* H_2O containing a reference protein *or* H_2O containing glutamate (to serve as a hydrolysis standard) is added onto each acidic droplet. Appropriate protein standards are 0.25–8 ng bovine serum albumin, and glutamate standards should be 1–5 pmol.

Comment: Example designs for oil-well ranks and general quantitative histochemical procedures (e.g., formulation of mineral oil, construction of pipettes) are given by Passonneau and Lowry (1993).

This first step is modified from the original method (Butcher and Lowry, 1976) in two ways. First, the assay is miniaturized by reduction of initial volumes. Second, because we found that nanoliter volumes of 6 N H_2SO_4 containing protein were especially difficult to deliver accurately, an aqueous protein sample is added to commensurately stronger H_2SO_4 to give the resulting droplet of appropriate protein and H_2SO_4 concentration.

Step 2: On the tip of a quartz fiber, a single cell (see Section II) is pushed through the oil and into contact with an H_2SO_4 blank droplet (Fig. 1).

Step 3: The oil-well rack is fitted with a spacer and a cover (Fig. 2) and is placed horizontally in a glass vessel. As a precaution against evaporation during the subsequent steps, 2 ml of water is placed in the vessel.

Comment: The vessel we have used (Outlaw et al., 1981) is a freeze-drying flask; the top and bottom are positioned to leave the orifice open, but the nipple itself is covered by a test tube. The entire assembly cannot be sealed tightly, but gaseous exchange between the inside and the outside of the vessel should be minimized.

Step 4: The assembly is placed inside a home-style pressure cooker (Fig. 2). Fitted with a 15-psi counterweight, the cooker is brought to pressure (120°C) and maintained for 6 h. With care to avoid excess outgassing, the 6-h hydrolysis period can proceed without interruption to replenish the cooker water.

Step 5: After the cooker returns to room temperature, the flask assembly is removed, and the oil-well cover and spacer are set aside. Thirty nanoliters 6 N NaOH is added to each droplet.

Step 6: Additional single-cell samples are added to reserved blank droplets. The oil-well rack is heated to 80°C for 20 min.

Comment: These droplets, to which cells were added after hydrolysis, will provide a "cell blank" that accounts for preexisting (nonprotein) glutamate.

IV. Analysis of Glutamate in the Protein Hydrolysate

This section describes glutamate measurement in two phases. In the first phase (Step 1), an enzymatic step couples glutamate reaction to reduction of NAD^+. In the second phase (Steps 2–4), formed NADH—stoichiometric with the picomole quantities of glutamate formed by protein hydrolysis—is quantified by enzymatic amplification techniques. A brief overview of the principles underlying metabolite measurement by enzymatic amplification is given by Outlaw (1980).

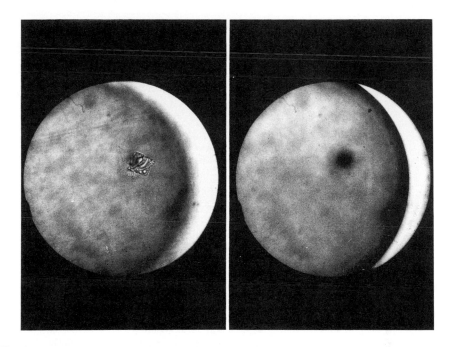

Fig. 1 Extraction of a single-plant-cell-sized sample in a microdroplet, which is contained within the oil of an oil well. One-half of a guard-cell pair was pushed through the oil and onto the reagent droplet. The tissue positions itself on the interface of the droplet and the oil. The focus for the photograph at left was on the sample (so the droplet periphery is unfocused). On the right, the periphery of the droplet is in focus and the sample is a dark blur. [This particular pair of photographs was taken during a series of guard-cell starch measurements. As epidermal cells contain little starch (Outlaw and Manchester, 1979), special care was not taken to ensure a high level of dissection precision, which is more aptly illustrated by Hampp and Outlaw (1987) and Harris *et al.* (1988). In addition, because starch and the analytical enzymes used to degrade it are macromolecules incapable of penetrating the guard-cell wall, atypically, the cells were cut to eliminate the wall barrier.]

Step 1: To each oil-well droplet, 0.5 μl of glutamate reagent is added. The reagent is 150 mM Tris-acetate (pH 8.4), 0.1 mM ADP, 0.5 mM NAD$^+$, and 150 μg · ml^{-1} glutamate dehydrogenase (EC 1.4.1.2).

Comment: Glutamate dehydrogenase is incorporated into the reagent just before use; inhibition by SO_4^{2-} (carried over from the hydrolysis step) explains the high concentration of this enzyme required.

Step 2: Following a 30-min incubation at room temperature, 0.5 μl 0.25 N NaOH is added to each well. The rack is heated to 80°C for 45 min.

Comment: This alkali/heating step destroys the NAD$^+$ of the glutamate reagent. NADH is unchanged by this step. Thus, after this step, NAD$_{total}$ equals NADH, which also equals glutamate originally present.

Step 3: An aliquot, 0.5 μl, of each droplet is transferred to a 10 × 75-mm borosilicate test tube that contains 50 μl of NAD cycling reagent. (The test tube

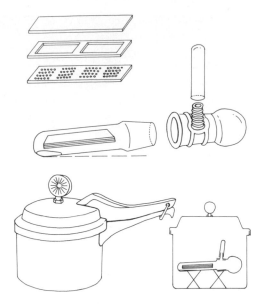

Fig. 2 Preparation of protein hydrolysate. The oil-well rack containing the 30-nl droplets of SO_4^{2-}/protein is covered by a spacer and Teflon top, which are loosely secured by small clips. Placed inside an unsealed vacuum flask containing a small amount of water, the oil-well racks are heated to 120°C for 6 h.

rack is kept on ice until transfers are complete.) The NAD cycling reagent is 100 mM Tris (pH 8), 300 mM ethanol, 1 mM DTT, 2 mM OAA, 0.02% (w/v) bovine serum albumen, 20 μg · ml^{-1} alcohol dehydrogenase (EC 1.1.1.1), and 2 μg · ml^{-1} malic dehydrogenase (EC 1.1.1.37). Incubation is for 60 min at 25°C. Incubation is terminated by submersion of the bottom of the test tube rack in 95°C water for 3 min, after which the rack is returned to a room-temperature bath.

Comment: The enzymes are added immediately before reagent use. The recommended enzyme concentrations are nominal. More or less of these cycling enzymes can be indicated after an initial trial assay. NAD contamination can be removed from alcohol dehydrogenase (Kato *et al.,* 1973) in the unlikely case that it presents an unacceptable blank.

In this step, NADH from Steps 1 and 2 is introduced in catalytic amounts to the cycling reagent. NADH is oxidized through the agency of malic dehydrogenase; resulting NAD$^+$ is, in turn, reduced back to NADH by alcohol dehydrogenase. Thus, for each "turn of the cycle," one molecule of OAA is lost, and one of malate, gained. (Similarly, ethanol is converted to acetaldehyde.) Overall, under fixed assay parameters, the amount of malate accumulated depends on the quantity of NADH transferred into the cycling reagent. As mentioned earlier, NADH itself is equal to glutamate originally present, so that the nanomole quantity of malate accumulated is proportional to the picomole quantity of glutamate that resulted from protein hydrolysis.

Step 4: To each test tube is added 1 ml of malate-indicating reagent (50 mM 2-amino-2-methyl propanol-Cl (pH 9.9), 0.2 mM NAD$^+$, 10 mM glutamate, 5 μg · ml^{-1} malic dehydrogenase, and 2 μg · ml^{-1} glutamic-oxalacetic transaminase (EC 2.6.1.1). After 15 min at room temperature, malate oxidation (and formation of fluorescent NADH) is complete. Sample fluorescence is determined.

Comment: NADH excitation maximum is 340 nm, but the spectrum is broad, and the isolated 365-nm-Hg-lines group works well; emission maximum is approximately 465 nm.

V. Advantages and Limitations

Figure 3 shows a typical, expected assay result that includes all experimentation vagaries, particularly lack of precision in the delivery of nanoliter volumes and the maintenance of a consistent volume during hydrolysis. In two randomly selected assays, the range ($n = 3$) of the four standards that bracketed the palisade cell protein content averaged 4.3% of the standard. Expressed as a percentage of the mean, the standard deviation of the protein content of palisade cells in a

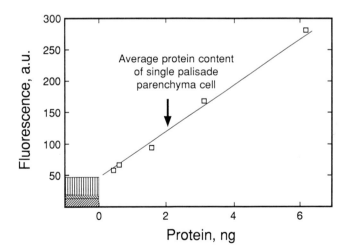

Fig. 3 Example results of one assay for the protein content of individual palisade cells (Outlaw and Manchester, unpublished). The histogram at left shows the total reagent blank, which has averaged about 1 ng protein. The final blank value is the sum of the blanks accumulated over the assay: hydrolysis and glutamate-specific step (▥), cycling step (▨), and malate-indicating step (▧). The malate-indicating step (essentially a within-assay constant) ranged from 30 to 60% of the overall blank. This blank can be diminished, should it be desired, by increasing the amplification in the cycling step.

typical assay was 36. As the method is destructive, this comparison shows that it would be pointless to seek improved precision.

Any method must be judged by the resources at hand and the purpose. The primary advantages of the glutamate method are (a) relatively high accuracy (using the Lowry as a standard) on animal (Butcher and Lowry, 1976) and plant (Outlaw et al., 1981) extracts, (b) stability and economy of stock reagents, (c) absence of toxic waste (e.g., radioactivity), and (d) use of only inexpensive and common instrumentation. The primary disadvantages are (a) the level of skill required for the micromanipulation and micropipetting, (b) the complicated protocol, and (c) the reliance on custom-fabricated tools. As mentioned in the introduction, several other methods deserve consideration. The o-phthalaldehyde-based microassay (Zawieja et al., 1984), in particular, is attractive for those who have access to and skill in the use of a high-quality quantitative fluorescence microscope. Although we have not used the microversion of the o-phthalaldehyde assay ourselves, our elaboration (Outlaw et al., 1985) and continued use (e.g., Tarczynski and Outlaw, 1993) of microdroplet fluorometry suggest the ease with which it might be done.

Acknowledgments

William H. Outlaw thanks the U.S. Departments of Energy and of Agriculture for support during the preparation of this review and A. B. Thistle for help in preparation of the manuscript.

References

Bradford, M. M. (1976). A rapid and sensitive method for the quantitation of microgram quantities of protein utilizing the principle of protein-dye binding. *Anal. Biochem.* **72,** 248–254.

Butcher, E. C., and Lowry, O. H. (1976). Measurement of nanogram quantities of protein by hydrolysis followed by reaction with orthophthalaldehyde or determination of glutamate. *Anal. Biochem.* **76,** 502–523.

Chang, J.-Y., Knecht, R., Jenoe, P., and Vekemans, S. (1989). Amino acid analysis at the femtomole level using dimethylaminoazobenzene sulfonyl chloride precolumn derivatization method: Potential and limitation. *In* "Techniques in Protein Chemistry" (T. E. Hugli, ed.), pp. 305–314. New York: Academic Press.

Ciesiolka, T., and Gabius, H.-J. (1988). An 8- to 10-fold enhancement in sensitivity for quantitation of proteins by modified application of colloidal gold. *Anal. Biochem.* **168,** 280–283.

Hames, B. D., and Rickwood, D. (1990). "Gel Electrophoresis of Proteins: A Practical Approach," 2nd Ed. Oxford: Oxford Univ. Press.

Hammer, K. D., and Nagel, K. (1986). An automated fluorescence assay for subnanogram quantities of protein in the presence of interfering material. *Anal. Biochem.* **155,** 308–314.

Hampp, R., and Outlaw, W. H., Jr. (1987). Mikroanalytik in der pflanzlichen Biochemie. *Naturwissenschaften* **74,** 431–438.

Harris, M. J., Outlaw, W. H., Jr., Weiler, E. W., and Mertens, R. (1988). Water-stress-induced changes in the abscisic acid content of guard cells and other cells of *Vicia faba* L. leaves, as determined by enzyme-amplified immunoassay. *Proc. Natl. Acad. Sci. U.S.A.* **85,** 2584–2588.

Hernández, M. J. M., Camañas, R. M. V., Cuenca, E. M., and Alvarez-Coque, M. C. G. (1990). Determination of the protein and free amino acid content in a sample using o-phthalaldehyde and N-acetyl-L-cycteine. *Analyst* **115,** 1125–1128.

Hunter, J. B., and Hunter, S. M. (1987). Quantification of proteins in the low nanogram range by staining with the colloidal gold stain AuroDye. *Anal. Biochem.* **164,** 430–433.

Kalinich, J. F., and McClain, D. E. (1992). An *in vitro* method for radiolabeling proteins with ^{35}S. *Anal. Biochem.* **205,** 208–212.

Kato, T., Berger, S. J., Carter, J. A., and Lowry, O. H. (1973). An enzymatic cycling method for nicotinamide-adenine dinucleotide with malic and alcohol dehydrogenases. *Anal. Biochem.* **536,** 86–97.

Krystal, G., MacDonald, C., Munt, B., and Ashwell, S. (1985). A method for quantitating nanogram amounts of soluble protein using the principle of silver binding. *Anal. Biochem.* **148,** 451–460.

Lee, T. T., and Yeung, E. S. (1992). Quantitative determination of native proteins in individual human erythrocytes by capillary zone electrophoresis with laser-induced fluorescence detection. *Anal. Chem.* **64,** 3045–3051.

Lowry, O. H., Rosebrough, N. J., Farr, A. L., and Randall, R. J. (1951). Protein measurement with the Folin phenol reagent. *J. Biol. Chem.* **193,** 265–275.

Merril, C. R. (1990). Silver staining of proteins and DNA. *Nature* **343,** 779–780.

Outlaw, W. H., Jr. (1980). A descriptive evaluation of quantitative histochemical methods based on pyridine nucleotides. *Annu. Rev. Plant Physiol.* **31,** 299–311.

Outlaw, W. H., Jr., and Manchester, J. (1979). Guard cell starch concentration quantitatively related to stomatal aperture. *Plant Physiol.* **64,** 79–82.

Outlaw, W. H., Jr., Manchester, J., and Zenger, V. E. (1981). The relationship between protein content and dry weight of guard cells and other single cell samples of *Vicia faba* L. *Histochem. J.* **13,** 329–336.

Outlaw, W. H., Jr., Springer, S. A., and Tarczynski, M. C. (1985). Histochemical technique. A general method for quantitative enzyme assays of single cell "extracts" with a time resolution of seconds and a reading precision of fmol. *Plant Physiol.* **77,** 659–666.

Passonneau, J. V., and Lowry, O. H. (1993). "Enzymatic Analysis. A Practical Guide." Totowa, New Jersey: Humana Press.

Peterson, G. L. (1979). Review of the Folin phenol protein quantitation method of Lowry, Rosebrough, Farr, and Randall. *Anal. Biochem.* **100,** 201–220.

Roth, M. (1971). Fluorescence reaction for amino acids. *Anal. Chem.* **43,** 880–882.

Sandwick, R. K., and Schray, K. J. (1985). Protein quantitation of as low as 10-ng/ml concentrations by competitive binding to polystyrene latexes. *Anal. Biochem.* **147,** 210–216.

Seitz, W. R. (1980). Fluorescence derivatization. CRC *Crit. Rev. Anal. Chem.* **8,** 367–405.

Smith, P. K., Krohn, R. I., Hermanson, G. T., Mallia, A. K., Gartner, F. H., Provenzano, M. D., Fujimoto, E. K., Goeke, N. M., Olson, B. J., and Klenk, D. C. (1985). Measurement of protein using bicinchoninic acid. *Anal. Biochem.* **150,** 76–85.

Stoscheck, C. M. (1987). Protein assay sensitive at nanogram levels. *Anal. Biochem.* **160,** 301–305.

Stoscheck, C. M. (1990). Quantitation of protein. *Methods Enzymol.* **182,** 50–68.

Stowell, C. P., Kuhlenschmidt, T. B., and Hoppe, C. A. (1978). A fluorescamine assay for submicrogram quantities of protein in the presence of Triton X-100. *Anal. Biochem.* **85,** 572–580.

Tarczynski, M. C., and Outlaw, W. H., Jr. (1987). Histochemical technique. Densitometry of nanogram quantities of proteins separated in one-dimensional microslab gels. *Plant Physiol.* **85,** 1059–1062.

Tarczynski, M. C., and Outlaw, W. H., Jr. (1993). The interactive effects of pH, malate, and glucose 6-P on guard-cell phosphoenolpyruvate carboxylase. *Plant Physiol.* **103,** 1189–1194.

Udenfriend, S., Stein, S., Bohlen, P., Dairman, W., Leimgruber, W., and Weigele, M. (1972). Fluorescamine: A reagent for assay of amino acids, peptides, proteins, and primary amines in the picomole range. *Science* **178,** 871–872.

Viets, J. W., Deen, W. M., Troy, J. L., and Brenner, B. M. (1978). Determination of serum protein concentrations in nanoliter blood samples using fluorescamine or *o*-phthalaldehyde. *Anal. Biochem.* **88,** 513–521.

Wallace, A., and Saluz, H. P. (1992). Ultramicrodetection of proteins in polyacrylamide gels. *Anal. Biochem.* **203,** 27–34.

Zawieja, D., Barber, B. J., and Roman, R. J. (1984). Analysis of picogram quantities of protein in subnanoliter-size samples. *Anal. Biochem.* **142,** 182–188.

CHAPTER 5

Photoaffinity Labeling and Strategies for Plasma Membrane Protein Purification

Joachim Feldwisch, Andreas Vente, Narciso Campos, Rolf Zettl, and Klaus Palme

Max-Planck Institut für Züchtungsforschung
Carl-von-Linné-Weg 10
D-50829 Köln, Germany

I. Introduction

Auxins play a major role in the regulation of plant cell elongation, cell division, morphogenesis, and differentiation. The molecular sequence of events responsible for the mode of action of auxins is still poorly understood. Receptor-like proteins that bind auxin and transmit the auxin signal have been postulated.

Due to the low concentration of most of these proteins, conventional approaches to identify and characterize them for molecular studies have been fraught with technical difficulties. To overcome problems associated with traditional auxin-binding studies (e.g., low receptor protein concentration, instability of auxin binding under experimental conditions, denaturation or loss of the auxin-binding proteins during purification) photolabile synthetic auxins may provide a valuable tool with which to tag auxin-binding proteins.

Photoaffinity probes have contributed greatly to the identification of receptor proteins and have aided in structural studies (Bayley and Knowles, 1977; Chowdhry and Westheimer, 1979). Upon illumination, photoaffinity ligands covalently label the ligand-binding polypeptide and allow receptor molecules to be followed throughout purification under both denaturing and nondenaturing conditions.

Photolabile synthetic auxins are currently widely used by several laboratories (for review see Jones, 1994; Palme, 1994). Here we describe recent progress in rapid and sensitive photoaffinity labeling of auxin-binding proteins. Since methods for the purification of plant proteins were recently reviewed by Weselake and Jain (1992), we will concentrate our discussion on strategies to purify membrane proteins photoaffinity-labeled with $5\text{-}N_3\text{-}[7\text{-}^3H]$indole-3-acetic acid ($[^3H]N_3IAA$).

II. Materials

A. Radiochemicals and Chemicals

$[^3H]N_3IAA$, 21.7 Ci/mmol (1 Ci = 37 GBq), was synthesized as described (Campos *et al.*, 1991). All chemicals were of analytical grade and purchased from Sigma (München, FRG) or Merck (Darmstadt, FRG), unless stated otherwise.

B. Plant Material

Seeds of *Zea mays* L. (cv. mutin 240; Kleinwanzlebener Saatzucht, Einbeck, FRG) were soaked in water overnight and grown on moist cotton for 3 days at 28°C in the dark. Coleoptiles, including the primary leaf, were harvested, frozen in liquid nitrogen, and stored at −70°C.

III. Methods

A. Preparation of Plasma Membranes

Plasma membranes were isolated by aqueous two-phase partitioning (see Larsson *et al.*, 1987, for a review). All procedures were performed at 4°C. Coleoptiles (200 g) including the primary leaves were homogenized using a Waring blendor or a Polytron (Kinematica, at speed setting 7–8) in 500 ml Buffer 1 (100 mM

Tris/citric acid pH 8, 500 mM sucrose, 1 mM EDTA, 5 mM ascorbic acid, 0.1 mM MgCl$_2$, 0.6% polyvinylpolypyrrolidone, 1 mM DTT, 3.5 μg/ml aprotinin, 0.1 μg/ml leupeptin). The homogenate was filtered through a nylon mesh (135 μm), and the particulate material retained by the mesh was reextracted using an equal volume of Buffer 1. The combined homogenates were centrifuged for 10 min at 10,000 $\times$ g. A microsomal pellet was obtained by centrifugation at 142,000 $\times$ g for 40 min [45 Ti Rotor (Beckman), 35,000 rpm, ~820 ml total volume of extract]. The microsomal pellet was suspended in Buffer 2 (5 mM potassium phosphate, pH 7.8, 330 mM sucrose) and adjusted to 18 g. One-half of this membrane suspension was then applied onto a "27-g system" (see protocols), resulting in a final weight of 36 g.

B. Protocol 1

1. Prepare two 27-g systems in 50-ml Falcon tubes (concentrations are calculated to the final weight of 36 g) consisting of 6.2% Dextran T 500 (11.16 g of 20% w/w in H$_2$O); 6.2% PEG 3350 (5.58 g of 40% w/w in H$_2$O); 0.33 M sucrose (3.05 g); 5 mM KPB (675 μl of 0.2 M KPB, pH 7.8; KPB comprises 1 part 0.2 M KHPO$_4$, pH 7.8; with 10 parts 0.2 M KHPO$_4$, pH 7.8; 4 mM KCl (108 μl of 1 M KCl); adjust with H$_2$O to 27 g (store at -20°C).

2. Apply each 9 g of the membrane suspension on a 27-g system and mix gently. After centrifugation for 10 min at 1500 $\times$ g in a swinging bucket rotor, an upper and a lower phase (U1 and L1) can be distinguished.

3. Carefully collect the upper phases (U1) and apply each on a new lower phase (L2, 10 ml). Each of the lower phases (L1) is reextracted with 10 ml fresh upper phase (U1'). Fresh lower and upper phases are produced from a 300-g system consisting of 6.2% Dextran T 500 (93 g of 20% w/w in H$_2$O); 6.2% PEG 3350 (46.5 g of 40% w/w in H$_2$O); 0.33 M sucrose (33,89 g); 5 mM KPB (7.5 ml of 0.2 M KPB pH 7.8); 4 mM KCl (1.2 ml of 1 M KCl); adjust with H$_2$O to 300 g. Mix thoroughly and incubate in a separation funnel overnight at 4°C. Collect the lower and the upper phases separately in 10-ml aliquots in 50-ml Falcon tubes (store at -20°C).

4. Repeat step 3 twice. Combine the fractions U4 and U4', dilute with at least three vol of Buffer 3 (250 mM sucrose in 10 mM Tris/Mes, pH 6.5), and centrifuge at 141,000 $\times$ g for 1 h (28,000 rpm, Beckmann SW28 rotor).

5. Decant the supernatant and drain the walls of the tube. Dissolve the membrane pellet in 1.2 ml Buffer 3, determine the protein concentration (Smith *et al.*, 1985), and store aliquots at -70°C.

C. Photoaffinity Labeling of Proteins

Samples containing 5–100 μg protein were incubated with 1 μM [^{3}H]N$_3$IAA for 10 min on ice in a total volume of 100 μl. Optionally, unlabeled auxins or related compounds were added for competition analysis. Photolysis was performed at either −196°C or 4°C for 10 min using a high-pressure mercury lamp (Philips HKP 125 W/L). This lamp emits visible light, UV light, and infrared light. To cool the lamp efficiently we used a water jacket of borosilicate glass (DURAN, spectral resolution 310 to 2200 nm). The lamp was placed at the bottom of the glass cylinder within a distance of about 5 cm to the samples. Illumination for 10 min is sufficient for complete photolytic degradation of [³H]N₃IAA (Campos *et al.,* 1991). After illumination, proteins were recovered either by centrifugation at 108,000 × *g* for 11 min using a TL-100 tabletop ultracentrifuge (Beckman) or by precipitation with 10% TCA. Pellets were dissolved in SDS loading buffer, heated to 70°C for 5 min, and separated by SDS–PAGE (12.5%). Fluorography was performed according to Skinner and Griswold (1983). Dried gels were exposed for 1 to 21 days to preflashed Kodak XAR-5 films.

D. Protocol 2

1. Prepare a mix containing membrane proteins (between 5 and 100 μg of protein), Buffer 3 and, optionally, 1 m*M* unlabeled auxins or related compounds as competitors. The final volume of this mix should be 90 μl. Keep the sample(s) on ice. Further work should be performed in the presence of dim red light.

2. Prepare a dilution of [³H]N₃IAA in Buffer 3 just before use. The concentration of [³H]N₃IAA should be 10 μ*M*.

3. Add 10 μl of the 10 μ*M* [³H]N₃IAA to each sample, mix briefly, and incubate for 10 min.

4. Pipette each sample into a lid (total lid volume is 260 μl) cut from a microcentrifuge tube and placed on a metallic block cooled on ice. Quickly freeze the samples by transferring the lids to the surface of a metallic block placed in liquid nitrogen (−196°C) or an ice-cold metallic block (4°C).

5. Place the lamp close to the sample block (distance approximately 5 cm). Switch on the lamp and illuminate for 10 min (this includes the heating time of the lamp).

E. Extraction of Membranes with Detergents

Membrane vesicles photoaffinity labeled with [³H]N₃IAA were subsequently extracted with Triton X-100 and with Triton X-114.

F. Protocol 3

1. Mix 200 μl plasma membrane vesicles containing 200 μg protein with 200 μl of Buffer 3 containing 2% Triton X-100.

2. Shake for 30–60 min at 4°C in an Eppendorf shaker.

3. Centrifuge for 11 min at 2°C at 108,000 × *g* (Tl 100.3 Rotor, Beckman).

4. Remove the supernatant and precipitate with 10% TCA (final concentration). This fraction contains proteins solubilized with Triton X-100.

5. Dissolve the insoluble pellet with 500 μl Buffer 4 (10 mM Tris/HCl, pH 7.5, 150 mM NaCl, 1% Triton X-114) and shake for 30–60 min at 4°C in an Eppendorf shaker.

6. Incubate samples for 5–10 min at 30°C to induce phase separation.

7. Centrifuge for 5 min at 2000 rpm in a microcentrifuge.

8. Collect the upper Triton X-114 depleted phase and precipitate with 10% TCA. The lower Triton X-114-enriched phase is diluted with 1200 μl H$_2$O and precipitated by the addition of 150 μl TCA (10% final concentration).

IV. Results and Discussion

A. Isolation of Plasma Membrane Vesicles

We have identified and purified several auxin-binding proteins from maize seedlings. As starting material we used either maize microsomal vesicles or plasma membrane-enriched vesicles. The purity of the plasma membrane-enriched vesicles was assessed previously by marker enzyme analysis (Feldwisch *et al.*, 1992). Analysis of organelle-specific marker enzymes revealed that these vesicles were essentially free of mitochondrial and tonoplast membranes. A 3.4-fold decrease in the endoplasmic reticulum specific marker enzyme cytochrome-*c* reductase and a 5.3-fold increase in activity of the plasma membrane-specific vanadate-sensitive ATPase suggested that these vesicles contained more than 90% plasma membranes. However, at present we cannot exclude minor contamination with organellar membranes (e.g., plastid membranes).

B. Photoaffinity Labeling

We established a method for photoaffinity labeling of proteins in our laboratory. It is easy to perform under dim red light conditions, even with a large number of samples (up to 40 per illumination) and the labeling parameters can easily be modified. The optimal labeling conditions, like illumination time and type of lamp, were as described by Campos *et al.* (1991). The degradation of the azido group of [^{3}H]N$_3$IAA was followed by measuring the decrease of absorbance at 244 nm after different lengths of illumination. After 9 min of illumination no further decrease in absorbance was observed, indicating that the photolytic degradation of [^{3}H]N$_3$IAA was already complete. Therefore, illumination for 10 min was used routinely in all further experiments. The use of a UV lamp (254 nm) instead of the high-pressure mercury lamp did not change the pattern of proteins labeled with [^{3}H]N$_3$IAA.

Using the approach described here we have identified several proteins in different fractions of maize seedlings that are labeled with [^{3}H]N$_3$IAA (Table

I). Some of these proteins were already detected in membrane preparations after fluorography for 1 day (e.g., pm60 and pm58), whereas other proteins could only be detected after fluorography for more than 14 days (pm23 and pm24). A critical aspect in using the photoaffinity labeling approach is to distinguish between specific and unspecific labeling. Specific photoaffinity labeling of proteins was assessed by labeling (*a*) at different temperatures, (*b*) in the presence of competitors, and (*c*) in the presence of a scavenger.

Routinely, we performed the photoaffinity labeling at 4°C and at −196°C in liquid nitrogen. For most of the proteins identified we achieved the highest labeling intensity at −196°C, whereas two proteins, ZmERabp1 and p23, were most strongly labeled at 4°C (Table I). p23 was the only protein that was not labeled at all at −196°C. The strong influence of the labeling temperature on the labeling specificity was best demonstrated by [^{3}H]N$_3$IAA labeling of plasma membrane proteins. In these experiments reduction of the labeling temperature drastically reduced the background labeling, resulting in labeling of only three proteins of 60 kDa (pm60), 58kDa (pm58), and 23 kDa (pm23) (Feldwisch *et al.*, 1992). The reason for this may be the reduced mobility of molecules at −196°C, so that only proteins that had already bound [^{3}H]N$_3$IAA to the binding site(s) at the time of shock-freezing were specifically labeled.

The labeling specificity was demonstrated by competition analysis. In these experiments the photoaffinity labeling was performed in the presence of a 1000-fold molar excess of unlabeled auxins or relevant analogs. Photoaffinity labeling with [^{3}H]N$_3$IAA of all proteins identified in maize was drastically reduced by IAA, 1-NAA, and 2,4-D. However, compounds with no auxin activity, like tryptophan and benzoic acid, did not reduce or only slightly reduced the labeling of these proteins, indicating that the labeling is due to specific interactions of [^{3}H]N$_3$IAA with the target proteins (Campos *et al.*, 1992; Feldwisch *et al.*, 1992; Feldwisch *et al.*, 1994).

Table I
Proteins Identified by Photoaffinity Labeling with [^{3}H]N$_3$IAA in Maize Seedlings

Protein	Labeling temperature (°C)	Labeling reduced by 10^{-3} *M* PABA[a]	Function
p60[b]	−196	Partially	ß-Glucosidase
p58[c]	−196	Partially	ß-Glucosidase
p23[d]	+4	Complete	Mn-SOD
ZmERabp1	+4	ND	?
pm60[e]	−196	Partially	ß-Glucosidase
pm58[f]	−196	Partially	ß-Glucosidase
pm24	−196	Partially	?
pm23	−196	Partially	?

[a] PABA, *p*-amino benzoic acid; ND, not determined. [b] Brzobohaty *et al.* (1993). [c] Zettl, unpublished results. [d] Feldwisch *et al.* (1995). [e] Feldwisch *et al.* (1994). [f] Feldwisch *et al.* (1994).

The most important experiment to prove labeling specificity is the use of a scavenger. Upon illumination the azido group of [^{3}H]N$_3$IAA is decomposed to a reactive nitrene. These nintrenes can undergo the following reactions: (*a*) insertion into C—H bonds to yield secondary amines, (*b*) cycloadditions to double bonds to form cyclic 3-member imines, (*c*) hydrogen atom abstractions followed by coupling of the free radicals, and (*d*) addition to nucleophiles (Glazer *et al.,* 1975). Due to these types of reactions, covalent linkage of the photolabile ligand is not restricted to the binding site of a target protein but may also occur at nonspecific sites on the target protein or even on other functionally irrelevant proteins. This pseudo-photoaffinity labeling can be reduced if a scavenger is included in the reaction. Scavengers should bind excess photogenerated reactive nitrenes thereby restricting labeling to the binding site of the target protein. We used *p*-amino benzoic acid (PABA) as a scavenger. With one exception (p23), labeling of all proteins identified in maize was only slightly reduced in the presence 1 m *M* PABA (Table I). Labeling of p23 (identified as Mn-SOD, Feldwisch *et al.,* unpublished work), however, seems to be a good example for pseudo-photoaffinity labeling. Sodium azide is a competitive inhibitor of Mn-SOD and Fe-SOD. As shown recently by X-ray structure analysis, azide is bound within the binding site of SOD to the open coordination position of the iron/manganese atom with a bond length of 2 Å (Stoddard *et al.,* 1990). This suggests that photoaffinity labeling by [^{3}H]N$_3$IAA of the maize Mn-SOD may be due to binding of the azido group and not to specific binding of the auxin molecule.

C. Strategies for Membrance Protein Purification

By photoaffinity labeling with [^{3}H]N$_3$IAA we have identified in different fractions of maize seedlings several proteins that differ in their membrane solubility. As shown in Fig. 1, p60, p58, ZmERabp1 and p23 were solubilized from microsomal membranes by acetone, *n*-butanol, and Triton X-100 extraction, whereas pm60, pm58, pm23, and pm24 were solubilized after extraction with Triton X-114 of either microsomal membranes or plasma membrane-enriched fractions. The purification strategies for all these proteins are outlined schematically in Fig. 1.

One prerequisite for membrane protein purification is the efficient solubilization of the protein from the membrane. Solubilization of membrane proteins is normally achieved by the use of nonionic, zwitterionic, or ionic detergents, organic solvents or high salt concentrations (Thomas and McNamee, 1990; Hjelmeland, 1990). The first step in developing a purification scheme should be the search for a detergent that not only gives the highest solubilization rate but also preserves the proteins of interest in its native structural and functional state. It is also important that the detergent be compatible with further purification steps, like ion-exchange chromatography or gel filtration, and does not interfere with enzymatic assays or binding assays used to identify the target protein. Due to the lack of a detergent of general use for the solubilization of membrane proteins,

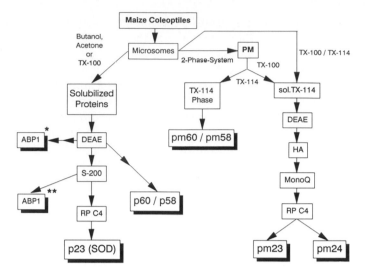

Fig. 1 Schematic representation of membrane-associated and membrane proteins from maize coleoptiles identified by photoaffinity labeling with [³H]N₃IAA. Proteins were purified either from microsomes or from a plasma membrane-enriched fraction (PM), as described previously [Napier *et al.,* 1988 (ABP1**); Palme *et al.,* 1990 (ABP1*), Campos *et al.,* 1992 (p60); Feldwisch *et al.,* 1992 (pm23/pm24); Feldwisch *et al.,* 1994 (pm60/pm58/p58)]. TX-100, Triton X-100 extraction; TX-114, Triton X-114 extraction; HA, hydroxylapatite chromatography.

as many different detergents as possible should be tested in the initial trials. Furthermore, the temperature-induced phase separation, a special characteristic of some detergents, should be considered when planning a purification scheme. Above a well-defined temperature, the cloud point (cp), some detergents, e.g., Triton X-100 (cp = 65°C) and Triton X-114 (cp = 22°C), form a two-phase system consisting of a detergent-enriched and a detergent-depleted phase. Using Triton X-114, hydrophobic integral membrane proteins partitioned into the lower detergent-enriched phase, whereas hydrophilic proteins partitioned into the upper detergent-depleted phase (Bordier, 1981). However, some integral membrane proteins, especially glycoproteins were also detected in the Triton X-114-depleted phase (Pryde, 1986), indicating that there is not a strict correlation between hydrophobicity of a protein and its partitioning in the two-phase system. In addition, temperature-sensitive membrane proteins can be isolated by using a mixture of Triton X-45 and Triton X-114, which allows the adjustment of the phase transition temperature anywhere between 0 and 22°C (Ganong and Delmore, 1991).

In our hands subsequent extraction of membrane vesicles with at least two different detergents turned out to be useful. The first detergent (Triton X-100) was used to solubilize most of the contaminating proteins. The second detergent (Triton X-114) was used to solubilize the remaining insoluble proteins.

Temperature-induced Triton X-114 phase separation was then used as an additional purification step. Using this strategy for the purification of pm23 and pm24, about 90% of contaminating proteins were removed by the subsequent detergent extractions (Feldwisch *et al.*, 1992). On the basis of protein determination, 73% of microsomal proteins were solubilized with Triton X-100, 17% were found in the Triton X-114-enriched phase, and 10% (including pm23 and pm24) were in the aqueous Triton X-114-depleted phase. Furthermore, Triton X-114 did not interfere with the photoaffinity labeling of pm23 and pm24. Therefore both proteins could be easily detected by [^{3}H]N$_3$IAA labeling in the following purification steps.

For the purification of membrane proteins solubilized with detergents, basically the same methods can be applied as those used to purify soluble proteins; these include gel filtration, ion-exchange, reversed-phase, and affinity chromatography (see Thomas and McNamee, 1990, for a recent review). In addition, chromatography on hydroxyapatite offers a further opportunity to separate proteins by their different types of adsorption to this matrix. Depending on the buffer system and salts used, hydroxyapatite columns can be optimized for the purification of either basic or acidic proteins (Gorbunoff, 1990).

As an example, Fig. 1 shows the series of separation steps that were used to purify pm23 and pm24 from maize membranes. After anion-exchange chromatography, hydroxyapatite chromatography, and reversed-phase chromatography a 7000-fold enrichment of these proteins was obtained. All chromatographic separations were performed in the presence of 0.06% (1.12 mM) Triton X-114. At this concentration Triton X-114 did not interfere significantly with UV monitoring. The purified proteins were further analyzed by amino acid sequence analysis. Synthetic oligonucleotides were then used to isolate the corresponding cDNAs (Feldwisch, Vente, Moore, and Palme, unpublished work).

VI. Conclusions

By photoaffinity labeling with [^{3}H]N$_3$IAA we have successfully identified several soluble and membrane-associated proteins. These proteins were purified and partial amino acid sequences determined. Although for some of these proteins enzymatic activities could already be predicted from similarities to known proteins (Brzobohaty *et al.*, 1993; Zettl *et al.*, 1993), no information is yet available for other proteins identified by this approach. Molecular cloning of their corresponding cDNA will assist in unravelling their function.

References

Bayley, H., and Knowles, R. (1977). Photoaffinity labeling. *In* "Methods in Enzymology: Affinity labeling" (W. B. Jakoby, M. Wilchek, eds.), Vol. 46, pp. 69–114. New York: Academic Press.

Bordier, C. (1981). Phase separation of integral membrane proteins in Triton X-114 solution *J. Biol. Chem.* **256**, 1604–1607.

Brzobohaty, B., Moore, I., Kristoffersen, P., Bako, L., Campos, N., Schell, J., and Palme, K. (1993). Release of active cytokinin by a β-glucosidase localized to the maize root meristem. *Science* **262**, 1051–1054.

Campos, N., Feldwisch, J., Zettl, R., Boland, W., Schell, J., and Palme, K. (1991). Identification of auxin binding proteins using an improved assay for photoaffinity labeling with 5-N$_3$-[7-^{3}H]-indole-3-acetic acid. *Technique* **3**, 69–75.

Campos, N., Laszlo, B., Feldwisch, J., Schell, J., and Palme, K. (1992). A protein from maize labeled with azido-IAA has novel β-glucosidase activity. *Plant J.* **1**, 675–684.

Chowdhry, V., and Westheimer, F. H. (1979). Photoaffinity labeling of biological systems. *Ann. Rev. Biochem.* **48**, 293–325.

Feldwisch, J., Zettl, R., Hesse, F., Schell, J., and Palme, K. (1992). An auxin-binding protein is localized to the plasma membrane of maize coleoptile cells. Identification by photoaffinity labeling and purification of a 23-kDa polypeptide. *Proc. Natl. Acad. Sci. U.S.A.* **89**, 475–479.

Feldwisch, J., Vente, A., Zettl, R., Bako, L., Campos, N., and Palme, K. (1994). Characterization of two membrane-associated β-glucosidases from maize coleoptiles. Submitted for publication.

Feldwisch, J., Zettl, R., Campos, N., and Palme, K. (1995). Identification of a 23 kDa protein from maize photoaffinity-labelled with 5-azido-[7-^{3}H]indol-3-ylacetic acid. *Biochem. J.* **305**, 853–857.

Ganong, B. R., and Delmore, J. P. (1991). Phase separation temperatures of mixtures of Triton X-114 and Triton X-45: Application to protein separation. *Anal. Biochem.* **193**, 35–37.

Glazer, A. N., DeLange, R. J., and Sigman, D. S. (1975). Chemical modification of proteins. Selected methods and analytical procedures. *In* "Laboratory Techniques in Biochemistry and Molecular Biology" (T. S. Work and E. Work, eds.), Vol. 4, pp. 167–179.

Gorbunoff, M. J. (1990). Protein chromatography on hydroxyapatite columns. *Methods Enzymol.* **182**, 329–338.

Hjelmeland, L. M. (1990). Solubilization of native membrane proteins. *Methods Enzymol.* **182**, 253–263.

Jones, A. M. (1994). Auxin-binding proteins. *Annu. Rev. Plant Physiol. Plant Mol. Biol.* **45**, 393–420.

Larsson, C., Widell, S., and Kjellbom, P. (1987). Preparation of high-purity plasma-membranes. *Methods Enzymol.* **148**, 559–568.

Napier, R. M., Venis, M. A. Bolton, M. A., Richardson, L. I., and Butcher, G. W. (1988). Preparation and characterisation of monoclonal and polyclonal antibodies to maize membrane auxin-binding protein. *Planta* **176**, 519–526.

Palme, K., Feldwisch, J., Hesse, T., Bauw, G., Puype, M., Vandekerckhove, J., and Schell, J. (1990). Auxin binding proteins from maize coleoptiles: Purification and molecular properties. *In* "Hormone Perception and Signal Transduction in Animals and Plants" (J. A. Roberts, C. Kirk, and M. Venis, eds.), Symp. Soc. Exp. Biol. XLIV, pp. 299–313.

Palme, K. (1994). Binding proteins to hormone receptors? *J. Plant Growth Regulat.* **12**, 171–178.

Pryde, J. G. (1986). Triton X-114: A detergent that has come from the cold. *TIBS* **11**, 160–163.

Skinner, M. K., and Griswold, M. D. (1983). Fluorographic detection of radioactivity in polyacrylamide gels with 2,5-diphenyloxazole in acetic acid and its comparison with existing procedures. *Biochem. J.* **209**, 281–284.

Smith, P. K., Krohn, R. I., Hermanson, G. T., Mallia, A. K., Gartner, F. H., Provenzano, M. D., Fujimoto, E. K., Goeke, N. M., Olson, B. J., and Klenk, D. C. (1985). Measurement of protein using bicinchoninic acid. *Anal. Biochem.* **150**, 76–85.

Stoddard, B. L., Ringe, D., and Petsko, G. A. (1990). The structure of iron superoxide dismutase from *Pseudomonas ovalis* complexed with the inhibitor azide. *Protein Eng.* **4**, 113–119.

Thomas, T. C., and McNamee, M. G. (1990). Purification of membrane proteins. *Methods Enzymol.* **182**, 499–520.

Weselake, R. J., and Jain, J. C. (1992). Strategies in the purification of plant proteins. *Physiol. Plant.* **84**, 301–309.

Zettl, R., Schell, J., and Palme, K. (1993). Photoaffinity labeling of *Arabidopsis thaliana* plasma membrane vesicles by 5-azido-[7-^{3}H]indole-3-acetic acid: Identification of a glutathione-S-transferase. *Proc. Natl. Acad. Sci. U.S.A.* **91**, 689–693.

CHAPTER 6

Cell Cycle Synchronization, Chromosome Isolation, and Flow–Sorting in Plants

Sergio Lucretti[*] and Jaroslav Doležel[†]

[*] ENEA Research Centre Casaccia
Biotechnology and Agriculture Sector
Via Anguillarese, 301
00060 S. M. di Galeria (Rome), Italy
[†] Norman Borlaug Centre for Plant Science
Institute of Experimental Botany
DeMontfort University
Sokolovska 6
CZ-77200 Olomouc, Czech Republic

METHODS IN CELL BIOLOGY, VOL. 50

I. Introduction

Plant cell populations synchronously traversing the cell division cycle or accumulated within a specific phase of the cycle represent a valuable model for studying a wide range of biological questions. Recent interest in synchronized cells has arisen not only in terms of studying the regulation of the cell division cycle, but also in terms of isolating chromosomes in suspension to permit purification, through flow-sorting, of individual chromosomes for molecular studies (Koch and Nasmyth, 1994; Doležel *et al.*, 1994).

Techniques for cell synchronization, for chromosome sorting, and for the production of chromosome-specific libraries were initially developed in mammalian species (for references, see Van Dilla and Deaven, 1990; Carter, 1993). The delay in developing equivalent techniques for the flow cytometric analysis and sorting of plant chromosomes was mainly due to difficulties in obtaining chromosome suspensions that contained intact and well-dispersed chromosomes and that were free of contaminating cellular debris.

The procedure for chromosome isolation involves four fundamental steps: (1) induction of cell cycle synchrony, (2) accumulation of cells in metaphase, (3) isolation of chromosomes from the cells, and (4) flow sorting of chromosomes.

A. Cell-Cycle Synchrony

Under normal conditions, cell-cycle synchrony occurs only rarely in the cells of plant organs (for a review, see Francis, 1992). Synchronizing techniques are therefore needed to make available large numbers of cells at defined phases of the cell cycle. Most of these techniques involve blocking DNA synthesis, which results in accumulation of cells at the G_1/S interface. Upon removal of the block, the cells synchronously traverse the subsequent phases of the cell division cycle. High degrees of synchrony can be obtained only if the treated populations contain a minimum of noncycling (resting) cells, as is the case of root tip meristems in actively growing roots (Van't Hof, 1973). Further advantages of root tip meristems compared to *in vitro* cultured cells include that they are karyologically stable and easier to handle.

Among a number of compounds that are known to inhibit DNA synthesis (Kornberg, 1980), only a few have been frequently used to synchronize cell cycle in root tips, including hydroxyurea (HU), 5-fluorodeoxyuridine (FudR), 5-aminouracil (5-AU), and aphidicolin (APC) (Eriksson, 1966; Scheuerman *et al.*, 1973; Sgorbati *et al.*, 1991). HU and FudR block DNA synthesis by inhibiting the formation of nucleotide precursors. 5-AU arrests DNA synthesis because it is a structural analogue of the pyrimidine bases. APC specifically inhibits DNA polymerase subunit α.

B. Metaphase Accumulation

Accumulation of cells in metaphase implies the use of some physical or chemical agent that inhibits mitotic spindle action. Those agents, if used at physiological concentrations, do not block cells indefinitely in metaphase; after a few hours, the chromosomes decondense, and form restitution nuclei and/or micronuclei. Colchicine (COL) is the most frequently employed chemical agent of its type. This alkaloid, extracted from *Colchicum autumnale*, binds to tubulin dimers, thereby preventing the formation of the mitotic spindle. COL is very effective in animal cell systems, where it is effective at micromolar levels, whereas with plant cells concentrations at the millimolar level are required for mitotic arrest (Morejohn et al., 1984a). Other compounds with significantly higher affinities to plant tubulins include the phosphoric amide herbicide amiprophos-methyl (APM), and the dinitroaniline herbicide oryzalin (ORY). APM and ORY appear especially promising as mitotic blocking agents for plant chromosome isolation. They have a high microtubule depolymerizing activity, and show visible antimicrotubular effects at low (micromolar) concentrations (Morejohn and Fosket, 1984b; Morejohn et al., 1987; Sree Ramulu et al., 1990; Doležel et al., 1992).

C. Chromosome Isolation

Methods for chromosome isolation were originally developed for animal and for human cells (Carrano et al., 1979). Chromosomes can be liberated after chemical or mechanical disruption of plasma membranes in a hypotonic buffer containing agents to weaken the membrane and to stabilize the released chromosomes (Gray and Langlois, 1986). For plant cells, these procedures must be modified through additional steps involving the use of polysaccharidases (pectinases, cellulases, hemicellulases) to digest the rigid cell wall (Mii et al., 1987). An alternative is to isolate the chromosomes directly by chopping of root tips using a scalpel; this requires mild prior fixation of the root tips with formaldehyde (Doležel et al., 1992). Recently, the technique has been successfully modified for the isolation of chromosomes from *A. cepa* (unpublished) and *P. sativum* (Gualberti et al., in preparation).

D. Flow-Karyotyping and Sorting

Flow cytometric analysis of suspensions of isolated chromosomes enables classification of chromosomes in the form of histograms (called flow-karyotypes) based on biochemical features such as their relative DNA content, base content, protein content, and various morphological parameters (DeLaat et al., 1986; Gray and Cram, 1990; Lucretti et al., 1993; Doležel et al., 1994). Flow-karyotyping has found applications in monitoring karyotype changes, for the detection of aberrant chromosomes, and for mutagenicity testing (Trask *et al.,* 1990; Gray *et al.,* 1988; Dietzel *et al.,* 1990). In humans, sorting of large quantities of chromosomes of a single type has been useful in the construction of chromosome-specific gene libraries (Van Dilla and Deaven, 1990), and in gene mapping (Cotter *et al.,* 1989). In plants, flow-karyotyping is made particularly difficult by a similarity of DNA content, base composition, and chromosome morphology among the individual chromosomes within a specific genome (Veuskens *et al.,* 1992). Flow-sorting is also affected by this lack of discrimination, since for most of the applications, such as gene mapping and construction of chromosome-specific gene libraries, sorted fractions of high purities are required. Recently, Wang *et al.* (1992) and Macas *et al.* (1993) reported the generation of a chromosome 4A-enriched library from a sorted suspension of *Triticum aestivum* chromosomes, and physical gene mapping utilizing sorted chromosome fractions of *Vicia faba,* respectively. To overcome the problem of chromosome similarity in *V. faba,* Macas *et al.* utilized genetically reconstructed karyotypes, where chromosomes are made morphologically distinct by known translocations (Schubert *et al.,* 1991).

II. Materials

A. Plant Material

Seeds of *V. faba* (2n = 12):

1. *V. faba* ssp. *minor* cv. Inovec were obtained from Dr. S. Ondro (Plant Breeding Station, SK-91624 Hornà Streda, Slovak Republic).

2. *V. faba* line with reconstructed karyotypes EF was kindly provided by Dr. I. Schubert (IPK, Gatersleben, Germany).

Because seed viability decreases with age, seed lots older than 2 years post-harvest should be avoided. One hundred seeds are used in one experiment described below.

B. Equipment and Other Materials

1. Flow cytometer/cell sorter
2. Fluorescence microscope
3. Refrigerator (+4°C)
4. Freezer (−20°C)
5. Biological incubator (heating/cooling, volume approx. 100 liters), internal temperature adjusted to 25 ± 0.5°C
6. Homogenizer (e.g., Ultra-Turrax T25 with GN2 generator)
7. Aquarium bubbler (regulated output, 10–100 ml air/min)
8. A plastic tray 1000 ml (e.g., 22 cm L · 7 cm W · 10 cm H) for germination of seeds

9. A plastic tray, internal volume approx. 1000 ml (e.g., 20 cm L × 14 cm W × 8 cm H) including an open-mesh basket to hold the germinated seeds
10. Chemical glassware
11. Perlite, approx. 3 liters (inert substrate for seed germination)
12. Glass petri dishes (diameter 60–90 mm)
13. Tweezers
14. Scalpels
15. Glass pipettes (1–10–25 ml)
16. Micro pipette with adjustable volume (1–1000 μl)
17. Nylon filters (51 μm, 10 μm), squares 4 × 4 cm
18. Polystyrene tubes (e.g., Falcon 2054)
19. Ice bucket + ice

C. Reagents and Stock Solutions

HU was obtained from Sigma (Catalog No. H8627); amiprophos-methyl (APM) was a gift from the Agriculture Chemical Division, Mobay Corp. (Kansas City); DAPI, was purchased from Molecular Probes (Eugene, OR).

1. Hoagland Nutrient Solution (Gamborg and Wetter, 1975; see also Chapter 24, Volume 49).

Prepare the solution as described by Doležel *et al.* (1992) just prior to use.

2. APM Stock Solution (200 mM)

Dissolve 60.86 mg APM in 10 ml of ice-cold acetone.

The APM stock solution can be maintained in 1-ml aliquots at −20°C for more than 1 year. When preparing the APM treatment solution, add the APM stock solution to the Hoagland solution with continuous stirring. The APM treatment solution must be prepared just before the use. APM is a mutagen, and must be handled with maximum care. Avoid skin contact; wear protective clothing, gloves, and eye/face protection. Do not breath the dust.

3. Tris Buffer

Dissolve in deionized H_2O (final concentrations are given in parentheses):

TRIS (10 mM)	0.606 g
Na$_2$EDTA (10 mM)	1.861 g
NaCl (100 mM)	2.922 g

Adjust volume to 500 ml with deionized H_2O, and the final pH to 7.5 using 1N NaOH.

4. LB01 Lysis Buffer

Dissolve in deionized H$_2$O (final concentrations are given in parentheses):

Tris (15 mM)	0.363 g
Na$_2$EDTA (2 mM)	0.150 g
spermine (0.5 mM)	0.020 g
KCl (80 mM)	1.193 g
NaCl (20 mM)	0.234 g
Triton X-100 (0.1%)	200 μl

Adjust volume to 200 ml with deionized H$_2$O, and final pH of the solution of 7.5 using 1N NaOH. Filter through a 0.22-μm filter to remove small particles. Add 220 μl mercaptoethanol (final conc. 15 mM) and mix well. Store at −20°C in 10-ml aliquots. More details on LB01 preparation can be found in Doležel *et al.* (1989).

5. DAPI Stock Solution (0.1 mg/ml)

Dissolve in deionized H$_2$O:
5 mg DAPI

Mix well for 60 min. Adjust volume to 50 ml with deionized H$_2$O. Filter through a 0.22-μm filter to remove small particles. Store at −20°C in 1-ml aliquots.

III. Procedure

Here we describe in detail a procedure for cell cycle synchronization, metaphase accumulation, chromosome isolation, and flow-sorting using *V. faba* root meristems. Prior the use, the temperature of all solutions must be adjusted to 25 ± 0.5°C. All incubations and treatments of seedlings are performed in a biological incubator at 25 ± 0.5°C, and all solutions are aerated.

A. Cell Cycle Synchronization and Accumulation of Metaphases

This procedure is based on a two-step blocking of the cell cycle: (1) synchronization at the G$_1$/S border with HU; (2) blocking at metaphase with APM (Fig. 1). Two different concentrations of HU may be used: 1.25 and 2.5 mM. With the lower concentration (1.25 mM) the peak of mitotic activity is observed approximately 2 h earlier. See Section IV for an explanation. The degree of synchrony should be checked on Feulgen-stained squash preparations and flowcytometrically by analysis of nuclear DNA content distributions.

1. Immerse seeds in distilled water (100 seeds in 500 ml H$_2$O) and leave them to imbibe for 24 h.

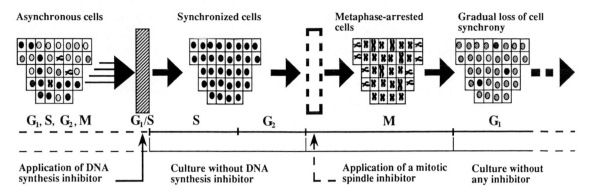

Fig. 1 A schematic representation of a synchronization procedure. This procedure is based on a reversible inhibition of DNA synthesis at the G_1/S interface. After the removal of the inhibitor, the cells traverse the cell cycle in a synchronous manner. In addition, synchronized cells can be treated with a mitotic spindle inhibitor that temporarily blocks cells at metaphase. After elimination of the mitotic spindle inhibitor, the cells can complete cytokinesis after which a gradual loss of cell cycle synchrony is observed.

2. Wet the perlite with the Hoagland nutrient solution and put it into a plastic tray (do not press!). Wash the seeds in distilled water and layer them onto the surface of the perlite. Cover the seeds with a layer of 1–2 cm of wet perlite (do not press!) and cover the tray with aluminium foil so that drying of perlite is minimized. Leave the seeds to germinate for 54 h.

3. Collect seedlings into distilled water, discard all nongerminating seeds and seedlings with primary roots shorter than ca. 1.5 cm.

4. Place the seedlings in plastic trays filled with the Hoagland solution containing 2.5 mM (1.25 mM) HU. Make sure that all roots are immersed in the solution. Incubate for 18 (18.5) h. Hydroxyurea treatment solution must be prepared just before use. HU is a mutagen, and must be handled with maximum care. Avoid skin contact; wear protective clothing, gloves, and eye/face protection. Do not breath dust.

5. Wash the roots briefly by immersing them in distilled water, and transfer the seedlings onto the trays filled with fresh Hoagland nutrient solution.

6. After a period of 6 (4.5) h following removal from the HU block, transfer the seedlings onto trays filled with the APM treatment solution, and incubate for 2 h before collecting root tips.

B. Preparation of Nuclear Suspensions

Nuclei from formaldehyde-fixed *V. faba* root tips are isolated by homogenization. This method is fast and effective, and allows monitoring of the degree of cell-cycle synchrony during the different treatments.

1. Cut the roots 1 cm from the root tip and transfer into distilled water.

2. Fix the roots by transferring them into a solution comprising 4% (v/v) formaldehyde in Tris buffer containing 0.1% Triton X-100, for 20 min at 5°C.

3. Wash the roots in Tris buffer three times, for 20 min at 5°C.

4. Exise five root tips (about 1.5–2 mm in length) and transfer them in 1 ml LB01 into a 5-ml polystyrene tube.

5. Isolate nuclei by homogenizing at 9500 rpm for 15 s at 5°C.

6. Filter the suspension through a 21-μm nylon mesh into a polystyrene tube.

7. Store the suspension at +5°C.

C. Preparation of Chromosome Suspensions

1. Immediately after APM treatment, cut the roots 1 cm from the root tip into distilled water.

2. Fix the roots in 4% (v/v) formaldehyde in tris buffer containing 0.1% Triton X-100 for 30 min at 5°C.

3. Wash the roots in Tris buffer three times for 20 min at 5°C.

4. Chop the meristem tips of 30 roots using a sharp scalpel in a glass petri dish containing 1 ml LB01 buffer. Avoid dispersion and/or drying.

5. Filter the suspension through a 51-μm nylon mesh into a polystyrene tube.

6. Store the suspension at +5°C.

D. Flow Cytometric Analysis and Sorting

1. Analysis

1. Prepare theoretical flow karyotypes, with both linear and logarithmic scales, using either a spreadsheet or dedicated computer software (Conia *et al.,* 1989; Dolezel, 1991b).

2. Just prior the analysis, syringe the suspension once carefully through a 22-gauge hypodermic needle.

3. Stain the chromosome suspension by addition of DAPI to a final concentration of 2 μg/ml.

4. Filter through a 10-μm nylon filter. Check the quality of suspension using a fluorscence microscope.

The experimental procedure described below was performed using a Becton-Dickinson FACStar[plus] flow cytometer and sorter (Becton Dickinson, San José, CA).

5. Switch on the laser (for DAPI analysis: the argon ion laser is turned at λ = 351–363 nm with a power output of 100 mW); let the laser stabilize according to the manufacturer's instructions.

6. Fill the sheath container with sheath fluid SF100.

KCl	80 mM, 5.965 g/liter
NaCl	20 mM, 1.169 g/liter

7. Leave the sheath fluid running to fill all the plastic lines and filters of the instrument.

8. Install the flow tip having the proper diameter (here 100 μm), and check for air bubbles by (*a*) visual inspection and (*b*) turning on the sorting unit and purging liquid into the flow cell chamber (the presence of air bubbles will cause alterations in the intensity of the piezo-derived sound emitted by the flow cell during purging).

9. Install the appropriate filter set for detection of the fluorescent signals (DAPI filter: LP400 or BP450/20).

10. Align the flow cytometer to obtain the lowest (most narrow) coefficient of variation (CV) on DNA fluorescence histograms, measuring fluorescence pulse height (FPH). Check the linearity of the measurement using an appropriate standard stained with DAPI. Chicken red blood cells (CRBCs) are good for this purpose, since they exhibit reasonably narrow CVs and, since they tend to aggregate, can also be used to assess machine linearity.

11. Set up the sweep trigger signal to DNA–FPH, in order to analyze all DNA particles in suspension, and determine the threshold level.

12. Let the sample stabilize at the appropriate flow rate; for chromosome analysis, this is 100–200 chromosomes/s.

13. Configure the pulse processing module (PPM) to measure the integral intensity of the DNA signals (area of the fluorescence pulse − FPA) and its length (width of the fluorescence pulse − FPW; this provides a means to discriminate aggregated particles from singlets having the same fluorescence intensity).

14. According to the known DNA content of CRBC nuclei (~2.3 pg/nucleus), set amplification and photomultiplier voltage values to place chromosome I of the *V. faba* standard karyotype (~4.2 pg) at channel 200 of the DNA–FPH histogram.

15. Run the chromosome sample, and adjust the photomultiplier voltage and amplification gains to match the theoretical distribution of nuclei and chromosomes in the logarithmic and linear DNA–FPA histogram distributions (Figs. 2A and 2B, respectively). First analyze the sample with logarithmic amplifiers to look at the relative fluorescence distribution of chromosomes and nuclei on the same histogram; then move to linear amplification and set up the analysis such that the chromosomes are evenly distributed along the DNA–FPA histogram, with chromosome I appearing at channel 200 (Fig. 3).

16. Set up conditions for biparametric analysis of DNA–FPA (DNA content) versus DNA–FPW (particle size), in order to obtain a distribution of clusters of signals of fluorescence; this will permit identification of chromosomes, their doublets, and chromatids (Fig. 4).

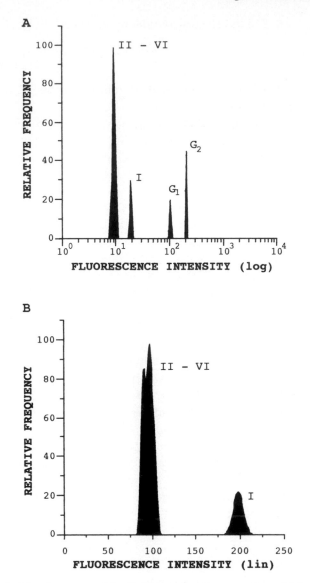

Fig. 2 Theoretical flow-karyotype of the *V. faba* standard line according to DNA content (Doležel, 1991b) calculated for a 3% coefficient of variation: (A) Logarithmic amplification; (B) Linear amplification.

2. Sorting

1. Switch on the sorting module, and adjust the drop drive frequency (DDF) and the drop drive amplitude (DDA) to generate the shortest breaking point distance (BPD), defined as the distance between the laser interception point with the stream and the first generated droplet; leave the piezo-

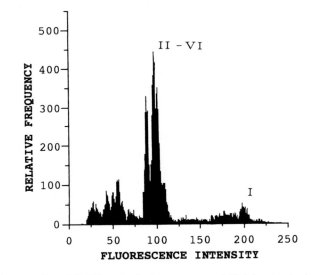

Fig. 3 Flow-karyotyping of DAPI stained chromosomes of *V. faba* standard line: metacentric chromosome I is located at about channel 200; acrocentrics and chromatids from chromosome I are located at channel 90–100.

electric crystal and deflection plates switched on. For the 100-μm flow tip, typical values are sheath fluid pressure (SFP), 17 psi; DDF, 23 kHz; DDA, 0.5 V.

2. Adjust the position and sharpness of the side streams by varying the drop drive phase (DDP) setting. This step is relevant to the final success of the sorting since sorted chromosomes must be all precisely deflected into the collection tube.

3. Calculate the average distance between 10 droplets (λ), define the drop delay (DD) setting (equal to BPD/λ), and enter this value into the DD counter of the flow cytometer. This establishes the correct time for droplet charging after detection of a specific chromosome.

4. Select the sorting envelope (SE: the number of droplets sorted together with the drop supposed to contain the selected chromosome) according to chromosome concentration, flow-stability, and desired final volume for the sorted fraction.

5. Define the regions for sorting on a biparametric FPA/FPW dot-plot, drawing rectangular regions around clusters of signals identifying specific chromosomes.

6. Sort a defined number of chromosomes (typically 50) onto a microscope slide, and check for yield (recovery), and purity by fluorescence microscopy. The anticoincidence circuitry is typically activated in order to prevent contamination with undesired particles. Continue the sort process, varying DD in fractions of droplets, until conditions for optimal recovery and purity are defined.

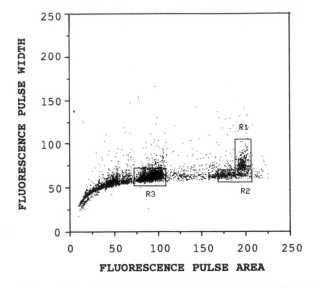

Fig. 4 Biparametric analysis (DOT-PLOT) of DNA-fluorescence pulse area versus DNA-fluorescence pulse width of chromosomes in suspension isolated from *V. faba* standard line root tips: R1 contains chromosome I; R2 contains doublets of acrocentrics and folded metacentrics; R3 contains acrocentrics (II–V) and metacentric chromatids.

7. Install the unit for collecting sorted fractions.

8. Check again for stability of the break-point and of the side streams.

9. Put the appropriate buffer for chromosome preservation into the collecting device (PCR buffer, LB01, etc.) install the device, and carry out the sorting.

IV. Critical Aspects of the Procedure

For isolation of chromosome suspensions, problems posed by the use of plant tissues include (*a*) difficulties in obtaining a high level of mitotic synchrony; (*b*) the variable degree of chromosomes stickiness and clumping observed after treatment with mitotic drugs; (*c*) a tendency for metaphase-blocked chromosomes to split into single chromatids; (*d*) the presence of the rigid plant cell wall which, in comparison to mammalian cells, increases debris and inhibits the release of intact chromosomes during homogenization. The use of DAPI, a fluorochrome that binds preferentially to A–T regions, for DNA staining and relative DNA fluorescence measurements allows, in comparison to other DNA stains such as propidium iodide, better discrimination of chromosomes and nuclei at different phases of the cell cycle and provides a lower background on DNA histograms (Otto, 1990).

The quality of chromosome suspensions depends heavily on the proportion of metaphase cells in the root meristems. Thus, special attention must be paid to the synchronization procedure. The procedure described here is relatively simple and usually no problems occur if it is performed carefully. Because cell-cycle kinetics depends on temperature (Murin, 1966), maximal attention should be paid to the temperature of the solutions (this must be adjusted prior to their use) and to the temperature inside the incubator. Special care must also be devoted to washing away residues of detergents and/or disinfectants from seeds and containers used for the experiments. Similarly, the length of all treatments must be scrupulously followed, the only exception being the length of germination in perlite, which may be extended up to 72 h.

A. Cell Cycle Synchrony

The effect of HU treatment depends critically on its concentration. Unnecessarily high concentrations may lead to a delayed recovery of the cells from the block and in poor synchrony. On the other hand, if HU is used in concentrations lower than optimal (e.g., 1.25–1.5 mM), cells escape from the block before the end of HU treatment. It is interesting to note that they resume DNA synthesis in a synchronous manner and that the degree of synchrony is fully comparable to that achieved using HU at 2.5 mM. Because the mitotic peak appears approximately 2 h earlier after the HU removal, this approach is appropriate if synchronization of cells in mitosis and/or accumulation of metaphases is required.

B. Metaphase Accumulation

The actual length of the APM treatment depends on the aim of the experiment. For chromosome isolation, we prefer to shorten the block to 2 h in order to decrease the proportion of single chromatids and to avoid chromosome decondensation. If just the highest number of metaphases is required, the APM treatment can be prolonged up to 4 h. The advantage of APM over other spindle poisons such as colchicine or oryzaline is that it is very effective and thus can be used in micromolar concentrations. Furthermore, in *V. faba,* the number of clumped ("ball") metaphases is lower after APM block compared to the situation observed using the other inhibitors (data not shown).

C. Nuclei and Chromosome Isolation

The release of nuclei and chromosomes from the cells is very sensitive to the concentration of formaldehyde, which is a volatile compound (Sgorbati *et al.,* 1986). Therefore, this chemical must be kept in a dark bottle tightly closed at room temperature. Since the formaldehyde concentration may vary between lots, it is advisable, particularly for chromosome isolation, to perform preliminary experiments to evaluate chromosome yield and morphology whenever changing the formaldehyde stock solution.

D. Flow–Karyotyping and Sorting

The purity of sorted fraction depends on the resolution of flow karyotypes. To achieve the highest resolution, the flow cytometer used for chromosome analysis must be well aligned (in our case full peak CV less than 2.5%), and it must be stable over long sorting periods.

Linearity of the measurements must be assessed with a proper standard such as CRBCs, since the peak position can be affected dramatically at the higher channel values. In this respect, optimizing the laser focus lens position is of major importance.

When setting the "threshold value" for sorting experiments based on DNA content, it is important that all DNA-containing particles be displayed on the DNA–FHP histogram; this is the condition for the instrument to recognize particles of no interest and avoiding contamination of sorted fractions. When the sorting module is on, the DDF and DDA can affect the signal acquisition by generating false signals. This problem can be avoided by (*a*) reducing the DDA voltage; (*b*) changing the DDF; (*c*) reducing the SFP; (*d*) tilting the fluorescence obscuration bar.

For sorting into small collection devices (e.g., PCR tubes), precise focusing of the sorted sample is of great importance. Adjusting the DDP allows precise timing between pulse charging and droplet formation, in order to avoid "fuzzy" sorting streams. The correct position for the sorting streams must be assessed by examining the sorting tube. During sorting, charged droplets of the same polarity repel each other. Therefore it is important to ground effectively the sorting tubes to prevent a dramatic loss in recovery of the sorted fraction.

V. Results and Discussion

We have found it useful to combine at least two methods to estimate the efficiency of the synchronization procedure: flow cytometry and karyological analysis. Flow cytometry allows rapid and precise analysis of nuclear DNA content in large populations of cells (Galbraith, 1990; Doležel, 1991a) and is used to monitor the progression of synchronized cells through the cell cycle and to estimate their position within the cell cycle and their proportion with respect to the total cell population. Analyses of mitotic synchrony as well as metaphase accumulation are performed on Feulgen-stained squash preparations. Other techniques such as immunolabeling of S-phase cells may be needed depending on the experimental design (Moretti *et al.*, 1992; Binarova *et al.*, 1993).

A. Cell Cycle Synchrony

The results of flow cytometric analysis of cell synchrony are summarized in Fig. 5. The HU treatment causes a block in DNA synthesis and accumulation

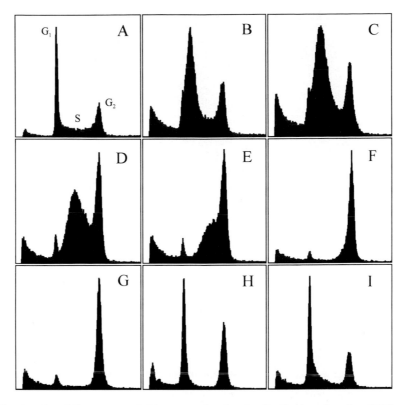

Fig. 5 A series of flow cytometric histograms showing the distribution of nuclear DNA contents in *V. faba* primary root tip cells (nuclei stained with DAPI). (A) Prior to HU treatment; in this population, 40, 37, and 23% of cells here in G_1-, $S-$, and G_2 phase, respectively. (B) Immediately after HU (2.5 mM/18 h) treatment and (C) 2 h; (D) 3 h; (E) 4 h; (F) 6 h; (G) 8 h; (H) 10 h; and (I) 12.5 h after HU removal. On the abscissa, the relative nuclear DNA content as fluorescence intensity (arbitrary units) of DNA is shown; on the ordinate, the number of analyzed nuclei is indicated.

of cells at the G_1/S interface and in early S phase. Consequently, mitotic activity is completely stopped. After release from the HU block, the meristem cells resume DNA synthesis and synchronously traverse S phase (Figs. 5C–5E). Six hours after the release, most of the synchronized cells are in G_2, with smaller fractions of cells either remaining in late S or already within G_1 (Fig. 5F). At this time, a gradual increase in the mitotic activity can be observed (Fig. 6). A maximal number of cells in G_2 is found 8 h after release of the block (Fig. 5G). At this time, a peak of mitotic activity is observed (Fig. 6). The degree of mitotic synchrony ranges typically between 45 and 55%, although mitotic indices up to 70% can be found in individual meristems (Fig. 7A). Synchronous mitosis followed by cell division leads to the increase in the frequency of cells in G_1 as observed 10 h after release of the block (Fig. 5H). Similarly, the number of

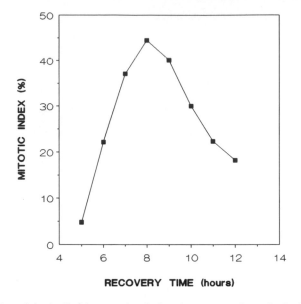

Fig. 6 Mitotic activity in *V. faba* root tips during the recovery from the hydroxyurea block.

mitotic cells decreases (Fig. 6). Twelve and a half hours after the release, synchronized cells start a second round of DNA replication (Fig. 5I).

B. Metaphase Accumulation

Treatment of populations of synchronized cells with APM for 2 h (4.5–6 h after HU removal) leads to accumulation of approximately 50% of the cells in metaphase. The frequency of cells blocked in metaphase depends on both the timing and the duration of the APM treatment. The highest frequency of metaphases (53.9%) was found when the root meristems were treated with 2.5 μM APM for 3 h, starting 6 h after removal of the hydroxyurea (Fig. 7B).

C. Nuclei and Chromosome Isolation

Nuclei can be isolated very efficiently by root tip homogenization; about 10^5 nuclei per root tip can be isolated, which is sufficient for flow cytometric analysis. The use of LB01 in respect to other isolation buffers [Tris, phosphate-buffered saline (PBS)] allows recovery of a higher number of morphologically well-preserved nuclei. Using the chopping procedure, the chromosome yield ranged from 4×10^5 to 1.6×10^6 per 30 root tips. A complex relationship between the duration of fixation and the yield of isolated nuclei and chromosomes was evident. On the one hand, a short fixation time did not preserve the morphology of nuclei and chromosomes. The chromosomes isolated from root tips fixed for 10 or 15 min were damaged and the chromosome suspensions contained a large amount

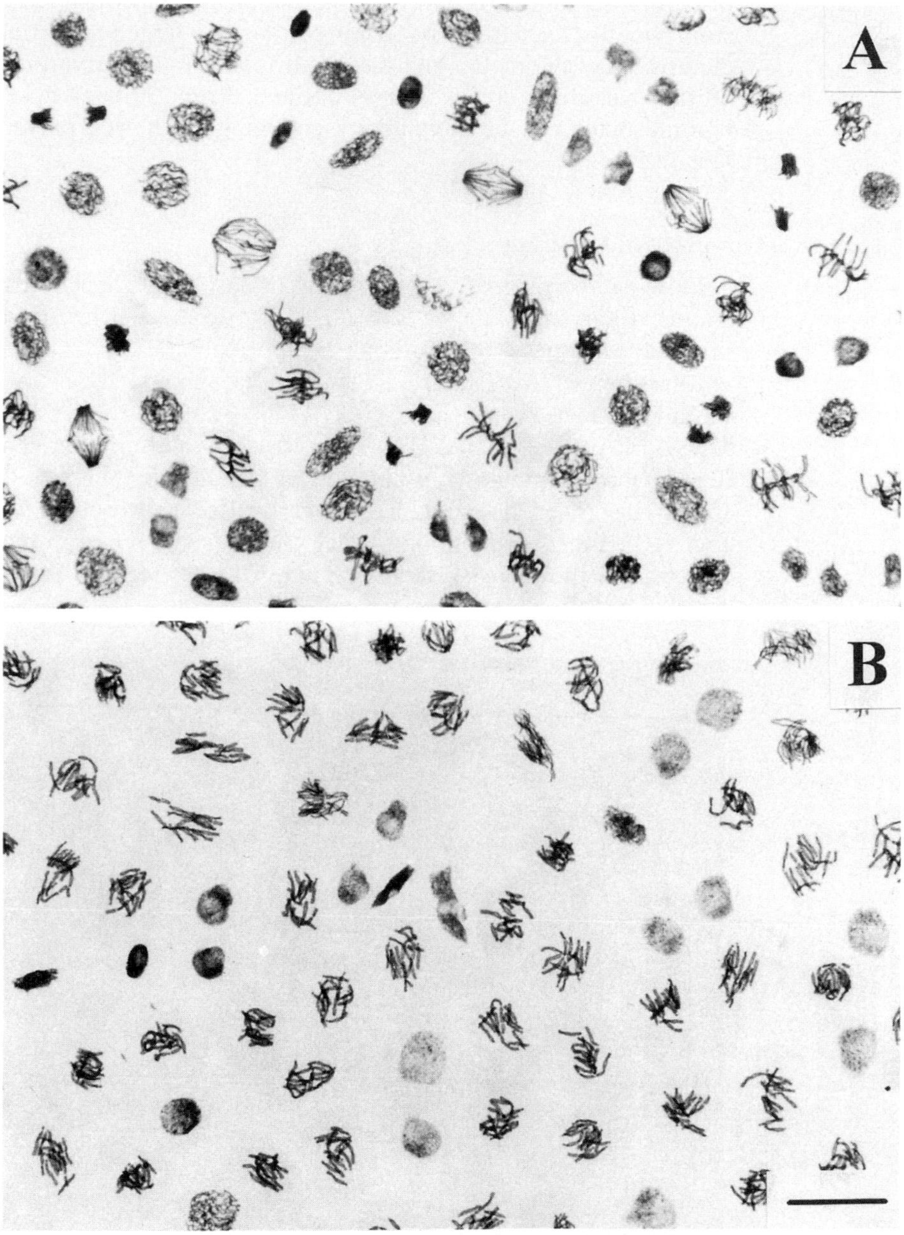

Fig. 7 (A) Mitotic activity of *V. faba* root tip cells 8 h after release from the HU block. (B) Metaphase cells in *V. faba* root tip accumulated after the treatment of synchronized cells with amiprophos-methyl (2.5 μM/3 h) (bar, 50 μm).

of chromosomal debris. On the other hand, chromosomal suspensions obtained from root tips fixed for 45 or 60 min contained a large proportion of chromosome clumps. Flow cytometric analyses confirmed these observations, and fixation for 30 min was therefore chosen as optimal. From 30 root tips, typically 1×10^6 chromosomes can be isolated by chopping, with well-preserved morphology (Fig. 8).

D. Flow-Karyotyping and Sorting

Flow-karyotyping of chromosomes in suspension isolated from the *V. faba* standard karyotype allows identification of metacentric chromosome I corresponding to a peak of fluorescence approximately at channel 200. All other chromosomes (II–VI) were grouped in a composite peak at channel 100 with an adjacent peak at channel 90 (Fig. 3). Using biparametric analysis of DNA–FPA versus DNA–FPW, it is possible to discriminate, according to their length, chromosome I from doublets of acrocentrics (Fig. 4: R1 and R2, respectively). In the reconstructed karyotype EF (for a theoretical flow-karyotype, see Fig. 9), peaks corresponding to each chromosome type can be discriminated on the basis of the DNA–FPA histogram of DAPI relative fluorescence intensity.

Fig. 8   A suspension of chromosomes isolated from *V. faba* standard line (bar, 50 μm).

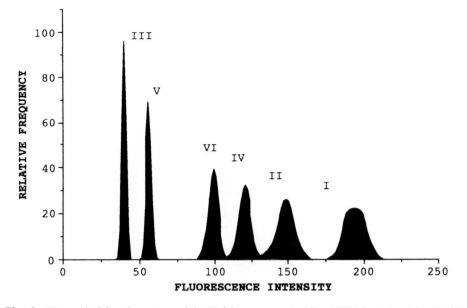

Fig. 9 Theoretical flow-karyotype of the *V. faba* reconstructed line "EF" (calculated for 3% CV; Doležel, 1991b).

Flow-sorting confirmed assignment of chromosomes I, II, IV, VI, V, and III, to peaks at channels 198, 146, 117, 84, 54, and 39, respectively (Fig. 10).

Flow-sorting of chromosomes in suspension can be performed, with different yields achieved according to the percentage of the chromosomal subpopulation that is sorted. From the standard karyotype of *V. faba,* chromosome I is usually recognizable in suspension as comprising 1.5–3% of the total chromosomal population; this low value is due to a tendency of this chromosome both to break at the centromeric region and to fold over, thus presenting an appearance similar to a couple of associated acrocentrics. It is possible to sort about 3×10^4 metacentric from 10^6 chromosomes in suspension (30 chopped root tips) with a purity better than 90% (Fig. 11).

VI. Conclusions and Perspectives

These methods have been used to obtain high metaphase indices for subsequent isolation of chromosome suspensions (Doležel *et al.,* 1992) and to study changes in cytoskeleton and phosphorylation events in individual phases of the cell cycle (Binarova *et al.,* 1993). Effective flow-sorting of pure single-type chromosome fractions (Lucretti *et al.,* 1993) was also achieved using this methodology. Sorted chromosomes have been used for physical gene mapping in *V. faba* (Macas *et al.,*

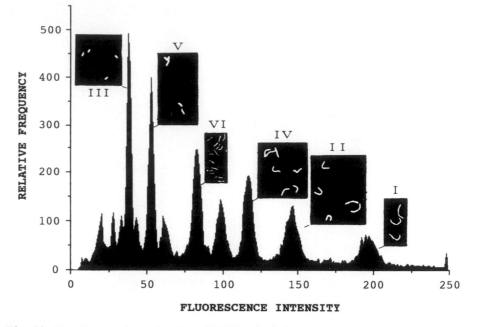

Fig. 10   Flow-karyotyping and sorting of DAPI-stained chromosomes in suspension isolated from reconstructed line "EF": each chromosome (I–VI) can be identified by a distinct peak, and subsequently can be sorted.

1993) and detailed cytogenetical studies (Schubert *et al.,* 1993). Sorted fractions of all the chromosomes from *V. faba* have been collected and are currently under analysis and amplification for the creation of chromosome-specific gene libraries (Macas *et al.,* work in progress). Recently, these procedures were found suitable for the study of the role of some plant hormones and their metabolites in the regulation of the cell cycle (Strnad and Doležal, pers. comm.).

The procedure for cell-cycle synchronization, metaphase blocking, and chromosome isolation can be modified for other species. The modifications concern the establishment of optimal HU concentration, selection of an appropriate mitotic blocking agent and its concentration, incubation times, and optimization of chromosome isolation procedure (concentration of formaldehyde and the length of fixation). Recently, we have modified the protocol described here for cell cycle synchronization and chromosome isolation in *Pisum sativum* (Gualberti *et al.,* in preparation), *Medicago sativa,* and *Zea mays* (unpublished).

Acknowledgement

We thank Dr. Jarmila Cihalikova and Dr. Guiliana Gualberti for their contribution to the development of the procedure for cell cycle synchronization and chromosome isolation in *Vicia faba*, Dr. I. Schubert and S. Ondro for supply of *V. faba* seeds, and Prof. A. Murin and Dr. K. Micieta for introduction to the model system of *V. faba* meristem root tips. Cell-cycle analysis was performed

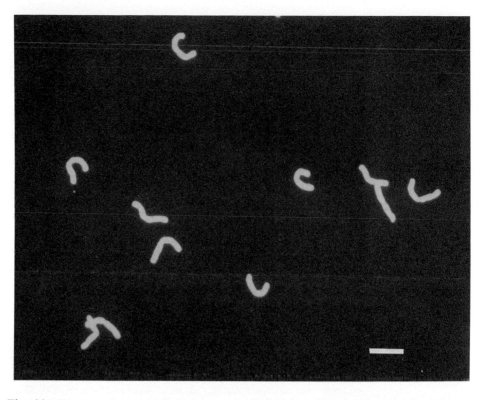

Fig. 11 Flow-sorted metacentric chromosome I obtained from chromosome suspensions isolated from *V. faba* standard line (bar, 10 μm).

using the PAS II flow cytometer kindly provided to one of us (J.D.) by Prof. Göhde (Partec, Germany). The gift of APM from the Agriculture Chemical Division, Mobay Corp. (Kansas City) is gratefully acknowledged. Part of this work was supported by grant 204/93/1097 from the Grant Agency of the Czech Republic. This is Publication 105 of ENEA, Biotechnology and Agriculture Sector.

References

Binarova, P., Cihalikova, J., and Doležel, J. (1993). Localization of MPM-2 recognized phosphoproteins and tubulin during cell cycle progression in synchronized *Vicia faba* root meristem cells. *Cell Biol. Int.* **17,** 847–856.

Carrano, A. V., Van Dilla, M. A., and Gray J. W. (1979). Flow cytogenetics: A new approach to chromosome analysis. *In* "Flow Cytometry and Sorting" (M. R. Melamed, P. F. Mullaney and M. L. Mendelsohn, eds.), pp. 421–451. New York: Wiley.

Carter, N. P. (1993). Gene mapping and PCR applications with flow-sorted chromosomes. *In* "Flow Cytometry, New Developments" (A. Jacquemin-Sablon, ed.), pp. 327–342. Berlin/Heidelberg: Springer-Verlag.

Conia, J., Muller, P., Brown, S., Bergounioux, C., and Gadal, P. (1989). Monoparametric models of flow cytometric karyotypes with spreadsheet software. *Theor. Appl. Genet.* **77,** 295–303.

Cotter, F., Nasipuri, S., Lam, G., and Young, B. D. (1989). Gene mapping by enzymatic amplification from flow-sorted chromosomes. *Genomics* **5,** 470–474.

De Laat, A. M. M., and Schel, J. H. N. (1986). The integrity of metaphase chromosomes of *Haplopappus gracilis* (Nutt.) Gray isolated by flow cytometry. *Plant Sci.* **47,** 145–151.

Dietzel, A., H., Hain, J., Virsik-Peuckert, R., P., and Harder, D. (1990). Algorithms for the evaluation of radiation induced chromosome aberration yields per cell from flow karyotypes. *Cytometry* **11,** 708–715.

Doleźel, J., Binarova, P., and Lucretti, S. (1989). Analysis of nuclear DNA content in plant cells by flow cytometry. *Biol. Plant.* **31,** 113–120.

Doleźel, J. (1991a). Flow cytometric analysis of nuclear DNA content in higher plants. *Phytochem. Anal.* **2,** 143–154.

Doleźel, J. (1991b). KARYOSTAR: Microcomputer program for modelling of monoparametric flow karyotypes. *Biológia* **46,** 1059–1064.

Doleźel, J., Cíhalíková, J., and Lucretti, S. (1992). A high-yield procedure for isolation of metaphase chromosomes from root tips of *Vicia faba* L. *Planta* **188,** 93–98.

Doleźel, J., Lucretti, S., and Schubert, I. (1994). Plant chromosome analysis and sorting by flow cytometry. *Crit. Rev. Plant Sci.* **13(3),** 275–309.

Eriksson, T. (1966). Partial synchronization of cell division in suspension cultures of *Haplopappus gracilis*. *Physiol. Plant.* **19,** 900–910.

Francis, D. (1992). The cell cycle in plant development. *New Phytologist* **122,** 1–20.

Galbraith, D. W. (1990). "Methods in Cell Biology" (Z. Darzynkiewicz and H. A. Grissman, eds.), pp. 549–562. San Diego: Academic Press.

Gamborg, O. L., and Wetter, L. R., eds. (1975). "Plant Tissue Culture Methods." Saskatoon, Canada: N.R.C. of Canada.

Gray, J. W., and Cram, L. S. (1990). Flow karyotyping and chromosome sorting. *In* "Flow Cytometry and Sorting." Second Edition (M. R. Melamed, T. Lindmo, and M. L. Mendelsohn, eds.), pp. 503–529. New York: Wiley-Liss.

Gray, J. W., and Langlois, R. G. (1986). Chromosome classification and purification using flow cytometry and sorting. *Annu. Rev. Biophys. Biophys. Chem.* **15,** 195–235.

Gray, J. W., Trask, B., Van den Engh, G., Silva, A., Lozes, C., Grell, S., Schonberg, S., Yu, L. C., and Golbus, M. S. (1988). Application of flow karyotyping in prenatal detection of chromosome aberratons. *Am. J. Hum. Genet.* **42,** 49–59.

Koch, C., and Nasmyth, K. (1994). Cell cycle regulated transcription in yeast. *Curr. Opinion Cell Biol.* **6,** 451–459.

Kornberg, A. (1980). "DNA Replication." San Francisco: Freeman.

Lucretti, S., Doleźel, J., Schubert, I., and Fuchs, J. (1993). Flow karyotyping and sorting of *Vicia faba* chromosomes. *Theor. Appl. Genet.* **85,** 665–672.

Macas, J., Doleźel, J., Lucretti, S., Pich, U., Meister, A., Fuchs, J., and Schubert, I. (1993). Localization of seed storage protein genes on flow-sorted field bean chromosomes. *Chromosome Res.* **1,** 107–115.

Mii, M., Saxena, P. K., Fowke, L. C., and King, J. (1987). Isolation of chromosomes from cell suspension cultures of *Vicia hajastana*. *Cytologia* **52,** 523–528.

Morejohn, L. C., Bureau, T. E., Tocchi, L. P., and Fosket, D. E. (1984a). Tubulins from different higher plant species are immunologically nonidentical and bind colchicine differentially. *Proc. Natl. Acad. Sci. U.S.A.* **81,** 1440–1444.

Morejohn, L. C., and Fosket, D. E. (1984b). Inhibition of plant microtubule polymerization *in vitro* by phosphoric amide herbicide amiprophos-methyl. *Science* **224,** 874–876.

Morejohn, L., C., Bureau, T. E., Mol-Bajer, J., Bajer, A. S., and Fosket, D. E. (1987). Oryzalin, a dinitroaniline herbicide, binds to plant tubulin and inhibits microtubule polymerization *in vitro*. *Planta* **172,** 252–264.

Moretti, F., Lucretti, S., and Doleźel, J. (1992). Plant cell cycle analysis on isolated nuclei using a monoclonal antibody against Burd Urd. *Eur. J. Histochem.* **36,** 367.

Murin, A. (1966). The effect of temperature on the mitotic cycle and its time parameters in root tips of *Vicia faba*. *Naturwissenschaften* **53,** 312–313.

Otto, F. J. (1990). DAPI staining of fixed cells for high-resolution flow-cytometry of nuclear DNA. *In* "Methods of Cell Biology" (Z. Darzynkiewicz and H. A. Crissman, Eds.), Vol. 33, pp. 105–110. San Diego: Academic Press.

Scheuermann, W., and Klaffke-Lobsien, G. (1973). On the influence of 5-aminouracil on the cell cycle of root tip meristems. *Exp. Cell Res.* **76,** 428–436.

Schubert, I., Doležel, J., Houben, A., Scherthan, H., and Wanner, G. (1993). Refined examination of plant metaphase chromosome structure at different levels made feasible by new isolation methods. *Chromosoma* **102,** 96–101.

Schubert, I., Riger, R., and Kunzel, G. (1991). Karyotype reconstruction in plants with special emphasis on *Vicia faba* L. *In* "Chromosome Engineering in Plants" (T. Tsuchiya T. and P. K. Gupta, eds.), part A, pp. 113–140. Amsterdam/New York: Elsevier.

Sgorbatti, S., Levi, M., Sparvoli, E., Trezzi, F., and Lucchini, G. (1986). Cytometry and flow cytometry of 4′,6-diamidino-2-phenylindole (DAPI)-stained suspensions of nuclei released from fresh and fixed tissues of plants. *Physiol. Plant.* **68,** 471–476.

Sgorbati, S., Sparvoli, E., Levi, M., Galli, M. G., Citterio, S., and Chiatante, D. (1991). Cell cycle kinetic analysis with flow cytometry in pea root meristem synchronized with aphidicolin. *Physiol. Plant.* **81,** 507–512.

Sree Ramulu, K., Verhoeven, H. A., Dijkhuis, P., and Gilissen, L. J. W. (1990). A comparison of APM-induced micronucleation and influence of some factors in various genotypes of potato and *Nicotiana. Plant Sci.* **69,** 123–133.

Trask, B., Van den Engh, G., Nussbaum, R., Schwartz, C., and Gray, J. (1990). Quantification of the DNA content of structurally abnormal X chromosomes and X chromosome aneuploidy using high resolution bivariate flow karyotyping. *Cytometry* **11,** 184–195.

Van Dilla, M. A., and Deaven, L. L. (1990). Construction of gene libraries for each human chromosome. *Cytometry* **11,** 208–218.

Van't Hof, J. (1973). The regulation of cell divisions in higher plants. *In* "Basic Mechanisms in Plant Morphogenesis" (P. Carlson, H. Smith, A. Sparrow, and J. Van't Hof, eds.), Vol. 25, pp. 152–165. Brookhaven, New York: Brookhaven National Laboratory.

Veuskens, J., Marie, D., Hinnisdaels, S., and Brown, S. C. (1992). Flow cytometry and sorting of plant chromosomes. *In* "Flow Cytometry and Cell Sorting" (A. Radbruch, ed.), pp. 177–188. Berlin/Heidelberg: Springer-Verlag.

Wang, M. L., Leitch, A. R., Schwarzacher, T., Heslop-Harrison, J. S., and Moore, G. (1992). Construction of a chromosome-enriched HpaII library from flow-sorted wheat chromosomes. *Nucleic Acids. Res.* **20,** 1897–1901.

CHAPTER 7

Principles and Applications of Recombinant Antibody Phage Display Technology to Plant Biology

William L. Crosby and Peter Schorr

Molecular Genetics Group
Plant Biotechnology Institute
National Research Council of Canada
Saskatoon, Saskatchewan, Canada S7N 0W9

I. Introduction

Since its inception (Kohler and Milstein, 1975), monoclonal antibody technology has found important applications in support of fundamental biological investigations. The availability of large quantities of monospecific antibody has provided research scientists with a variety of diagnostic, biochemical, and molecular tools with which to address their particular biological problems.

The past decade has seen a dramatic increase in our understanding of the molecular genetic bases of antibody expression in mammalian systems. In particular, the application of recombinant DNA techniques has made possible the precise manipulation of those genetic segments encoding the vast combinatorial repertoire of antibody structures that are characteristic of the animal immune response. Corresponding advances have been made in our understanding of the structure of immunoglobin gene family hierarchies, particularly from murine and human sources (Williams and Winter, 1993; Tomlinson *et al.,* 1993). Likewise, a considerable body of information has accumulated pertaining to the cloning and expression of V-gene repertoires from cDNA and germline sources. Recombinant V-gene repertoires have been explored as a source of antibody segments with defined antigen-binding properties, and have been expressed in a variety of transgenic hosts including bacteria, fungi, baculovirus, plant, and animal systems (Plückthun, 1992).

Almost a decade ago, the potential of filamentous bacteriophage systems for the expression of oligopeptides and recombinant antibody molecules was recognized (Smith, 1985). Phage display techology has since found widespread application for the cloning and structural presentation of antibody repertoires in prokayotic hosts (Chester *et al.,* 1994; Nissim *et al.,* 1994; Griffiths, 1993; Barbas, III *et al.,* 1991; Clackson *et al.,* 1991; Kang *et al.,* 1991; Marks *et al.,* 1991). Until recently, such libraries achieved recombinant diversities of generally 10^6–10^8 independent clones—a figure significantly less than the approximate 10^7–10^9 or 10^{10}–10^{12} antibody diversity presented at any one time by the native murine or human immune system, respectively. Although such libraries have served as "single-pot" sources of antibodies against different antigens (Nissim *et al.,* 1994), the avidity of antibodies recovered has been low—generally in the micromolar K_D range. More recently, an *in vivo* "combinatorial infection" strategy (Waterhouse *et al.,* 1993) has been employed for the generation of very high diversity (approximatcly 10^{11}) V-gene repertoires from PCR-amplified human germline segments (Griffiths *et al.,* 1994).

II. Construction of High-Diversity, Naive Recombinant Antibody Libraries in Bacteriophage fd

The construction and screening of a high-diversity human Fab phage display library have been described in detail elsewhere and is briefly reviewed here for clarity of discussion. V-gene repertoires for use in library construction have been amplified by the polymerase chain reaction (PCR) either from single-strand cDNA isolated from a lymphocyte population following immunization or from naive germline sources. The options open to the investigator with respect to template, source organism, and cloning strategy are beyond the scope of this article. Rather, the reader is directed to any of several reviews that highlight the relative merits of different cloning and expression strategies (Winter, 1993; Winter and Harris, 1993; Hoogenboom *et al.*, 1992; Plückthun, 1992).

A. Amplification and Assembly of High-Diversity Human V Gene Repertoires

In the "combinatorial infection" strategy (Waterhouse *et al.*, 1993), a human heavy-chain repertoire was cloned using PCR techniques from 49 independent V_H segments incorporating a randomized 4–12 amino acid CDR3 domain. This V_H repertoire was cloned as the N-terminal fusion with bacteriophage fd gIII protein in a colE1 pUC-based "donor" expression plasmid. The heavy-chain gene was expressed with an N-terminal *pel*B leader sequence designed to direct its excretion to the host periplasm. Furthermore, the V_H gene was flanked by wild-type *lox*P and mutant *lox*P511 recombination sites (Waterhouse *et al.*, 1993). The resulting plasmid-born V_H repertoire was transformed and maintained in a F^+ *E. coli* strain (e.g., TG1). The κ and λ light-chain repertoires incorporating randomized CDR3 domains were amplified as independent libraries from germline $V_κ$ and $V_λ$ segments and cloned into a bacteriophage fd "acceptor" vector incorporating an N-terminal phage fd gIII transit peptide. V_H and V_L segments were randomly combined by infection of F^+ *Escherichia coli* host cells harboring the donor V_H repertoire with aliquots of either phage fd light-chain library, and coinfected with phage P1 to provide *Cre* recombinase in *trans*. The coresidence of the two episomes in *E. coli* results in a high frequency of recombination–cointegration events involving the heavy-chain plasmid and incoming light-chain phage fd chromosome. Upon subsequent resolution of the cointegrants, approximately half the phage chromosomes will carry a Fab operon exhibiting a new juxtaposition of heavy- and light-chain genes. Recombined phage are grown and recovered from coinfective centers, propagated through one round of amplification in a bacterial host that does not express *cre* recombinase, and concentrated by PEG/NaCl precipitation to high titer ($1–4 \times 10^{12}$ phage/ml); aliquots are stored at $-90°C$, where we have measured a titer half-life in excess of 3 months (Schorr and Crosby, unpublished).

B. Rationale for High-Diversity Libraries

Consistent with previous theoretical predictions (Perelson and Oster, 1979), the high-diversity recombinant human Fab library has proven to be a useful source of antibodies against diverse hapten and polypeptide (e.g., thyroglobulin) antigens, including a growing list of human "self" and "foreign" antigens. Comparative panning using a low-diversity subset of the same library resulted in the recovery of a relatively restricted population of binding antibodies that exhibited reduced binding affinities to several hapten molecules (Griffiths *et al.*, 1994). Thus, greater combinatorial diversity equates to an enhanced utility of the library as a potential "single-pot" source of recombinant antibodies capable of recognizing a variety of different antigens.

As a single-pot source of recombinant antibodies, such libraries avoid the need to assemble a new library from immunized animal tissues in order to have a high probability of recovering antibodies against each new antigen under study. Furthermore, the large size of the repertoire results in the recovery of phage populations that exhibit a correspondingly more diverse content of high-affinity recombinant antibodies (Griffiths *et al.,* 1994).

III. Panning and Characterization of Recombinant Antibodies from Phage Libraries

The process of recovering phage displaying a recombinant antibody by "biological panning" entails the indirect immobilization of phage to a solid surface or recoverable particle via an antibody–antigen interaction. In one such approach, the antigen is immobilized to a solid surface such as polystyrene petri dishes or Immunotubes (Nunc). In a second approach, the antigen is biotinylated and presented to the phage library in solution, followed by recovery of the antigen–antibody complex using streptavidin-coated paramagnetic beads (Hawkins *et al.,* 1992). In both protocols, phage become indirectly immobilized to the solid surface or bead via an antigen–antibody interaction. After immobilization, washing, and recovery, phage are eluted or cleaved from the solid support, pooled, and amplified in preparation for further rounds of affinity enrichment. The process of phage "panning" mimicks important aspects of the *in vivo* immune response in animals in that (*a*) antibodies are "presented" on the surface of a biologically replicating entity (B-lymphocytes versus phage fd) and (*b*) the biological vector displaying the desired antibody is selectively propagated and amplified in response to antigen, leading to preferential enrichment of binding versus nonbinding entities in the population.

A. Panning Using Immunotubes

1. Coat 5-ml Immunotubes (Nunc) by adding 4 ml of a fresh 100 μg/ml solution of antigen in sterile phosphate-buffered saline (PBS) (1) and incubate overnight at room temperature (RT). At the same time, initiate a 10- to 20-ml overnight preculture of F^+ *E. coli* (e.g., strain TG1) in 2× TY medium (1% yeast extract, 1.6% tryptone, 0.5% NaCl, pH 7.1) at 37°C.

2. The next day, drain the tubes and wash them three times in quick succession by filling with sterile PBS from a new wash bottle, and emptying with a vigorous wrist action. Tubes are blocked by filling to the brim (5 ml) with 2% (w:v) skim milk powder in sterile PBS (2% MPBS), and incubated standing at 37°C for 2 h.

3. After blocking, tubes are washed 3× in quick succession with sterile PBS as in (2) above. To each tube, add 2 ml of 4% MPBS, followed by 1-ml aliquots each of the V_κ and V_λ-recombined phage diplay library. Seal the tubes well using

stoppers or Parafilm and incubate for 30 min with gentle tumbling, followed by 1.5 h standing at RT. Dilute the overnight *E. coli* preculture 1:100 into a 250-ml flask containing 20 ml of prewarmed 2× TY medium, and incubate with vigorous shaking (250–300 rpm) at 37°C.

4. After the binding incubation, decant and discard the phage/MPBS solution and wash the tubes in quick succession, as in (2) above, 5× with sterile PBS + 0.05% (v:v) Tween 20, followed by 5× with sterile PBS. Drain the tubes well, but do not allow to dry.

5. To each tube, add 1 ml of *fresh* 100 mM triethylamine (TEA) in H$_2$O, seal, and incubate with gentle tumbling for 10 min at RT. Remove the 1 ml phage-containing TEA and neutralize by adding the phage-containing solution to 0.5 ml of 1.0 M Tris–HC1, pH 7.5, prealiquoted to microfuge tubes. Store the eluted phage on ice. Note: phage should not be neutralized in the antigen-coated Immunotube in order to avoid the potential for rapid reimmobilization of phage.

6. Check the OD$_{590}$ of the subcultured *E. coli* host cells, and confirm that the OD is between 0.6 and 0.9 (mid-log) after about 2 h of culture. Combine 1 ml of eluted, neutralized phage with 10 ml of the mid-log culture in disposable 15-ml culture tubes, mix well, and place in a 37°C water bath for 30 min without agitation. After incubation and infection, determine the number of phage as tranducing particles by plating 10^{-1} through 10^{-5} dilutions to 2× TY medium containing 15 μg/ml tetracycline (2× TY + Tet). Centrifuge the remaining 10-ml mixture of phage-infected cells at 5000 × *g* for 10 min in a clinical or preparative centrifuge, resuspend to 1 ml 2× TY and plate evenly on 243 × 243-mm-square petri dishes (e.g., Bioassay Dish, Nunc) containing 2× TY + Tet plus 1.5% agar. Dry the plates in a laminar flow hood, invert, and incubate overnight at 37°C (10–12 h) or 30°C (15 h or longer).

7. The next morning, harvest the lawn of *E. coil* colonies by scraping evenly (we use a stout, bent glass rod) into 10 ml of 2× TY + Tet. An even suspension is more important to achieve than total recovery at this point, and you will likely recover about 6–7 ml. Scrape the cells toward one edge of the dish, then tilt to one corner and wash by successive pipetting with a 10-ml pipette until the suspension is homogeneous. Inoculate the recovered cells to 200 ml of liquid 2× TY + Tet in a 1L flask and incubate 6–8 h at 30°C with vigorous aeration. Count the overnight dilution plates, and calculate the previous round's total number of eluted phage. Seal the dilution plates with film, and store at 4°C for possible later use in phage ELISA assays (Section C). Note that if this section of the protocol is started sufficiently early in the day (typically, 6:30 AM), there is enough time to complete the subsequent phage preparation and a round of panning on the same day (by 9:00 PM). In this way, 1 round per day can be completed. If this is not convenient, incubate the amplification culture overnight at 30°C.

8. Following incubation of the amplification culture at 30°C, centrifuge the culture at 5000 × *g,* 10 min at 4°C. To effectively reduce the potential for phage

cross-contamination between selection rounds, use centrifuge bottles that have been treated in 0.1 N NaOH, followed by thorough rinsing with sterile distilled water and autoclaving. Decant the phage-containing supernatant to a sterile flask, add one-fifth volume of a PEG–NaCl solution (2.5 M NaCl + 20% (w:v) PEG-6000; BDH Catalog No. B44391-36) and incubate on ice for 30 min to 1 h.

9. Following incubation on ice, pellet the aggregated phage by centrifuging at 12,000 $\times$ g for 20 min, 4°C. Discard the supernatant and drain well, taking care to wipe excess traces of NaCl/PEG solution from the tube. The phage pellet will appear as a tan to opalescent pellet extending broadly up the wall of the centrifuge bottle. Using a pipette, gently resuspend the phage pellet in 10 ml sterile PBS, and transfer to a graduated disposable 50-ml conical tube.

10. Dilute the phage suspension to 40 ml with PBS and reaggregate by adding 8 ml of NaCl/PEG solution. Incubate on ice 15 min, followed by centrifugation as in (9) above. Drain well, removing all traces of NaCl/PEG. Resuspend the phage pellet to 1.5 ml sterile PBS, and use 1 ml in subsequent rounds of panning.

B. Solution Panning Using Paramagnetic Beads

1. In this protocol (Hawkins *et al.,* 1992), the chosen antigen (polypeptide) is covalently attached to a biotin linker arm containing a thiol-cleavable bridge (ImmunoPure NHS-SS-Biotin, Pierce Chemical Co.) according to the manufacturer's instructions.

2. The modified and washed (or dialyzed) antigen is diluted to the desired concentration (typically 50–100 nM final concentration) in 0.5 ml of 6% MPBS, 10 μl Tween 20, and added to 1 ml of phage from the library, or from a previous round of selection. Incubate for 1 h in a 15-ml disposable tube with gentle tumbling. Dilute the overnight *E. coli* preculture 1:100 to 20 ml prewarmed 2× TY medium and incubate at 37°C with vigorous aeration.

3. Initially using 250 μl Streptavidin M-280 Dynabeads (Dynal, Inc., Lake Success, NY) per selection, block the beads by incubating in 5 ml of 4% MPBS followed by incubation with gentle tumbling for 5 min at RT. Separate the beads using a magnet, decant, and discard the bead-free solution. Add 5 ml 4% MPBS and incubate at RT for 1 h with gentle tumbling. Separate on a magnet, decant, and discard the bead-free solution and resuspend the beads in 300 μl 2% MPBS.

4. To the phage/antigen mixture, add 300 μl of blocked Dynabeads, and incubate with gentle tumbling for 15 min at RT. Separate on a magnet, decant, and discard the bead-free solution, and resuspend the beads in 1 ml of 2% MPBS before transferring to a 1.5-ml microfuge tube. Wash the beads 15 times with PBS or 2% MPBS in the sequence 2× PBS followed by 1× 2% MPBS.

5. Resuspend the washed beads in 300 μl of freshly prepared 50 mM dithiothreitol in PBS, and stand at RT for 5 min. Separate beads on a magnet.

6. Add 150 μl of phage-containing bead-free solution to 10 ml of mid-log phase *E. coli* host cells in a 15-ml culture tube, mix well, and incubate without

shaking at 37°C for 30 min. Plate dilution aliquots to 2× TY + Tet plates for calculation of the number of phage-infected cells. Pellet the remaining infective centers and amplify as for Immunotubes (6–10) above.

7. Calculate the number of infective centers from the plate counts, and store the plates at 4°C for possible later use in phage ELISA assays (Section C).

C. Phage ELISA Assay for Recombinant Antibodies

1. Starting with a round of selection where a significant increase in phage numbers was observed, and extending to the previous and succeeding rounds, innoculate 94 individual colonies from the stored infective center dilution plates to 200 μl 2× TY + Tet per well using (initially) one 96-well polycarbonate microtiter plate per round. Leaving two wells per plate as uninoculated blanks, culture the microtiter plates with shaking overnight at 30°C in a closely fitting plastic box or other sealed container to maintain high humidity. Take care to leave the microtiter plate lid ajar to permit aeration; we usually rest the lid on a few sterile toothpicks laid across the microtiter plate. Coat ELISA assay plates (Falcon No. 3912) with 100 μl of a 10 μg/ml solution of antigen in PBS, and stand overnight at RT.

2. Wash the coated ELISA assay plates 3× in PBS; we invert and "slap" the plate on the wall or ledge of a sink, followed by submersion in a 4L plastic tub filled half-full of PBS. Take care to remove bubbles during submersion, and do not be afraid to slap the plates around to help empty the wells between washes—they do not break easily. To block, the plates are inverted and tamped to a paper towel, and 200 μl of 2% MPBS is added to all wells followed by incubation for 2 h standing at 37°C.

3. After blocking, coated ELISA assay plates are washed 3× in PBS as in (2) above, and 50 μl of 4% MPBS to is added each well. Microtiter plates containing individual phage cultures are centrifuged at 3000 rpm (about 2500 × g) for 15 min in a benchtop clinical centrifuge equipped with a microtiter plate adaptor. Following centrifugation, 50 μl of phage-containing supernatant is added to the ELISA assay plates containing 4% MPBS, and the mixture incubated for 1 h standing at RT.

4. Following phage binding/incubation, the contents of the ELISA assay plates are discarded and the plates washed 3× in PBS + 0.05% Tween 20, followed by 3× in PBS. The plates are tamped to remove excess PBS, followed by addition of 100 μl of an appropriate dilution of rabbit anti-fd or anti-M13 polyclonal antisera in 2% MPBS, and incubated for 30 min at RT. Note: polyclonal serum raised against bacteriophage M13 or fd is commonly of high titer; our antiserum was raised in rabbits against phage M13K07 (Pharmacia) and is used at 1 : 12,000 dilution. For each batch of antiserum, an optimal dilution should be determined to maintain a final assay blank value of ≤0.05.

5. The ELISA assay plates are washed as in (4) above, and an appropriate dilution in 2% MPBS of secondary antibody horseradish peroxidase (HRP) conjugate (e.g., goat anti-rabbit; Sigma Catalog No. A9169; we currently are using Lot No. 121H4817 at 1:8000 dilution), followed by incubation at RT for 30 min.

6. Assay plates are washed as in (4) above, and tamped in preparation for development of the HRP signal. For each 96-well assay plate freshly prepare 10 ml of HRP development solution (10 ml of 0.1 M Na-acetate, pH 6, plus 100 μl of TMB substrate solution [10 mg/ml 3,3′,5,5′-tetramethylbenzidine–HCl in 50% (v:v) DMSO], Sigma Catalog No. T8757) plus 4 μl 30%H_2O_2. Note that TMB stock solutions should be stored at 4°C in the dark. Using a multichannel pipettor, aliquot 100 μl of development solution to each column of wells at timed intervals (e.g., 15 s) and incubate 10 min at room temperature (standard reaction), or until a signal-over-blank blue color develops. Stop the action by adding 25 μl of 2 M H_2SO_4, and read the endpoint yellow color in a microtiter plate reader at $OD_{450-650}$.

D. Soluble Subcloning for Excreted, Recombinant Fabs

The direct kinetic characterization of phage-borne recombinant antibodies is complicated by the fact that the Fab heavy chain is expressed as a gene-III fusion in variable multiple-copy number of 3–5 per phage. The recombinant Fab antibodies can be subcloned from phage chromosomal DNA preparations to bacterial plasmid expression vectors as a pool using PCR techniques already described (Griffiths *et al.,* 1994). During subcloning, the heavy chain is removed from its phage gene-III fusion context, and cloned instead as a fusion with a murine *myc* epitope. As in the phage format, both light and heavy chains are expressed as fusions with heterologous leader peptides (gIII and *pel*B, respectively) designed to direct their secretion to the periplasm of host *E. coli* cells, where subunit polypeptides are assembled to form a functional, monovalent antibody molecule. Expressed as a frequency of the phage binder population, the frequency of plasmid subclones that are competent for secretion and assembly of soluble antibodies are a small subset of phage binders—usually 5–40% at any given round of selection. At least two possibilities may account for the lower frequency of soluble binders vis-a-vis phage binders in the population: (*a*) polyvalent binding in the phage format may constitute an artificially stable antibody–antigen interaction, which does not score positive for univalent low-affinity soluble molecules and (*b*) conformational changes imparted to the heavy chain upon removal from its gene-III fusion context may impair its secretion and/or its ability to assemble properly with its corresponding light chain to form a functional soluble antibody molecule.

ELISA assays for soluble recombinant antibody expression are performed as for phage ELISAs (described above) with some minor changes. Host cells carrying recombinant plasmid Fab constructs under the direction of a *lac*Z promoter are grown to mid-log phase in 96-well microtiter plates and induced overnight in the presence of 1 mM IPTG at 30°C. During the extended culture, soluble antibodies that are correctly secreted and assembled will accumulate in the culture medium. Cell-free aliquots of the culture medium are subsequently assayed as for phage ELISAs, except the primary (anti-M13) antibody is replaced by an equal mixture of rabbit polyclonal anti-Human κ chain and anti-Human λ chain antisera. The anti-κ and anti-λ antibodies (Sigma Catalog No. K-4502, Lot No. 111H4823 and Catalog No. L-8141, Lot No. 122H4809, respectively) are both used at 1:16,000 dilution, and the ELISA signal is developed as described for the phage assay. The relative intensities of ELISA signal from soluble antibody assays will vary, but are routinely lower compared to the phage ELISA reaction for at least two reasons: (*a*) as already mentioned, there is the potential for more stable, high-affinity binding through polyvalency in the phage format and (*b*) there are probable signal enhancements inherent to the phage ELISA assay itself, since each binding phage presents some 2000 copies of the fd or M13 coat lipoprotein antigen that is subsequently amplified and detected by the rabbit polyclonal antiserum.

E. Comparative Merits of Immunotube versus DynaBead Panning

Depending upon the experimental objectives involved, high-diversity single-pot prokaryotic antibody repertoires can present enormous practical and biological advantages over conventional antibody technologies. Phage display libraries can generally be screened against relatively low quantities of purified antigen, and existing panning strategies deliver recombinant antibodies as phage binders in a matter of 5–7 days, and at a fraction of the expense of an equivalent monoclonal antibody experiment. The procedures themselves require no specialized media or cell biology expertise beyond that normally found in a typical molecular biology laboratory. Like cloned hybridoma cell lines, each individual phage binder presents a monospecific antibody that recognizes a single epitope. Alternatively, phage may be combined as a pool and used as a "polyclonal" collection of binders recognizing multiple epitopes. Interestingly, panned and selected phage can display a high degree of specificity, as evidenced by recently published data where polyclonal phage pools distinguished between closely related members of the Kringle serine protease family (Griffiths *et al.,* 1994).

The use of prokaryotic antibody expression libraries may pose an added benefit in that the use of animals and associated immunization regimes can be entirely avoided. Furthermore, since the recombinant antibodies are expressed and assembled outside a eukaryotic cell context, they may not be subjected to the same biological constraints of immune "tolerance" and immune "education" toward "self" antigens intrinsic to the animal system. As such, prokaryotic expression libraries may prove to be valuable sources of recombinant antibodies capable of binding conserved self antigens such as murine BIP, calnexin, or calreticulin, and recently published data suggest that this is indeed the case (Nissim *et al.*, 1994; Griffiths *et al.*, 1994).

In principle, each round of Immunotube selection and amplification will result in enrichment of the phage population in favor of those that bind the immobilized antigen. Perhaps not surprisingly, the conduct of a given experiment with respect to the kinetics of enrichment and recovery of recombinant antibody binders at each round of panning can vary widely between different antigens and different selection protocols. Such variability can be attributed to several factors inherent in the panning procedure including (*a*) different binding capacity and loading of individual antigens to the solid support (Immunotube) surface, (*b*) the number of antigen-binding phage initially present in the library population (*c*) the affinity of phage binders for the selected antigen, and (*d*) the stringency (volume and number) of washing conditions used prior to phage elution and amplification. In a "typical" Immunotube panning experiment a total of about 1×10^3–5×10^4 phage are eluted at round 1 (R1). If the numbers of eluted phage are significantly below this value (e.g., 500 or less) one might be dealing with a situation of poor binding affinity of the antigen for the solid support, in which case biotinylation and solution panning protocols using Dynabeads may present a viable alternative (see below). On the other hand, low elution numbers at R1 might also signify that the library content of antigen binders was low or nonexistent, although this has not been our experience in experiments involving more than 17 different polypeptide and nucleic acid antigens to date. Normally by R3–R4 the numbers of total phage eluted between rounds of panning are seen to increase—sometimes dramatically, such that 1–2 orders of magnitude increases are common. A sustained increase in eluted phage numbers through successive rounds of panning generally indicates successful enrichment for a genuine antigen-binding phage population. In this case, the investigator is justified to proceed to phage ELISAs, where the percentage of phage binders in the population is determined directly.

In comparison to the solid-phase panning protocol involving Immunotubes, the use of paramagnetic beads in solution presents significant advantages as well as some minor disadvantages. In this protocol (Hawkins *et al.*, 1992), the antigen is covalently attached to biotin via a thiol-cleavable linker molecule and subsequently washed or purified free of biotinylating reagent. In practice the requirement for biotinylation has not been a limitation involving a wide variety of protein and nucleic acid antigens. Biotinylation can be readily extended to include different lipid and carbohydrate attachment chemistries, and is therefore applicable to most biological antigens. This may not be correspondingly true for the binding properties of Immunotubes where, as has been already mentioned, surface retention and loading may vary enormously between individual antigens, or under different binding conditions of pH and ionic strength. Although a more tedious series of washing steps is involved for the magnetic bead protocol, in general it is more effective in terms of the kinetics of appearance of binders— possibly due to the relatively high surface area presented by microscopic beads for the capture of antigen–antibody/phage complexes. Unlike the Immunotube solid surface, where the concentration of surface-bound antigen is variable, the solution panning protocol permits careful control over, and may present a higher absolute concentration of, biotinylated antigen during panning. Thus it is possible to skew the recovery of phage binders in favor of relatively high-affinity interactions by lowering the concentration of antigen during panning.

Both the Immunotube and Dynabead panning procedures can be modified to select for phage that recognize epitopes that are unique to one antigen among two or more closely related antigens. For example, the library phage pool can be preexposed to a competing concentration of common epitopes (or antigen) before panning to the immobilized or biotinylated antigen carrying the epitope of interest. In principle, phage that recognize epitopes unique to the immobilized target antigen will be selected and enriched under these conditions. By way of example, Table I shows the results of parallel Immunotube panning experiments against a *mal*E :: ALS fusion protein (Bekkaoui *et al.,* 1993) under noncompetitive conditions, versus conditions where the phage were precompeted against an excess of *mal*E carrier protein. Under noncompetitive conditions, binders specific for the ALS peptide domain were recovered at a frequency of about 2% of the phage at R3 and R4, whereas under competitive panning conditions the frequency of ALS-specific phage rose an order of magnitude to occupy approximately 20% of the recovered phage population.

Up to this point, it has been convenient to consider the situation where phage pools are panned and enriched, either competitively or noncompetitively, against purified (usually homogeneous) preparations of antigen. However, the competitive panning strategy can be equally applied to more complex mixtures that differ by some specific antigen or epitope of interest. For example, using competi-

Table I
Enrichment for Recombinant Antibodies Against a *mal*E-ALS Fusion Protein Using Noncompetitive and Competitive Panning Protocols[a]

Panning round	Noncompetitive protocol			Competitive protocol		
	Total phage eluted	% *mal*E-ALS binders	% binders specific for ALS	Total phage eluted	% *mal*E-ALS binders	% Binders specific for ALS
1	3.5×10^5	<1	<1	—	—	—
2	2.0×10^6	<1	<1	2.3×10^5	44	8
3	1.7×10^6	51	2	1.8×10^8	99	17
4	4.4×10^8	87	3	NT	NT	NT

[a] Phage were panned using standard Immunotube protocols as described in the text. Under noncompetitive conditions, panning tubes were coated in PBS containing 100 μg/ml of the 78-kDa fusion protein, comprising the 42-kDa *mal*E domain fused to a 36-kDa ALS domain from *Brassica napus* ALS1 gene (Wiersma *et al.,* 1989; Bekkaoui *et al.,* 1993). For competitive panning, amplified phage from the first round of noncompetitive panning were incubated in PBS containing 0.5 m*M* purified *mal*E protein for 60 min at 37°C, prior to panning against immobilized *mal*E :: ALS fusion protein (NT, not tested).

tive panning strategies we have successfully recovered phage antibodies specific for recombinant proteins expressed in unfractionated extracts of vaccinia virus-infected mammalian cells *in vitro,* following competitive panning against the corresponding nonrecombinant extracts (Loh and Crosby, unpublished).

It is important to recognize that the panning process involves a biological amplification step at each round, with the attendant risk of selecting for phage that display an inherent growth advantage (including nonrecombinant phage). Thus, extended rounds of panning may have the effect of increasing the percent of phage binders in the population to very high levels (near 100%) at the expense of population diversity, such that a high frequency of identical clones are recovered. The optimal degree of panning and population enrichment can therefore be seen as striking a balance between the *frequency* of binders versus diversity of the population. In the event that panning delivers a population of phage binders of low diversity, that diversity can in principal be reintroduced by several different approaches. For example, a low-diversity phage population can be resubmitted to "combinatorial infection" in *cre* recombinase-expressing host cells carrying the donor heavy-chain repertoire, resulting in the reshuffling and diversification of the population's V_H repertoire. The resulting phage can then be subsequently repanned for binders to the antigen, with the likelihood that new $V_H + V_L$ combinations would emerge. A second approach involves the introduction of random structural diversity by propagation of the phage population in F^+ mutator strains of *E. coli.* Mutator alleles such as *mut*D preferentially introduce random base substitution mutations at an approximate frequency of 10^{-4} per nucleotide, or nearly 4 orders of magnitude greater spontaneous mutation frequency than the wild type (Cox and Horner, 1986). Since the average panning experiment involves on the order of 10^{12} phage, it is clear that an enormous molecular diversity will be represented in the mutagenized population that can be panned. Although the random mutational approach doubtless introduces deleterious mutations that compromise the replicational ability, selectable marker, or other biological property of the phage fd vector, these will not be overly apparent in the population, which is subsequently *functionally* amplified, repanned, and enriched for binders to the chosen antigen. Indeed, the random mutational approach might be held dearest to the geneticist's heart in that it makes no assumptions about the structure–function basis of antibody–antigen interaction.

IV. Application of Recombinant Antibody Technology to Plant Biology

To the plant biologist, it may be worthwhile to regard recombinant antibody phage display technology in its simplest terms—as a powerful, convenient and cost-effective source of defined protein ligands where both the genotypic and the phenotypic basis for the protein interaction are manifest in a single, replicating biological vector.

It is certainly true that most applications mentioned above could be profitably used in the area of plant biology. Virtually any aspect of the application of immunological techniques can benefit from recombinant antibody expertise including immunolocalization studies, immunocytochemistry, immunoaffinity purification, and protein–protein interaction applications. However, while recognizing the importance of diagnostic and biochemical applications, it is perhaps the competitive panning strategies involving complex targets that deserves more careful consideration.

Just as competitive panning strategies can be applied to complex mixtures of antigens, they can, in principle, be applied to the competitive panning of intact organelles, cells, or organs representing different states of genetic competence or developmental totipotency. For example, surface epitopes unique to *Brassica* SI (self-incompatibility phenotype) versus non-SI pollen could be targeted by differential panning of pollen harvested from different genetic backgrounds. Instead of recovering and amplifying R1 phage as a pool, infective centers could be aliquoted and maintained as an ordered array of low-diversity pools in microtiter plates. Such ordered arrays of recombinant antibodies, representing binders to all eipitopes presented on a complex surface, could be subsequently screened for binders that distinguish the functional or developmental state under study. The differential antigens involved need not be restricted to polypeptides, and the recombinant antibodies can be used directly as immunoaffinity purification reagents for purification of the specific epitope involved. A second related, and perhaps equally intriguing, application involves the assay of ordered phage pools for their ability to inhibit directly or stimulate some biochemically assayable function *in vitro*. For example, phage pools could be surveyed for their ability to interdict *in vitro* preprotein uptake and maturation by organelles, receptor/effector interaction or cell/pathogen recognition. Among the bioactive recombinant antibodies recovered, one would reasonably expect that some would recognize antigens that were directly involved in mediating the function under assay, and to serve as useful reagents for their biochemical and molecular characterization.

At the moment, plant biologists and geneticists alike suffer from the lack of an effective gene "knockout" capability of the sort that has been used so profitably in bacteria, fungi, and many higher eukaryotic systems for establishing clear gene–function relationships. Although antisense, ribozyme, and cosuppression strategies have been useful in this regard, their utility has not been universal. An alternative strategy has emerged involving the post-translational interdiction of plant gene product function via the transgenic expression of recombinant antibodies directed to specific antigens. For example, a recent published report has described the interdiction of a plant virus assembly by expression of a recombinant murine monoclonal-derived single-chain Fv construct (scFv) directed to the viral coat protein, resulting in a virus-tolerant phenotype (Taviadoraki *et al.,*

1993). At another level, the regulation of plant metabolism may well involve the establishment and maintenance of defined protein–protein interactions, and such interactions are being studied using approaches such as the yeast 2-hybrid system, described elsewhere in this volume. Antibody phage display libraries, in particular those of the single-gene scFv format, may provide significant advantages for this kind of metabolic interdiction approach, since they rapidly deliver the structural genes that encode recombinant antibodies with defined antigen-binding properties.

Acknowledgments

The authors are indebted to Dr. A. D. Griffiths, Professor Greg Winter, FRS, and other members of the MRC Centre for Protein Engineering, Cambridge, for their patient introduction to phage display technology during a study leave by W.L.C. to the Centre in 1993. We are grateful for stimulating discussions with regional colleagues including Drs. S. M. Hemmingsen (PBI), J. S. Lee and E. B. Waygood (Department of Biochemistry, University of Saskatchewan), and D. Galbraith (Department of Plant Sciences, University of Arizona, Tucson). This publication is NRCC No. 38456.

References

Ausubel, F. M., Brent, R., Kingston, R. E., Moore, D. D., Seidman, J. G., Smith, J. A., and Struhl, K. (1989). "Current Protocols in Molecular Biology." New York: Wiley–Interscience.

Barbas, C. F., III, Kang, A. S., Lerner, R. A., and Benkovic, S. J. (1991). Assembly of combinatorial antibody libraries on phage surfaces: The gene III site. *Proc. Natl. Acad. Sci. U.S.A.* **88,** 7978–7982.

Bekkaoui, F., Schorr, P., and Crosby, W. L. (1993). Acetolactate synthase from *Brassica napus:* Immunological characterization and quaternary structure of the native enzyme. *Physiol. Plant.* **88,** 475–484.

Chester, K. A., Begent, R. H. J., Robson, L., Keep, P., Pedley, R. B., Boden, J. A., Boxer, G., Green, A., Winter, G., Cochet, O., and Hawkins, R. E. (1994). Phage libraries for generation of clinically useful antibodies. *Lancet* **343,** 455–456.

Clackson, T., Hoogenboom, H. R., Griffiths, A. D., and Winter, G. (1991). Making antibody fragments using phage display libraries. *Nature* **352,** 624–628.

Cox, E. C., and Horner, D. L. (1986). DNA sequence and coding properties of mutD(dnaQ) a dominant Escherichia coli mutator gene. *J. Mol. Biol.* **190,** 113–117.

Griffiths, A. D. (1993). Building an in vitro immune system—Human antibodies without immunisation from phage display libraries. *Ann. Biol. Clin. Paris* **51,** 554.

Griffiths, A. D., Williams, S. C., Hartley, O., Tomlinson, I. M., Waterhouse, P., Crosby, W. L., Kontermann, R. E., Jones, P. T., Low, N. M., Allison, T. J., Prospero, T. D., Hoogenboom, H. R., Nissim, A., Cox, J. P. L., Harrison, J. L., Zaccolo, M., Gherardi, E., and Winter, G. (1994). Isolation of high affinity human antibodies directly from large synthetic repertoires. *EMBO J* **13,** 3245–3260.

Hawkins, R. E., Russell, S. J., and Winter, G. (1992). Selection of phage antibodies by binding affinity. Mimicking affinity maturation. *J. Mol. Biol.* **226,** 889–896.

Hoogenboom, H. R., Marks, J. D., Griffiths, A. D., and Winter, G. (1992). Building antibodies from their Genes. *Immunol. Rev.* **130,** 41–68.

Kang, A. S., Barbas, C. F., Janda, K. D., Benkovic, S. J., and Lerner, R. A. (1991). Linkage of recognition and replication functions by assembling combinatorial antibody Fab libraries along phage surfaces. *Proc. Natl. Acad. Sci. U.S.A.* **88,** 4363–4366.

Kohler, G., and Milstein, C. (1975). Continuous cultures of fused cells secreting antibody of predefined specificity. *Nature* **256,** 495–497.

Marks, J. D., Hoogenboom, H. R., Bonnert, T. P., McCafferty, J., Griffiths, A. D., and Winter, G. (1991). By-passing immunization. Human antibodies from V-gene libraries displayed on phage. *J. Mol. Biol.* **222,** 581–597.

Nissim, A., Hoogenboom, H. R., Tomlinson, I. M., Flynn, G., Midgley, C., Lane, D., and Winter, G. (1994). Antibody fragments from a "single pot" phage display library as immunochemical reagents. *EMBO J* **13,** 692–698.

Perelson, A. S., and Oster, G. F. (1979). Theoretical studies of clonal selection, minimal antibody repertoire size and reliability of self-nonself discrimination. *J. Theor. Biol.* **81,** 645–670.

Plückthun, A. (1992). Monovalent and bivalent antibody fragments produced in *Escherichia coli*— Engineering, folding and antigen binding. *Immunol. Rev.* **130,** 151–188.

Smith, G. P. (1985). Filamentous fusion phage: Novel expression vectors that display cloned antigens on the virion surface. *Science* **228,** 1315–1317.

Taviadoraki, P., Benvenuto, E., Trinca, S., DeMartinis, D., Cattaneo, A., and Galeffi, P. (1993). Transgenic plants expressing a functional single-chain Fv antibody are specifically protected from virus attack. *Nature* **366,** 469–472.

Tomlinson, I. M., Walter, G., Marks, J. D., Llewelyn, M. B., and Winter, G. (1993). The repertoire of human germline V_H sequences reveals about fifty groups of V_H segments with different hypervariable loops. *J. Mol. Biol.* **227,** 776–798.

Waterhouse, P., Griffiths, A. D., Johnson, K. S., and Winter, G. (1993). Combinatorial infection and in vivo recombination: a strategy for making large phage antibody repertoires. *Nucleic Acids Res.* **21,** 2265–2266.

Wiersma, P. A., Schmiemann, M. G., Condie, J. A., Crosby, W. L., and Moloney, M. M. (1989). Isolation, expression and phylogenetic inheritance of an acetolactate synthase gene from Brassica-napus. *Mol. Gen. Genet.* **219,** 413–420.

Williams, S. C., and Winter, G. (1993). Cloning and sequencing of human immunoglobulin-V-lambda gene segments. *Eur. J. Immunol.* **23,** 1456–1461.

Winter, G. (1993). Immunological techniques. *Curr. Opin. Cell. Biol.* **5,** 253–255.

Winter, G., and Harris, W. J. (1993). Humanized antibodies. *Trends Pharmacol. Sci.* **14,** 139–143.

CHAPTER 8

Isolation and Characterization of Plant Nuclei

Tom J. Guilfoyle

Department of Biochemistry
University of Missouri
Columbia, Missouri 65211

I. Introduction

A number of procedures and variations on these procedures have been described for isolating plant nuclei (reviewed by Watson and Thompson, 1986). A wide range of homogenization buffers containing stabilizing agents such as sucrose, glycerol, or hexylene glycol and divalent cations (e.g., Mg^{2+}, Ca^{2+}, or polyamines) have been employed along with filtration protocols and centrifugation strategies for isolating this plant organelle. It is now common practice to purify plant nuclei on Percoll gradients using protocols similar to those described by Luthe and Quatrano (1980) or Hagen and Guilfoyle ((1985). Two additional treatments are often employed to aid in the isolation and purification of plant nuclei. One of these treatments involves immersing embryos, whole seedlings, or excised organs for a few minutes in cold diethylether and subsequent removal of this solvent (Hamilton *et al.,* 1972). This treatment appears to facilitate the disruption of tissues and cells, but also effects the integrity and stability of nuclei released during cell disruption. The second treatment employs a nonionic detergent such as Triton X-100 for lysing contaminating organelles. This detergent treatment does, however, damage the membranes of plant nuclei (Guilfoyle *et al.,* 1986; Watson and Thompson, 1986).

A number of criteria have been used to assess the purity, integrity, and composition of isolated nuclei, including light and electron microscopy, protein/RNA/DNA ratios, chromatin structure, specific activity and duration of *in vitro* RNA or DNA synthesis, and analysis of specific *in vitro* run-on transcripts. The most common method used to assess purity and integrity of isolated nuclei is light microscopy of the isolated organelles after staining with a dye that reacts with nucleic acids and/or basic proteins. Nomarski optics are also commonly employed. SDS or acid–urea–polyacrylamide and agarose gel electrophoresis can be combined with microscopy to better characterize the integrity of nuclear components (i.e., DNA, RNA, and protein). The integrity of specific DNA or RNA species can be assessed by hybridization, and specific proteins can be analyzed by western blotting. Assays for specific enzymes (e.g., RNA polymerases, DNA polymerases, topoisomerases, protein kinases) have been used to characterize the metabolic state and capacity of purified nuclei. A routine method for the purification and assessment of purification for plant nuclei is described below, and protocols are outlined for determining DNA content and RNA polymerase activities in the purified organelles.

II. Materials

Diethylether

Nuclei isolation buffer I: 20 mM Tris–HCl (pH 7.2), 10 mM MgCl$_2$, 2 M sucrose, 1 mM dithiothreitol

Nuclei isolation buffer II: 20 mM Tris–HCl (pH 7.2), 2 mM *EDTA, 0.5 mM* EGTA, 0.5 mM spermidine, 0.15 mM spermine, 1 mM dithiothreitol, 40% glycerol

Percoll (Pharmacia LKB Biotechnology, Piscataway, NY) buffers: 50% Percoll (v/v) in nuclei isolation buffer I or II; 25% Percoll (v/v) in nuclei isolation buffer I or II

Nuclei storage buffer I: 20 mM Tris–HCl (pH 7.2), 10 mM MgCl$_2$, 1 mM dithiothreitol, 50% glycerol

Nuclei storage buffer II: 20 mM Tris–HCl (pH 7.2), 2 mM EDTA, 0.5 mM EGTA, 0.5 mM spermidine, 0.15 mM spermine, 1 mM dithiothreitol, 50% glycerol

Nuclei stain buffer: 0.1% Azure C in 20 mM Tris–HCl (pH 7.2), 10 mM MgCl$_2$, 250 mM sucrose

2× transcription buffer: 40 mM Hepes (pH 7.9), 0.2 mM EDTA, 2 mM dithiothreitol, 20 mM MgCl$_2$, 20% glycerol, 0.5 mM ATP, CTP, and GTP, 0.05 mM UTP (0.1–1 μCi [α-^{32}P]UTP/50 μl, New England Nuclear, Boston, MA)

1 M KCl stock solution

1 mg/ml α-amanitin stock solution

DE81 paper (Whatman LabSales, Inc., Hillsboro, OR)

0.5 M sodium phosphate buffer (pH 7)

95% ethanol

Distilled H$_2$O

Wheaton Dounce tissue grinder (Fisher Scientific, St. Louis, MO)

Cheesecloth

Miracloth (CalBiochem, La Jolla, CA)

Polytron PT20ST (Brinkman, Westbury, NY)

Refrigerated centrifuge with JS13.1 rotor (Beckman, Palo Alto, CA)

50-ml polypropylene centrifuge tubes (Fisher Scientific, St. Louis, MO)

15-ml Corex centrifuge tubes (Fisher Scientific, St. Louis, MO)

A light microscope with 10×, 20×, and 40× objectives (Nikon, Garden City, NY)

Pipetman (20 μl; Rainin Instrument Co., Woburn, MA)

Microcentrifuge tubes (1.5 ml: Fisher Scientific, S. Louis, MO)

Diphenylamine reagent: A diphenylamine stock solution is prepared by dissolving 1.5 g of diphenylamine in 100 ml glacial acetic acid, to which 1.5 ml of concentrated sulfuric acid is added. Just prior to use, 0.1 ml of aqueous acetaldyde (1 ml of cold acetaldehyde/50 ml distilled H$_2$O) is added per 20 ml of diphenylamine stock solution.

Calf thymus DNA standard solution: DNA is dissolved at 0.3 mg/ml in 5 mM NaOH. Standards are prepared by dissolving a given volume of stock solution in an equal volume of 1 N perchloric acid. The standard solutions are heated to 70°C for 15 min.

Spectrophotometer, DU-50 (Beckman, Palo Alto, CA)

All chemicals unless indicated otherwise are the highest grade available from Sigma Chemical Co. (St. Louis, MO).

========= **III. Methods**

A. Isolation of Plant Nuclei

The method described below has proven effective for the isolation of nuclei from a variety of plant tissues, including raw germ (e.g., wheat, rice), germinating seedlings (e.g., wheat, rice, soybean, mung bean, pea), plumules (e.g., pea, soybean, mung bean, green bean), curd (i.e., cauliflower), shoots and roots (e.g., soybean, pea, mung bean, maize), and leaves (e.g., turnip, tobacco, *Arabidopsis*). Plant material is cooled to 4°C, and all isolation procedures are carried out at this same temperature. The plant seedlings or organs are weighed, chopped into small pieces with a razor blade, and submersed in cold diethylether for 2–5 min. The ether is poured off, and any remaining ether is blown off with a stream of air in a hood. Five volumes of nuclei isolation buffer I or II per tissue weight is added to the plant material (e.g., 50 ml of isolation buffer per 10 g soybean plumules; the procedures that follow are based on a 10-g sample of plant material), and the mixture is homogenized at medium speed for 0.5–1 min in a Polytron PT20ST. If nuclei isolation buffer I is used at this step, then buffer I is used throughout the purification. If buffer II is used at this step, then buffer II is used throughout the purification. The homogenate is filtered through four layers of cheesecloth, followed by two layers of miracloth. The filtrate is poured into 50-ml centrifuge tubes and subjected to centrifugation at 1000–4000 rpm (depending on the source of the nuclei; pelleting of nuclei versus contaminating organelles should be monitored by microscopy; see below) for 10 min in a JS13.1 swinging bucket rotor. The supernatant is poured off, and a spatula is used to remove the upper gelatinous pellet from the firm white starch layer on the very bottom of the tube. The gelatinous pellet is suspended with a Dounce tissue grinder in nuclei isolation buffer at 25% of the original volume used for homogenization. The mixture is spun at 1000–4000 rpm for 5 min. The gelatinous pellet is collected as described above and suspended with a Dounce tissue grinder in 5 ml (or 10% of the original volume used for homogenization) nuclei isolation buffer. The mixture is carefully layered on top of 3 ml 25% Percoll in nuclei isolation buffer which, in turn has been layered over 3 ml 50% Percoll in nuclei isolation buffer (in a 15-ml Corex tube). Care must be take not to disturb the Percoll layers. The Percoll gradient tubes are spun at 7000 rpm for 20 min in a JS13.1 rotor. A Pasteur pipette is used to remove the upper layer, the 25% Percoll interface, and most of 25% Percoll layer. The 50% Percoll interface, containing the nuclei, is collected and suspended in 10 vol of nuclei isolation uffer with a Dounce tissue grinder. The mixture is spun at 4000 rpm for 10 min in a JS13.1 rotor and the pellet is collected and subjected to another wash in nuclei isolation buffer using a Dounce tissue grinder. The nuclei are repelleted by centrifugation at 4000 rpm for 10 min. The pellet is suspended with a Dounce tissue grinder in a small volume (e.g., 500 μl per 10 gram of plant tissue used in the original homogenization) of nuclei storage buffer I or II, and the purified nuclei are stored in 1.5-ml microcentrifuge tubes at −80°C. The nuclei can be stored frozen for years at −80°C in storage buffer and subjected to thawing and refreezing many times without observable effects on integrity and enzymatic activities.

B. Monitoring the Purity of Nuclei

At each step in the purification procedure described above (starting with the filtered homogenate), the nuclei should be examined for purity and integrity by using light microscopy. As a general rule, 10 μl of nuclei is mixed in 20 μl of Azure C stain buffer using a 20-ml pipetman. The nuclei are stained pink to purple and are observed as round or oval-shaped structures with darker purple round nucleoli within. Electron microscopy can be employed for greater resolution of nuclear structure. Some representative light and electron micrographs of purified plant nuclei are shown in Fig. 1.

C. DNA Determinations

DNA content of nuclei is determined using the diphenylamine assay described by Burton (1968). Aliquots of nuclei are incubated at 70°C for 10 min in 0.5 N perchloric acid. Dilutions of the nuclear extracts 0.5 N perchloric acid are mixed with 2 vol of diphenylamine reagent in glass tubes. Standard calf thymus DNA and blank tubes are prepared in the same manner at the same time, and all tubes are incubated at 25–30°C for 15–17 h. Absorbance at 600 nm is measured with a spectrophotometer. For additional details and troubleshooting, Burton (1968) should be consulted.

D. *In Vitro* Transcription

In vitro transcription assays are carried out with 10 μl of purified nuclei (approximately 10–50 μg DNA, as determined by the diphenylamine assay) in a total volume of 100 μl containing 20 mM Hepes (pH 7.9), 0.1 mM EDTA, 1 mM dithiothreitol, 10 mM MgCl$_2$, 50 or 300 mM KCl, 10% glycerol, 0.25 mM ATP, CTP, and GTP, 0.025 mM UTP (0.1–1 μCi of [α-^{32}P]UTP; [^{3}H]UTP or [^{14}C]UTP may be substituted for [^{32}P]UTP and any other nucleotide triphosphate may be substituted for UTP) and 0, 1, or 100 μg/ml α-amanitin. Stock solutions of 2× transcription buffer, 1 M KCl, and 1 mg/ml α-amanitin are made at appropriate dilutions into the transcription mixture to attain the final concentrations. Distilled H$_2$O is used to adjust the final volume to 100 μl. RNA polymerase activities are measured in 50 mM KCl at 0, 1, and 100 μg/ml α-amanitin as well as 300 mM KCl and 0, 1, and 100 μg/ml α-amanitin. Assays are carried out at 25–30°C for 10–15 min. Alternatively, the kinetics of RNA synthesis can be measured by removing aliquots at 5, 10, 20, 30, and 60 min. Assays are terminated by spotting 10-μl aliquots on Whatman DE81 paper and washing the

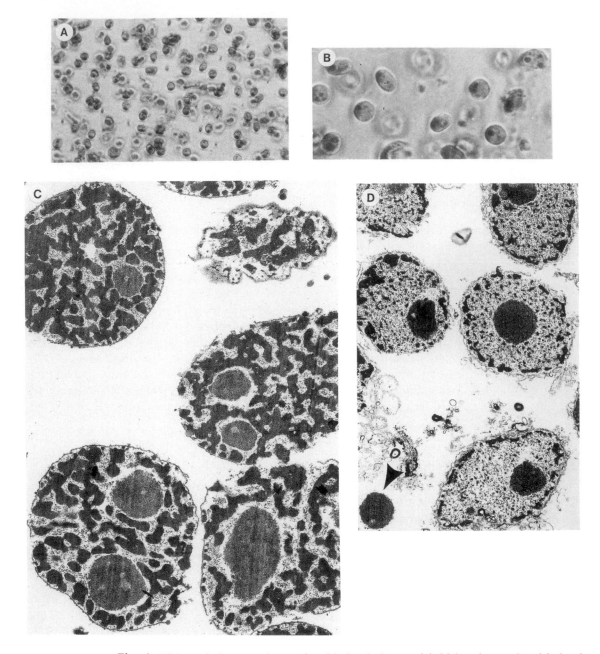

Fig. 1 Light and electron micrographs of isolated plant nuclei. Light micrographs of isolated wheat germ nuclei (A) 80× and (B) 200×. Electron micrographs of nuclei purified from wheat germ (C; 5000×) and etiolated soybean plumules (D;10,000×). Arrow in (D) indicates the position of a free nucleolus.

paper 5× for 5 min each in 0.5 M sodium phosphate buffer (pH 7), 2× in H_2O, and 2× in 95% ethanol. The paper is air-dried and radioactivity incorporated into RNA is determined by liquid scintillation counting. Specific activity is reported as picomoles UMP incorporated into RNA per 100 μg DNA per minute or per 10 or 15 min.

IV. Results and Discussion

A. Choice of Plant Material

Different plant organs and tissues present different levels of difficulty in purifying intact nuclei. The plant cell wall presents a substantial barrier to isolating fragile organelles such as nuclei. In some cases (e.g., wheat and rice germ), the nuclei appear to be relatively stable and can be easily isolated without rupture. In other cases (e.g., soybean hypocotyl or plumule and cauliflower curd), the nuclei are fragile, and these plant materials generally require an ether treatment to isolate intact, relatively undamaged nuclei. Plant materials that are rich in nuclei and lack a large central vacuole (e.g., germ, etiolated plumules, curd) are generally more amenable to the purification of nuclei than mature tissues that contain large vacuoles (e.g., maturing and mature leaves, stems, and roots). In general, nuclei are easier to purify from etiolated tissues compared to green tissues that contain mature chloroplasts. Chloroplasts can present a real challenge to nuclear purification, especially if detergents are omitted from the purification procedure. With green tissues, it is generally necessary to use a detergent such as Triton X-100 (0.1–1%) to purify nuclei away from lysed chloroplasts, and this detergent treatment results in some nuclear membrane disintegration (Watson and Thompson, 1986; Guilfoyle *et al.,* 1986.)

It is not uncommon to freeze plant tissues in liquid N_2 and store these tissues at −80°C prior to isolating nuclei for nuclear run-on *in vitro* transcription experiments. Although this may not present a major problem with *in vitro* transcription in some cases, in other cases prior freezing of plant tissues may be detrimental for maximum run-on transcription rates (Guilfoyle, unpublished results with auxin-induced transcripts in soybean plumules). Freezing of plant tissues can also be detrimental to nuclear integrity. Although wheat germ may be frozen without much apparent effect on the integrity and activity of isolated nuclei, nuclei prepared from frozen cauliflower curd or soybean plumules are clearly damaged (by microscopic observation) and have substantially lower levels of transcriptional activity.

Protoplasts serve as a good source for nuclei because enzymatic degradation of the cell wall facilitates the release of largely undamaged nuclei upon lysis of the plasma membrane. When protoplasts are used for nuclei isolation, the ether step is omitted. With protoplasts and tissue cultured cells (treated with cell wall degrading enzymes), nuclei can be purified by cell lysis using tissue grinders or by passing the protoplasts or cells through a 25-gauge needle several times. The nuclei are subsequently purified by centrifugation on Percoll gradients (Will-mitzer and Wagner, 1981; Hicks and Raikhel, 1993).

B. Choice of Nuclei Isolation Buffer and Purification Strategy

It is a general procedure to use Mg^{2+} as a divalent cation to stabilize chromatin and nuclear proteins during purification of nuclei. In some cases, however, it is beneficial to substitute polyamines (i.e., spermidine and spermine) in the presence of metal chelators (i.e., EDTA and EGTA) for the divalent cation. In this case, the polyamines act to stabilize the chromatin and nuclear proteins, and at the same time, the metal chelators act to inhibit deoxyribonuclease and phenolic oxidases (Willmitzer and Wagner, 1981). The inhibition of these enzymes can be important in the preparation of high-molecular-weight nuclear DNA and in protecting nuclear enzymes from oxidation.

The pH of the homogenization buffer can also be important. In a number of cases, low pH buffers, in the range pH 5.5–6, have been found to increase the yield, integrity, and purity of the isolated nuclei (Chen *et al.*, 1975; Saxena *et al.*, 1985).

Percoll concentrations may have to be adjusted differently than those described for isolating nuclei from certain plant tissues. Although the procedure described is generally applicable to a variety of plant tissues, the bulk of nuclei from some plant tissues will band on top of the 25% Percoll interface as opposed to the 50% interface. It is also possible that nuclei from some plant tissues will penetrate into the 50% Percoll layer and even pellet. Thus, it is important to examine each interface as well as the pellet in Percoll gradients for nuclei. For best results, it is recommended that Percoll concentrations be adjusted by trial and error to conditions that give the maximum yield of nuclei at a specific Percoll interface for a given plant tissue. An alternative strategy with Percoll is to suspend the nuclei in the 50% Percoll buffer and layer the 25% Percoll on top of this followed by buffer alone. When these gradients are subjected to centrifugation, the nuclei will float to the top of the 50% layer. In some cases, this strategy can improve purification.

C. Microscopic Observation

Microscopic observation reveals that ether treatment of cauliflower curd results in better yield and isolation of round, apparently undamaged, nuclei (Guilfoyle, unpublished results). In contrast, cauliflower curd nuclei isolated in the absence of an ether treatment are clearly ruptured and exhibit a variety of shapes in the microscope. The isolation of nuclei from wheat and rice germ does not appear to be substantially improved by an ether treatment, however. Although ether treatment may not always improve the purification and yield of nuclei, it is generally not detrimental to nuclear purification and should be included in routine nuclei purification protocols. Addition of detergents to nuclei isolation buffers can have some detrimental effects (as mentioned above). Although detergents such as Triton X-100 do not appear to alter the shape and apparent integrity of nuclei observed with the light microscope, electron microscopy reveals that the nuclear envelope is severely damaged or totally removed by this treatment (Guilfoyle *et al.*, 1986).

D. DNA Content and Composition of Isolated Nuclei

To standardize nuclear preparations used in further analysis such as enzymatic activity measurements, the DNA content per given volume of purified nuclei is usually determined. The most common procedure used to determine DNA content in isolated nuclei is the diphenylamine procedure described by Burton (1968). The integrity of the DNA can be assessed by agarose gel electrophoresis (Watson and Thompson, 1986). In some cases, a count of nuclei per given volume (using a hemocytometer) is used in place of DNA content determinations to standardize numbers of nuclei in enzymatic assays. Protein content is also occasionally used to standardize nuclear preparations, but protein contents in isolated nuclei can vary widely (Chen *et al.*, 1975). Although the protein content of isolated nuclei provides little information, the protein composition of isolated nuclei can provide information on the composition, integrity, and purity of isolated nuclei. Protein composition can be assessed by SDS–polyacrylamide gel electrophoresis and western blotting. Histones, which characteristically migrate in the range 10–20 kDa on 12 to 15% SDS gels, should be the dominant proteins observed with highly purified nuclei (Spiker *et al.*, 1983).

E. *In Vitro* Transcription

To evaluate preparations of isolated nuclei for metabolic competence, it is common to measure the specific activity of RNA polymerase in the purified organelles. In some instances, only total activity is measured, which includes RNA polymerases I, II, and III; however, a more thorough assessment of RNA polymerase activity distinguishes among RNA polymerase I, II, and III or between RNA polymerases I + III and RNA polymerase II. In plant nuclei, RNA polymerases I and II are generally the dominant activities, and RNA polymerase III may be difficult to detect. RNA polymerase I, II, and III or I + III and II activities are distinguished by including α-amanitin in the *in vitro* transcription reaction. Plant RNA polymerase I is refractory to α-amanitin, up to 2 mg/ml of this toxin (Guilfoyle, 1983). All plant RNA polymerase II enzymes examined are 50% inhibited at about 0.05 μg/ml α-amanitin (Guilfoyle, 1983). Plant RNA polymerase III shows an intermediate sensitivity to this toxin, ranging from 50% inhibition at 5 μg/ml in wheat germ to greater than 2 mg/ml in cauliflower (Guilfoyle, 1983). When nuclei are incubated in 1 μg/ml α-amanitin, RNA polymerase II activity is inhibited by greater than 95%; therefore, activity measured at this concentration of toxin represents almost exclusively RNA polymerase

I + III. If RNA polymerase activity observed at 1 μg/ml α-amanitin is subtracted from total RNA polymerase activity observed with no inhibitor, the value obtained is equal to RNA polymerase II activity. When wheat germ nuclei are incubated in 100 μg/ml α-amanitin, RNA polymerase II and III are strongly inhibited, and the remaining activity is largely due to RNA polymerase I. By subtracting RNA polymerase activity observed at 100 μg/ml α-amanitin from total activity, the value obtained is equivalent to RNA polymerase II + III activity, and if RNA polymerase II activity (determined at 1 μg/ml α-amanitin) is subtracted from this value, RNA polymerase III activity is obtained. It is not possible to distinguish RNA polymerase I and III activities in cauliflower nuclei, however, because both RNA polymerases I and III in cauliflower are almost totally refractory to 1 mg/ml of the toxin (inhibition observed beyond this high toxin concentration may not be meaningful).

It should be noted that types and concentrations of monovalent and divalent cations can greatly impact on RNA polymerase activity in isolated nuclei. In many cases, higher activities are observed if 1 mM MnCl$_2$ is substituted for or included with 10 mM MgCl$_2$. The concentration of monovalent cation can also greatly effect the nuclear RNA polymerase activities. Although it is not always the case, RNA polymerase I and III activity are generally most active at low concentrations of monovalent salt (e.g., 50–300 mM KCl), but RNA polymerase II may be most active at high concentrations of salt (e.g., 300–500 mM KCl). The monovalent salt effect on RNA polymerase II activity probably results from disruption of chromatin structure rather than any intrinsic property of the enzyme, since purified RNA polymerase II is most active at low salt concentrations when assayed on a deproteinized DNA template (Guilfoyle, 1983). It is advisable to determine optimal conditions for RNA polymerase I, II, or III activities before carrying out specific *in vitro* assays (i.e., *in vitro* run-on experiments with specific gene probes). When making comparisons from one nuclei preparation to another, it is important to note that specific activities for total or individual RNA polymerase I, II, and III activities in isolated nuclei can vary widely, depending on such things as the state of the tissue chosen for nuclei purification (dividing versus nondividing; dry versus imbibed; light grown versus dark grown, etc.) as well as the method chosen for *in vitro* assay (divalent and monovalent salt concentrations, nucleotide concentrations, assay temperature, etc.). Some representative data on specific activities, α-amanitin inhibition, and salt effects on nuclear RNA polymerase activities are shown in Table I.

V. Conclusions and Perspectives

The isolation of pure, intact, metabolically active plant nuclei is important to many fields of plant molecular and cellular biology. In molecular biology, purified nuclei are important for demonstrating transcriptional regulation of specific genes (i.e., *in vitro* run-on transcription), RNA-processing, isolation of transcription

Table 1

Specific Activities and Ratios of RNA Polymerases I + III and II in Purified Plant Nuclei[a]

Tissue source	Activity total	Activity + α-amanitin	Activity RNAP I + III	Activity RNA PII	Ratio
SB-P-LS	13.4	7.2	7.2	6.2	1.2
SB-P-HS	30.4	14.6	14.6	15.8	0.9
P-P-LS	5.4	4.3	4.3	1.1	3.9
P-P-HS	6.2	3.7	3.7	2.5	1.5
C-LS	47.2	30	30	17.2	1.7
C-HS	80.8	26.8	26.8	54	0.5
W-LS	2.3	2.2	2.2	0.1	22
W-HS	3.5	1.7	1.7	1.8	0.9

[a] Nuclei were isolated from 8-day-old etiolated soybean plumules (SB-P), 10-day-old etiolated pea plumules (P-P), cauliflower curd (C), and 18-h imbibed wheat embryo axes using nuclei isolation buffer I. Assays were carried out for 10 min at 25°C at low salt (LS, 50 mM KCl) or high salt (HS, 300 mM KCl) in the presence or absence of 1 μg/ml α-amanitin. Specific activities are reported in picomoles UMP incorporated into RNA/100 μg DNA in a 10-min assay period. RNA polymerase I + III and RNA polymerase II activities were calculated as described in the text. The ratio is RNA polymerase I + III/RNA polymerase II activities.

factors and nuclear proteins. In cell biology, these isolated organelles are important for studying nuclear protein targeting and retention, the nuclear matrix, and chromatin structure.

Acknowledgments

Research in the author's laboratory on nuclei isolation, characterization, and *in vitro* transcription was supported from grants from NSF PCM 8208496 and NIH GM24096 and GM37950.

References

Burton, K. (1968). Determination of DNA concentration with diphenylamine. *Methods Enzymol.* **12B,** 163–168.

Chen, Y., Lin, C. Y., Chang, H., Guilfoyle, T. J., and Key, J. L. (1975). Isolation and properties of nuclei from control and auxin-treated soybean hypocotyl. *Plant Physiol.* **56,** 78–82.

Guilfoyle, T. J., Suzich, J., and Lindberg, M. (1986). Synthesis of 5S rRNA and putative precursor tRNAs in nuclei isolated from wheat embryos. *Plant Mol. Biol.* **7,** 95–104.

Guilfoyle, T. J. (1983). DNA-dependent RNA polymerases of plants and lower eukaryotes. *In* "Enzymes of Nucleic Acid Synthesis and Modification. RNA Enzymes" (S. T. Jacob, ed.), Vol. 2, pp. 1–42. Boca Raton, FL: CRC Press.

Hagen, G., and Guilfoyle, T. J. (1985). Rapid induction of selective transcription by auxins. *Mol. Cell. Biol.* **5,** 1197–1203.

Hamilton, R. H., Kunsch, U., and Temperli, A. (1972). Simple procedures for the isolation of tobacco leaf nuclei. *Anal. Biochem.* **49,** 48–57.

Hicks, G. R., and Raikhel, N. V. (1993). Specific binding of nuclear localization sequences to plant nuclei. *Plant Cell* **5,** 983–994.

Luthe, D. S., and Quatrano, R. S. (1980). Transcription in isolated wheat nuclei. Isolation of nuclei and elimination of endogenous ribonuclease activity. *Plant Physiol.* **65,** 305–308.

Saxena, P. K., Fowke, L. C., and King, J. (1985). An efficient procedure for the isolation of nuclei from plant protoplasts. *Protoplasma* **128,** 184–189.

Spiker, S., Murray, M. G., and Thompson, W. F. (1983). DNase I sensitivity of transcriptionally active genes in intact nuclei and isolated chromatin of plants. *Pro. Natl. Acad. Sci. U.S.A.* **80,** 815–819.

Watson, J. C., and Thompson, W. F. (1986). Purification and restriction endonuclease analysis of plant nuclear DNA. *Methods Enzymol.* **118,** 57–75.

Willmitzer, L., and Wagner, K. G. (1981). The isolation of nuclei from tissue-cultured plant cells. *Exp. Cell Res.* **135,** 69–77.

CHAPTER 9

Isolation of Nuclei Suitable for *in Vitro* Transcriptional Studies

John C. Cushman

Department of Biochemistry and Molecular Biology
Oklahoma State University
Stillwater, Oklahoma 74078

I. Introduction

The isolation of intact nuclei from plant sources has been used for a variety of experimental purposes including the isolation of high-molecular-weight DNA (Watson and Thompson, 1986), *in vitro* DNA synthesis (Roman *et al.,* 1980), isolation of labeled transcripts for differential screening of cDNA libraries (Somssich *et al.,* 1989), preparation of nuclear extracts for *in vitro* transcription systems

(Roberts and Okita, 1991), isolation of nuclear proteins (Harrison *et al.,* 1992), and studies of protein targeting to the nucleus (Hicks and Raikel, 1993). One of the most important uses of isolated nuclei, however, involves the *in vitro* transcription run-on assay (Luthe and Quatrano, 1980a,b; Gallagher and Ellis, 1982; Silverthorne and Tobin, 1984; Hagen and Guilfoyle, 1985; Guilfoyle *et al.,* 1986).

The accumulation of mRNA can be determined by either increased rates of transcription or decreased rates of degradation. To distinguish between these two possibilities, *in vitro* transcription run-on techniques using isolated nuclei were developed to measure directly transcription rates from specific genes (McKnight and Palmiter, 1979). In this technique, isolated nuclei are allowed to elongate transcripts initiated *in vivo* in the presence of one or more radiolabeled ribonucleoside triphosphates. Specific radiolabeled transcripts can then be quantitated by hybridization to complementary DNA probes. Isolated nuclei from plants have been used extensively for *in vitro* transcription run-on experiments to demonstrate the transcriptional control of specific genes in response to a wide variety of abiotic and biotic stimuli and by plant growth regulators and developmental signals.

II. Isolation of Nuclei

Many protocols for isolating nuclei from leaf tissue, plant embryos (Luthe and Quatrano, 1980a,b), cultured plant cells (Willmitzer and Wagner, 1981; Keim, 1987; Wollgiehn, 1991), and protoplasts (Saxena *et al.,* 1985) have been described. The basic steps involved in the process of nuclei isolation include (1) tissue disruption and breaking of cell walls, (2) separation of the nuclei from the crude cellular lysate by filtration, (3) differential centrifugation, (4) solubilization of the plasma membrane and other subcellular organelles with nonionic detergents, and finally, (5) the purification of the intact nuclei from cellular contaminants by continuous or discontinuous density gradient centrifugation. For certain purposes, such as the isolation of crude nuclear protein extracts, differential centrifugation provides nuclei preparations of sufficient purity. For the isolation of nuclei for use in transcriptional run-on studies, however, it is essential to remove other cellular contaminants by density gradient centrifugation. The successful isolation of plant nuclei from certain plant tissue sources is complicated by contamination from abundant, transcriptionally active, subcellular organelles (i.e., plastids and mitochondria). Such contamination can be minimized by selective disruption of the organelles using low concentrations of nonionic detergents and the addition of a variety of substances that limit particle aggregation.

The disruption of tissues and breaking of cell walls can be accomplished by a variety of mechanical methods with blenders, polytron mixers, mortar and pestles, and Potter–Elvehjem tissue homogenizers usually yielding the best results (Willmitzer and Wagner, 1981). To enhance the yield of intact nuclei by reducing the damage caused by mechanical disruption procedures, partial enzymatic digestion of cell walls from tissue or tissue-cultured suspension cells can be used to facilitate cell wall disruption (Willmitzer and Wagner, 1981; Wollgiehn, 1991). Pretreatment of tissues with cold ether apparently enhances cell disruption presumably through disruption of the outer cell membrane and partial removal of cuticular waxes (Watson and Thompson, 1986). Disruption of tissues is often carried out in extraction media containing membrane stabilizing agents. One popular agent, hexylene glycol, has been used for the isolation of plant nuclei from leaf tissue (Watson and Thompson, 1986). Alternatively, low-molecular-weight osmotica, such as sucrose or glycerol are used because of their relatively high osmotic potential. This favors the disruption of other cellular organelles while leaving nuclei relatively intact. Replacement of sucrose with glycerol reportedly reduces the loss of soluble proteins and small-molecular-weight RNAs (Guilfoyle *et al.,* 1986).

Cellular disruption entails the breaking of other subcellular compartments, such as vacuoles, which can be rich sources of proteases and nucleases. In order to minimize the detrimental effects of exposure to vacuolar sap, homogenization of tissues is carried out in large volumes (typically 5–10 times the weight of the original tissue) and is followed immediately by centrifugation. Homogenization buffers commonly contain divalent cations (Mg^{2+}) to stabilize the nuclei. Alternatively, polyamines are used in conjunction with EDTA. Chelating agents such as EDTA are added to inhibit DNases and phenol oxidases. Polyamines help stabilize the nuclei by counteracting the chromatin condensation effects of EDTA, resulting in increased nuclear yields (Willmitzer and Wagner, 1981). Buffer type and pH can have a profound effect on nuclear yield. Tris- or Hepes-based homogenization buffers having a pH range of 7.2–7.8 have been commonly used when isolating nuclei from leaf tissue (Hagen and Guilfoyle, 1985). MES–KOH buffers with a relatively low pH, however, apparently improve overall nuclear yields from plant protoplasts (Saxena *et al.,* 1985). In addition, low pH buffers (pH 5.2–6) have been reported to yield more stable nuclei than pH ranges (pH 7.3–8) more suitable for animal nuclei (Willmitzer and Wagner, 1981; Saxena *et al.,* 1985).

After cellular disruption, the tissue brei is filtered, typically through several layers of cheesecloth or Miracloth, to remove cellular debris. The cellular membrane components remaining in the filtrate are often solubilized by low concentrations of nonionic detergents, such as Triton X-100 or Nonidet P-40. Detergents are also useful for preventing the aggregation of the nuclei. Disrupting other transcriptionally active cellular components such as plastids and mitochondria while preserving the structure of intact nuclei is a challenge and requires careful attention to isolation conditions. Relatively high concentrations of Triton X-100 (0.1–1%) will disrupt nuclear membranes, but nuclei will appear essentially intact upon examination by light microscopy (Mösinger and Schafer, 1984). Nuclei exposed to Triton will retain native chromatin structure and RNA polymerase activity despite the fact that membrane damage will occur as a result of the exposure to this detergent (Watson and Thompson, 1986). It is important to note, however, that the synthesis of pre-tRNA and 5S rRNA transcripts can be severely limited when nonionic detergents are used (Guilfoyle *et al.,* 1986).

Final purification of nuclei from other cellular contaminants is accomplished by density centrifugation on continuous or discontinuous gradients of Percoll, a colloidal silica coated with poly(vinylpyrolidone). The low viscosity and osmotic potential yet relatively high density of Percoll make it an ideal gradient material for nuclei isolations. Percoll gradients underlain with a 2 *M* sucrose cushion were first used by Luthe and Quatrano (1980a) for the isolation of nuclei and have become the method of choice for obtaining intact nuclei relatively free of cellular contaminants, particularly ribonuclease activity. Step gradients of Percoll are often preferable to continuous gradients because of the heterogenous size and density of nuclei commonly encountered in certain plants due to systemic endo-polyploidy (De Rocher *et al.,* 1990). If one requires large amounts of nuclei, a batch format can be adopted to save time when processing large amounts of tissue homogenates. Using this format, high concentrations of Percoll are used to float the nuclei and allow separation of starch granules, followed by lower Percoll concentrations to float cell wall debris and sediment nuclei (Willmitzer and Wagner, 1981). Alternatively, cytoplasmic contaminants can be removed by sedimenting nuclei through Percoll, followed by flotation of the nuclei on a Percoll pad (Wollghein, 1991).

A. Materials

Mortar and pestle (Coors)

Polytron mixer, PT-10-35 (Brinkmann)

Nylon meshes of 54 and 25 μm (Nitex)

Miracloth (CalBiochem)

Centrifuge, Sorvall RC-5B superspeed centrifuge (DuPont) or equivalent

Rotor, swinging bucket, HB-4, Sorvall (DuPont)

30-ml Corex tubes (Corning)

Pipetman, P-200 (Rainin)

Ethyl ether (anhydrous; Fisher E138-1)

Percoll (Sigma P-1644)

Triton X-100 (Rohm and Haas; Sigma T-6878)

Spermidine, free base (Sigma, S-0266). Make 0.1 *M* stock in diH$_2$O and aliquot into 500-μl portions. Store -20°C,

Spermine, tetrahydrochloride (Sigma, S-1141). Make 0.1 *M* stock in diH$_2$O and aliquot into 500-μl portions. Store -20°C,

Nuclei Isolation Buffer I (NIB I):

1 M sucrose, 10 mM Tris–HCl, pH 7.2, 5 mM MgCl$_2$, 10 mM 2-mercaptoethanol. Autoclave and store at 4°C. Add 2-mercaptoethanol immediately prior to use.

Nuclei Isolation Buffer II (NIB II):

250 mM sucrose, 10 mM morpholinoethanesulfonic acid (MES)–KOH, pH 5.4, 10 mM NaCl, 10 mM KCl, 2.5 mM EDTA, 0.1 mM spermine, 0.5 mM spermidine, 10 mM 2-mercaptoethanol. Autoclave and store at 4°C. Add 2-mercaptoethanol immediately prior to use.

Nuclei Storage Buffer (NSB):

20% glycerol, 20 mM N-2-hydroxyethylpeperazine-N'-2-ethanesulfonic acid (Hepes)–KOH, pH 7.2, 5 mM MgCl$_2$, 1 mM dithiothreitol. Store −20°C.

10% Triton X-100

B. Methods

The methods for isolating nuclei from plant tissues outlined below rely upon Percoll density step gradient centrifugation for the purification of nuclei. The first protocol is based on the procedure of Hagen and Guilfoyle (1985), for isolating nuclei from hypocotyls and etiolated plumules of soybean. The homogenization buffer contains a high concentration of sucrose at pH 7.2 and uses Mg^{2+} to stabilize the nuclei. This method has been adapted for use with green tissues (Cushman *et al.*, 1989). The second method is based upon procedures described for the isolation of nuclei from plant protoplasts (Saxena *et al.*, 1985; Hicks and Raikel, 1993) and has been adapted here for intact plant tissues. This method employs a homogenization buffer having a low pH and polyamines to stabilize the nuclei. Both methods yield nuclei suitable for *in vitro* transcription run-on assays; however, the nuclei obtained using Method II appear more intact. Modifications for isolating nuclei from green versus nongreen tissue and small versus large-scale preparations are included in the protocols. Precool all glassware, solutions, and utensils to 4°C. Prepare Percoll step gradients prior to beginning the procedure. All centrifugation steps use a swinging bucket rotor to minimize mechanical stress on the nuclei.

1. Method I

1. Collect leaves, wash with water if needed, and weigh them.

2. Optional: In a fume hood, transfer each batch of leaves to a glass beaker on ice. Treat leaves with ice cold diethyl ether (3–5 ml/g fresh weight of tissue) for 3–5 min. Drain ether off and remove remaining ether by blotting leaves with paper towels. Rinse leaves once with ice-cold nuclear isolation buffer (NIB I) (3 ml/g fresh weight). Discard rinse. All subsequent manipulations are performed on ice in a cold room at 4°C.

3. Transfer (ether-treated) leaves to a precooled mortar and pestle and grind tissue (10 g) in 10 vol (100 ml) of ice-cold NIB I until completely disrupted. For

processing larger amounts of tissue, a blender (Waring) or polytron mixer (PT 10-35, Brinkmann) may be used; however, it is important to keep frothing to a minimum.

4. Filter homogenate through two layers of Miracloth and then through a series of nylon meshes (i.e., 54- and 25-μm) of decreasing size into a glass beaker sitting on ice. The meshes are fitted over the mouth of plastic reagent bottles of decreasing size so they may be nested one inside the other. The smallest has a diameter that will fit above the beaker (250 ml) being used to collect the filtrate.

5. Measure volume of the filtrate. If green tissue is used, gradually add 10% Triton X-100 dropwise with stirring to a final concentration of 0.4%.[1] If nongreen tissue is used, proceed to step 7.

6. Incubate the lysate on ice for 5 min to allow solubilization of chloroplast membranes.

7. Carefully overlay the lysate onto a 2-step gradient consisting of 25/75% Percoll prepared in NIB I without Triton X-100. Prepare step gradients in 30-ml Corex tubes with each step consisting of 5 to 7.5 ml.

7A. Large-scale alternative. For processing larger amounts of plant material, the crude lysate is spun at 4000 rpm ($2600 \times g$)[2] for 10 minutes in a Sorvall HB-4 rotor. Carefully decant the supernatant and gently resuspend the pellet(s) using a wide-bore serological pipette in 15 mls NIB I (per gradient) without Triton X-100. The resuspended crude nuclei are then loaded onto the two-step gradient as above.

8. Centrifuge at 6500 rpm ($7000 \times g$) for 30 min in a swinging bucket rotor (Sorvall HB-4).

9. Remove upper part of the gradient with a serological pipette and carefully remove the nuclei, which will appear as a white fluffy band at the 25/75% Percoll interface with a Pasteur pipette. Transfer nuclei to a new Corex tube and dilute with 3 volumes NIB I.

10. Centrifuge 4000 rpm ($2600 \times g$) for 10 min in a swinging bucket rotor (Sorvall HB-4).

11. The nuclei pellet is suspended in 20 ml of NIB I and pelleted again by centrifugation at 2500 rpm ($1000 \times g$) for 10 min to remove residual Percoll.

12. The final nuclear pellet is resuspended in 250 μl NSB I using a pipetman with wide bore (cut off) tip.

13. Aliquot nuclei (50 μl) into prechilled microfuge tubes.

14. Nuclei may be used immediately or frozen in liquid N_2 and stored at $-80°C$ until use. Nuclei may be stored at $-80°C$ for many months without significant loss of activity.

---

[1] The optimal concentration of detergent for elimination of membrane-bound organelle contamination must be determined empirically.

[2] The g values refer to the maximum centrifugal field of the rotor.

15. Nuclei yield may be estimated by counting in a hemocytometer using a light microscope with phase contrast or Nomarski optics. If possible, it is useful to stain the nuclei preparation with either 1% DAPI (4',-diamidino-2-phenylindole, Sigma D-9542) or 1–5 μg/ml Hoechst dye No. 33258 (Polysciences, Inc.), and visualize using fluorescence optics.

2. Method II

1. Collect, wash, and weigh leaves as described in Method I. Be sure to conduct all subsequent manipulations on ice in a cold room at 4°C.

2. Grind leaves in ice-cold nuclear isolation buffer II (NIB II) as described in Method I,

3. After the homogenate has been filtered (and detergent-treated), carefully overlay the lysate onto a two-step gradient consisting of 25/75% Percoll prepared in NIB II.

3A. Large-scale alternative. For processing larger amounts of plant material, the crude lysate may be spun at 2500 rpm (1000× g) for 10 min in a Sorvall HB-4 rotor. Carefully decant the supernatant and gently resuspend the pellet(s) using a wide-bore pipette or paint brush in 15 ml NIB II (per gradient) without Triton X-100. The resuspended crude nuclei are then loaded onto the two-step gradient as above.

4. Centrifuge step gradients at 3500 rpm (2000× g) for 20 min in a swinging bucket rotor (Sorvall HB-4).

5. Remove the upper part of the gradient with a serological pipette and then carefully remove the nuclei which will appear as a white fluffy band at the 25/75% Percoll interface with a Pasteur pipette. Transfer nuclei to a new Corex tube and dilute with 3 vol NIB II.

6. Centrifuge 2500 rpm (1000× g) for 10 min in a swinging bucket rotor (Sorvall HB-4).

7. The nuclei pellet is suspended in 20 ml of NIB II and pelleted again by centrifugation at 2500 rpm (1000× g) for 10 min to remove residual Percoll.

8. The final nuclear pellet is resuspended in NSB (250 μl), aliquoted, frozen, and stored as described in Method I.

III. *In Vitro* Transcription Run–On Assays

The following protocol is based on the *in vitro* transcription run-on procedures of Chappell and Hahlbrock (1986), Hagen and Guilfoyle (1985), and Wollgiehn (1991). RNA polymerase requires both monovalent and divalent cations. In general, RNA polymerase activity is stimulated by increasing amounts of monovalent cations [with $(NH_4)_2SO_4$ being more stimulatory than KCl]; however, nuclei will lyse at cation concentrations greater than about 125 mM (Luthe and Quatrano, 1980b; Gallagher and Ellis, 1982). Therefore, $(NH_4)_2SO_4$ concentrations of 50–75 mM are typically used. $MgCl_2$ also stimulates RNA polymerase activity and it is typically used at concentrations from 5 to 15 mM. The *in vitro* incorporation by nuclei is dependent upon the addition of exogenous ribonucleoside triphosphates. Isolated nuclei will usually display linear rates of incorporation for only about the first 15–20 min, but will continue to incorporate radiolabeled UTP for several hours (Silverthorne and Tobin, 1984; Chappell and Hahlbrock, 1986).

A. Materials

Microfuge, Eppendorf 5415c (Brinkmann)
Minifold II Slot blot apparatus (Schleicher & Schull, Inc.)
2.4 cm-DE81 filter discs (Whatman)

RNasin (Promega)
Ribonucleoside triphosphates (Boehringer Mannheim)
$[\alpha\text{-}^{32}P]$UTP (3000 Ci/mmol, ICN)
DNase, RNase-free (RQ1, Promega)
tRNA, *E. coli*, RNase free (Boehringer Mannheim)

5× Transcription buffer:
100 mM Hepes–KOH, pH 7.8; 375 mM $(NH_4)_2SO_4$, 25 mM $MgCl_2$

Transcription extraction buffer:
7.5 M urea, 100 mM LiCl, 20 mM EDTA, 5% SDS, 10 mM ATA

B. Methods

1. On ice, compose the transcription reaction in a 100-μl reaction volume, adding the components in the order listed below. Add the nuclei last and gently mix without introducing air bubbles.

Final concentration	Stock solution	For 100 μl
1× transcription buffer	5×	20 μl
10% v/v glycerol	—	10 μl
0.5 mM each ATP, CTP, GTP	100 mM	1.5 μl
1 mM dithiothreitol	100 mM	1.0 μl
100 μCi $[\alpha\text{-}^{32}P]$UTP (3000 Ci/mmol, ICN)	10 μCi	10 μl
100 U RNasin (Promega)	40 U/μl	5 μl
Sterile diH$_2$O	—	5 μl
Nuclei (1–2 × 10^7)	—	50 μl

2. Incubate at 30°C for 15–20 min.

3. If incorporation rates of $[\alpha\text{-}^{32}P]$UTP into radiolabeled transcripts are to be

followed, remove duplicate 2- to 5-μl samples at appropriate time intervals and spot onto 2.4-cm DE81 filter discs (Whatman) on aluminum foil. After 30–60 s, drop discs into a beaker containing 5–10 ml per disc of 5% Na_2HPO_4. Wash filters five times for 5 min each time in 5% Na_2HPO_4. Wash twice for 1 min in diH_2O followed by two washes in 95% EtOH. Air-dry filters and determine incorporation rates by liquid scintillation counting.

4. Terminate reaction by adding 2 μg DNaseQ (Promega Biotech, RNase free) and 50 μg tRNA (*E. coli,* RNase free, Boehringer Mannheim). Incubate for an additional 15 min at 30°C.

5. Extract radiolabeled transcripts from the reaction by adding 200 μl transcription extraction buffer followed by 300 μl phenol–chloroform–isoamyl alcohol (25:24:1).

6. Vortex for 1 min and centrifuge 14,000 rpm for 10 min in a microfuge.

7. Transfer aqueous phase to a fresh tube. Repeat extraction of aqueous phase with another 300 μl phenol–chloroform–isoamyl alcohol (25:24:1).

8. Transfer aqueous phase to a fresh tube. Add NH_4^+-acetate (DEPC treated) to a final concentration of 2 *M* and 2.5 vol of absolute EtOH.

9. Mix well and allow to precipitate overnight at −20°C.

10. Centrifuge at 14,000 rpm for 30 min at 4°C. Decant supernatant and carefully rinse pellet twice with 70% ethanol. Dry pellet in speed vac. Resuspend the pelleted nucleic acids in 1 ml (pre)hybridization fluid (50% formamide, 6× SSC,[3] 5× Denhardt's reagent,[4] 100 μg/ml denatured, fragmented salmon sperm DNA).

IV. Detection of Labeled Transcripts

The detection of radiolabeled transcripts relies upon hybridization to saturating levels of DNA specific for the gene of interest. The target or probe DNA is linearized by restriction digestion, denatured, and transferred by Southern blotting of plasmid DNA inserts or slot blotting onto nitrocellulose or nylon membranes using a slot-blot manifold such as the Minifold II (Schleicher & Schuell).

A. Materials

Nitrocellulose (NitroBind, MSI)
Nylon filters (Magna NT, MSI)
Minifold II (Schleicher & Schuell).
Heat-sealable polyester plastic bags (281-616, Curtin Matheson).

[3] 1× SSC contains 0.15 *M* NaCl, 0.015 sodium citrate, pH 7.
[4] 1× Denhardt's reagent contains 0.2% w/v Ficoll (type 400, Pharmacia, 0.02% w/v polyvinylpyrrolidone, 0.02% w/v BSA (Fraction V, Sigma).

Impulse sealer, TISH 300 (TEW Electric Sealing Equipment Co., Curtin Matheson)

X-ray film (Kodak, X-OMAT AR-5)

Autoradiography cassette (Fisher, AC-810)

Intensifying screens (Fisher, IS-810)

Polyvinylpyrollidone (Sigma, P-5288)

Ficoll 400 (Sigma F-2637)

BSA, fraction V (Fisher BP-1600)

DEPC (diethyl pyrocarbonate, Sigma D-5758)

(Pre)Hybridization fluid:
50% formamide, 6× SSC, 5× Denhardt's reagent, 100 μg/ml denatured, fragmented salmon sperm DNA.

Blotto (Pre)Hybridization fluid:
50% formamide, 6× SSC, 0.25% non-fat dry milk, (Carnation)

Wash solution I:
6× SSC, 0.1% SDS

Wash solution II:
0.1× SSC, 0.1% SDS

B. Methods

1. Preparation of DNA Slot Blots

Applying DNAs applied to membranes using a slot-blot device facilitates quantifying the hybridization signal. Nitrocellulose (Nitrobind, MSI) or nylon filters (Magna, NT, MSI) may be used. Nylon filters are preferred because they are more durable and can be reprobed with transcripts from various batches of nuclei.

1. Quantify plasmid DNAs containing the target genes of interest. Linearize the DNA by digestion with an appropriate restriction endonuclease. Dilute DNA in sufficient volume of 1× TE to allow ample volumes of solution for applying to the slot blot apparatus (300-500 μl).

2. Denature plasmid DNA by treatment with 0.3 M NaOH at 65°C for 1 h.

3. Neutralize DNA by adding NH_4^+-acetate(pH 7) to a final concentration of 0.5 M. Add 20× SSC to achieve a final concentration of 10× SSC.

4. Assemble slot blot apparatus according to manufacturer's instructions.

5. Apply 5 μg DNA to each slot. Apply vacuum to draw samples onto the filter. Apply an equal volume of 10× SSC to each slot and apply vacuum again.

6. Disassemble the slot-blot apparatus and carefully remove filter. UV cross-link filter or bake filter for 2 h at 80°C in a vacuum oven.

7. Store filter in a heat-sealable plastic bag.

2. Detection of Labeled Transcripts

1. Prehybridize DNA slot-blot filters overnight 42°C in 50–100 μl (pre)hybridization fluid/cm^2 of filter in a heat-sealable plastic bag. Blotto (pre)hybridization fluid can be substituted for the Denhardt's containing (pre)hybridization fluid with excellent results (Cushman *et al.*, 1989), but it may contain RNase activity from the nonfat dry milk. Therefore, it is highly recommended that Blotto (pre)-hybridization fluid be treated with 0.25% v/v DEPC (diethyl pyrocarbonate) before use.

2. Decant fluid and replace with hybridization fluid containing radiolabeled transcripts (50–100 μl/cm^2 filter).

3. Hybridize 48 h at 42°C.

4. Wash filters with wash solution I two times for 30 min each at room temperature.

5. Wash filters with wash solution II two times for 30 min each at 55°C.

6. Expose filters to X-ray film for 12–72 h in an X-ray cassette with intensifying screens at −80°C.

7. X-ray film images can be quantified using scanning densitometry. Alternatively, the hybridization signals can be measured directly by liquid scintillation counting of individual slots or using a betascope or phosphorimaging system.

IV. Critical Aspects of the Procedure

A. Nuclei Isolation

The quality and state of starting material are critical to the overall yield and transcriptional activity. Plant tissues can typically yield 10 times less nuclei per gram fresh weight than cultured cells (Harrison *et al.*, 1992). The use of fresh tissue will yield 5- to 10-fold more nuclei than frozen tissue (Harrison *et al.*, 1992). In addition, nuclei from fresh tissue will have 5- to 10-fold more transcriptional activity than those from frozen tissue (Hagen and Guilfoyle, 1985). Also, when homogenizing the plant material, it is important not to use too much tissue. Use of more than 20% w/v of tissue to homogenization buffer can result in overloading the capacity of the gradients, which in turn will result in more contaminated nuclei preparations. Finally, the percentage of Percoll in the step gradients may be altered depending on the size of nuclei being isolated. Lower percentages may be used if smaller nuclei are being isolated. The size of the nuclei is dependent on the tissue source being used.

Detergent concentration and type are also critical factors in determining the quality of nuclei preparations. For nongreen tissue sources, it is possible to avoid the use of detergents altogether. For *in vitro* studies involving transcription of small molecular weight RNAs, it is recommended that detergents be avoided altogether (Guilfoyle *et al.*, 1986). However, nuclei preparations from green tissues require detergent use to reduce or eliminate plastid contamination. Depending on the amount and source of plant tissue being used, the concentration of detergent may be varied. For example, lowering the concentration of Triton X-100 below 0.1% may improve the integrity of the nuclei. The degree of contamination, however, from chloroplasts (and other subcellular organelles) will increase if detergent concentrations are too low. It is best to determine empirically the lowest amount of detergent that is sufficient to yield intact, yet pure nuclei. If chloroplast contamination is perceived to be a problem, the nuclei may be treated with higher concentrations of detergent and repurified on a fresh gradient after collection from the step gradient interface. Triton X-100 is the most commonly used detergent; however, various researchers have reported success employing other nonionic detergents, such as Nonidet P-40 (Willmitzer and Wagner, 1981; Wollgiehn, 1991), or Surfonyl 485 (Nothacker and Hildebrant, 1985).

B. *In Vitro* Transcription Run-on Assays

The exact conditions for optimal transcription rates by isolated nuclei vary depending on the origin of the starting material. The most important aspect controlling RNA polymerase activity is the proper balance of mono- and divalent cations used. Salt concentrations that yield optimal polymerase activity may not be optimal for run-on activity of specific genes transcribed by RNA polymerase II (Chappell and Hahlbrock, 1986). In addition, prediction of salt effects on mRNA versus rRNA transcription based on incorporation studies does not necessarily correspond to transcripts quantified by hybridization to immobilized DNA probes (Wollgiehn, 1991). Therefore, it is important that transcription conditions for a particular gene of interest be optimized empirically using hybridization to immobilized DNA probes. Normally, "balanced" *in vitro* transcription conditions are chosen such that both RNA polymerase I and II are equally active (Wollgiehn, 1991). To preserve the integrity of transcripts, incubation times should be kept to less than 30 min and an RNase inhibitor should be included in the reaction mixture.

C. Detection of Labeled Transcripts

Immobilization of DNA target sequences provides a reliable means of quantifying specific target DNAs. Slot blots are preferred because they are easier to prepare and quantify than Southern blots of insert DNAs. It is important to bind sufficient amounts of target insert DNA to provide excess hybridization sites for the radiolabeled transcripts. As little as 200 ng of insert has been found to be saturating (Chappell and Hahlbrock, 1984). The absolute transcription rates of various nuclei preparations from different treatments or experimental conditions can be quite variable making accurate quantitation of a particular transcript a challenge. Even if care is taken to use equivalent amounts of nuclei and hybridize with equivalent amounts of radiolabeled transcripts, it is often difficult to obtain equivalent levels of hybridization signal. Therefore, the addition of target DNA sequences to slot blots that are known to be constitutively expressed in a particular experimental system is often desirable. As shown in Fig. 2, hybridization signals to genes encoding 18S rRNA, actin, or tubulin can be used to normalize absolute transcriptional activities. The level of transcript abundance for a particular message of interest can then be expressed in relative terms.

V. Results and Discussion

Figure 1 shows the morphology of nuclei isolated using Method II. Although the nuclei appear intact, they have most likely lost some nuclear membrane integrity due to exposure to Triton X-100 (Gallagher and Ellis, 1982; Mösinger and Schäfer, 1984). In general, the degree of intactness of nuclei isolated using Method I is lower than for nuclei isolated with Method II as assessed by light and fluorescence microscopy. Furthermore, the nuclei isolated with Method I

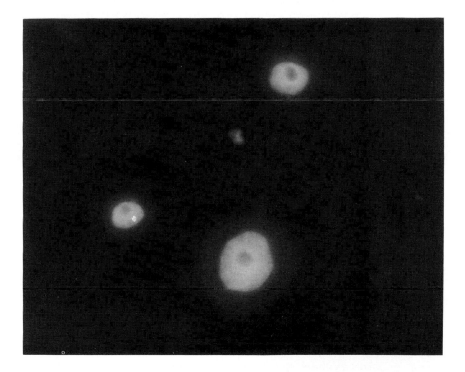

Fig. 1 Photomicrograph of isolated ice plant nuclei. Nuclei were isolated using method II from 6-week-old ice plant leaves (*Mesembryanthemum crystallinum*), stained with 5μg/ml Hoechst dye No. 33258 (Polysciences, Inc.), and photographed at 625× magnification under UV-illumination (350 nm excitation/420 nm emission) using fluorescence optics on an Olympus model BH-2S microscope. Note the size heterogeneity of the nuclei indicative of systemic endopolyploidy.

tend to have a greater amount of contamination, mainly cell wall debris. The yield of nuclei is also greater when Method II is used. Despite these limitations, we have obtained excellent *in vivo* transcription run-on results with nuclei isolated using Method I as illustrated in Fig. 2.

Figure 2 shows a typical result of transcripts generated by nuclei isolated according to Method I. Nuclei were isolated from *Mesembryanthemum crystallinum* at various time points during a cycle of 12 h of illumination followed by 12 h of darkness. The transcription rate of the chlorophyll a/b binding gene (*Cab*) is dramatically up-regulated before and during the illuminated period in a diurnal fashion. The important feature to note in this figure is the use of internal control genes that do not change their transcription rates significantly (i.e., actin and tubulin) during the same time period. These genes can be used as internal controls with which to compare the expression of target genes of interest. The expression of 18S ribosomal RNA transcripts produced by RNA polymerase I may also be used as an internal control. The use of such internal controls can become important when the overall transcriptional activity of the isolated nuclei varies. Variation in activity can be quite dramatic depending on the phenomenon being studied. For example, when nuclei are isolated from ice plants stressed with 0.5 *M* NaCl to induce CAM gene expression, overall transcription rates tend to decline, whereas transcription of salt-induced genes can increase dramatically (Cushman *et al.*, 1989).

VI. Conclusions and Perspectives

The use of isolated nuclei for *in vitro* transcription run-on assays has played an important role in analyzing the transcriptional regulation of many plant genes. Since the first demonstration of transcriptional activation of a specific plant genes

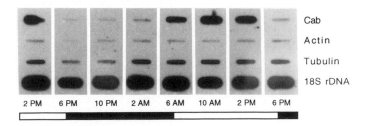

Fig. 2 Diurnal transcriptional regulation of Cab gene expression in ice plant. *In vitro* transcription run-on assays were performed with nuclei isolated using method I from 6-week-old ice plants. Radiolabeled transcripts were hybridized with slot-blotted DNA (5 μg per slot) from plasmids containing the Chloropyll a/b binding *Cab*, actin, and tubulin genes from ice plant (unpublished), and 18S rDNA gene from soybean (Eckenrode *et al.*, 1985). Nuclei were harvested at the times indicated from plants grown under a 12-h light (6 AM to 6 PM)/dark (6 PM to 6 AM) cycle. White and black bars indicate periods of illumination and darkness, respectively.

by light (Gallagher and Ellis, 1982), the method has proved useful for analyzing a wide range of developmental, environmental, biological, and plant growth regulator effects on plant gene expression. The synthesis and accurate run-on transcription of genes within isolated nuclei have led to our ability to sort out the contribution of transcriptional activation events from other regulatory mechanisms, such as mRNA stability, that ultimately govern gene expression. Despite their great importance in transcription run-on assays, isolated nuclei have traditionally been demonstrated to only have the ability to complete the synthesis of nascent RNA chains. To study the function of plant *trans*-acting factors *in vitro,* the development of nuclear-based transcription systems capable of accurately initiating transcripts would be of great utility. Recent reports of transcriptional reinitiation in isolated nuclei using incorporation of [γ-S]- or [β-S]-containing ribonucleotides and organomercurial chromatography have shown that reinitiation events are likely to occur (Mennes *et al.,* 1992), but can be nonspecific and difficult to distinguish due to the activity of nuclear RNA kinases (Schweitzer and Hahlbrock, 1993). In the future, with improvements in nuclear isolation and transcript detection techniques, isolated nuclei could potentially play an important role in furthering our understanding of transcriptional activation events in plants.

Acknowledgments

I thank Dr. Nick Cross for the use of his fluorescent microscope and Dr. Jay De Rocher for his help in preparing Fig. 2. I also thank Mary Ann Cushman, and Nancy Forsthoefel for helpful comments on the manuscript.

References

Chappell, J., and Hahlbrock, K. (1984). Transcription of plant defense gene in response to UV light or fungal elicitor. *Nature* **311,** 76–78.

Chappell, J., and Hahlbrock, K. (1986). Salt effects on total and gene-specific in vitro transcriptional activity of isolated nuclei. *Plant Cell Rep.* **5,** 398–402.

Cushman, J. C., Meyer, G., Michalowski, C. B., Schmitt, J. M., and Bohnert, H. J. (1989). Salt stress leads to the differential expression of two isogenes of phospho*enol*pyruvate carboxylase during Crassulacean acid metabolism induction in the common ice plant. *Plant Cell* **1,** 715–725.

De Rocher, E. J., Harkins, K. R., Galbraith, D. W., and Bohnert, H. J. (1990). Developmentally regulated systemic endopolyploidy in succulents with small genomes. *Science* **250,** 99–101.

Eckenrode, V. K., Arnold, J., and Meagher, R. B. (1985). Comparison of the nucleotide sequence of soybean 18S rRNA with the sequences of other small-subunit rRNAs. *J. Mol. Evolution* **21,** 259–269.

Gallagher, T. F., and Ellis, R. J. (1982). Light-stimulated transcription of genes for two chloroplast polypeptides in isolated pea nuclei. *EMBO J.* **12,** 1493–1498.

Guilfoyle, T. J., Suzich, J., and Lindberg, M. (1986). Synthesis of 5S rRNA and putative precursor tRNAs in nuclei isolated from wheat embryos. *Plant Mol. Biol.* **7,** 95–104.

Hagen, G., and Guilfoyle, T. J. (1985). Rapid induction of selective transcription by auxins. *Mol. Cell. Biol.* **5,** 1197–1203.

Harrison, M. J., Choudhary, A. D., Lawton, M. A., Lamb, C. J., and Dixon, R. A. (1992). Analysis of defense gene transcriptional regulation. *In* "Molecular Plant Pathology. A Practical Approach"

(S. J. Gurr, M. J. McPherson, D. J. Bowles, eds.), Vol. I, pp. 147–162. London/New York: Oxford Univ. Press.

Hicks, G. R., and Raikel, N. V. (1993). Specific binding of nuclear localization sequences to plant nuclei. *Plant Cell* **5,** 983–994.

Keim, P. (1987). Isolation of nuclei from soybean suspension cultures. *Methods Enzymol.* **148,** 535–541.

Luthe, D. S., and Quatrano, R. S. (1980a). Transcription in isolated wheat nuclei. I. Isolation of nuclei and elimination of endogenous ribonuclease activity. *Plant Physiol.* **65,** 305–308.

Luthe, D. S., and Quatrano, R. S. (1980b). Transcription in isolated wheat nuclei. II. Characterization of RNA synthesized in vitro. *Plant Physiol.* **65,** 306–313.

McKnight, G. S., and Palmiter, R. D. (1979). Transcriptional regulation of the ovalbumin and conalbumin genes by steroid hormones in chick ovaduct. *J. Biol. Chem.* **254,** 9050–9058.

Mennes, A. M., Quint, A., Gribnau, J. H., Boot, C. J. M., van der Zaal, E. J., Maan, A. C., and Libbenga, K. R. (1992). Specific transcription and reinitiation of 2,4-D-induced genes in tobacco nuclei. *Plant Mol. Biol.* **18,** 109–117.

Mösinger, E., and Schafer, E. (1984). In vivo phytochrome control of in vitro transcription rates in isolated nuclei from oat seedlings. *Planta* **161,** 444–450.

Nothacker, K. D., and Hildebrandt, A. (1985). Isolation of highly purified and more native nuclei of *Physarum polycephalum* utilizing Surfonyl, hexylene glycol and Percoll. *Eur. J. Cell Biol.* **39,** 278–282.

Roberts, M. W., and Okita, T. W. (1991). Accurate *in vitro* transcription of plant promoters with nuclear extracts prepared from cultured plant cells. *Plant Mol. Biol.* **16,** 771–786.

Roman, R., Caboche, M., and Lark, K. G. (1980). Replication of DNA by nuclei isolated from soybean suspension cultures. *Plant Physiol.* **66,** 726–730.

Saxena, P. K., Fowke, L. C., and King, J. (1985). An efficient procedure for isolation of nuclei from plant protoplasts. *Protoplasma* **128,** 184–189.

Schweitzer, P., and Hahlbrock, K. (1993). Post-transcriptional transfer of γ-thio affinity label to RNA in isolated parsley nuclei. *Plant Mol. Biol.* **21,** 943–947.

Silverthorne, J., and Tobin, E. M. (1984). Demonstration of transcriptional regulation of specific genes by phytochrome action. *Proc. Natl. Acad. Sci. U.S.A.* **81,** 1112–1116.

Somssich, I. E., Bollmann, J., Hahlbrock, K., Kombrink, E., and Schulz, W. (1989). Differential early activation of defence-related genes in elicitor-treated parsley cells. *Plant Mol. Biol.* **12,** 227–234.

Watson, J. C., and Thompson, W. F. (1986). Purification and restriction endonuclease analysis of plant nuclear DNA. *Methods Enzymol.* **118,** 57–75.

Willmitzer, L., and Wagner, K. G. (1981). The isolation of nuclei from tissue-cultured plant cells. *Exp. Cell Res.* **135,** 69–77.

Wollgiehn, R. (1991). Conditions for gene-specific transcription of isolated nuclei from tomato cell cultures. *Biochem. Physiol. Pflanzen.* **187,** 305–315.

CHAPTER 10

Analysis of the H+-ATPase and Other Proteins of the Arabidopsis Plasma Membrane

G. Eric Schaller* and Natalie D. DeWitt[†]

* Department of Botany and
† Department of Horticulture
University of Wisconsin
Madison, Wisconsin 53706

I. Introduction

The plasma membrane serves as a boundary between the plant cell and its environment, and is the primary location for proteins involved in transport and signal transduction (Sussman, 1994). Plasma membrane proteins have a wide range of structural and functional characteristics, with no single defining feature other than location. For example, the plasma membrane H^+-ATPase is an integral membrane protein that spans the lipid bilayer at least eight times. Most of the protein faces the cytoplasm or is embedded in the membrane, with only a minimal extracellular portion. In contrast, the receptor-like protein kinases contain only a single transmembrane domain, but have large cytoplasmic and extracellular domains (Schaller and Bleecker, 1993). Other proteins are more peripherally associated with the plasma membrane, such as some cytoskeletal proteins, calcium-dependent protein kinases, and calmodulin. Lacking transmembrane domains, they may contain hydrophobic or lipidated regions capable of membrane association, or may instead interact with membrane-anchored proteins.

Although proteins of the plasma membrane play a crucial role in cell growth and maintenance, altogether they probably constitute less than 3% of the total cellular protein (Sussman, 1994). For this reason, the biochemical analysis of plasma membrane proteins has traditionally taken place with plant species that satisfy two criteria: first, that the plants are of sufficient size and abundance to permit the isolation of large amounts of plasma membrane; second, that they exhibit relatively low proteolytic activity. The importance of this last criteria cannot be underestimated, for the large central vacuole of the plant cell contains a high concentration of proteases that are released following homogenization of the plant tissue. Proteolysis occurs rapidly, severely hampering biochemical characterization.

Activity of the H^+-ATPase in the purified plasma membranes is typically used as a criteria for the relative integrity of the membrane preparation. In good preparations of plasma membrane from oat, tomato, and beet, the specific activity of the H^+-ATPase is 1 to 2 μmol phosphate released per milligram protein per minute when assayed at 30°C. In other plants, the specific activity may be orders of magnitude less, depending upon species and isolation conditions. Plasma membrane and H^+-ATPase isolation has been extensively reviewed for several of these biochemically tractable plant species (Briskin *et al.*, 1987; Larsson *et al.*, 1987; Widell and Larsson, 1990; Sandelius and Morré, 1990).

Ease of genetic analysis, not biochemical purification, has established *Arabidopsis thaliana* as an important model system for the study of plant development (Meyerowitz, 1987). Arabidopsis has a short life cycle and one of the smallest genome sizes of any plant, which facilitates the generation and screening of mutant populations, as well as the cloning of genes. The small size of the plant itself permits large mutant screens in relatively small lab areas, as opposed to fields or greenhouses. Together classical and reverse genetic approaches have identified numerous novel proteins, as well as homologues for proteins previously identified in other plant species. Thus our clearest picture of any single plant species is rapidly emerging from these studies on Arabidopsis. However, this picture is sorely lacking in biochemical information on how these cloned gene products function in Arabidopsis. Usually they are assigned a function based

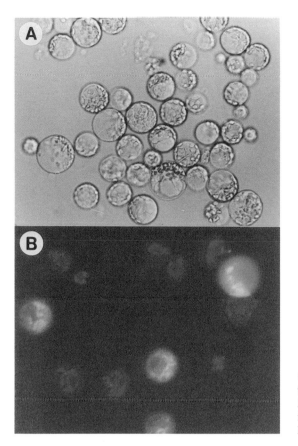

Color Plate 1 (Chapter 1) (A) Maize protoplasts prior to transfection, observed under Nomarski illumination. Magnification, ×300. (B) Maize protoplasts observed under epifluorescence illumination 18 hr following transfection. Magnification, ×400.

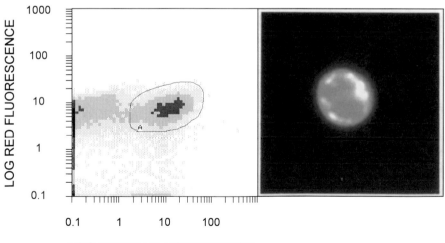

Color Plate 2 (Chapter 1) Illustration of the sort windows employed for the transfected protoplasts. (Right) Protoplast observed under epifluorescence illumination following sorting. Magnification, ×1000.

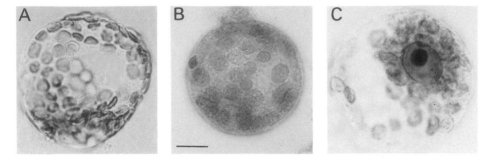

X-gluc   DAPI

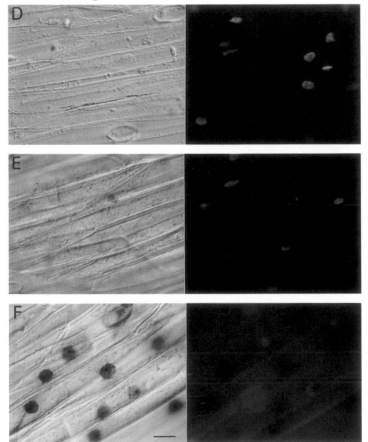

Color Plate 3 (Chapter 20) *In situ* histochemical localization of GUS activity in transfected protoplasts and transgenic plants. (A–C) Protoplasts transfected in the absence of DNA (panel A), or with plasmids encoding GUS (panel B) or GUS/NIa (panel C), were reacted with X-gluc. Bar in (B) equals 20 μm. (D–F) Epidermal leaf strips from nontransformed plants (panel D), or from transgenic plants expressing GUS (E) or GUS/NIa (F), were reacted with X-gluc and DAPI stain. The X-gluc reaction product was visualized using DIC microscopy (left), and the DAPI stain was visualized by UV fluorescence microscopy (right). Bar, 20 μm.

upon analogy to similar proteins in other systems, and seldom upon analysis of the actual protein in Arabidopsis. The reason for this is simple: Arabidopsis is not a good model system for biochemical studies. It is too small for the ready isolation of large amounts of material, and it does not appear to be particularly low in proteolysis. However, with the increasing focus of research upon Arabidopsis, there is growing incentive to begin a biochemical analysis of the proteins under study, although such studies still lag behind what has been accomplished in more amenable plant systems.

II. Growth of Arabidopsis

Arabidopsis plants can be grown in a variety of ways to facilitate the later isolation of protein and membranes. Plants can be grown in pots containing soil or vermiculite according to standard protocols. A 3:1 mixture of potting soil to perlite is suitable, with optimal growth conditions usually being 20–22°C with 50–75% humidity and 100–200 μE m^{-2} s^{-1} of light. Plants should be watered with a nutrient solution such as 10% (v/v) Hoagland's solution or 0.5× Murashige and Skoog (MS) salts (Sigma). To increase vegetative growth the plants can be grown with 8-h days, which delays bolting. Growing Arabidopsis in pots is probably the best approach if a particular tissue is desired. However, for general membrane isolation, where the tissue type is not of great importance, two alternative growth procedures can often be useful. (1) growth in liquid culture and (2) growth on petri plates. Both these growth procedures yield plants with turgid tissues that can be homogenized more easily and completely than plants grown in soil. Furthermore, the plants are harvested without the need for removal of soil and potentially contaminating algae or fungi.

For either liquid culture or petri dishes, the Arabidopsis seed is first sterilized. Seed is placed in a microfuge tube and incubated for 1 min with 70% (v/v) ethanol. The ethanol is removed and the seeds are then incubated for 15 to 30 min in a freshly made solution of 30% (v/v) Clorox bleach, and 1% (w/v) Triton X-100. At this point seeds should be moved to a sterile hood and all the remaining work done under sterile conditions. The Clorox solution is removed, and the seed washed at least five times with 1-ml quantities of sterile water, until the solution does not foam when shaken vigorously.

The liquid media used for plant growth contains MS salts, 2% (w/v) sucrose, and B5 vitamins consisting of inositol (100 mg/ml), nicotinic acid (1 mg/ml), pyridoxine HCl (1 mg/ml), and thiamine HCl (10 mg/ml). B5 vitamins can be made up as a 1000-fold concentrated stock solution and filter-sterilized. The liquid media is adjusted to pH 5.6–5.8 with KOH, and autoclaved in Erlenmeyer flasks capped with aluminum foil: 50 ml media in a 125-ml flask; 100 ml in a 500-ml flask. Sterile seed is added directly to the media. Typically 10 to 50 seeds are used per flask. Higher seed number gives proportionately more root growth relative to leaves. Flasks are placed on a gently rotating platform shaker. Plant tissue is usually ready for harvest between 2 and 4 weeks, at which point the leaves should be green and turgid (beyond 4 weeks, the plants often begin to show signs of senescence). Roots or leaves can be separated, or the entire tissue mass can be harvested. This method is often particularly useful for isolation of root material that cannot be obtained easily from plants grown in soil.

For growth on petri plates a similar medium is used, containing $0.5\times$ MS salts, B5 vitamins, and 0.8% (w/v) agar. Sucrose is not necessary, but may be added, usually to 2% (w/v). The pH is adjusted to 5.6–5.8 with KOH, the media is autoclaved, and plates are poured. Deep petri plates (20 mm) should be used in order to leave room for plant growth. Arabidopsis seed can be plated out directly upon the agar, but for later harvesting of small seedlings, it is sometimes easier to grow the plants upon a layer of cellophane. Circles of permeable cellophane (Idea Scientific) are autoclaved in water, and then laid upon the agar plates. Seeds are plated upon the cellophane, and when ready for harvesting the plants can be easily scraped off the cellophane, which their roots do not penetrate. For plating, the seed is drawn up in a 1-ml pipet tip in a small quantity of sterile water, and is then deposited upon the plate. The plates are allowed to stand uncovered in the sterile hood to allow excess water to evaporate from the surface. In order to increase and coordinate the seed germination, plates containing Arabidopsis seed are stratified by wrapping the plate in parafilm and placing it in the refrigerator for 4–7 days. Following stratification, the parafilm is removed, and the plates are placed under lights for growth.

III. Rapid Extraction Procedures for Protein

In most plant species, the biochemical characterization of a protein has traditionally preceded the cloning of its gene. However, one usually proceeds backward with Arabidopsis, from cloning the gene to biochemical analysis of the gene product. Therefore antibodies against the protein of interest are usually obtainable, in the form of antisera prepared against homologous proteins of other organisms, or against bacterially expressed gene products. Specifically, portions of the coding sequence for the gene of interest are expressed in *Escherichia coli,* and the purified protein is used for antibody production. Little additional effort is required to generate polyclonal antibodies against two separate regions of the protein. Since both antibodies should recognize the same polypeptide in plant extracts or tissue sections, these independent antisera serve as a control to evaluate spurious crossreaction with unrelated proteins.

Antibodies can be used to determine the correct size of the protein of interest, prior to going through further purification. This is often important information, given the proteolytic degradation problems in plants. Total protein can be extracted from plant tissue, separated by polyacrylamide gel electrophoresis in the presence of denaturing sodium dodecyl sulfate (SDS–PAGE), and the molecular weight of the specific protein then determined from Western blots. For this to be successful, the antibody must be of a high enough titer to detect the protein in a crude extract and specific enough that it does not recognize a wide variety of unrelated proteins. Furthermore, the extraction procedure used should denature proteins as rapidly as possible, thereby inactivating proteases. Several extraction procedures have proven useful in this respect, employing either sodium dodecyl sulfate (SDS) or trichloroacetic acid as the denaturing agents.

In the first protein extraction method, homogenization of the plant tissue is performed directly in loading buffer for SDS–PAGE [125 mM Tris, pH 6.8, 4% (w/v) SDS, 20% (v/v) glycerol, 10% (v/v) β-mercaptoethanol, 0.01% (w/v) bromphenol blue]. Freshly harvested plant material is added to loading buffer, and homogenization performed in microcentrifuge tubes, using plastic pestles (e.g., Kontes). Typically 1 mg (fresh weight) of plant material will yield from 30 to 100 μg of total protein, enough for at least one lane on SDS–PAGE. Based on these yields, it is convenient to homogenize using 20 μl of loading buffer per milligram of plant material. A minimum volume of at least 60 μl should be used, with several hundred microliters being optimal. Plant material should be thoroughly homogenized and briefly centrifuged to pellet the plant debris, then the supernatant containing the protein is transferred to a new tube. The material is then ready for SDS–PAGE and Western blotting. Some investigators have suggested that plant material be flash frozen in liquid nitrogen prior to homogenization in boiling loading buffer. However, this is not necessarily advisable for membrane proteins, since integral membrane proteins such as the H⁺-ATPase aggregate upon heating and then cannot be resolved on SDS–PAGE.

In the second method, homogenization of the plant tissue is performed in cold 10% tricholoracetic acid, which serves to denature and precipitate the plant proteins. Homogenization is carried out in the same manner as with loading buffer. Following homogenization, the sample is centrifuged in a microfuge for 15 min at 4°C. The supernatant is removed and discarded, then the pellet centrifuged again briefly, and any remaining supernatant removed. It is important to remove as much of the trichloroacetic acid as possible. The large pellet will contain both plant debris and the acid-precipitated protein. Loading buffer for SDS–PAGE is added to the pellet, and the pellet brought into solution by pipetting up and down, vortexing, and using a bath-sonicator if available. The bromphenol blue dye in the loading buffer will turn yellow due to the acid, and should be neutralized by adding microliter amounts of 2 M Tris base, until the loading buffer turns blue again. The microfuge tube should then be briefly centrifuged to pellet the insoluble debris, and the supernatant loaded onto a polyacrylamide gel for SDS–PAGE.

Both these methods are relatively rapid methods for protein isolation. However, no single method works for all proteins, and some proteolysis can occur no matter how rapid the extraction procedure. It is recommended that both methods be tried with the protein of interest, as this involves little additional work.

IV. Isolation of Membranes

All homogenization and isolation procedures for Arabidopsis should take place at 4°C with chilled equipment. Arabidopsis tissue (1–100 g fresh weight) is cut into smaller pieces with a razor blade or sharp scissors, then homogenized with a mortar and pestle. The homogenization buffer contains 250 mM Tris (pH 8.5 at 4°C), 290 mM sucrose, 25 mM EDTA, 75 mM β-mercaptoethanol, and protease inhibitors. Instead of β-mercaptoethanol, 1–10 mM dithiothreitol (DTT) can be used as a reducing agent. For each gram of tissue, 1–2 ml of homogenization buffer is used, and the material ground until the tissue in homogeneous. The homogenate is then filtered through Miracloth (Calbiochem) or several layers of cheesecloth.

In order to remove insoluble plant debris and organelles, the homogenate is centrifuged at low speed. This is typically accomplished by centrifugation at 10,000× g for 15 min. Some membranes also pellet in this low-speed centrifugation step, so if the membrane fraction is later to be analyzed on a sucrose density gradient, this low-speed centrifugation step can be reduced to 5000× g for 5 min to minimize membrane losses. A large portion of the plant debris will still be removed, but not as large percentage of the microsomes. Any remaining plant debris and organelles will pellet during later centrifugation on the sucrose density gradient.

The supernatant from the low-speed centrifugation step is centrifuged at 100,000× g for 30 min to pellet the microsomes. The microsomes are then resuspended in a small volume of buffer. The volume and composition of this buffer depend upon how the microsomes will next be used. For two-phase purification of plasma membranes, the microsomes should be resuspended in 0.33 M sucrose, 3 mM KCl, 5 mM potassium phosphate, pH 7.8, and then immediately processed. If the microsomes are to be run on a sucrose density gradient, they can be resuspended in homogenization buffer at 0.5 ml buffer/10 g fresh weight and run on a gradient immediately or stored at −80°C. If no further purification of the microsomes is desired, they can conveniently be resuspended in 50 mM Tris (pH 7.5), 1 mM EDTA, 1 mM DTT, 10% (v/v) glycerol at 0.5 ml buffer/g fresh weight and stored at −80°C. When isolated in this manner, microsomes of Arabidopsis exhibit vanadate-sensitive ATPase activity ranging from 75 to 400 nmol phosphate released/mg protein/min when assayed at 37°C, with a K_m for ATP of 0.6 mM. Highest specific activities for ATPase activity were observed with young plants grown on petri plates.

All buffers used in the isolation of plant membranes should contain protease inhibitors. As a minimum, 1 mM phenylmethylsulfonyl fluoride (PMSF) should be used, which is one of the most effective inhibitors against plant proteases. For added protection, 10 μg/ml leupeptin, 1 μg/ml pepstatin, and 10 μg/ml aprotinin can also be included. Macroglobulin (1 unit/ml) has also proved quite effective as a plant protease inhibitor, and is useful if proteolytic problems are particularly acute. Macroglobulin is sensitive to reducing agents and should not be used in the presence of β-mercaptoethanol and DTT. Stock solutions of protease inhibitors should be prepared as follows: 100 mM PMSF in isopropanol; 1 mg/ml leupeptin in water; 1 mg/ml pepstatin in methanol; 5 mg/ml aprotinin in water; and 100 units/ml macroglobulin in water. Protease inhibitor stocks should be stored frozen, except for PMSF which is stored at 4°C. Protease inhibitors should be added to buffers just prior to use. This is particularly important in the case of PMSF, which rapidly loses activity in aqueous solutions.

V. Subfractionation of Membranes

Crude microsomal fractions can be used for many experiments, particularly when one is attempting to limit manipulations of the material prior to analysis. However, in many cases plasma membrane vesicles must be purified, either to enrich for a plasma membrane protein or to determine if a protein is localized to the plasma membrane.

A. Isolation of Plasma Membrane

Two-phase partitioning in aqueous Dextran–polyethylene glycol has become the accepted method for the isolation of high-purity plasma membranes. In this method plasma membranes are separated from other membranes based upon differences in their surface charge rather than size or density, usually allowing greater purification than is possible by density gradient centrifugation. In particular, the two-phase method cleanly resolves chloroplast thylakoid membranes away from the plasma membranes, which cannot be accomplished on a density gradient. The two-phase method has an additional advantage in that the isolated plasma membrane vesicles are predominantly sealed and oriented right-side-out, unlike the density gradient where isolated plasma membrane vesicles are of mixed orientation. This allows the vesicles to be used in studies on membrane transport or membrane protein topology.

Details of the two-phase method for plasma membrane purification have been reviewed (Briskin *et al.*, 1987; Larsson *et al.*, 1987; Sandelius and Morré, 1990), and the reader is referred to these for experimental details. Most recently, a modified two-phase procedure has been published that has advantages over the original procedure in that it is simpler, is less time-consuming, and uses less material (Shimogawara and Usuda, 1993). Several considerations should be kept in mind when isolating Arabidopsis plasma membranes by the two-phase procedure when using these methods. It is important when using the standard procedure to start with enough plant tissue, for the yield of plasma membranes rapidly decreases if not enough tissue is used. A minimum of 60 g of tissue by fresh weight should be used, with 100 g of tissue being optimal. The procedure can be scaled down; however, results from scaling down can be erratic. It is also important to check the purity of isolated plasma membrancs by using marker assays (Section VI), although an initial estimation of how well the separation is working can be accomplished by following partitioning of the green thylakoid membranes into the lower phase. Finally, enrichments for Arabidopsis plasma membranes using the two-phase procedure have not been as high as enrichments found with some other plant species, with specific activities of Arabidopsis plasma membrane markers usually being 3- to 5-fold higher than in microsomes (Meyer et al., 1989), compared to 10-fold enrichments commonly observed with oat, soybean, and beet (Sandelius and Morré, 1990).

B. Separation of Membranes by Density-Gradient Centrifugation

Membrane vesicles derived from various organelles have characteristic densities, which permit their separation using equilibrium density (isopycnic) centrifugation. Membrane vesicles sedimented in a density gradient equilibrate at their buoyant densities, which are determined by their protein/lipid ratio. This approach has been used to purify distinct vesicle populations such as secretory vesicles (Taiz et al., 1983), and to identify the cellular location of putative auxin and fusicoccin receptors (Ray, 1977; Meyer et al., 1989), and a variety of other enzymatic and binding activities.

After density-gradient centrifugation, fractions are collected and assayed for the presence of membrane markers, which identify vesicles from the various organelles. Ideally organellar membranes separate into discrete bands according to their buoyant density, and are identified by specific membrane markers. In reality, gradients rarely resolve cellular membranes so neatly, and are subject to significant heterogeneity in the behavior of the membranes. This heterogeneity may reflect lateral heterogeneity of membranes from a single organelle, with vesicles sedimenting at several densities. Alternatively, organellar membranes from different cell types may have different buoyant densities. In addition, most membrane markers are far from ideal, and may be unevenly distributed throughout a membrane or be active in more than one membrane. Therefore, multiple marker assays should be conducted on each gradient.

Bearing in mind these caveats, subfractionation of membranes is extremely useful for isolating a particular membrane population, or for identifying the subcellular location of a particular protein, binding property, or enzymatic activity. Membrane fractionation procedures vary widely depending on their intended purpose, and one may choose from a variety of gradient compositions (e.g., sucrose, dextran, or renografin), gradient configurations (e.g., continuous linear gradients or discontinuous "step" gradients), and centrifugation parameters (e.g., differential, rate zonal, or equilibrium density regimes). Below is described the subfractionation of Arabidopsis membranes by buoyant density centrifugation over a continuous sucrose gradient.

All solutions should be chilled to 4°C, and manipulations done on ice or in a cold room. Two sucrose solutions are prepared, consisting of 20% (w/w) and 45% (w/w) sucrose, each dissolved in centrifugation buffer (10 mM Tris, pH 7.5, 1 mM EDTA, and 1 mM DTT). Protease inhibitors may be included if needed. To form the gradient, a gradient marker such as a Hoefer Scientific SG series is loaded with each of the two sucrose solutions (shallower or deeper gradients may be prepared by modifying starting sucrose concentrations). First, a 1-ml cushion of 45% (w/w) sucrose is formed in the bottom of 17-ml Beckman ultra-clear centrifuge tubes. Then a continuous (linear) gradient from 20% (w/w) to 45% (w/w) sucrose is formed. A 17-ml gradient will clearly resolve microsomes from about 10 g fresh weight of Arabidopsis tissue. If one increases the amount of microsomes run on the gradient, there is a risk of overloading the gradient and losing resolution.

Before loading onto the gradient, total microsomes (~0.5 ml) from the differential centrifugation step (Section IV) are thoroughly resuspended using several passes of a glass homogenizer to ensure uniform resuspension of pellet. Microsomes are then carefully pipetted onto the surface of the gradient. Sucrose gradients are centrifuged in a swinging bucket rotor at $150,000 \times g$, for 4 to 5 h at 4°C. Fractions (1 ml) are then collected using a fraction-collecting device or by punching a hole in the bottom of the tube. These can be frozen at −80°C or immediately used for marker assays. Sucrose concentrations of each fraction are determined with a refractometer.

Topological orientation of the vesicles (i.e., sidedness) is affected by freeze/thawing. Therefore, marker assays should be immediately performed with enzymes that depend on detergent stimulation because of an active site hidden inside a vesicle (e.g., Triton-stimulated UDPase activity).

VI. Membrane Markers

A wide variety of markers have been used to identify membranes from different organelles. Few such marker enzymes can be considered perfect, as they often have a primary location in one membrane and a secondary location in another, or they are unevenly distributed throughout a given membrane. In the discussion below, a number of marker assays are evaluated with respect to Arabidopsis membranes, with particular attention to markers for the plasma membrane. This discussion is by no means inclusive, and the reader is referred to two excellent evaluations of plant membrane markers for a more thorough treatment of the subject (Hall, 1983; Widell and Larsson, 1990), and for descriptions of marker assays not included here.

A. Plasma Membrane

One of the most distinct and frequently used markers for the plasma membrane is the H⁺-ATPase. The H⁺-ATPase is an electrogenic proton pump, and uses the chemical energy of ATP to drive the extrusion of protons into external medium (Sussman, 1994). This generates a proton electrochemical gradient across the plasma membrane that is used to energize solute uptake systems. The H⁺-ATPase belongs to the family of "P-type" ATPases, which have a catalytic subunit of M_r 100,000, are inhibited by vanadate, and form an acyl-phosphate intermediate during the reaction cycle. Antibodies directed against the plant H⁺-ATPase serve as excellent markers for the plasma membrane. Vanadate-sensitive ATPase activity has also been used for localization of the plasma membrane, but cannot be considered unequivocal since activity is often observed in other membrane fractions. Other plasma membrane markers that have been demonstrated effective in Arabidopsis include fusicoccin-binding and glucan synthase II activity.

1. Plasma Membrane H⁺-ATPase Immunoblotting

Antibodies directed against the Arabidopsis plasma membrane H⁺-ATPase provide an excellent tool for identifying plasma membrane vesicles. Polyclonal antisera directed against the carboxy terminus of the AHA2 isoform has been demonstrated to label specifically the plasma membrane using immunogold electron microscopy and immunofluorescence (DeWitt, 1994). When Arabidopsis membranes are fractionated by sucrose density-gradient centrifugation, this anti H⁺-ATPase sera labels a 100-kDa protein with a peak around 38% (w/w) sucrose (Fig. 1), which corresponds to previously reported densities for plasma membranes from Arabidopsis and several other plant species (Chanson *et al.,* 1984; Hager *et al.,* 1991; Meyer *et al.,* 1989).

Aliquots of each gradient fraction are solubilized in SDS–PAGE loading buffer [125 mM Tris, pH 6.8, 4% (w/v) SDS, 20% (v/v glycerol, 10% (v/v) β-

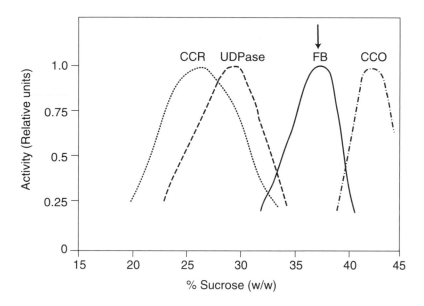

Fig. 1 Distribution of membrane markers following fractionation of Arabidopsis membranes over a sucrose density gradient. Shown are peaks of activity for NADH cytochrome *c* reductase (CCR), a marker for the endoplasmic reticulum; latent UDPase (UDPase), a marker for Golgi membranes; fusicoccin-binding (FB), a marker for the plasma membrane; and cytochrome *c* oxidase (CCO), a marker for the mitochondrial inner membrane. The peak of immunodecorated H⁺-ATPase (arrow) is coincident with fusicoccin binding.

mercaptoethanol, 0.01% (w/v) bromphenol blue], and loaded on an 8–10% (w/v) acrylamide gel. (Note: to avoid aggregation of H⁺-ATPases, samples should not be heated prior to loading.) SDS–PAGE, electroblotting, and antibody incubations are conducted using standard procedures. We have found that chemiluminescent detection systems based on horseradish peroxidase-coupled secondary antibodies (e.g., Amersham, or Kirkegaard and Perry) are ideal for this purpose, as they are extremely sensitive, and allow one to strip and reuse blots for probing with different antibodies.

One must bear in mind that there are at least 10 P-type ATPase isoproteins in Arabidopsis, with diverse cellular and possibly organellar locations (Sussman, 1994). Because of their structural similarities, antiseras directed against the plasma membrane H⁺-ATPase often crossreact with multiple isoproteins. In particular, it is known that the AHA2 antisera crossreacts with multiple P-type H⁺-ATPase isoproteins found in different cell types (DeWitt, 1994).

2. Vanadate–Sensitive ATPase Activity

P-type ATPases such as the plasma membrane H⁺-ATPase are characterized by vanadate inhibition. Thus, vanadate-inhibited ATPase activity is frequently used for identification of plasma membranes. When Arabidopsis membranes are fractionated by sucrose density-gradient centrifugation, a primary peak of activity is noted at 38% (w/w) sucrose, coincident with other plasma membrane markers and previous reports from other plant species. However, a secondary peak of vanadate-inhibited ATPase activity is also frequently observed at about 30% (w/w) sucrose, perhaps due to secretory vesicle ATPases or a lighter population of plasma membrane vesicles (Chanson *et al.,* 1984; Widell and Larsson, 1990).

a. Assay for ATPase Activity

The assay for the vanadate-sensitive H⁺-ATPase is carried out in 0.5 ml of 5 mM ATP, 5 mM MgCl$_2$, 10 mM Pipes, adjusted to pH 6.7 with Tris. In order to reduce the background of other ATPase activities, 5 mM NaN$_3$, 0.1 mM sodium molybdate, and 100 mM KNO$_3$ are also included in the assay to inhibit mitochondrial, soluble, and vacuolar ATPases, respectively. KNO$_3$ provides potassium needed for maximal activity of the H⁺-ATPase, but if KNO$_3$ is not desired, the assay should contain 50 mM KCl instead, which will still stimulate the plasma membrane ATPase but not inhibit the vacuolar ATPase. This assay buffer can be made up as a 10-fold concentrated stock and stored frozen until ready to use.

A large percentage of the ATPase activity is latent because the active site is hidden inside right-side-out vesicles (Larsson *et al.,* 1987). This is especially true with plasma membrane isolated by the two-phase procedure, where the majority of vesicles are right-side-out. On buoyant density gradients, populations of both inside-out and right-side-out vesicles are found. For access of the assay mix to the vesicle interior, ATPase assays are run in the presence of either 0.1 mg/ml lysophosphatidylcholine (from 10 mg/ml stock) or 0.015% (w/v) Triton X-100, added to the assay buffer immediately prior to assay. However, detergent concentrations over 0.02% (w/v) Triton X-100 seem to inhibit ATPase activity.

To assay vanadate-sensitive ATPase activity, duplicate samples are set up in plastic tubes in the presence and absence of 100 μM Na_3VO_4. Assays are typically performed at room temperature, 30°C, or 37°C. (Approximately twice as much activity is observed at 37°C than at room temperature.) The assay is started by the addition of the membranes, and the reaction allowed to proceed for 15 to 60 min, depending upon enzymatic activity. Usually 10–50 μg protein is suitable for the assay when using the molybdate/semidine method of phosphate determination, less when using the malachite green method described below.

b. Determination of Released Phosphate

There are several alternative methods for determining released phosphate. Two methods are described below: a standard assay using molybdate/semidine as the color reagent (Dryer *et al.*, 1957), and a more sensitive assay using molybdate/ malachite green as the color reagent (Lanzetta *et al.*, 1979).

Molybdate/Semidine Method. The reaction is stopped by adding to the ATPse assay mix: 0.1 ml 50% (v/v) trichloroacetic acid, immediately followed by 2.5 ml 10 mM ammonium molybdate. The samples are vortexed and allowed to stand at room temperature for 5–10 min, then 2 ml of semidine solution is added. The semidine (*N*-phenyl-*p*-phenylenediamine monohydrochloroide) stock solution is made by dissolving 0.5 g semidine in 5 ml ethanol, to which is added 10 g $NaHSO_3$ in 1 liter water. The solution is extensively stirred then filtered. Both semidine and molybdate stock solutions should be stored in dark bottles. Following the addition of semidine to the assay, color is allowed to develop for approximately 15 min and the absorbance is read at 710 nm. A phosphate standard for the ATPase assays is prepared by adding 1 μmol of K_2HPO_4 from a stock solution to an assay tube. A blank assay with no additions should also be run to subtract from the phosphate standard.

Malachite Green Method. The malachite green method of released phosphate detection offers the advantage of extremely high sensitivity (in the nanomolar range). Relatively small reaction volumes are used (50–100 μl), so this method is particularly useful for assaying frequently low yields of membranes in Arabidopsis preparations. To prepare the color reagent, 3 parts 0.045% (w/v in water) malachite green/HCl, are mixed with 1 part 4.2% (w/v in 4 N HCl) NH_4 molybdate. This is covered with foil, stirred for 30 min, then filtered through two No. 1 Whatman filters into a polyethylene container. For 10 ml of this mixture, add 200 μl 2% (w/v) tergitol NP, and store covered at 4°C for up to 1 week. To determine released phosphate in the ATPase assay, add 800 μl of the above solution to 50–100 μl ATPase assay volume in a microfuge tube. Wait 1 min, then add 100 μl of 34% (w/v) sodium citrate-2H_2O. Phosphate standards are prepared containing 0 to 30 nmol P_i from potassium phosphate. After 0.5–2 h of incubation, the absorbance is read at 630 nm.

3. Fusicoccin Binding

The fungal toxin fusicoccin rapidly stimulates the plasma membrane H^+-ATPase through an unknown mechanism, probably first interacting with a receptor located in the plasma membrane. Fusicoccin binding sites have been detected in fractionated Arabidopsis membranes at 38% (w/w) sucrose, the same density as a large peak of vanadate-sensitive ATPase activity (Meyer et al., 1989). Thus, competable binding of [³H]fusicoccin serves as an alternate and specific marker for Arabidopsis plasma membrane (Fig. 1).

Assays are conducted in microfuge tubes. Membrane fractions are added to 0.9 ml binding buffer (10 mM MES–Tris, pH 6, 1 mM CaCl$_2$, 1 mM MgSO$_4$, 1 mM EDTA, 1 mM KF, 2.6 mM DTT) containing 10 nM [³H]fusicoccin (35 Ci/mM). The same volume of membrane fractions is added to duplicate tubes containing 10 $\mu$$M$ nonradioactive fusicoccin in addition to 10 nM radioactive fusicoccin in binding buffer. After a 1-h incubation at room temperature, tubes are centrifuged in a microfuge for 1 h at 4°C. The supernatant is discarded, and tubes are inverted to drain all traces of supernatant from the pellet. The bottom of each tube containing the drained pellet is cut off, placed in scintillation fluid, and sonicated and vortexed until the pellet dissolves. Radioactivity is measured using a scintillation counter. Fusicoccin binding to membranes is calculated as the amount of [³H]fusicoccin binding that is competed off the membranes with 10 $\mu$$M$ nonradioactive fusicoccin.

4. Glucan Synthase II

1,3-β-Glucan synthase (glucan synthase II) is frequently used as a plasma membrane marker. However, when used to assay fractionated Arabidopsis membranes, the standard assay procedure gives a primary peak of activity at the density of plasma membrane vesicles (38% w/w), and a secondary peak of activity in the approximate density of Golgi apparatus membrane vesicles (31% w/w). This is perhaps unsurprising since 1,3-β-glucan synthase activity is notoriously difficult to distinguish from the 1,4-β-glucan synthase activity of the Golgi, and has been also observed in plasma membrane-associated structures, such as coated vesicles (Widell and Larsson, 1990). This assay is nonetheless reproducible and has pronounced activity in the plasma membrane, although further optimization of this assay may be needed to improve its specificity for Arabidopsis plasma membranes.

Glucan synthase II is assayed in 100-μl reaction volumes, containing 50 mM Hepes–KOH, pH 7.25, 0.8 mM spermine, 16 mM cellobiose, 4 mM EGTA/ 4 mM CaCl$_2$, 0.5 mM DTT, 0.010% (w/v) digitonin, and an aliquot of membrane fraction (usually 50 μl for sucrose gradient fractions). The reaction is started by adding unlabeled UDP–glucose containing 0.15 μCi[UDP-¹⁴C]glucose to a final concentration of 0.5 mM. The reaction is incubated at 25°C for 30 min and is stopped by placing tubes in boiling water for 5 min. The samples are transferred to Whatman filter paper discs and are washed 2 × 1 h in 0.35 mM ammonium acetate, pH 3.6, 30% (v/v) ethanol. The filters are dried and the radioactivity is counted on a scintillation counter.

B. Thylakoid Membranes

Chlorophyll A and B pigments reside specifically in thylakoid membranes, so simple measurements of their absorbances provide an excellent marker for thylakoid membranes from disrupted chloroplasts (intact chloroplasts are removed in the initial centrifugation step). Arabidopsis thylakoid membranes sediment predominantly at 40% (w/w) sucrose. To assay, an aliquot of each fraction is added to 750 μl of 95% (v/v) ethanol, and $A_{664.2}$ and $A_{648.6}$ are measured spectrophotometrically. Total chlorophyll a and b content is then determined by inserting the measured absorbance values into the equation $C_{a+b} = 5.24A_{664.2} + 22.24A_{648.6}$, with concentrations given as micrograms per milliliter plant extract solution (Lichtenthaler, 1987).

C. Endoplasmic Reticulum

Antimycin-insensitive NAD(P)H cytochrome c reductase is the most commonly used marker for the endoplasmic reticulum, although significant activity is also detected in the plasma membrane when assayed in the presence of detergent (Askerlund $et\ al.$, 1991). In sucrose gradients of Arabidopsis membranes assayed in the absence of detergent, its activity prominently peaks at 26% (w/w) sucrose (Fig. 1), so its major activity is coincident with the density previously reported for endoplasmic reticulum membranes (Hager $et\ al.$, 1991; Jacobs and and Hertel, 1978; Lord, 1987; Perlin and Spanswick, 1980). This density of endoplasmic reticulum membranes is characteristic for those isolated under conditions with low magnesium, which strips off ribosomes. High magnesium isolation conditions will yield endoplasmic reticulum membranes containing ribosomes, which sediment at a slightly higher density, in addition to smooth endoplasmic reticulum membranes (Lord, 1987).

NADH cytochrome c reductase activity is assayed at room temperature in 1-ml reaction volumes, containing 20 mM potassium phosphate, pH 7.2, 0.2 mM NADH, 0.02 mM cytochrome c, 10 mM KCN, and 1 μM antimycin A (Lord, 1987). A stock solution of antimycin A is made up in ethanol. The reaction is started by the addition of protein, and activity is recorded as the rate of ΔA_{550}. The amount of cytochrome c reduced is estimated using an extinction coefficient for cytochrome c of 18.5 mM^{-1} cm^{-1} at 550 nm.

D. Mitochondrial Membranes

Cytochrome c oxidase is one of the few markers whose subcellular localization has been clearly localized, in this case, to the inner membrane of the mitochondria (Widell and Larsson, 1990). In sucrose gradients of Arabidopsis membranes, a single peak of activity is observed at 42% (w/w) sucrose (Fig. 1). The assay is similar to NADH cytochrome c reductase activity, although in this case the oxidation of cytochrome c is monitored after addition of membranes. First, a 45 mM stock solution of cytochrome c is reduced by adding sodium dithionite until the A_{550}/A_{565} is between 5 and 12. The reaction buffer contains 3 ml 50 mM KPO$_4$, 0.01% (w/v) digitonin to which protein is added. To start the reaction, 0.1 ml of 45 mM reduced cytochrome c is then added, and activity determined as described for NADH cytochrome c reductase assay (Hodges and Leonard, 1974).

E. Golgi Membranes

The Golgi apparatus is a heterogeneous structure that is polarized with respect to secretion, and that contains both cisternal and vesicular elements. Latent NDPase activity is frequently used as a marker for Golgi membranes (Widell and Larsson, 1990). Latent IDPase and UDPase activities peak at about 28–30% (w/w) in sucrose gradients of Arabidopsis membranes (Fig. 1), which is consistent with previously reported densities for Golgi membranes of other plant species (Chanson et al., 1984; Perlin and Spanswick, 1980; Nagahashi and Kane, 1982). Earlier protocols involved activating the latent enzyme activity by long (e.g., 4- to 6-day) incubations of membranes before performing the assay. Nagahashi and Kane's use of detergent to activate the latent UDPase activity significantly expedites and improves the assay, and so is described here. Freezing and thawing of membranes may activate latent enzymatic activity, so this assay should be performed on fresh membranes for maximal detergent activation.

Add 1–5 μl of a membrane fraction to 50–100 μl assay buffer, consisting of 3 mM UDP, 3 mM MnSO$_4$, 30 mM MES–TRis, pH 6.5. A duplicate reaction with the addition of 0.03% (w/v) Triton X-100 should also be prepared. Incubate at 37°C for 20 min. Released phosphate is then determined using the malachite green method (Section VI,A,2,a). If other methods of phosphate determination are used, proportionately higher volumes of protein should be assayed (e.g., 10–100 μl of protein to 0.9 ml of assay buffer). Triton-activated UDPase activity is determined by subtracting the activity in absence of detergent from that in presence of detergent.

F. Protein Determination

When assaying membrane fractions for protein content, detergents should be included in the assay to expose portions of the proteins that are buried in the membrane. If detergents are not included, protein concentration will be severely underestimated. Because detergents needed for solubilization of membrane proteins can interfere with Bradford protein assays, the Lowry assay (Lowry et al., 1951) has frequently been used for determination of membrane protein concentration. Although differences in sucrose concentrations, such as those encountered in sucrose density-gradient centrifugation, are reported to affect the Lowry assay, this is only a problem if using fraction volumes greater than 5% of the Lowry assay volume (Gerhardt and Beevers, 1968).

Three stock reagents are used for the Lowry solution. Reagent A is 2% (w/v) Na_2CO_3 in 0.1 N NaOH. Reagent B is 2% (w/v) NaK tartrate tetrahydrate. Reagent C is 1% (w/v) $CuSO_4 \cdot 5H_2O$. These can be prepared in advance and stored indefinitely. Just before the assay, reagent D is freshly prepared by mixing 50 ml reagent A, 0.5 ml reagent B, and 0.5 ml reagent C, added in this order. Reagent E is made by freshly diluting a stock solution of 2 N Folin and Ciocalteu's phenol reagent to 1 N in water. Protein samples are added to 1 ml 0.4% (w/v) deoxycholate. The assay is initiated by adding 3 ml reagent D to each sample at timed intervals and mixing. After exactly 10 min, 0.3 ml reagent E is added to the tubes and mixed. The color is allowed to develop for about 45 min and absorbance is read at 710 nm. Bovine serum albumin is usually used in setting up a standard curve, with the maximum amount of protein being 100 μg per sample. Samples can be assayed in smaller sample volumes by reducing the amounts of all reagents proportionately in the assay.

VII. Conclusions and Perspectives

The methods described here represent the application of traditional biochemical techniques to the investigation of proteins in the Arabidopsis plasma membrane. Based on this analysis, Arabidopsis cellular membranes share many biochemical features with those of other plant species that have been more rigorously characterized. For example, the Arabidopsis plasma membrane is enriched for by two-phase partitioning, although the process is potentially more arduous than with bulkier plants. In addition, when examined by sucrose density-gradient centrifugation, the Arabidopsis plasma membrane vesicles band with a peak around 38% (w/w) sucrose. This is similar to what has been observed with plasma membrane vesicles of other plant species, and indicates that the Arabidopsis plasma membrane has biochemical properties, such as its overall lipid and protein composition, in common with these plants. Equally important is the finding that other membrane systems of Arabidopsis band at densities different from that of the plasma membrane, permitting resolution of different membranes and identification of proteins belonging to different membrane systems.

A limitation to this type of biochemical analysis arises from the finding that many proteins are encoded by multigene families. This is the case for both the Arabidopsis H^+-ATPase and the calmodulin-domain (formerly called calcium-dependent) protein kinases, both of which have been found associated with the plasma membrane (Sussman, 1994; Schaller et al., 1992). Members of multigene families encode similar proteins that are often immunologically and biochemically indistinguishable by conventional means. However, despite their similarities, differences in their subcellular localization, activity, regulation, and associations with cellular components may have important consequences in the cells in which they exist. Because of their structural similarities, antibodies prepared against one isoprotein often recognize other members of the family. As a result, one ends up characterizing not one protein, but a family of isoproteins. To circumvent this problem in the study of specific H$^+$-ATPase isoforms, molecular genetic approaches have been successfully applied to the study of the Arabidopsis plasma membrane H$^+$-ATPase. These methods include epitope tagging and expression in yeast of specific H$^+$-ATPase genes.

Epitope tagging involves introducing a small antigenic determinant from a foreign protein into a protein of interest. The nucleotide sequence encoding the epitope is cloned into a cDNA or genomic clone encoding the protein, and this construct is then expressed in living cells (see Chapter 10, this volume). Two commonly used epitopes are c-myc, which is derived from a nuclear protein, and HA, which is derived from an influenza hemagglutinin (Kolodziej and Young, 1991). Epitope tagging of yeast and mammalian proteins has become common-place, and more recently has been used with a number of plant proteins, including the H$^+$-ATPase of Arabidopsis.

The H$^+$-ATPase isoproteins share between 70 and 94% amino acid identity (Sussman, 1994). Some isoprotein family members have been demonstrated to be differentially expressed in a tissue-specific manner, with two being expressed predominantly in roots (Harper et al., 1990) and one expressed specifically in phloem (DeWitt et al., 1991). Further defining their function, localization, and associations requires isoprotein-specific antibodies for immunolocalization and immunoprecipitation studies. However, it is difficult to generate antibody probes that can discriminate between the different isoproteins. To overcome this problem, the method of epitope tagging was employed. Specifically, a genomic clone of a H$^+$-ATPase isoform, under control of its own promoter, was tagged with nucleotide sequences encoding either the c-myc or HA epitopes (DeWitt, 1994). The modified genes were then expressed in transgenic plants, and the epitope-tagged proteins monitored using monoclonal antisera specific for each introduced epitope. The c-myc epitope tag proved superior, providing strong signals without crossreaction to endogenous plant antigens. The epitope-tagged H$^+$-ATPase was readily detectable in cell extracts on Westerns, and on plant sections using immunofluorescence. This method provides an important experimental advantage in that a genetic negative control is available in the form of untransformed plants, helping to distinguish bona fide signals from spurious antibody crossreaction, often a serious problem in immunolocalization.

Epitope tagging will ultimately permit the unambiguous cellular and subcellular localization of each H$^+$-ATPase isoprotein. One can then address questions regarding the heterogeneity of ATPase activity that has previously been observed in fractionated cell membranes (Larsson et al., 1987). This heterogeneity has been attributed to different cell types having plasma membranes with different buoyant densities, so that cell-specific ATPases are resolved. It also has been proposed that the plasma membrane may contain microdomains of different buoyant densities (Lutzelschwab et al., 1989), like the distinct apical and basolateral domains of mammalian epithelial cells. Another possibility is that P-type ATPases reside in membranes other than the plasma membranes. Indeed, a recently cloned P-type ATPase is localized to the chloroplast envelope (Huang *et al.,* 1993). Using immunoaffinity chromatography and precipitation techniques, an epitope-tagged isoprotein can be isolated without contamination from other family members. Such purification approaches could also be used to isolate membrane vesicle populations from different cell types, organelles, or even membrane domains in which the tagged protein resides.

A second method by which an individual isoprotein can be examined biochemically is by expression in a heterologous system, particularly a system suited to high-level expression. Although *E. coli* is frequently used for the expression of eukaryotic proteins, they are often produced in denatured inactive forms. Membrane proteins are particularly difficult to express in *E. coli,* because of low protein yields and toxic effects upon the bacteria. Yeast, on the other hand, is a eukaryote and employs a secretory system similar to that of plants. Several different isoforms of the H^+-ATPase from Arabidopsis have been expressed in yeast with functional activity (Palmgren and Christensen, 1994). The expressed H^+-ATPase is membrane-associated, hydrolyzes ATP, and pumps protons. Perhaps due in part to high levels of expression, most of the Arabidopsis ATPase does not reach the plasma membrane, but is trapped at an earlier stage of the secretory pathway in the endoplasmic reticulum. Taking advantage of this localization, the expressed H^+-ATPase from Arabidopsis can be purified in milligram quantities from yeast without contamination by the endogenous yeast H^+-ATPase of the plasma membrane. The functional equivalency of the plant H^+-ATPase with the yeast H^+-ATPase has been demonstrated by complementation of a mutated yeast ATPase with the AHA2 isoform of the Arabidopsis ATPase (Palmgren and Christensen, 1993). Interestingly, this required truncation of the last 92 amino acids of AHA2 for targeting of the ATPase to the plasma membrane.

These molecular genetic and biochemial methods should soon permit elucidation of differences in the spatial organization and function of each Arabidopsis H^+-ATPase isoprotein, and should be applicable to the study of other plasma membrane proteins as well. In particular, the described molecular genetic approaches will permit examination of individual isoproteins that were previously immunologically and biochemically indistinguishable. Long-standing questions of plant biology now can be addressed from a different perspective, with a heightened awareness of the complexity and diversity of molecular players.

Acknowledgments

The authors acknowledge Michael R. Sussman for useful discussions and critical reading of the manuscript. This work was supported by a grant to the University of Wisconsin from the DOE/ NSF/USDA Collaborative Program on Research in Plant Biology (Grant No. BIR 92-20331).

References

Askerlund, P., Laurent, P., Nakagawa, H., and Kader, J.-C. (1991). NADH-ferricyanide reductase of leaf plasma membranes. *Plant Physiol.* **95,** 6–13.

Briskin, D. P., Leonard, R. T., and Hodges, T. K. (1987). Isolation of the plasma membrane: Membrane markers and general principles. *Methods Enzymol.* **148,** 542–558.

Chanson, A., McNaughton, E., and Taiz, L. (1984). Evidence for a KCl-stimulated, Mg^{2+}-ATPase on the Golgi of corn coleoptiles. *Plant Physiol.* **76,** 498–507.

DeWitt, N. D., Harper, J. F., and Sussman, M. R. (1991). Evidence for a plasma membrane proton pump in phloem cells of higher plants. *Plant J.* **1,** 121–128.

DeWitt, N. D. (1994). Epitope tagging and reporter gene studies identify an Arabidopsis plasma membrane proton pump specific for phloem companion cells. PhD thesis.

Dryer, R. L., Tammes, A. R., and Routh, J. I. (1957). The determination of phosphorus and phosphatase with N-phenyl-*p*-phenylenediamine. *J. Biol. Chem.* **225,** 177–183.

Gerhardt, B., and Beevers, H. (1968). Influence of sucrose on protein determination by the Lowry procedure. *Anal. Biochem.* **24,** 337–352.

Hager, A., Debus, G., Edel, H. G., Stransky, H., and Serrano, R. (1991). Auxin induces exocytosis and the rapid synthesis of a high-turnover pool of plasma-membrane H+-ATPase. *Planta* **185,** 527–537.

Hall, J. (1983). Plasma Membranes. *In* "Isolation of Membranes and Organelles from Plant Cells" (J. L. Hall and A. L. Moore, eds.), pp. 55–81. London: Academic Press.

Harper, J. F., Manney, L., DeWitt, N. D., Yoo, M. H., and Sussman, M. R. (1990). The *Arabidopsis thaliana* plasma membrane H+-ATPase multigene family. *J. Biol. Chem.* **265,** 13601–13608.

Hodges, T. K., and Leonard, R. T. (1974). Purification of a plasma membrane-bound adenosine triphosphatase from plant roots. *Methods Enzymol.* **32,** 392–406.

Huang, L., Berkelman, T., Franklin, A. E., and Hoffman, N. E., (1993). Characterization of a gene encoding a Ca^{+2}-ATPase-like protein in the plastid envelope. *Proc. Natl. Acad. Sci. U.S.A.* **90,** 10066–10070.

Jacobs, M., and Hertel, R. (1978). Auxin binding to subcellular fractions from *Cucurbita* hypocotyls: *In vitro* evidence for an auxin transport carrier. *Planta* **142,** 1–10.

Kolodziej, P., and Young, R. A. (1991). Epitope tagging and protein surveillance. *Methods Enzymol.* **194,** 508–519.

Lanzetta, P. A., Alvarez, L. J., Reinach, P. S., and Candia, O. A. (1979). An improved assay for nanomole amounts of inorganic phosphate. *Anal. Biochem.* **100,** 95–97.

Larsson, C., Widell, S., and Kiellbom, P. (1987). Preparation of high-purity plasma membranes. *Methods Enzymol.* **148,** 558–568.

Lichtenthaler, H. K. (1987). Chlorophylls and carotenoids: Pigments of photosynthetic biomembranes. *Methods Enzymol.* **148,** 350–382.

Lord, J. M. (1987). Isolation of endoplasmic reticulum: General principles, enzymatic markers, and endoplasmic reticulum-bound polysomes. *Methods Enzymol.* **148,** 542–558.

Lowry, O. H., Rosebrough, N. J., Farr, A., and Randall, R. J. (1951). Protein measurement with the Folin phenol reagent. *J. Biol. Chem.* **193,** 265–275.

Lutzelschwab, M., Asard, H., Ingold, U., and Hertel, R. (1989). Heterogeneity of auxin-accumulating membrane vesicles from *Cucurbita* and *Zea:* A possible reflection of cell polarity. *Planta* **177,** 304–311.

Meyer, C., Feyerabend, M., and Weiler, E. W. (1989). Fusicoccin-binding proteins in *Arabidopsis thaliana* (L.) Heynh. *Plant Physiol.* **89,** 692–699.

Meyerowitz, E. M. (1987). *Arabidopsis thaliana. Annu. Rev. Genet.* **21,** 93–111.

Nagahashi, J., and Kane, A. P. (1982). Triton-stimulated nucleoside diphosphatase activity: Subcellular localization in corn root homogenates. *Protoplasma* **112,** 167–173.

Palmgren, M. G., and Christensen, G. (1994). Functional comparisons between plant plasma membrane H+-ATPase isoforms expressed in yeast. *J. Biol. Chem.* **269,** 3027–3033.

Palmgren, M. J., and Christensen, G. (1993). Complementation *in situ* of the yeast plasma membrane H+-ATPase gene *pma1* by an H+-ATPase gene from a heterologous species. *FEBS Lett.* **317,** 216–222.

Perlin, D. S., and Spanswick, R. M. (1980). Labeling and isolation of plasma membranes from corn leaf protoplasts. *Plant Physiol.* **65,** 1053–1057.

Ray, P. M. (1977). Auxin-binding sites of maize coleoptiles are localized on membranes of the endoplasmic reticulum. *Plant Physiol.* **59,** 594–599.

Sandelius, A. S., and Morré, D. J. (1990). Plasma membrane isolation. *In* "The Plant Plasma Membrane: Structure, Function, and Molecular Biology" (C. Larsson and I. M. Møller, eds.). Berlin: Springer-Verlag.

Schaller, G. E., Harmon, A. C., and Sussman, M. R. (1992). Characterization of a calcium- and lipid-dependent protein kinase associated with the plasma membrane of oat. *Biochem.* **31,** 1721–1727.

Schaller, G. E., and Bleecker, A. B. (1993). Receptor-like kinase activity in membranes of Arabidopsis thaliana. *FEBS Lett.* **333,** 306–310.

Shimogawara, K., and Usuda, H. (1993). A concentrating two-phase partitioning: Its application to isolation of plasma membrane from maize roots. *Anal. Biochem.* **212,** 381–387.

Sussman, M. R. (1994). Molecular analysis of proteins in the plant plasma membrane. *Annu. Rev. Plant Physiol. Plant Mol. Biol.* **45,** 211–234.

Taiz, L., Murray, M., and Ronbinson, D. G. (1983). Identification of secretory vesicles in homogenates of pea stem segments. *Planta* **158,** 534–539.

Widell, S., and Larsson, C. (1990). A critical evaluation of markers used in plasma membrane purification. *In* "The Plant Plasma Membrane: Structure, Function, and Molecular Biology" (C. Larsson and I. M. Møller, eds.). Berlin: Springer-Verlag.

CHAPTER 11

Isolation and Functional Reconstitution of the Vacuolar H+-ATPase

John M. Ward[*] and Heven Sze[†]

[*] Department of Biology
University of California, San Diego
La Jolla, California 92093-0116
[†] Department of Plant Biology
University of Maryland
College Park, Maryland 20742

I. Introduction

Vacuolar proton-translocating ATPases (V-type H+-ATPases) acidify the lumen of internal organelles of eukaryotic cells including those of the endocytic pathway, the Golgi network, lysosomes, vacuoles, and secretory vesicles (for review, see Forgac, 1989). In plants, V-type H+-ATPases are known to be present in the tonoplast (vacuole membrane), the Golgi membrane, and the endoplasmic reticulum (Herman et al., 1994). Studies at the molecular level suggest the presence of several isoforms that may be expressed in a tissue-specific or developmentally specific pattern (Sze *et al.,* 1992b). These pumps have a central role in physiologically important processes such as the regulation of cell turgor, the secondary transport of ions and metabolites, and the regulation of cytoplasmic Ca^{2+} concentration and pH (for review: Sze *et al.,* 1992a). Knowledge of the structure, transport properties, and regulation of plant H^+-ATPases is important for our understanding of the physiological function of this class of transporters in plant cells. Much of our information concerning plant V-type ATPases has come from studying the transport properties of the enzyme in native vesicles and the protein structure and ATP hydrolytic activity of the detergent solubilized and purified H^+-ATPases. The incorporation of purified H^+-ATPases into sealed phospholipid vesicles, referred to as reconstitution, has now enabled studies of the H^+-transport activity of plant enzymes of this class in the absence of other ion-conducting membrane proteins (Ward and Sze, 1992b; Warren *et al.,* 1992; Yamanishi and Kasamo, 1994). This approach is useful since purified preparations of H^+-ATPases, of known subunit composition, can be assayed for the complete enzymatic reaction: coupled ATP hydrolysis and H^+-translocation. The reconstitution procedure presented here has been used to study the subunit composition and regulation of the oat root V-type H^+-ATPases and may have further utility in the identification and characterization of additional H^+-ATPase complexes from other plant tissues or membrane fractions. These pumps may have distinct structural and functional features related to their specific physiological roles.

II. Materials

Acridine orange was purchased from Eastman Kodak (Rochester, NY). Crude phospholipids were obtained from Avanti Polar Lipids (Pelham, AL). Q-Sepharose FF and Sephacryl S400HR were from Pharmacia LKB Biotechnology, Inc. (Piscataway, NJ). Bio-beads SM-2 were purchased from BioRad (Richmond, CA). Bafilomycin A_1 was a gift from Dr. Karlheinz Altendorf (University of Osnabruck, Germany). All other reagents were obtained from Sigma (St. Louis, MO).

III. Purification and Reconstitution of the V-Type H^+-ATPase

A. Isolation of Low-Density Membrane Vesicles

This procedure for the isolation of low-density vesicles by differential and gradient centrifugation, enriched in tonoplast, was originally described by Churchill and Sze (1983) and modified by Randall and Sze (1986, 1987) and Ward and Sze (1992a). Membrane vesicles prepared using this method are enriched in vacuolar H^+-ATPase activity; the specific activity is normally between 0.3 and 0.4 μmol/min/mg protein (Randall and Sze, 1986; Ward and Sze, 1992a). Approximately 60% of the total ATP hydrolysis activity is inhibited by Bafilomycin A_1 (Ward and Sze, 1992b) in the absence of other inhibitors. This membrane fraction contains vesicles derived from mature vacuoles, provacuoles, and probably endoplasmic reticulum and Golgi (Sze, 1985; Herman *et al.*, 1994).

Oat seeds (*Avena sativa*) are first imbibed for several hours under running tap water. The seeds are then spread in a monolayer between two layers of cheesecloth suspended on a plastic grid and germinated in the dark over an aerated 0.5 mM CaSO$_4$ solution. After 4 days the roots are harvested and rinsed twice with deionized H$_2$O at 4°C. The oat roots (typically 90 g) are added to homogenization buffer (Table I), cut into 1-cm sections, ground using a mortar and pestle chilled to 4°C at a medium-to-tissue ratio of 2 ml/g fresh weight, and strained through cheesecloth. Whole cells and mitochondria are removed by centrifugation for 15 min at 13,000× g. The supernatant is then subjected to centrifugation at 60,000× g (Beckman SW 28, r_{max}) for 30 min to collect microsomal membranes. The microsomal pellets are suspended in a total of 24 ml of resuspension buffer (Table I) and four 6-ml aliquots are layered over 10 ml of 6% (w/v) Dextran (MW 80,000) in resuspension buffer in 16.5-ml tubes. After centrifugation for 2 h at 70,000× g the turbid layer containing low-density vesicles is collected (1 ml per tube) at the Dextran interface. Normally between 3 and 4 mg of vesicle protein are obtained from 90 g of oat roots. All procedures are conducted at 4°C. Vesicle protein may be stored for several weeks at −80°C.

B. Solubilization and Purification of the Vacuolar H⁺-ATPase

1. Solubilization of Membrane Protein

Low-density vesicles are concentrated by 10-fold dilution into resuspension buffer, pelleted at 85,000× g (Beckman SW 28, r_{max}) and resuspended to 5 mg protein/ml in resuspension buffer. The vacuolar H⁺-ATPase is solubilized and purified using methods modified from Randall and Sze (1986, 1987) and Parry

Table I
Solutions Used for Isolation of Low-Density
Vesicles from Oat Roots

Homogenization buffer	Resuspension buffer
50 mM Hepes–BTP, pH 7.4	25 mM Hepes–BTP, pH 7.2
250 mM sorbitol	250 mM sorbitol
6 mM EGTA	1 mM DTT
0.2% (w/v) BSA	0.1 mM PMSF
1 mM DTT[a]	
0.1 mM PMSF[b]	

[a] DTT, dithiothreitol.
[b] PMSF, phenylmethylsulfonyl fluoride.

et al. (1989) as described in Ward and Sze (1992a). Low-density vesicles (10 to 20 mg protein) are solubilized by the addition of an equal volume of solubilization buffer (Table II, final Triton X-100 concentration of 5%) with mixing. The presence of glycerol during solubilization and chromatography is required to maintain the H^+-ATPase in functional form (Randall and Sze, 1986; Parry *et al.,* 1989; Ward and Sze, 1992a). In the original procedure of Randall and Sze (1986, 1987), ATP was present in the solubilization buffer. In subsequent experiments, MgATP was demonstrated to enhance the dissociation and inactivation of the ATPase complex (see Parry *et al.,* 1989; Ward *et al.,* 1992) and was therefore omitted from the solubilization buffer. After a 30-min incubation at 4°C the solubilization solution is centrifuged at 180,000× *g* (Beckman TY 65, r_{max}) for 1 h. Approximately 60% of the membrane protein and NO_3-sensitive ATP hydrolysis activity is solubilized and recovered in the clear supernatant. Due to the large size of the vacuolar H^+-ATPase complex (650 kDa; Ward and Sze, 1992a), centrifugation following solubilization is particularly important to remove high-molecular-weight, nonsolubilized complexes that otherwise migrate in the void volume during the subsequent gel filtration step, close to the migration of the ATPase.

2. Gel Filtration Chromatography

Based on its large size, the vacuolar H^+-ATPase can be separated from other solubilized proteins by gel filtration chromatography. The solubilized membrane proteins are loaded on a Sephacryl S400HR column (1.6 × 100 cm) equilibrated with gel filtration elution buffer (Table II) at 4°C. The column is eluted at 20 ml/h and 2.7-ml fractions are collected. Samples (20 to 30 μl) of fractions are assayed immediately for NO_3-sensitive ATPase activity in the presence of 1 mg/ml soybean phospholipids as described (Randall and Sze, 1986; Ward and Sze, 1992a).

Table II
Solutions Used for the Purification of H^+-ATPase

Solubilization buffer	Q buffer	Gel filtration elution buffer
10% (v/v) Triton X-100	10 mM Tris–Cl, pH 6	5 mM Tris–Mes, pH 8
20% (v/v) glycerol	20% (v/v) glycerol	10% (v/v) glycerol
2 mM EDTA–Tris, pH 7.2	0.3% (v/v) Triton X-100	0.3% (v/v) Triton X-100
8 mM MgSO$_4$	0.05 mg/ml soybean phospholipids	0.05 mg/ml soybean phospholipids
	1 mM EDTA–Tris, pH 6	1 mM EDTA–Tris, pH 8
	4 mM MgSO$_4$	4 mM MgSO$_4$
	1 mM DTT	1 mM DTT
	0.1 mM PMSF	0.1 mM PMSF

3. Anion Exchange Chromatography

Membrane proteins solubilized using a nonionic detergent, in this case Triton X-100, may be separated based on differences in charge by ion exchange chromatography. In this procedure, the H^+-ATPase is further purified based on its affinity for the positively charged, quaternary amine support Q-Sepharose. The fractions from the leading half of the ATPase activity peak eluted from the gel filtration column contain the highest specific activity and are pooled (6 to 8 ml), diluted with an equal volume of Q buffer (Table II) and slowly adjusted to pH 6 with 0.1 N HCl. This pH adjustment of the pooled gel filtration fractions is important since the ATPase was found to be denatured on Q-Sepharose columns when the pH was buffered at 7 or higher (Ward and Sze, unpublished data) possibly due to a stronger interaction with the positively charged column matrix. A Q-Sepharose FF anion exchange column (1 × 2.5 cm) is equilibrated with Q buffer at 4°C. The pooled fractions are then applied and the column is washed with 20 ml of 0.1 M KCl in Q buffer at 30 ml/h. The purified H^+-ATPase is eluted with a step gradient of 0.1 to 0.3 M KCl in Q buffer and, under these conditions, is recovered in a single 3.5-ml fraction.

C. Reconstitution of the Purified H⁺-ATPase into Phospholipid Vesicles

1. Preparation of Sonicated Phospholipids

Sonicated soybean or *Escherichia coli* phospholipids are prepared using a method described by Ambudkar and Maloney (1986). *E. coli* phospholipids, purchased in solution in chloroform (acetone–ether precipitated), are aliquoted (100 mg) and first dried under N_2. The phospholipids are then solubilized in 2 ml of diethyl ether containing 5 mM 2-mercaptoethanol and dried to a thin film under N_2. Residual ether is removed by incubation for 30 min under vacuum. Two milliliters of resuspenson buffer is added and the tube sealed under N_2. The phospholipids are sonicated to clarity in a bath sonicator (Bransonic 2000) in three 10-min cycles at 18 to 20°C. The phospholipids are stored on ice and used for experiments on the same day. Soybean phospholipids (100 mg), purchased in lyophilized form, are solubilized in ether and sonicated to clarity as described above. Sonicated soybean phospholipids, present in ATP hydrolysis assays, during solubilization, and in gel filtration elution and Q buffers (Table II) are stored at −80°C prior to use.

2. Reconstitution Procedure

The reconstitution of detergent solubilized membrane proteins into phospholipid vesicles involves the removal of the detergent in the presence of sonicated phospholipids. In this procedure, Triton X-100 is removed using Bio-beads SM-2, which are first treated with methanol and washed extensively with deionizedH_2O as described by Holloway (1973). Ten milligrams (200 μl) of sonicated

E. coli phospholipids is added to 1-ml aliquots of purified H$^+$-ATPase (100 to 120 μg protein/ml) and this mixture is incubated at 4°C for 20 min. Bio-beads SM-2 (0.3 g) are then added and the mixture is incubated with gentle rocking at 20°C for 30 min. The protein–phospholipid suspension is removed from the Bio-beads by pipette, added to a fresh 0.3-g aliquot of Bio-beads, and incubated for an additional 30 min as described above. We estimate that the 0.6 g of SM-2 beads used in this procedure is sufficient to reduce the concentration of Triton X-100 from 0.3% (the concentration in Q buffer) to the critical micelle concentration of about 0.005% (Holloway, 1973). The suspension is removed by pipette and an additional 10 mg of sonicated *E. coli* phospholipids is added. This mixture is incubated at 4°C for 20 min and then added to 19 ml of loading buffer (Table III) at 20°C. This step is used to load the proteoliposomes (sealed phospholipid vesicles containing membrane protein) with K$^+$ and to further reduce the concentration of Triton X-100. After a 20-min incubation at 20°C, proteoliposomes are collected by centrifugation at 185,000× g (Beckman TY 65, r_{max}) for 1 h at 4°C. The translucent pellets are resuspended in a total of 400 μl loading buffer, stored on ice, and assayed for H$^+$-pumping and ATP hydrolysis activities immediately.

D. Proton Transport Assays

The formation of a pH gradient (inside acidic) is used to assay the H$^+$-transport activity of the purified and reconstituted H$^+$-ATPase. The fluorescent probe acridine orange is a permeant weak base and accumulates in acidic compartments. The accumulation of acridine orange results in a quenching of the fluorescence signal, which is monitored using an excitation wavelength of 495 nm and an emission wavelength of 525 nm, at 22°C. In our experiments a Farrand System 3 spectrofluorometer (Ward and Sze, 1992b) was utilized for these measurements.

1. Assay for Sealed Liposomes or Proteoliposomes

To assay ion transport in a reconstituted system it is necessary to work with proteoliposomes with a low background permeability to ions, referred to here

Table III
Solutions Used for the Reconstitution and Assay of Transport Activity

Loading buffer	Proton pumping assay buffer
150 mM K$_2$SO$_4$	50 mM K$_2$SO$_4$
50 mM Hepes–BTP, pH 7	166 mM sorbitol
1 mM EDTA–Tris, pH 7	33 mM Hepes–BTP, pH 7
1 mM DTT	0.33 mM EDTA–Tris, pH 7
0.1 mM PMSF	1 mM ATP–BTP
	0.5 μM acridine orange

as sealed proteoliposomes. During development of the reconstitution procedure, we determine whether the liposomes or proteoliposomes formed were tightly sealed using a method described by D'Souza *et al.* (1987). Following sonication, 200 μl liposomes (10 mg) are diluted 20-fold into loading buffer (Table III) at 20°C, incubated for 20 min at 20°C, and collected by centrifugation at 185,000× g (Beckman TY 65, r_{max}) for 1 h at 4°C. This serves to load the liposomes with K^+. The clear pellets are suspended in 200 μl of loading buffer. The liposomes or proteoliposomes are diluted into K^+-free assay buffer containing 2.5 mM Hepes–BTP, pH 7, 250 mM sorbitol, and 8 μM acridine orange in a stirred fluorescence cuvette. If the addition of K^+-loaded liposomes to assay media lacking K^+ resulted in an immediate and rapid quenching of fluorescence (Fig. 1A) the liposomes were considered leaky to both K^+ and H^+. The interpretation of this result was that an outward-directed leak of K^+ created a negative (inside) membrane potential, which drove H^+ accumulation, resulting in fluorescence quenching. However, if liposomes are tightly sealed their addition into a K^+-free assay media does not result in fluorescence quenching (Figs. 1B and 1C). Nigericin (0.25 μM final concentration), a H^+/K^+ exchange ionophore, will cause rapid acidification of the liposomes (Fig. 1B). In contrast, the addition of valinomycin (0.5 μM, final concentration), which transports K^+ selectively, results in less rapid

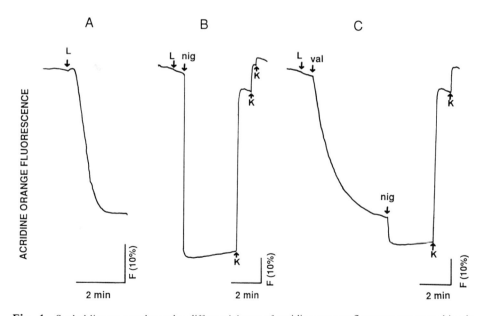

Fig. 1 Sealed liposomes showed a differential rate of acridine orange fluorescence quenching in the presence of valimonycin and nigericin. Demonstration of leaky (A) and sealed (B, C) liposomes. Potassium-loaded liposomes were diluted into assay buffer containing 2.5 mM Hepes–BTP, pH 7, 250 mM sorbitol, and 8 μM acridine orange in a final volume of 2 ml. L, val, nig, and K represent the additions of 10 μl of liposomes, 5 μl of 100 μM valinomycin, 5 μl of 50 μM nigericin, or 10 μl of 2 M KCl, respectively.

fluorescence quenching (Fig. 1C). Following the addition of nigericin, the pH gradient is dissipated by the addition of KCl to a final concentration of 20 to 30 mM. This assay is useful for the testing of different batches of phospholipids which may vary considerably and which, in our experience, should not be stored more than several months. In addition, this assay is used to determine whether sealed proteoliposomes are formed following reconstitution of purified H$^+$-ATPase prior to H$^+$-transport assays.

2. Assay for Proton Pumping

Proteoliposomes (10 or 20 μl, 1.5 to 3 μg protein) are diluted into proton-pumping assay buffer (Table III) in a final volume of 1 ml in a stirred fluorescence cuvette. This produces an initial three-fold transmembrane K$^+$ gradient (higher inside) that is sufficient to dissipate the positive (inside) membrane potential generated by electrogenic H$^+$-pumping in the presence of valinomycin (0.5 μM final concentration), a K$^+$ ionophore. The reaction was started by the addition of 5 μl of 1 M MgSO$_4$ (5 mM final concentration).

IV. Results and Discussion

The purification procedure results in the isolation of concentrated (100 to 120 μg protein/ml) H$^+$-ATPase suitable for reconstitution experiments. The specific activity is typically 2.1 to 3 μmol/min/mg protein at 22°C. The ATP hydrolysis activity of the purified enzyme is stimulated twofold by additional soybean phospholipids (Randall and Sze, 1986; Ward and Sze, 1992a). The purified oat root H$^+$-ATPase has a relative molecular mass of 650 kDa, as determined by gel filtration chromatography, and contains 10 subunits with molecular masses of 70, 60, 44, 42, 36, 32, 29, 16, 13, and 12 kDa (Fig. 2, Ward and Sze, 1992a,b). Phospholipids extracted from E. coli were found to produce more tightly sealed proteoliposomes than those from soybean when using the assay described here, and were therefore used for our reconstitution experiments. However, H$^+$-transport by the purified H$^+$-ATPase was also achieved using soybean phospholipids in a number of experiments (Ward and Sze, unpublished data). The reconstitution procedure is based on the ability of hydrophobic SM-2 Bio-beads to adsorb Triton X-100 in the presence of sonicated phospholipids. One advantage of this procedure is that Triton X-100 is an inexpensive, nonionic detergent compatible with several chromatography procedures used to prepare purified vacuolar H$^+$-ATPase. This procedure was found to be highly specific for the H$^+$-ATPase. In particular, the vacuolar H$^+$-translocating pyrophosphatase was not efficiently reconstituted by this procedure (Ward and Sze, unpublished data). Another distinct procedure was required to reconstitute this H$^+$-pump (Britten et al., 1992). When a crude Triton X-100-solubilized membrane fraction is subjected to this reconstitution procedure, purified and transport-competent H$^+$-ATPase

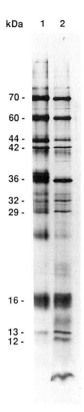

Fig. 2 The purified and reconstituted vacuolar H⁺-ATPase contains 10 subunits. The purified (lane 1, 2.5 μg protein) and reconstituted (lane 2) ATPase, containing similar levels of NO_3 sensitive ATPase activity (5.2 nmol/min), were subjected to SDS–polyacrylamide electrophoresis. The gel was silver stained.

can be recovered (Ward and Sze, unpublished data). The specificity of this procedure may be due to distinct structural features of this enzyme, which is composed predominantly of a large hydrophilic domain. Another reconstitution procedure, based on the removal of octy-β-D-glucopyranoside by dilution (Ambudkar and Maloney, 1986) was also used successfully to reconstitute transport activity by the oat root H⁺-ATPase in crude form (Ward and Sze, unpublished data). This latter procedure is more generally applicable to other membrane proteins and has been used to reconstitute H⁺/Ca^{2+} exchange activity from tonoplast membranes of oats (Schumaker and Sze, 1990). Purification using octy-β-D-glucopyranoside was not attempted due to the relatively high cost of this detergent.

Using the procedure described here, between 30 and 50% of the ATP hydrolytic activity of the purified preparation is typically recovered following reconstitution. The ATP hydrolytic activity of the reconstituted H⁺-ATPase is not stimulated by he addition of soybean phospholipids, but is stimulated 2.5-fold by gramicidin, which dissipates the ATP-dependent H^+ gradient, allowing maximal hydrolysis activity. The addition of Mg^{2+} and ATP to the reconstituted H^+-ATPase results in the generation of a pH gradient (acid inside), as demonstrated by the fluorescence quenching of acridine orange (Fig. 3). The rate of fluorescence quenching is stimulated 20-fold by the addition of valinomycin (Fig. 3A). This result indicates that, in the absence of valinomycin, H^+-transport generates a positive membrane potential that limits proteoliposome acidification, as presented in Fig. 4. The addition of valinomycin dissipates the membrane potential by allowing the efflux of K^+ and stimulates the rate of H^+-transport and vesicle acidification. In the absence of Mg^{2+}, valinomycin has no effect on acridine orange fluorescence under these conditions, where only a 3-fold transmembrane K^+ gradient is present (Fig. 3B). Bafilomycin A_1 is a specific inhibitor of vacuolar H^+-ATPases. The addition of 15 nM Bafilomycin completely inhibits H^+-transport by the ATPase (Fig. 3B). In addition, H^+-pumping activity of the reconstituted

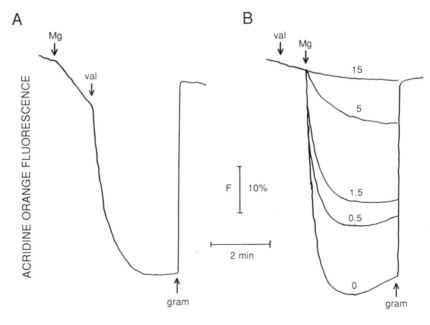

Fig. 3 The purified and reconstituted H^+-ATPase is a functional H^+ pump. Proteoliposomes (10 μl, NO_3 sensitive ATPase activity, 3.5 nmol/min) were diluted into assay buffer as described for H^+-pumping assays. To start the reaction, 5 μl of 1 M MgSO$_4$ was added. When present, Bafilomycin A_1 in dimethyl sulfoxide (DMSO) was added (0.5% DMSO final) and incubated at 22°C for 10 min prior to the assay val and gram refer to the additions of 5 μl of 100 μM valinomycin or 5 μl of 1 mg/ml gramicidin D, respectively. (Reproduced from Ward and Sze, 1992b, with permission from the American Society of Plant Physiologists).

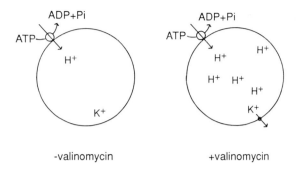

Fig. 4 Valinomycin stimulates the rate of proteoliposome acidification by dissipation of the membrane potential. The positive (inside) membrane potential generated by the H⁺-ATPase limits the rate of H⁺-transport in the absence of valinomycin. The addition of valinomycin, which mediates the efflux of K⁺, dissipates the membrane potential and allows rapid H⁺ transport.

ATPase is inhibited by NO_3^- and DCCD, inhibitors of V-type ATPases (Ward and Sze, 1992b). The reconstituted ATPase is not inhibited by orthovanadate, an inhibitor of P-type ATPases such as the plasma membrane H⁺-pump, or by azide, an inhibitor of F-type, coupling factor ATPases of the mitochondria and chloroplast. These results indicate that the H⁺-ATPase is an electrogenic pump and that the proteoliposomes were sealed to K⁺ and H⁺.

The reconstitution procedure was developed to determine the structure and to elucidate the regulation of the oat root vacuolar H⁺-ATPase. For example, a primary goal was to determine the minimal subunit composition required for both ATP hydrolytic activity and H⁺ transport. Here we show that the purified vacuolar H⁺-ATPase lacks a 100-kDa subunit (Fig. 2) that is part of the purified H⁺-ATPase from red beet storage tissue (Parry *et al.*, 1989). Interestingly, the red beet H⁺-ATPase, purified and reconstituted using the procedures described here, contained subunits identical to those previously reported including a 100-kDa polypeptide (Parry *et al.*, 1989) and was also functional in H⁺-pumping (Ward and Sze, unpublished data). These results, as well as those of Warren *et al.* (1992), demonstrate that a 100-kDa subunit may not be required for H⁺-pumping activity. However, because the 100-kDa subunit may be highly susceptible to proteolytic degradation the presence of a 100-kDa subunit in the native membrane-bound enzyme of oats cannot be excluded by this approach.

V. Conclusions and Perspectives

Reconstitution of the purified vacuolar H⁺-ATPase provides a valuable system for studying the protein structure, regulation, and transport properties of this H⁺-pump. The procedure described here is rapid and provides high recovery of the ATPase, making it suitable for the structural characterization of other possible isoforms of plant vacuolar ATPases (for review see Sze *et al.,* 1992b). Recently, it was demonstrated that the oat root V-type ATPase, dissociated into a large hydrophilic domain and a membrane integral domain, can be reassembled into a functional H^+-pump (Ward *et al.,* 1992). This suggests that the reassembly of a functional H^+-ATPase from purified components could yield valuable information concerning the role of individual subunits in the regulation, catalytic activity, and transport properties of vacuolar ATPases.

References

Ambudkar, S. V., and Maloney, P. C. (1986). Bacterial anion exchange. *J. Biol. Chem.* **261,** 10079–10086.

Britten, C. J., Zhen, R.-G., Kim, E. J., and Rea, P. A. (1992). Reconstitution of transport function of vacuolar H^+-translocating inorganic pyrophosphatase. *J. Biol. Chem.* **267,** 21850–21855.

Churchill, K. A., and Sze, H. (1983). Anion-sensitive, H^+-pumping ATPase of in membrane vesicles from oat roots. *Plant Physiol.* **71,** 610–617.

D'Souza, P. M., Ambudkar, S. V., August, J. T., and Maloney, P. C. (1987). Reconstitution of the lysosomal proton pump. *Proc. Natl. Acad. Sci. U.S.A.* **84,** 6980–6984.

Forgac, M. (1989). Structure and function of vacuolar class of ATP-driven proton pumps. *Physiol. Rev.* **69,** 765–796.

Herman, E. M., Li, X., Su, R. T., Larsen, P., Hsu, H.-T., and Sze, H. (1994). Vacuolar-type H^+-ATPases are associated with the endoplasmic reticulum and provacuoles of root tip cells. *Plant Physiol.* **106,** 1313–1324.

Holloway, P. W. (1973). A simple procedure for removal of Triton X-100 from protein samples. *Anal. Biochem.* **53,** 304–308.

Parry, R. V., Turner, J. C., and Rea, P. A. (1989). High purity preparations of higher plant vacuolar H^+-ATPase reveal additional subunits. *J. Biol. Chem.* **264,** 20025–20032.

Randall, S. K., and Sze, H. (1986). Properties of the partially purified tonoplast H^+-pumping ATPase from oat roots. *J. Biol. Chem.* **261,** 1364–1371.

Randall, S. K., and Sze, H. (1987). Purification and characterization of the tonoplast H^+-translocating ATPase. *Methods Enzymol.* **148,** 123–132.

Schumaker, K. S., and Sze, H. (1990). Solubilization and reconstitution of the oat root vacuolar H^+/Ca^{2+} exchanger. *Plant Physiol.* **92,** 340–345.

Sze, H. (1985). H^+-translocating ATPases: Advances using membrane vesicles. *Annu. Rev. Plant Physiol.* **36,** 175–208.

Sze, H., Ward, J. M., and Lai, S. (1992a). Vacuolar H^+-translocating ATPases from plants: Structure, function, and isoforms. *J. Bioenerg. Biomembr.* **24,** 371–381.

Sze, H., Ward, J. M., Lai, S., and Perera, I. (1992b). Vacuolar-type H^+-translocating ATPase in plant endomembranes: Subunit organization and multigene families. *J. Exp. Biol.* **172,** 123–135.

Ward, J. M., and Sze, H. (1992a). Subunit composition and organization of the vacuolar H^+-ATPase from oat roots. *Plant Physiol.* **99,** 170–179.

Ward, J. M., and Sze, H. (1992b). Proton transport activity of the purified vacuolar H^+-ATPase from oats: Direct stimulation by Cl^-. *Plant Physiol.* **99,** 925–931.

Ward, J. M., Reinders, A., Hsu, H.-T., and Sze, H. (1992). Dissociation and reassembly of the vacuolar H^+-ATPase complex from oat roots. *Plant Physiol.* **99,** 161–169.

Warren, M., Smith, J. A. C., and Apps, D. K. (1992). Rapid purification and reconstitution of a plant vacuolar ATPase using Triton X-114 fractionation: Subunit composition and substrate kinetics of the H^+-ATPase from the tonoplast of *Kalanchoë diagremontiana. Biochem. Biophys. Acta* **1106,** 117–125.

Yamanishi, H., and Kasamo, K. (1994). Effects of cerebroside and cholesterol on the reconstitution tonoplast H^+-ATPase purified from mung bean (*Vigna radiata* L.) hypocotyls in liposomes. *Plant Cell Physiol.* **35,** 655–663.

CHAPTER 12

Isolation and Fractionation of Plant Mitochondria and Chloroplasts: Specific Examples

José Manuel Gualberto, Hirokazu Handa, and Jean Michel Grienenberger

Institut de Biologie Moléculaire des Plantes CNRS
Université Louis Pasteur
12, rue du Général Zimmer
67084 Strasbourg, France

I. Introduction

Plant organelles have been studied for their unique metabolisms. It is not necessary to emphasize the importance and complexity of the metabolic pathways that specifically take place in plant organelles, but we want to underline the peculiar mechanisms involved in the biosynthesis of mitochondria and chloroplast in plants, where the productive interaction of the three cellular genomes (nuclear, mitochondrial, and chloroplast) is required. Gene expression in chloroplasts and mitochondria is far from being simple, involving complex mechanisms of cis- and trans- RNA splicing and extensive RNA editing (Gray *et al.,* 1992; Gruissem and Tonkin, 1993). The study of these complex mechanisms is difficult, mostly because few genetic tools can be applied and the purification of sufficient amounts of purified plant organelles is time- and labor-consuming. These problems have impaired most efforts in obtaining specific organellar *in vitro* systems and are a major problem when the purification of minor organellar proteins is attempted.

The purpose of this chapter is to describe general methods for the isolation and fractionation of plant mitochondria and chloroplasts. The methods have to be adapted for each specific purpose, and we discuss some of our experience with problems derived from the use of different plant material. Even when working with a same plant species, for each specific purpose (protein purification, nucleic acids extraction, *in organello* synthesis, import studies etc.) the appropriate starting material must be chosen. We derive most of our experience from our studies on plant mitochondria, and discuss in more detail the problems involved with the isolation of that type of organelle. For the isolation of chloroplasts, we refer to widely used general purification methods. Detailed protocols can also be found in other reviews (Leaver *et al.,* 1983; Douce *et al.,* 1987; Walker *et al.,* 1987; Robinson and Barnett, 1988; Schuster *et al.,* 1988).

II. General Considerations

Both mitochondria and chloroplasts are isolated from total plant extracts by differential centrifugation. The material obtained is contaminated with soluble hydrolases, other cellular fractions, membrane fragments, bacteria and, in the case of mitochondria isolation, the presence of plastid material. Further purification by density gradient centrifugation is therefore advisable for most subsequent studies. Sucrose or Percoll gradients are routinely used.

Extraction buffers used for the purification of mitochondria or chloroplast must contain an osmoticum to prevent organelle burst due to hypotonic shock. Mannitol, sorbitol, and sucrose are the most commonly used osmotic agents. In theory, mannitol and sorbitol should be preferred, because they are not usually metabolized. For the short purification times usually required in plant mitochondria isolation, we have not seen any inconvenience in using sucrose. It is, however, advisable to test the buffer composition that gives better results, if mitochondria is isolated from a tissue not previously tested. For chloroplast isolation, sorbitol is the most frequently used osmoticum. For certain species, the osmolarity conditions must be individually optimized.

Another problem that has to be considered when working with plant tissues is the high contamination with secondary products released from the vacuole. The strong shearing forces required to break the plant cell wall result in the inevitable disruption of the vacuole and release of its content. Hydrolytic enzymes, phenolic compounds, terpenes, and other secondary products released from the vacuole are detrimental to many components of organellar membranes and if certain precautions are not taken, the extraction yield and the quality of the material obtained are poor. This problem can be circumvented if mitochondria and chloroplasts are prepared from protoplasts (Nishimura *et al.,* 1982), but this alternative is only practical to obtain limited amounts of high-quality plant organelles. To minimize the effect of the detrimental compounds released during homogenization, different protecting agents are included in extraction buffers, such as PVP, BSA, and reducing agents (cysteine, β-mercaptoethanol, ascorbate). Addition of EDTA to the extraction buffer also protects the organellar membranes from ion-dependent lipases and phospholipases (Douce *et al.,* 1987).

III. Purification of Plant Mitochondria

Plant mitochondria can be routinely prepared from a variety of plant tissues. Storage organs (potato, sugar beet), green leaves (tobacco, pea, petunia), etiolated seedlings (wheat, maize, mung bean), tissue cultures, and other sources of plant material have been used successfully for the purification of plant mitochondria (Leaver *et al.,* 1983). If specific studies are to be pursued, it is necessary to test several species and tissues for the best yield and activity required. Protocols for mitochondria purification can be adapted without major problems to different plant materials, apart from differences in homogenization conditions required to disrupt the plant cells. Homogenization can be accomplished with a Polytron homogenizer, with a Waring blender, or with a Moulinex mixer, or by grinding the tissue in a mortar and pestle. In general, milder homogenization conditions result in better quality of isolated mitochondria.

Below we describe the protocol that we routinely utilize in our laboratory to isolate mitochondria from wheat seedlings or embryos. For most purposes, we use mitochondria purified from 7-day-old wheat seedlings: good mitochondria extraction yields are easily obtained and mitochondrial RNA and DNA isolated from seedling mitochondria seem to be least contaminated with nucleases. On the other hand, wheat embryos are very rich in mitochondria that prove to be of better quality for many biochemical studies: it has been the material of choice used to prepare the first plant mitochondrial transcription, processing, and editing *in vitro* systems (Hanic-Joyce *et al.,* 1990; Hanic-Joyce and Gray, 1991; Araya *et al.,* 1992) and, according to our experience, it is a better material for the study of mitochondrial ribosomes (see Section IV).

A. Isolation of Mitochondria from Wheat Embryos or Etiolated Seedlings

1. Preparation of Plant Material

To prepare mitochondria from wheat seedlings, seeds are surface-sterilized with sodium hypochlorite (10 min in a 3% solution), washed several times with sterile water, and allowed to germinate on vermiculite or other appropriate support. Green or etiolated seedlings are typically grown for 4–8 days, respectively, under the light or in the dark.

Viable wheat embryos are a good starting material for mitochondria extraction and can be prepared as previously described (Johnson and Stern, 1957). In brief, wheat grains are ground in a Waring blender and the material is sieved through 2-, 1-, and 0.6-mm meshes respectively. The material remaining on the upper sieve is recovered for an additional grinding cycle. Free embryos are retained in the lower sieve and are further purified by blowing, to eliminate the contaminating aleurones (we use a small ventilator, set at low speed, or a hair drier), followed by two rounds of flotation in a 25:10 carbon tetrachloride/cyclohexane solution. Purified embryos (about 3–5 g for kilogram of wheat grains) are dried under the hood and are stored in the cold room under vacuum. To circumvent most of the labor-intensive (and dirty) steps of seed grinding and sieving, a convenient alternative is to use an already ground fraction from a commercial mill. We use a fraction resulting from the first industrial grinding and sieving (smaller than 2 mm) that is mostly free of flour and is enriched in free embryos. With that fraction, we process three times less material to obtain the same amount of embryos. Before mitochondrial extraction, embryos are surface-sterilized by immersion 3 min in 0.1% sodium hypochlorite, washed four to six times with sterile water and allowed to imbibe 16 to 24 h over water-saturated filter papers in petri dishes (5 g of embryos on a 13-cm-diameter petri dish). If bacterial contaminations are persistent, 1 mg/ml lysozyme and 50 μg/ml of cefotoxin can be included in the solution.

2. Isolation of Washed Mitochondria

All solutions and glassware used are autoclaved. During all procedures, temperature is kept below 5°C. *Extraction buffer* (0.4 M sucrose, 50 mM Tris–HCl, pH 7.5, 3 mM EDTA, 0.2 mM EGTA) is added with 0.1% BSA (fraction V) and 4 mM β-mercaptoethanol just before use. Harvest 500 g of 7-day-grown wheat seedlings and cut them into small pieces of about 1 cm long. Wash the material with cold sterile water and homogenize in 2 to 4 v/w of extraction buffer (large ratios of extraction buffer to plant material normally give better extraction yields). Homogenization is conducted in a precooled Waring blender at low speed by bursts of 5 s, as many times as required to obtain a creamy past (usually three to seven times, according to the tissue and buffer volume). Long homogenization times result in better cell disruption, but also in mitochondria loss. The homogenate is filtered through nylon meshes (95–25-μm pore sizes) or alternatively, through two layers of Miracloth (Calbiochem, USA). The brei is squeezed to recover most of the fluid.

For mitochondria extraction from wheat embryos, overnight imbibed embryos (20 g) are ground by batches of 5 g in 100 ml of extraction buffer either with a mortar and pestle or with a Polytron (Kinematica GmbH, Switzerland) set at

40% power for 10 s. The homogenate is filtered as described above. A reextraction of the solid fraction can slightly increase the final yield.

The extract is centrifuged 10 min at 1000× g (2500 rpm in a Beckman JA10 rotor) to pellet most cell debris, starch, and nuclei. If working with green tissue, most chloroplasts are pelleted by a second centrifugation at 3000× g. The supernatant is recovered into new centrifuge bottles and the mitochondria fraction is pelleted by centrifugation at 10,000× g (7000 rpm in a Beckman JA10 rotor) for 20 min. The supernatant is discarded and the mitochondrial pellet is resuspended in the remaining buffer with a small clean brush. Additional buffer is added (5–10 ml of buffer for mitochondrial pellet) and homogenization is continued with a Dounce homogenizer with a loose pestle. The mitochondrial suspension is adjusted to one-fourth of the initial volume and centrifuged again 5 min at 1000× g. The supernatant is transferred to new centrifuge tubes and mitochondria are pelleted by centrifugation at 10,000× g for 15 min. The washed mitochondrial fraction is resuspended in a small volume and layered over sucrose or Percoll gradients for further purification.

B. Gradient Purification of Mitochondria

Mitochondria purified by differential centrifugation are still contaminated with other membrane fractions (broken mitochondria, membrane vesicles, plastids, peroxisomes), soluble hydrolytic enzymes, and bacteria (Douce *et al.*, 1987). The intact mitochondria must be purified by isopycnic centrifugation in sucrose or Percoll gradients. Sucrose step gradients are routinely used to purify mitochondria from nongreen tissues. They have, however, the inconvenience that mitochondria can be damaged by the osmotic shock caused by the high sucrose concentrations (Douce *et al.*, 1987). Dilution to isoosmotic conditions must proceed slowly to avoid a too-rapid inner membrane expansion.

Percoll gradients allow rapid purification of mitochondria in isoosmotic and low-viscosity conditions and should be used if functional mitochondria are to be obtained (Neuburger *et al.*, 1982). Percoll has the additional advantage over sucrose that, in mitochondrial preparations from green tissues, a good gradient purification from contaminating the thylakoid membranes can be obtained. It also seems that Percoll gradients achieve a better purification of mitochondria from contaminating hydrolases: Neuburger *et al.* (1982) reported that the integrity of Percoll-purified mitochondria can be maintained up to 3 days. There are, however, two situations where we found that sucrose gradient purification is to be preferred: when bacterial contamination is a problem, sucrose gradients allow better purification from contaminating bacteria (Leaver *et al.*, 1983), and when purifying mitochondrial RNA or DNA, the nucleic acids extracted from sucrose gradient-purified mitochondria are usually of better quality than those from Percoll-purified mitochondria (possibly because of the contaminating Percoll that is difficult to eliminate completely from the mitochondrial pellet). After gradient purification, mitochondria from most frequently studied species can be used for nucleic acids isolation, respiration studies, *in organello* protein synthesis, etc. (Leaver *et al.*, 1983; Douce *et al.*, 1987; Schuster *et al.*, 1988).

1. Sucrose Gradient Purification of Mitochondria

The mitochondrial suspension obtained by differential centrifugation is layered onto sucrose step gradients (in a 40-ml SW27 rotor tube, 9 ml of 0.9 M sucrose, 10 ml of 1.6 M sucrose, and 9 ml of 1.8 M sucrose in 50 mM Tris–HCl, pH 7.5, 3 mM EDTA, 0.2 mM EGTA, and 0.1% BSA). For mitochondria extracted from 500 g wheat seedlings or from 20 g wheat embryos, prepare four 40-ml gradient tubes. Spin the gradients in a SW27 rotor or equivalent at 25,000 rpm for 1 h at 4°C. The light brown mitochondrial band at the 0.9–1.6 M interface is carefully collected with a sterile Pasteur pipette and is very slowly diluted with at least 3 vol of wash buffer. To avoid osmotic damage of mitochondria, the dilution takes about 20–30 min. The purified mitochondria are pelleted by centrifugation at 10,000× g for 15 min.

2. Percoll Gradient Purification of Mitochondria

Continuous, self-generating and discontinuous Percoll gradients can be used. We use a discontinuous gradient system (Jackson *et al.*, 1979) for most purification procedures. Percoll cannot be autoclaved in the presence of salts or sugars. Percoll solutions are prepared from 100% Percoll, 2× concentrated extraction buffer, and sterile water. In a 40-ml centrifuge tube, the step gradient comprises three layers of, respectively, 12 ml of 13% Percoll, 12 ml of 21% Percoll, and 6 ml of 45% Percoll. The hydrophobicity of the 13 and 21% layers is increased by the addition of 100 mM propane-1,2-diol. This permits organelle separation also according to the surface properties of the membranes (Jackson *et al.*, 1979). Mitochondria from green or nongreen tissues are layered on top of the gradient, which is centrifuged in a Beckman JA20 rotor (or equivalent) at 13,000× g for 30 min. Thylakoid membranes band at the 13–21% Percoll interface, whereas mitochondria band at the 21–45% interface. Avoid contamination with the fraction that is denser than 45% Percoll, which mostly contains peroxisomes (Neuburger *et al.*, 1982). The mitochondria are collected with a Pasteur pipette, diluted with at least 10 vol of extraction buffer and centrifuged at 10,000× g for 15 min. The loose mitochondrial pellet is resuspended in extraction buffer and recentrifuged, to eliminate most of the remaining Percoll. The washing is repeated if Percoll has not been fully eliminated.

For mitochondria purification from green tissue, an improved protocol of Percoll gradient purification has been reported that results in a better separation of mitochondria from thylakoid membranes (Douce *et al.*, 1987). The method employs, in a single step, a self-generating gradient of Percoll combined with a linear gradient of PVP. Two solutions of 28% Percoll, 1× extraction buffer are prepared, containing, respectively, 0 and 10% PVP25,000. PVP will take some time to dissolve, so the solutions must be prepared in advance. Using a gradient former, linear gradients are prepared in 40-ml polycarbonate centrifuge tubes. The mitochondrial suspension is layered on top of the gradients, which are centrifuged at $40,000\times g$ for 45 min (18,000 rpm in a Beckman JA20 rotor). Mitochondria form very fine discrete bands in the lower part of the gradient.

C. Fractionation of Mitochondria

Fractionation of mitochondria into a soluble fraction (matrix and intermembrane fractions) and a membrane fraction is eventually necessary, for instance, to study the function of unidentified mitochondrial polypeptides and as a primary step in the purification of mitochondrial enzymes. Mitochondria can be lysed without the use of detergents by sonication. The mitochondria pellet is resuspended in 10 mM Tris–HCl, pH 8, 0.15 M NaCl to a protein concentration of 1 mg/ml and sonication is conducted till the suspension clears (about 20 s). A 5-min centrifugation at $12,000\times g$ removes unbroken mitochondria, and a centrifugation at $40,000\times g$ for 20 min (18,000 rpm in a Beckman JA20 rotor) yields the soluble fraction and the total membrane fraction. Proteins loosely associated with the membranes can be solubilized by resuspension of the membrane pellet in 0.1 M Na$_2$CO$_3$, to a final concentration of 1 mg protein/ml (Fujiki *et al.*, 1982). After incubation on ice for 30 min, with occasional resuspension with a glass potter homogenizer, the stripped membranes are pelleted by centrifugation. For more precise characterization of mitochondrial membrane fractions, detailed protocols have been published (Mannella, 1987; Møller *et al.*, 1987).

D. Assessing the Quality of Isolated Mitochondria

The quality of purified plant mitochondria preparations can be assessed by evaluating several parameters. These include the ADP/O ratios, the dependence of respiratory rates by exogenous ADP, the permeability of the outer membrane to cytochrome *c,* and the appearance in electron micrographs (Douce *et al.*, 1987). We define "practical purity" as the degree of mitochondrial purity and integrity required for a specific purpose. Mitochondria preparations used for biophysical and enzymatic experiments must be of the highest possible quality, which is usually obtained in detriment of yield. On the other hand, rather crude mitochondrial fractions can be used for certain applications. As an example, membrane fractions obtained from wheat germ are a rich source of functional membrane-associated mitochondrial enzymes such as cytochrome *c* oxidase (Peiffer *et al.*, 1990), even if intact mitochondria cannot be obtained from the material.

Measurement of respiration rates requires an oxygen electrode that is not standard equipment of most laboratories. We routinely evaluate the quality of our mitochondrial preparations in two ways: When analyzing mitochondrial gene expression, the quality of mitochondrial RNA isolations is evaluated in denaturing formaldehyde gels. If the integrity of the outer membrane is critical, the permeability of the outer membrane to added cytochrome *c* mitochondrial RNA is rapidly degraded by contaminating RNAses that seem to be attached to the outer membrane and as a consequence RNA isolated from poor mitochondria preparations (with many broken mitochondria and resealed vesicles) is mostly degraded.

Damage of the mitochondrial outer membrane is conveniently evaluated by following the reduction of exogenous cytochrome *c* by succinate : cytochrome *c* oxidoreductase, which is localized in the external face of the inner membrane. To 1 ml of respiration buffer (20 mM K-phosphate buffer, pH 7.2, 0.3 M sucrose, 1 mM EDTA, 5 mM glycine) in a plastic spectrophotometer cuvette, the following solutions are added: 5 μl KCN, 0.2 M, 5 μl ATP, 0.2 M, 5 μl cytochrome *c*, 16 mM, and a mitochondria aliquot of about 100 μg. The rate of reduction of cytochrome *c* is measured by change in absorbance at 550 nm (extinction coefficient: 21 mM^{-1} cm^{-1}) after addition of 10 μl of 1 M succinate. A second aliquot of mitochondria is ruptured by osmotic shock by addition of 0.5 ml of cold sterile water. Osmolarity is reestablished by the addition of 0.5 ml of 2$\times$ respiration buffer and the reduction of cytochrome *c* is followed as described above. The percentage of intact mitochondria in the preparation is given by the relation $\% = 100 \times (1 - P_1/P_0)$ where P_1 and P_0 are the slopes of the $A_{550\ nm}$ measurements for untreated (P_1) and osmotically shocked (P_0) mitochondria.

IV. Specific Example: Isolation of Mitochondria Free from Cytosolic Ribosomes

For specific purposes, a further purification step of mitochondria or a judicious choice of starting materials is required to minimize contamination from other cellular components. In the process of developing a procedure for the isolation of pure mitochondrial ribosomes, we found that mitochondrial RNA isolated from wheat etiolated seedlings were largely contaminated with cytosolic ribosomal RNAs (Fig. 1). The contaminating cytosolic rRNA could not be removed even if sucrose or Percoll gradients were used for the purification of mitochondria. This result indicated that wheat mitochondria from etiolated seedlings are not contaminated with free cytosolic rRNA molecules but with cytosolic ribosomes themselves. As has been shown for yeast mitochondria (Kellems and Butow, 1974), it seems that a large number of cytosolic polysomes are tightly bound to the mitochondrial surface and synthesize mitochondrial proteins encoded by nuclear genes. In order to reduce the contamination with cytosolic ribosomes, we introduced two alternative steps into our mitochondrial purification procedure: "mitoplast"-making and puromycin treatment.

Mitoplasts are mitochondria lacking outer membranes to which cytosolic polysomes are tightly attached. For mitoplast-making, the mitochondrial pellet is

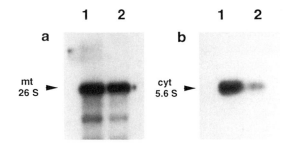

Fig. 1 Northern blot analysis of mitochondrial and mitoplast RNAs (lanes 1 and 2, respectively) isolated from etiolated seedlings. RNAs were size-fractionated on formaldehyde agarose gels and transferred to Nylon membranes. The filters were hybridized with antisense oligonucleotides specific for (a) mitochondrial 26S rRNA and (b) for cytosolic 5.8S rRNA.

suspended in phosphate buffer (10 mM sodium phosphate, pH 7.4, 1 mM EDTA, 30 min on ice) to swell inner membranes osmotically and to rupture and strip the outer membrane. After the osmotic treatment, mitoplasts are separated from outer membranes and broken mitochondrial fractions by centrifugation through a discontinuous sucrose gradient of 22, 33, and 47% sucrose (Zhuo and Bonen, 1993). The mitoplast fraction is recovered from the 33–47% sucrose interface, slowly diluted with extraction buffer, and recovered by centrifugation as described above.

Puromycin is an inhibitor of protein synthesis that, by acting as an analogue of aminoacyl-tRNA, induces a premature discharge of nascent polypeptides from ribosomes. It is supposed that the interaction between cytosolic polysomes and the mitochondrial surface mainly involves the nascent polypeptide chains specifically bound to protein import sites on the mitochondrial surface (Kellems *et al.*, 1975). Therefore, the puromycin-induced discharge of nascent polypeptides should release cytosolic polysomes from the mitochondrial surface. We incubated mitochondria with 1 mM puromycin in the presence of 300 mM KCl to prevent an ionic interaction between the ribosomes and the membrane. After puromycin treatment, mitochondria were washed and resuspended with extraction buffer.

To assess the effectiveness of mitoplast-making and puromycin treatment in reducing cytosolic ribosomes contamination, Northern blots of mitochondrial RNA were hybridized with antisense oligonucleotide probes specific for mitochondrial or cytosolic rRNAs. Mitoplasts prepared from etiolated seedlings were much less contaminated with cytosolic rRNAs than untreated mitochondria. As shown in Fig. 1, when same amounts of RNA are loaded on agarose gels, the amount of mitochondrial 26S rRNA present in mitoplasts is nearly the same as that found in untreated mitochondria, but contaminating nuclear 5.8S rRNA in the mitoplast fraction is reduced more than 60%. Although this procedure is effective in reducing the levels of contaminants, about one-third of contaminating cytosolic ribosomes was still present, still too much if pure mitochondrial ribosomes are to be analyzed. Puromycin treatment had efficiency in removing contaminants from mitochondria similar to that of mitoplast-making (about 60% reduction, data not shown).

In such a case, the second important factor that was considered to minimize contaminants in the mitochondrial sample was the type of tissues used as a starting material for mitochondrial extraction. As mentioned before, wheat embryos are very rich in mitochondria, although their preparation is labor- and time-consuming. When we compared mitochondrial RNA isolated either from wheat germinated embryos or seedlings, we found a large difference between the ratios of cytosolic rRNA contamination. As shown in Fig. 2, mitochondrial RNA isolated from wheat embryos had over 90% less contaminating cytosolic 18S rRNA than that found in mitochondria from etiolated seedlings. This result suggested that the actual number of cytosolic ribosomes attached to the mitochondrial surface depends on the tissue used for mitochondrial extraction and/or the metabolic state at the time of extraction. In yeast, such a diversity in the number of cytosolic ribosomes bound to mitochondria has also been observed; mitochondria of stationary cells had 40–60% fewer cytosolic ribosomes than log-phase cells (Kellems and Butow, 1974).

Mitoplast-making or puromycin treatment were also tested on mitochondria isolated from wheat embryos. The effect was nearly the same as in the case of mitochondria isolated from etiolated seedlings; 50–60% of the contaminating cytosolic 18S rRNA was removed from the embryo mitochondrial fraction. Finally, using the combination of wheat embryo as a starting material and additional purification by mitoplast-making or puromycin treatment, we obtained a mitochondrial RNA sample containing about 5% of contaminating cytosolic rRNA molecules when compared with the extract from etiolated seedlings (Data not shown). Our results strongly demonstrate that for each specific purpose, for the best purity and yields desired, the appropriate plant material must be chosen and the isolation procedures adopted with care.

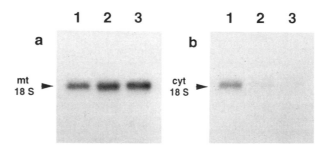

Fig. 2 Northern blot analysis of mitochondrial and mitoplast RNAs isolated from wheat embryo. Lane 1 contains mitochondrial RNA isolated from etiolated seedlings as a reference. Lanes 2 and 3 contain mitochondrial and mitoplast RNAs isolated from wheat embryo, respectively. Filters were hybridized with antisense oligonucleotide probes specific (a) for mitochondrial 18S rRNA and (b) for cytosolic 18S rRNA.

V. Purification of Chloroplasts

Chloroplasts can be isolated from many types of plant tissues by differential centrifugation followed by Percoll gradient purification of intact chloroplasts (Walker *et al.,* 1987). Most studies on chloroplast biogenesis have used chloroplasts purified from spinach or pea leaves because high yields of intact chloroplasts can be isolated routinely from these plants. Chloroplasts have successfully been isolated from many other higher plant species, from algae, and from *Euglena gracilis,* although with variable efficiency. The isolated chloroplasts can be used for protein transport studies, for *in organello* protein and RNA synthesis, for the preparation of *in vitro* transcription and RNA maturation systems and, in scale-up preparation, for the purification of chloroplast proteins. We describe here basic procedures for chloroplast purification, since excellent protocols for *in vitro* chloroplast assays have been described in other reviews, including in this series (Ellis and Hartley, 1982; Mishkind *et al.,* 1987; Walker *et al.,* 1987; Perry *et al.,* 1991). The method that we outline below for the isolation of chloroplasts from pea is an adaptation of the method outlined by Ellis and Hartley (1982) and can also be used for chloroplast isolation from other dicot and monocot plants. Low chloroplast yields obtained from other plant tissues are possibly related to higher production of starch granules that rupture the chloroplast envelope membrane during chloroplast isolation.

A. Isolation of Chloroplasts from Pea Seedlings

Conditions of plant growth are critical for reproducible chloroplast isolation. Pea seedlings are grown for 10–15 days on vermiculite in controlled environment chambers at 20–25°C and under a photoperiod of 12 h. The day before chloroplast isolation, plants are kept in the dark until harvesting, to deplete chloroplasts of accumulated starch grains. Most active chloroplasts are obtained when tissue homogenization is performed at low temperature and chloroplasts are separated from the homogenate as rapidly as possible.

The expanded top leaves are harvested and mixed with cold *grinding buffer* (0.33 *M* sorbitol, 50 m*M* Hepes–KOH, pH 7.3, 2 m*M* EDTA, and 0.1% BSA) in a bucket on ice. For 20 g of plant tissue, use 100 ml grinding buffer. The leaves are homogenized with a Polytron (Kinematica GmbH, Switzerland) set at 70% power by three bursts of 5 s each. The homogenate is filtrated through a 0.95-μM nylon mesh (or through two layers of Miracloth) into precooled polycarbonate centrifuge tubes. Chloroplasts are pelleted by centrifugation at 3000× *g* for 5 min (4000 rpm in Beckman JA20 rotor). The supernatant is discarded and chloroplasts are gently suspended in the remaining buffer with a fine brush. The volume is increased to 30 ml, the suspension is transferred to sterile Corex tubes, and chloroplasts are pelleted at 3000× *g* for 5 min. An alternative procedure is to underlay the homogenate with about one-fourth volume of 40% Percoll solution and centrifuge at 3000× *g* for 5 min. Intact plastids pellet while most broken plastids collect at the buffer–Percoll interface (Mishkind *et al.,* 1987). Chloroplasts from this crude fraction can be used for the preparation of stromal and thylakoid fractions. If intact chloroplasts are required, a Percoll gradient centrifugation is desirable.

B. Percoll Gradient Purification of Chloroplasts

Linear self-generating Percoll gradients are prepared before tissue homogenization. In 40-ml polycarbonate tubes, mix 13 ml of Percoll and 13 ml of $2\times$ concentrated grinding buffer. Centrifuge 30 min at $40,000\times g$ (18,000 rpm in a Beckman JA20 rotor) for the gradient to form. Allow the rotor to stop without brake and store in the cold until use. On top of each gradient, layer 2–3 ml of the chloroplast suspension. Centrifuge at $8000\times g$ for 10 min in a swing bucket rotor (7000 rpm in a Beckman JS13 rotor). Allow the rotor to stop without brake. The upper green band is constituted of broken chloroplasts. The lowest green band of intact chloroplasts is gently recovered using a sterile Pasteur pipette and diluted with 3–5 vol of the appropriate buffer containing 0.33 M sorbitol and 50 mM Hepes–KOH, pH 8, for protein import experiments or pH 7.5 for envelope preparation (Robinson and Barnett, 1988). Chloroplasts are pelleted by centrifugation at $3000\times g$ for 5 min, resuspended in a small volume of buffer, and kept on ice until use, in tubes covered with aluminium foil.

C. Assessing the Quality of Chloroplast Preparations

Criteria for the integrity of chloroplasts rely on assays in which the substrate does not penetrate the chloroplast envelope. Several of these assays (such as reduction of ferricyanide and measurement of CO_2 assimilation rates) are described elsewhere (Walker *et al.,* 1987). The quality of the chloroplasts can also be estimated by determining the ratio of intact versus broken chloroplasts by phase-contrast microscopy. For practical purposes, determination of the chlorophyll concentration is convenient, since it can be correlated to the total chloroplast protein concentration and/or to the number of chloroplasts. For that, a small aliquot is diluted with 1 ml of 80% acetone. Insoluble material is removed by centrifugation and the optical density of the supernatant is determined at 652 nm. The concentration in chlorophyll is given by: chlorophyll (mg/ml) = $OD_{652} \times 0.02899$/sample volume. Generally, for intact chloroplast preparations, there are about 10^9 chloroplasts/mg of chlorophyll.

D. Fractionation of Chloroplasts

For specific studies, such as localization of imported proteins, chloroplasts can be fractionated into the constituent fractions, which include the soluble fraction (stroma) and thylakoid membranes. Envelope membranes, which constitute only about 1% of the chloroplast membranes, can be isolated from scaled-up chloroplast preparations (Douce and Joyard, 1982).

To prepare the stromal fraction, chloroplasts are lysed by hypotonic shock, by resuspension in 25 mM Hepes–KOH, pH 8, 4 mM MgCl$_2$, followed by incubation on ice for 10 min. The complete lysis of chloroplasts can be checked by centrifugation of an aliquot through a 1-ml cushion of 40% Percoll, at 3000× g for 5 min. Unbroken chloroplasts will pellet at the bottom of the tube. The chloroplast suspension can be centrifuged at 40,000× g for 20 min (18,000 rpm in a Beckman JA20 rotor) to yield the soluble stromal fraction and a pellet of total chloroplast membranes. To purify the thylakoid fraction, the membrane pellet is resuspended in 25 mM Hepes–KOH, pH 8, 0.3 M sucrose, 4 mM MgCl$_2$ and layered on top of a discontinuous sucrose gradient constituted by equal volumes of 1.2 M sucrose, 1 M sucrose, and 0.6 M sucrose in 25 mM Hepes–KOH, pH 8, 4 mM MgCl$_2$. After centrifugation at 90,000× g for 1 h (27,000 rpm in a Beckman SW40 rotor), the purified thylakoid fraction is found at the bottom of the tube. A fraction of envelope membranes can be recovered at the 0.6–1 M sucrose interface. More easily, thylakoid membranes can be purified by simple washing: after three to four rounds of resuspension and pelleting by centrifugation, most contaminating membranes are removed.

VI. Conclusions

This chapter describes the current methods used in our laboratory for the extraction and purification of mitochondria and chloroplasts. These methods were derived and adapted from the numerous original papers that have been published in the literature as described in the text. Our experience is mainly based on studies on plant mitochondrial genome structure and expression. In preparing the extraction and purification procedure, it is necessary to keep in mind the specific goal that is to be achieved. An extraction procedure will not be exactly the same if one wants to study an enzyme with full activity, a multisubunit complex, or pure nucleic acids. The extraction procedure will have to be adapted for each case. The specific example described in this chapter, the extraction and purification of plant mitochondrial ribosomes, indicates also that one of the most important choices to be made is the determination of the best plant material from which the mitochondria will be obtained. In this case, it is clear that mitochondria obtained from wheat embryos are much less contaminated by cytoplasmic ribosomes than mitochondria obtained from wheat etiolated seedlings. Therefore any new project involving extraction and purification of plant cell organelles should consider the initial plant tissue, its metabolic state (i.e., a green or nongreen tissue), and the specificity of the final molecule to be studied.

References

Araya, A., Domec, C., Bégu, D., and Litvak, S. (1992). An *in vitro* system for the editing of ATP synthase subunit-9 messenger RNA using wheat mitochondrial extracts. *Proc. Natl. Acad. Sci. U.S.A.* **89,** 1040–1044.

Douce, R., Bourguignon, J., Brouquisse, R., and Neuburger, M. (1987). Isolation of plant mitochondria: General principles and criteria of integrity. *Methods Enzymol.* **148,** 403–414.

Douce, R., and Joyard, J. (1982). Purification of the chloroplast envelope. *In* "Methods in Chloroplast Molecular Biology" (M. Edelman, R. B. Hallick, and N. H. Chua, eds.), pp. 239–256. New York: Elsevier Biomedical Press.

Ellis, J. R., and Hartley, M. R. (1982). Preparation of higher plant chloroplasts active in protein and RNA synthesis. *In* "Methods in Chloroplast Molecular Biology" (M. Edelman, R. B. Hallick, and N. H. Chua, eds.), pp. 169–188. New York: Elsevier Biomedical Press.

Fujiki, Y., Fowler, S., Shio, H., Hubbard, A. L., and Lazarow, P. B. (1982). Polypeptide and phospholipid composition of the membrane of rat liver peroxisomes: Comparison with endoplasmic reticulum and mitochondrial membranes. *J. Cell Biol.* **93,** 103–110.

Gray, M. W., Hanic-Joyce, P. J., and Covello, P. S. (1992). Transcription, processing and editing in plant mitochondria. *Annu. Rev. Plant Physiol. Plant Mol. Biol.* **43,** 145–175.

Gruissem, W., and Tonkin, J. C. (1993). Control mechanisms of plastid gene expression. *Crit. Rev. Plant Sci.* **12,** 19–55.

Hanic-Joyce, P. J., and Gray, M. W. (1991). Accurate transcription of a plant mitochondrial gene *in vitro. Mol. Cell. Biol.* **11,** 2035–2039.

Hanic-Joyce, P. J., Spencer, D. F., and Gray, M. W. (1990). *In vitro* processing of transcripts containing novel tRNA-like sequences ("t-elements") encoded by wheat mitochondrial DNA. *Plant Mol. Biol.* **15,** 551–559.

Jackson, C., Dench, J. E., Hall, D. O., and Moore, A. L. (1979). Separation of mitochondria from contaminating subcellular structures utilizing silica sol gradient centrifugation. *Plant Physiol.* **64,** 150–153.

Johnson, F. B., and Stern, H. (1957). Mass isolation of viable wheat embryos. *Nature* **179,** 160–161.

Kellems, R. E., and Butow, R. A. (1974). Cytoplasmic type 80S ribosomes associated with yeast mitochondria. III. Changes in the amount of bound ribosomes in response to changes in metabolic state. *J. Biol. Chem.* **249,** 3304–3310.

Kellems, R. E., Allison, V. F., and Butow, R. A. (1975). Cytoplasmic type 80S ribosomes associated with yeast mitochondria. IV. Evidence for ribosomes binding sites on yeast mitochondria. *J. Cell Biol.* **65,** 1–14.

Leaver, C. J., Hack, E., and Forde, B. G. (1983). Protein synthesis by isolated plant mitochondria. *Methods Enzymol.* **97,** 476–484.

Mannella, C. A. (1987). Isolation of the outer membrane of plant mitochondria. *Methods Enzymol.* **148,** 453–464.

Mishkind, M. L., Greer, K. L., and Schmidt, G. (1987). Cell-free reconstitution of protein transport into chloroplasts. *Methods Enzymol.* **148,** 274–294.

Moller, I. M., Lidén, A. C., Ericson, I., and Gardeström, P. (1987). Isolation of submitochondrial particles with different polarities. *Methods Enzymol.* **148,** 442–453.

Neuburger, M., Journet, E. P., Bligny, R., Carde, J. P., and Douce, R. (1982). Purification of plant mitochondria by isopycnic centrifugation in density gradients of Percoll. *Arch. Biochem. Biophys.* **217,** 312–323.

Nishimura, M., Douce, R., and Akazawa, T. (1982). Isolation and characterization of metabolically competent mitochondria from spinach leaf protoplasts. *Plant Physiol.* **69,** 916–920.

Peiffer, W. E., Ingle, R. T., and Ferguson-Miller, S. (1990). Structurally unique plant cytochrome c oxidase isolated from wheat germ, a rich source of plant mitochondrial enzymes. *Biochem.* **29,** 8696–8701.

Perry, S. E., Li, H. M., and Keegstra, K. (1991). *In vitro* reconstitution of protein transport into chloroplasts. *Methods Cell Biol.* **34,** 327–344.

Robinson, C., and Barnett, L. (1988). Isolation and analysis of chloroplasts. *In* "Plant Molecular Biology: A Practical Approach" (C. H. Shaw, ed.), pp. 67–78. Oxford/Washington, DC: IRL Press.

Schuster, W., Hiesel, R., Wissinger, B., Schobel, W., and Brennicke, A. (1988). Isolation and analysis of plant mitochondria and their genomes. *In* "Plant Molecular Biology: A Practical Approach" (C. H. Shaw, ed.), pp. 79–102. Oxford/Washington, DC: IRL Press.

Walker, J. D. A., Cerovic, Z. G., and Robinson, S. P. (1987). Isolation of intact chloroplasts: General principles and criteria of integrity. *Methods Enzymol.* **148,** 145–156.

Zhuo, D. G., and Bonen, L. (1993). Characterization of the S7 ribosomal protein gene in wheat mitochondria. *Mol. Gen. Genet.* **236,** 395–401.

CHAPTER 13

Isolating the Plant Mitotic Apparatus: A Procedure for Isolating Spindles from the Diatom *Cylindrotheca fusiformis*

Harrison Wein, Barbara Brady, and W. Zacheus Cande

Department of Molecular and Cell Biology
University of California
Berkeley, California 94720

I. Introduction

Mitotic spindles isolated from diverse species have reproduced anaphase-like movements *in vitro* (Masuda *et al.,* 1990; Rebhun and Palazzo, 1988; Palazzo *et al.,* 1991; Sakai *et al.,* 1976), but among plants, the only mitotic spindles that have demonstrated anaphase movement outside the cell have been prepared from the diatoms *Stephanopyxis turris* (Cande and McDonald, 1985, 1986; McDonald *et al.,* 1986; Baskin and Cande, 1988) and *Cylindrotheca fusiformis* (Hogan *et al.,* 1993a). Elongation of these isolated diatom spindles closely approximates *in vivo* anaphase B movement. Mitotic spindles have also been isolated from the tobacco BY-2 cell culture line, but to our knowledge these spindles are not functional for anaphase A (chromosome-to-pole movement) or anaphase B (spindle elongation) *in vitro* (Yasuhara *et al.,* 1992; Kakimoto and Shiboaka, 1988; Nagata *et al.,* 1992).

The chief advantage of using diatoms for spindle isolation is the structure of their spindles. In diatoms, the chromosomes and kinetochore microtubules involved in anaphase A are spatially distinct from the central spindle, a highly ordered paracrystalline array of microtubules that is responsible for spindle elongation (McDonald *et al.,* 1977, 1979; McIntosh *et al.,* 1979; McDonald and Cande, 1989; Pickett-Heaps and Tippet, 1978). The microtubules of each half-spindle are approximately uniform in length and interdigitate to form a well-defined zone of antiparallel microtubule overlap. The high level of organization in *S. turris* spindles was exploited to demonstrate that the force driving spindle elongation during anaphase B is generated by mechanochemical enzymes in the zone of microtubule overlap (Cande and McDonald, 1985, 1986).

Although *S. turris* provided useful physiological and structural information, the cells could not be grown to a density sufficient for the bulk isolation of spindles. To obtain the large number of spindles necessary for biochemical analysis, Hogan *et al.* (1992, 1993a) developed an isolated spindle model using cells of the pennate marine diatom *C. fusiformis.* Cells of this smaller diatom can be grown to a density over three orders of magnitude higher than that of *S. turris* and can be synchronized so that up to 75% of the population is in metaphase or early anaphase. *C. fusiformis* spindles are more robust than those from *S. turris* or from most other organisms; even after being subjected to shearing or high centrifugal forces, these spindles can still be reactivated for anaphase B movement with the simple addition of ATP. Hogan *et al.* (1992, 1993b) used isolated spindles and a permeabilized cell model from *C. fusiformis* to obtain evidence suggesting that a kinesin-related protein in the diatom central spindle generates the force that drives anaphase spindle elongation.

Recently, our lab has made several changes to improve the *C. fusiformis* spindle isolation procedure. The new procedure is more consistent and reliable than that published previously (Hogan *et al.,* 1992, 1993a). In addition, the spindle yield has been increased without loss of purity. Most significantly, however, this procedure opens the way for further purification steps.

II. Methods

A. Spindle Isolation

Culturing of cells, synchronization, and collection were performed as described by Hogan *et al.* (1993a) with slight modifications. *C. fusiformis* cultures (provided by Dr. Ben E. Volcani, Scripps Institute for Oceanography, La Jolla, CA) were maintained at 22°C in natural sea water supplemented with F/2 medium (Guillard, 1975). To synchronize cells, several (typically 12) 8-liter polycarbonate carboys were filled with 7 liters of sterilized sea water, supplemented with F/2, and inoculated to a density of 2.5×10^4 cells/ml. Cells were grown under wide spectrum fluorescent lights (GE F40PL/AQ, General Electric) with bubbled 3% CO_2 and vigorous stirring for approximately 36 h. Lights were extinguished for 24 h and turned on again to obtain a population of synchronously dividing cells. Seven hours after the initiation of the second light period, with cell density at approximately 5×10^5 cells/ml, nocodozole was added to 0.1 μM. At this low concentration of nocodozole, interphase microtubule arrays disappear but metaphase central spindles still form and existing spindles do not depolymerize. Cells were harvested 3 to 4 h after drug treatment with a tangential flow filtration system containing 0.45-μm Durapore filters (Millipore Corp., Bedford, MA).

Figure 1 is a flow chart outlining the spindle isolation procedure. Synchronized cells were concentrated to 1 liter by tangential flow filtration, then pelleted at $2500 \times g$ for 5 min (Sorvall H-1000B rotor, Sorvall RT6000B tabletop centrifuge,

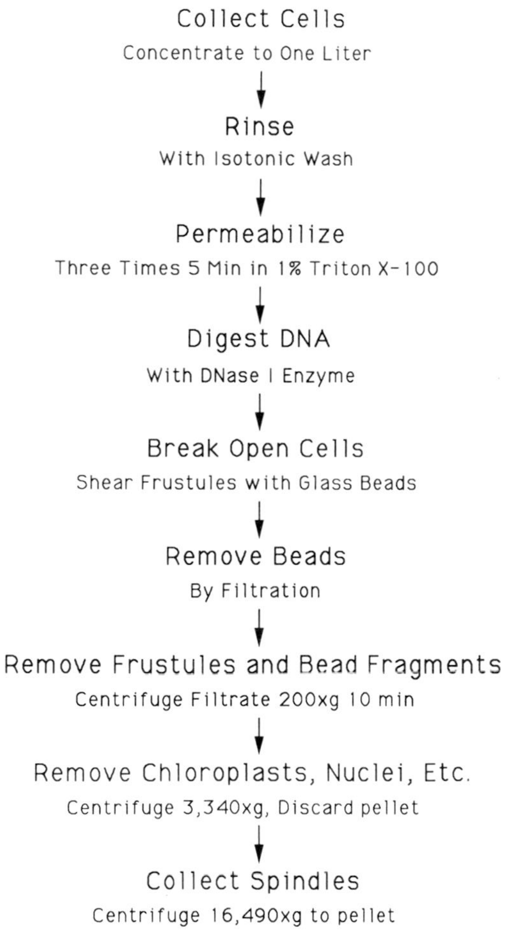

Collect Cells
Concentrate to One Liter

↓

Rinse
With Isotonic Wash

↓

Permeabilize
Three Times 5 Min in 1% Triton X-100

↓

Digest DNA
With DNase I Enzyme

↓

Break Open Cells
Shear Frustules with Glass Beads

↓

Remove Beads
By Filtration

↓

Remove Frustules and Bead Fragments
Centrifuge Filtrate 200xg 10 min

↓

Remove Chloroplasts, Nuclei, Etc.
Centrifuge 3,340xg, Discard pellet

↓

Collect Spindles
Centrifuge 16,490xg to pellet

Fig. 1   General outline of the *C. fusiformis* mitotic spindle isolation procedure.

Sorvall, DuPont, Wilmington, DE) and rinsed in 1 liter of an isotonic buffer [100 mM NaCl, 5 mM KCl, 0.1 M tris(hydroxylmethyl)aminomethane (Tris), 0.02 M ethylene glycol-bis(b-aminoethyl ether)N,N,N',N'-tetraacetic acid (EGTA), 10 mM trans-1,2-diaminocyclohexane-N,N,N',N'-tetraacetic acid (CDTA), pH 6] to lower the divalent cation concentration surrounding the cells. From this point on, samples were kept at or below 4°C. Cells were pelleted and resuspended in 240 ml of cold permeabilization buffer [50 mM 1,4-piperazinediethane sulfonic acid (Pipes), 5 mM MgSO$_4$, 5 mM EGTA, 40 mM β-glycerophosphate, 1 μM rac-6-hydroxy-2,5,7,8-tetramethylchromane-2-carboxylic acid (TROLOX), 1% Triton X-100, 3% dimethyl sulfonic acid (DMSO), 1 mM dithiothreitol (DTT), 1 mM phenylmethylsulfonyl fluoride (PMSF), and a proteinase inhibitor cocktail {1 μg/ml leupeptin, 1 μg/ml pepstatin A, 10 μg/ml Nα-benzoyl-L-arginine methyl ester (BAME), 10 μg/ml Na-p-tosyl-L-arginine methyl ester (TAME), 10 μg/ml N-tosyl-L-phenylalanine chloromethyl ketone (TPCK), 10 μg/ml soybean trypsin inhibitor, 1 μg/ml aprotinin (Masuda et al., 1988)}, pH 7] to solubilize lipids and proteins. After rocking for 5 min, the cells were pelleted and resuspended once more in permeabilization buffer. They were subjected to a total of three 5-min permeabilizations. Thus, including the spins to pellet them, the cells remained in the permeabilization buffer for approximately 30 min.

After the last permeabilization, cells were pelleted and resuspended in 450 ml of DNase buffer [50 mM Pipes, pH 7, 2.5 mM MgCl$_2$, 0.5 mM MnCl$_2$, 40 mM β-glycerophosphate, 1 μM TROLOX, 0.02% Triton X-100, 3% DMSO, 1 mM DTT, 1 mM PMSF, proteinase inhibitor cocktail] with 80 units DNase I enzyme (Boehringer Mannheim, Indianapolis, IN) per milliliter of buffer. The suspension was rocked for 20 min and pelleted once more.

Cells were resuspended in a small volume of bead beat (BB) buffer [50 mM Pipes, 5 mM MgSO$_4$, 10 mM EGTA, 40 mM β-glycerophosphate, 1 μM TROLOX, 0.02% Triton X-100, 3% DMSO, 1 mM DTT, 1 mM PMSF, proteinase inhibitor cocktail, pH 7]. An 85-ml bead beat chamber (Biospec Products, Bartlesville, OK) was prepared by filling one-third its volume with 0.1-mm glass beads moistened with fresh BB buffer. As cell breakage is more effective in a dilute suspension, half the cells were transferred to the chamber and the chamber was filled with fresh BB buffer. The diatom shells (frustules) were broken by homogenization on ice with three 20-s bursts and 20-s rests between each burst.

The other half of the cells were subjected to the same treatment and, after both fractions were bead-beaten, the homogenate was filtered through a 30-μm Nitex filter (Tetco, Inc., Elmsford, NY) to remove glass beads and debris. The filtrate's volume was then raised to 310 ml with BB buffer. Increasing the volume at this point is crucial; the proper fractionation of spindles and contaminants by centrifugation depends on a dilute suspension.

The diluted filtrate was transferred to seven 50-ml centrifuge tubes and centrifuged for 10 min at 200×g to remove the diatom frustules and remaining glass beads. The pellet resulting from this spin was quite loose, so the supernatant was removed very carefully with a minimum of jostling using a 25-ml pipette.

The supernatant was measured and divided equally into ten 40 ml-polycarbonate tubes. The sample in each tube was underlain with 3 ml of a 40% sucrose solution in PMEG [50 mM Pipes, pH 7, 5 mM MgSO$_4$, 5 mM EGTA, 40 mM β-glycerophosphate, 1 μM TROLOX, 1 mM DTT, 1 mM PMSF, proteinase inhibitor cocktail] and centrifuged for 15 min at 3340×g in a swinging bucket rotor (Beckman JS13.1 or Sorvall HB-4).

The supernatant, including sucrose cushion, was retained. The discarded pellet, rich in nuclei and chloroplasts, appeared green and was quite firm. The supernatant was transferred to 10 fresh 40-ml polycarbonate tubes and underlain once more with 3 ml of the 40% sucrose cushion. The samples were centrifuged at 16,490×g for 15 min to pellet spindles. Finally, the supernatant was discarded.

The final spindle pellets appeared yellowish brown. If they were still green, the pellets were resuspended and combined into a total volume of 30 ml of permeabilization buffer, then centrifuged as in the previous step through a 3-ml 40% sucrose cushion.

Whether or not the last rinse was performed, the final pellet(s) was resuspended in a small volume of BB buffer (typically about 6 ml, but volume depends on the intended use) and frozen in liquid N$_2$ for long-term storage. Isolated spindles remain functional for at least 1 year when stored this way.

B. Determination of Spindle Purity

An estimate of spindle purity was obtained using a visual assay. Samples of permeabilized cells and all spindle fractions were centrifuged (2500×g for 5 min or 10,000×g for 10 min, respectively) through 3 ml of BB buffer onto poly-L-lysine-coated coverslips and fixed for 10 min in 0.1% glutaraldehyde, 0.05% paraformaldehyde in PME [50 mM Pipes, 5 mM MgSO$_4$, 5 mM EGTA]. Spindles were visualized with a monoclonal anti-sea urchin α-tubulin (provided by Dr. David Asai, Purdue University) and FITC-conjugated secondary antibody (Sigma Chemical Co., St. Louis, MO). The number of spindles in five 400× fields was counted and averaged. Protein concentrations (fractions were prepared as described in Section C) were measured with the BCA protein assay (Pierce, Rockford, IL). Finally, the relative spindle purity of each fraction (Table I) was estimated by dividing the average number of spindles in a 400× field by the total amount of protein in the sample used to make the coverslip.

C. Preparation of Protein Fractions for Analysis

Protein fractions were prepared with the goal of obtaining the highest possible yield of total protein. Whole cells and permeabilized cells (A,B in Table I and Fig. 3) were pelleted (2500×g for 5 min), resuspended in PBS + 0.5% SDS, vortexed vigorously with an equal volume of 0.1-mm glass beads, and centrifuged briefly to remove the beads. The spindle fractions (C–F) were pelleted (16,000×g for 15 min) and resuspended in PBS + 0.5% SDS. All samples were boiled for

Table I
Analysis of Spindle Purity

	Stage in isolation	No. spindles/total protein (sp/mg)	Improvement over previous step	Improvement over whole cell extract
A	Whole cell	3,200	—	—
B	Permeabilized	7,100	2.2×	2.2×
C	DNased, bead beat	28,900	4.1×	9.0×
D	After 200× g spin	21,700	.75×	6.8×
E	After 3340× g spin	26,700	1.2×	8.3×
F	After collection	40,100	1.5×	12.5×

Note. The average number of spindles in one 400× field was divided by the amount of protein loaded onto the corresponding coverslip as described in Section II,B. The final fraction is enriched more than 12-fold for mitotic spindles over the starting material.

5 min and pulsed briefly in a microfuge to clear the solutions; the supernatants were then removed for further analysis. To obtain a purer preparation of spindle proteins (G), an isolated spindle fraction was pelleted and treated with PMEG + 0.2 M KI for 20 min to depolymerize microtubules and extract spindle proteins. The solution was cleared of nonsoluble components by centrifugation ($40,000 \times g$ for 20 min) and dialyzed in three changes of PBS for 1 h each.

III. Results and Discussion

Figure 2 shows images of permeabilized cells (a,b,c), freshly broken cells (d,e,f), and the final isolated spindles (g,h,i). One indication of contamination is the presence of a strong bright red background caused by photosynthetic pigments. Black-and-white reproductions cannot capture the dramatic decrease in this contamination between the broken cell fraction and the final isolated spindles. Nonetheless, both the elimination of diatom frustules and the reduction of other contaminating materials are obvious in these images, as is the virtually complete digestion of DNA. The main contaminant in the final fraction is an uncharacterized flocculant material, and we are currently experimenting with methods of eliminating it. The spindles in the final fraction appear thicker and more brightly stained by anti-tubulin antibodies than those prepared using the previous spindle isolation procedure (Hogan *et al.,* 1992, 1993a). This enhanced staining suggests that spindle integrity is better preserved with the new procedure.

Determining the biochemical purity of a mitotic spindle preparation can be difficult. The most abundant spindle protein, tubulin, is not present exclusively in the spindle. Tubulin exists both in a large soluble pool and polymerized into non-spindle-associated cytoplasmic microtubules. Thus, quantitative assays based

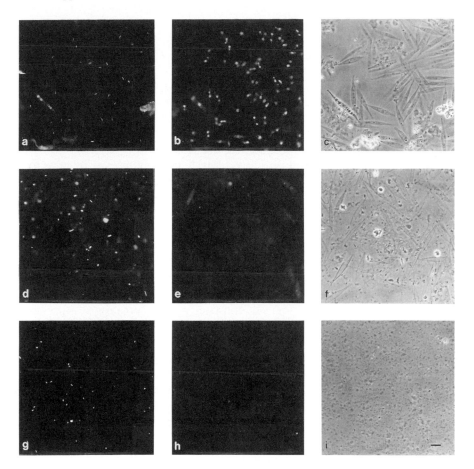

Fig. 2 Micrographs of samples from three stages in the spindle isolation procedure show a clear reduction in contamination. Permeabilized cells (a,b,c), freshly broken cells (d,e,f), and the final isolated spindles (g,h,i) are shown. Anti-α-tubulin immunofluorescence (a,d,g), DAPI DNA staining (b,e,h), and phase contrast (c,f,i) images are included. Bar, 16 μm.

on antibody recognition of tubulin (such as ELISAs) can not accurately determine the degree of spindle protein purity. To quantitate the degree of spindle purity in our preparation, we compared the number of spindles at a particular point in the preparation to the amount of protein present at that step (Table I). This method of determining spindle purity relies on a visual assay rather than on an antibody that reacts with both spindle and non-spindle tubulin.

Many spindles are lost in the two centrifugation steps following cell breakage. However, these steps are necessary because they rid the suspension of frustules, chloroplasts, nuclei, and other contaminants. The level of spindle purity gained by centrifugation easily compensates for the loss in yield. Our estimates are that, at the end of the spindle isolation procedure, about 30% of the spindles are retained and that the final fraction is enriched 12-fold for spindle proteins.

Figure 3 shows a Coomassie blue-stained gel and a tubulin Western blot of samples taken at each step in the spindle isolation procedure; a salt-extracted spindle fraction is also included. An equal amount of protein (5 μg) was loaded in each lane and then subjected to 10% SDS–PAGE. Since a subunit of ribulose-1,5-bisphosphate carboxylase, an abundant photosynthetic protein, and tubulin migrate to approximately the same point by electrophoresis, tubulin levels must be assessed by Western blots. The tubulin level relative to the total amount of protein drops during permeabilization (lane A to B) when the pool of soluble tubulin is lost. In subsequent steps, the tubulin level increases or remains constant relative to total protein (lanes B to F).

A KI salt extraction (lane G) enriches for spindle proteins and yields a different protein profile than the other fractions. Assuming that all spindle proteins in the final fraction are extracted by this salt treatment (no spindles are visible in the pellet), this extract represents approximately a 70-fold enrichment in spindle-associated proteins over the starting material. Starting with 94 liters of cells, the final yield of this salt-extracted spindle protein is approximately 0.5 mg.

IV. Critical Aspects of the Procedure

One of the main difficulties in obtaining a good spindle preparation is synchronizing a population of cells so that a large number of dividing cells can be collected. Our strategy, which employs a light–dark cycle coupled with the micro-

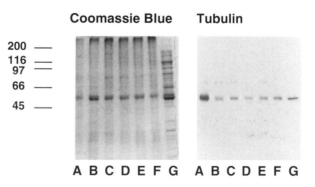

Fig. 3 The final isolated spindle fraction shows an enrichment in spindle associated tubulin. Five micrograms of protein was loaded in each lane and samples were subjected to 10% SDS–PAGE. Fraction letters A–F are as assigned in Table I, whereas sample G is a salt-extracted isolated spindle fraction (described in Section II,C). The gel on the left has been stained with Coomassie blue, while the immunoblot on the right has been probed with an anti-α-tubulin antibody and developed by ECL (Amersham Life Science, Arlington Heights, IL).

tubule inhibiting drug nocodozole, is effective in synchronizing diatoms. With *C. fusiformis* cultures, we typically obtain a population with 65–75% of the cells in metaphase or early anaphase. In tobacco BY-2 cell culture lines, an alternative strategy involving the use of aphidicolin, a DNA α-polymerase inhibitor, and propyzamide, a microtubule-inhibiting drug, yields an 80–90% mitotic index (Kakimoto and Shiboaka, 1988; Nagata *et al.*, 1992). No other plant cell culture line has demonstrated this cell line's high levels of synchrony; indeed, few preparations of any cell type can achieve such high synchrony.

Throughout the diatom spindle isolation procedure, the phosphorylated state of the spindle must be maintained in order to preserve the capacity for *in vitro* spindle elongation. This was demonstrated for spindles prepared from both the large centric diatom *S. turris* (Wordeman and Cande, 1987) and the pennate diatom *C. fusiformis* (Hogan *et al.*, 1993a). All buffers, therefore, contain the competitive phosphatase inhibitor β-glycerophosphate.

Early in the procedure, unbroken cells were pelleted and resuspended three times in fresh permeabilization buffer in order to remove detergent-soluble material. In the previous *C. fusiformis* spindle isolation procedure (Hogan *et al.*, 1992, 1993a), this was accomplished by subjecting isolated spindles to Minitan tangential flow filtration (Millipore Corp., Bedford, MA) in a detergent buffer after cell breakage and centrifugation. Spindles were often disrupted by the long tangential flow filtration step, as they were constantly in motion and under pressure. The new technique removes soluble components faster and with less agitation.

Another new step was added to the procedure before cell breakage: the digestion of DNA with DNase I enzyme. In addition to reducing chromatin-associated contaminants in the final fraction, this step had two unforeseen benefits. First, DNA digestion allowed for more consistent results in later centrifugation steps. In the old preparation, DNA was disrupted during bead-beating, leaving spindles with random fragments of attached chromatin (confirmed by electron microscopy, data not shown). Because of the varied amounts of associated DNA, spindles in the old procedure were heterogeneous with respect to their centrifugation properties. When a DNA digestion step was incorporated into the procedure before cell breakage, isolated spindles in later steps behaved with more consistency. Second, an early DNA digestion step made the final spindle pellets easier to resuspend. Pellets that contained isolated spindles from the old preparation were extremely difficult to resuspend without disrupting spindle integrity; they could be resuspended only with vigorous shearing through a syringe. In the old preparation, Minitan tangential flow filtration was used not only to remove detergent-soluble components, but also because the technique allowed for the collection of spindles without pelleting. We found that pellets containing spindles from which the chromatin had been digested were considerably easier to resuspend.

Freeing the spindle of its surrounding cell is a particularly difficult step in the isolation of any mitotic apparatus. To break the silicon-based shells of *C. fusi-*

formis, shearing forces were generated using glass beads and rapid agitation. The more fragile shells of *S. turris* were broken with several strokes of a Dounce homogenizer. On the other hand, cellulose-based cell walls such as those that surround BY-2 cells must be enzymatically digested to free the mitotic spindles within.

Finally, 40% sucrose cushions are used in two different centrifugation steps. In the first, the density of the sucrose cushions helps keep spindles in suspension during a low-speed spin that removes contaminants. Spindles are then pelleted at a higher speed through a second sucrose cushion for separation from lipids and soluble proteins in the supernatant, and also for collection. As does the earlier DNA digestion step, requiring spindles to travel through a 40% sucrose cushion eases the eventual resuspension of the spindle pellet. DNA digestion and the use of 40% sucrose cushions have allowed for the complete elimination of Minitan tangential flow filtration from the procedure. Spindles can now be collected without a significant disruption of spindle integrity, opening the way for further purification steps.

V. Conclusions and Perspectives

Genetic systems have provided good evidence linking particular proteins to mitotic spindle functions, but mutant phenotypes in these organisms are often difficult to analyze biochemically and cytologically. Biochemical studies are required to confirm genetic predictions and reveal the specific functions of individual proteins. Genetic approaches also may not identify all proteins crucial for spindle function because the genes encoding for some of these proteins may be essential for viability. Furthermore, certain genes may be functionally redundant and thus difficult to isolate using genetic screens. For these reasons, a biochemical approach is still necessary to decipher the molecular events behind mitotic spindle movements.

The diatom *C. fusiformis* provides both a relatively pure isolated mitotic spindle preparation and a reliable *in vitro* assay for anaphase B function. Biochemical preparations will help identify previously undiscovered spindle proteins, whereas the *in vitro* functional assay will allow for the identification of proteins involved in spindle elongation. We now use this improved procedure routinely in our laboratory. Spindle yield is quite high, and the spindles are intact as judged by immunofluorescence. We are currently experimenting with a variety of biochemical methods to further purify mitotic spindle proteins.

References

Baskin, T. I., and Cande, W. Z. (1988). Direct observation of mitotic spindle elongation *in vitro*. *Cell Motil. Cytoskeleton* **10**, 210–216.

Cande, W. Z., and McDonald, K. L. (1985). *In vitro* reactivation of anaphase spindle elongation using isolated diatom spindles. *Nature* **316:6024,** 168–170.

Cande, W. Z., and McDonald, K. L. (1986). Physiological and ultrastructural analysis of elongating mitotic spindles reactivated *in vitro*. *J. Cell Biol.* **103**, 593–604.

Guillard, R. R. L. (1975). Culture of phytoplankton for feeding marine invertebrates. *In* "Culture of Marine Invertebrate Animals" (W. L. Smith and M. H. Chanley, eds.), pp. 29–60. New York: Plenum.

Hogan, C. J., *et al.* (1992). Physiological evidence for the involvement of a kinesin-related protein during anaphase spindle elongation in diatom central spindles. *J. Cell Biol.* **119:5**, 1277–1286.

Hogan, C. J., *et al.* (1993a). The diatom central spindle as a model system for studying antiparallel microtubule interactions during spindle elongation in vitro. *In* "Motility Assays for Motor Proteins" (J. M. Scholey, ed.), Methods in Cell Biology, Vol. 39, pp. 277–292. San Diego: Academic Press.

Hogan, C. J., *et al.* (1993b). Inhibition of anaphase spindle elongation *in vitro* by a peptide antibody that recognizes kinesin motor domain. *Proc. Natl. Acad. Sci. U.S.A.* **90**, 6611–6615.

Kakimoto, T., and Shiboaka, H. (1988). Cytoskeletal ultrastructure of phragmoplast-nuclei complexes isolated from cultured tobacco cells. *Protoplasma Suppl.* **2**, 95–103.

Masuda, H., *et al.* (1988). The mechanism of anaphase spindle elongation: Uncoupling of tubulin incorporation and microtubule sliding during *in vitro* spindle reactivation. *J. Cell Biol.* **107**, 623–633.

Masuda, H., *et al.* (1990). In vitro reactivation of spindle elongation in fission yeast *nuc2* mutant cells. *J. Cell Biol.* **110**, 417–425.

McDonald, K. L., *et al.* (1977). On the mechanism of anaphase spindle elongation in *Diatoma vulgare*. *J. Cell Biol.* **74**, 377–388.

McDonald, K. L., *et al.* (1979). Cross-sectional structure of the central mitotic spindle of *Diatoma vulgare*. Evidence for specific interactions between antiparallel microtubules. *J. Cell Biol.* **83**, 443–461.

McDonald, K. L., *et al.* (1986). Comparison of spindle elongation *in vivo* and *in vitro* in *Stephanopyxis Turris*. *J. Cell Sci. Suppl.* **5**, 205–227.

McDonald, K. L., and Cande, W. Z. (1989). Diatoms and the mechanism of anaphase spindle elongation. *In* "Algae as Experimental Systems" (A. W. Coleman, L. J. Goff, J. R. Stein-Taylor, eds.), pp. 3–18. New York: Alan R. Liss, Inc.

McIntosh, J. R., *et al.* (1979). Three-dimensional structure of the central mitotic spindle of *Diatoma vulgare*. *J. Cell Biol.* **83**, 428–442.

Nagata, T., *et al.* (1992). Tobacco BY-2 cell line as the "HeLa" cell in the cell biology of higher plants. *Int. Rev. Cytol.* **132**, 1–30.

Palazzo, R. E., *et al.* (1991). Reactivation of isolated mitotic apparatus: Metaphase versus anaphase spindles. *Cell Motil. Cytoskeleton* **18**, 304–318.

Pickett-Heaps, J. D., and Tippit, D. H. (1978). The diatom spindle in perspective. *Cell* **14**, 455–467.

Rebhun, L. I., and Palazzo, R. E. (1988). *In vitro* reactivation of anaphase B in isolated spindles of the sea urchin egg. *Cell Motil. Cytoskeleton* **10**, 197–209.

Sakai, H., *et al.* (1976). Induction of chromosome motion in the glycerol-isolated mitotic apparatus: Nucleotide specificity and effects of anti-dynein and myosin sera on the motion. *Dev. Growth Differ.* **18:3**, 211–219.

Wordeman, L., and Cande, W. Z. (1987). Reactivation of spindle elongation *in vitro* is correlated with the phosphorylation of a 205 kd spindle-associated protein. *Cell* **50**, 535–543.

Yasuhara, H., *et al.* (1992). ATP-sensitive binding to microtubules of polypeptides extracted from isolated phragmoplasts of tobacco BY-2 cells. *Plant Cell Physiol.* **33:5**, 601–608.

CHAPTER 14

Chromoplasts

C. A. Price, Noureddine Hadjeb, Lee A. Newman, and Ellen M. Reardon

Waksman Institute
Rutgers University
Piscataway, New Jersey 08855-0759

I. Introduction

What is a chromoplast? Plastids have many differentiation states, including proplastids, chloroplasts, amyloplasts, etioplasts, gerontoplasts, and chromoplasts. These various forms of the organelle play very different roles and are mostly interconvertible. An exception are gerontoplasts, the terminal state of chloroplasts in senescent leaves. Chromoplasts are plastids in which exceptional amounts of carotenoids accumulate (Sitte *et al.,* 1980). As a consequence, chromoplasts confer bright colors to plant tissues: shades of orange, yellow, and red.

Chromoplasts, therefore, are the quintessential plastids of flowers and fruits, attracting insects for pollination and animals for seed dispersal.

Origin of chromoplasts: Chromoplasts may arise from chloroplasts, proplastids, or leucoplasts. Immature fruits of wild-type tomato (*Lycopersicon esculentum*) and pepper (*Capscium annuum*), for example, are typically green with normal-appearing chloroplasts. As they ripen, the chlorophyll disappears while carotenoids accumulate. In fruits of some mutants (see II,C), however, the immature fruits are white and become red during ripening. In other mutants the carotenoids accumulate normally, but the chlorophyll remains; the ripe fruits are brown. The essential character of chromoplasts, the accumulation of carotenoids, is independent of the presence or absence of chlorophyll.

Function of chromoplasts: The function of chromoplasts is esthetic, making plant organs attractive to animals such as ourselves. They are not, therefore, essential to the life of the plant.

In contrast to chloroplasts, which contain a limited set of carotenoids that vary little across the plant kingdom, chromoplasts have been the sites of widespread experimentation by nature in the varieties as well as the amounts of carotenoids. The carotenoids that accumulate to such high levels in chromoplasts may be normal intermediates of carotenoid biosynthesis, such as lycopene in tomato (*L. esculentum*) or β-carotene in carrot (*Daucus carota*). They may also be genus-specific derivatives of the standard array of chloroplast carotenoids (Goodwin, 1980), such as the keto-carotenoids in red fruits of *C. annuum* (bell pepper) or escholtzxanthin, a 3,3′-dihydroxy*retro*-β-carotene in the flowers of *Escholtzia californica* (California poppy).

As we note below, plastids are the exclusive sites of carotenoid biosynthesis in plants (see II,A).

II. Characterization of Chromoplasts

A. Activities of Chromoplasts

Carotenoid biosynthesis: Carotenoid biosynthesis originates as a branch of the isoprenoid pathway (Britton, 1993). The very low activity in plastids of enzymes leading to the synthesis of isopentenyldiphosphate has led to the view that the early steps in isoprenoid synthesis occur in the cytoplasm (Camara *et al.*, 1983; Kreuz and Kleinig, 1984). Starting from isopentenyldiphosphate, however, plastids appear to be the sole sites of isopentenyl pyrophosphate isomerase, geranylgeranyl pyrophosphate synthase, and phytoene synthase (Camara *et al*, 1989). Intact plastids from *N. pseudonarcissus* convert exogenous isopentenyl diphosphate to α- and β-carotene without the accumulation of intermediates (Beyer, 1989). Activities of isopentenyl pyrophosphate isomerase, geranylgeranyl pyrophosphate synthase, and phytoene synthase, desaturases, and cyclization reactions can be demonstrated in plastid extracts (Beyer and Kleinig, 1989; Camara *et al.*, 1989).

While chromoplasts are a rich source of these enzymes, qualitatively identical reactions almost certainly occur in chloroplasts.

Protein synthesis in chromoplasts: Carde *et al.* (1989) reported the absence of ribosomes in electron photomicrographs of chromoplasts in fruits of *C. annuum.* Hadjeb (Hadjeb, 1993), however, found vigorous incorporation of labeled amino acids into protein *in organello.* In contrast, protein synthesis by chloroplasts isolated from immature fruits proceeded much less rapidly.

Import of cytoplasmically synthesized proteins: The fragility of tomato chromoplasts prevented Lawrence *et al.,* from direct measurements of the uptake of cytoplasmically synthesized proteins. They were, however, able to demonstrate the transport into pea chloroplasts of several proteins from tomato that increase during fruit ripening. Together with the identification of nuclear-encoded proteins that are specific to chromoplasts (below), these data make it almost certain that chromoplasts import cytoplasmically synthesized proteins by mechanisms similar to those found in chloroplasts.

B. Molecules Specific to Chromoplasts

Small molecules: Chromoplast carotenoids have provided a rich substrate for organic chemistry with many chemical species serving as botanical species markers (Goodwin, 1980; Goodwin, 1988; Karrer and Jucker, 1948).

Isolation of ChrA and ChrB: We analyzed preparations of pure chromoplasts and chloroplasts from fruits of *C. annuum* (see Section III) by SDS gel electrophoresis for proteins that might be specific for chromoplasts (Hadjeb *et al.,* 1988). We found ChrA, an integral membrane protein of 58 kDa, which accounted for half of the total protein of the plastids, and ChrB, a peripheral membrane protein of 35 kDa, which accounted for 10%.

Distribution of ChrA and ChrB in *C. annuum:* Using antisera raised against ChrA we determined that the protein occurs only in tissues of *C. annuum* that contain chromoplasts (Newman, 1993; Newman *et al.,* 1989), and is particularly rich in fruits of cultivars selected for paprika production (Covello and Gray, 1993). The occurrence of ChrA correlates exclusively with the y^+ genotype (Oren-Shamir *et al.,* 1993) (Fig. 1). This gene corresponds in turn to the occurrence of the keto-carotenoids, capsanthin and capsorubin, and to the phenotype of red and orange fruits (Table I). We note that these keto-carotenoids are peculiar to *Capsicum* spp. and that no immunological signals for ChrA have been detected in species other than *Capsicum.*

As in the case of ChrA, antisera to ChrB detect this protein only in chromoplast-containing tissues, but ChrB occurs in pepper chromoplasts of all genotypes (Oren-Shamir *et al.,* 1993).

A family of ChrB proteins: Proteins from carrot root and from flowers of cucumber (*Cucumis sativus*), daffodil (*Narcissus psuedonarcissus*), and *Nicotiana alata* have similar apparent molecular weights and cross-react with antisera to ChrB (Newman et al., in preparation). No signals were detected in leaf tissue

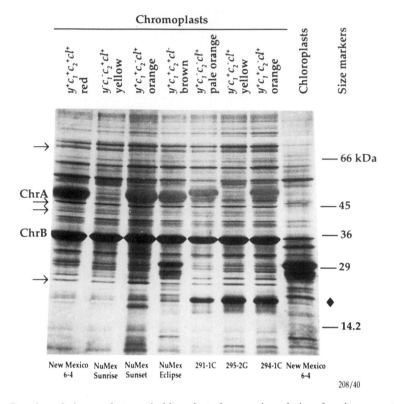

Fig. 1 Proteins of chromoplasts and chloroplasts from various fruits of various genotypes of *Capsicum annuum* (from Oren-Shamir *et al.*, 1993). Proteins from isolated plastids were separated by SDS–gel electrophoresis. The cultivars (bottom labels) represented genotypes (top labels) that differ in y, c_1, and c_2, the three genetic loci known to affect carotenoid accumulation in mature fruits (cf. Table I). Note that the occurrence of ChrA correlates exclusively with the genotype y^+, whereas ChrB occurs in chromoplasts of all genotypes.

from any of these plants. Negative responses were obtained with proteins from fruits of tomato and from petals from any of five varieties of rose. [Note: Although we detected no immunological cross-reaction in tomato, Bathgate *et al.* (1985) reported the appearance of a protein of molecular weight similar to that of ChrB during ripening.]

The case of cucurbit flowers is particularly interesting. Antiserum to ChrB raised against the *Capsicum* protein detected a protein of similar molecular weight in cucumber, but not in petals of squash (*Cucurbita pepo*). Antiserum raised against the cucumber protein, however, detected the protein strongly in squash and weakly in *Capsicum* (Vainstein *et al.*, 1994). The differences between the pepper and cucumber proteins led the Vainstein group to name the cucumber protein ChrC.

Nasturtium majus is another interesting case: the petals (but not leaves) of red and orange varieties contained proteins of high molecular weight that cross-

Table I
Genes Affecting Chromoplast Development in Pepper (*Capsicum annuum*)[a]

y	c_1	c_2	cl	Fruit color
+	+	+	−	Red
+	−	+	−	Light red
+	−	−	−	Orange
−	+	+	−	Orange-yellow
−	+	−	−	Pale orange-yellow
−	−	+	−	Lemon-yellow
−	−	−	−	White
+	+	+	+	Brown
−	+	+	+	Olive-green

[a] From Smith (1950) and Hurtado-Hernandez and Smith (1985).

Note. The genes listed here control pigmentation in the mature chromoplast. A separate series, G_1, G_2, determine whether the immature fruit contains leucoplasts or chloroplasts (Odland and Porter, 1938).

reacted with antiserum to ChrB; yellow petals and green leaves displayed very weak signals.

Role of ChrA: ChrA occurs as a carotenoid-protein complex (Cervantes-Cervantes *et al.*, 1990), the first carotenoid-binding protein to be discovered in higher plants. This observation and the correlation of ChrA with the occurrence of keto-carotenoids leads might suggest a role for ChrA in the conversion of epoxy-carotenoids to the corresponding ketones.

Role of ChrB: In *C. annuum* ChrB occurs in loose association with chromoplast membranes. ChrC, the cucumber variant of ChrB, occurs as a carotenoid-protein complex (Vainstein *et al.*, 1994). Deruère *et al.* (1994a) find a close association between ChrB (which they call fibrillin) and chromoplast fibrils; they also show that these fibrils can be reconstituted by mixing ChrB with lipids extracted from the membranes or with purified carotenoids. The widespread distribution of this family of proteins leads us to speculate that it may play a role in chromoplast differentiation, at least with the tubulous type of chromoplasts.

Other chromoplast-specific proteins: Carotenoid-binding proteins in higher plants have also been reported from carrot (Milicua *et al.*, 1991). ELIPs (early light-induced proteins) (Grimm *et al.*, 1989) were identified as chlorophyll-binding proteins that appear early in plastid development, but Cbr, an algal homolog (Lers *et al.*, 1991), has been found to bind carotenoids (Levy *et al.*, 1993).

C. Chromoplast-Specific Genes

Plastid genome: Plastid DNA persists in chromoplasts and is identical, at least in sequence, to the plastid DNA of chloroplasts (Gounaris *et al.*, 1986; Iwatsuki *et al.*, 1985; Thompson, 1980). [Akazawa's laboratory has reported differential methylation of chromoplasts (Ngernprasirtsiri *et al.*, 1988), but this appears not to be a general phenomenon.] It may be that chromoplasts retain their DNA in order to be able to redifferentiate into chloroplasts, as occurs in some plants, such as *Citrus* spp.

Nuclear genes: All genes known to affect chromoplast development and all genes known to be involved in carotenoid synthesis in higher plants are nuclear. The best characterized genes affecting chromoplast development are in pepper and tomato (Tables I and II), but none have been isolated. As detailed below, the only genes whose expression is clearly chromoplast-specific are *ChrA* (the gene encoding ChrA) and *ChrB* (encoding ChrB). The gene encoding GGPP synthase has been isolated (Kuntz *et al.*, 1992), and it *may* be a chromoplast-specific gene.

ChrA: Hadjeb (1993) determined the N-terminal amino acid sequences of ChrA and ChrB, designed oligomeric probes based on these sequences, and probed a genomic library of *C. annuum*. He obtained several genomic clones for *ChrA*, one of whose sequences corresponded exactly to the determined amino acid sequence of the mature protein and contained an in-frame transit peptide and several hundred bases upstream (AC = M95788).

ChrB: Hadjeb (1993) also determined the N-terminal sequence of ChrB. R. Schantz and M. Kuntz obtained a match between this sequence and the sequence of a previously unidentified clone in a cDNA library generated from ripe fruits of *C. annuum*. They probed RNAs from developing fruits with this cDNA (AC = X71952) and found transcripts accumulating to a high level late in ripening (Deruère *et al.*, 1994a). The genomic sequence has been reported (AC = X77290; Deruére *et al.*, 1994b).

M. Kuntz kindly provided us with the ChrB cDNA. In our hands it produced similar northern blots (Fig. 2) and is consistent with a protein that appears early in chromoplast development (Newman, 1993). Newman (unpublished) found, however, that *ChrB* cross-hybridizes with the first exon of *ChrA*. She selected a 3′-fragment of *ChrB* that did *not* cross-hybridize; the northern blots obtained did not differ substantially from those probed with the entire cDNA.

Genes encoding enzymes of carotenoid biosynthesis: A number of genes encoding enzymes in the pathway of carotenoid biosynthesis have been isolated recently (Table III); the products of all or most of them are probably plastidic. Transcripts of a gene encoding GGPP synthase, isolated from *C. annuum* (Kuntz *et al.*, 1992), are much more abundant in chromoplast-containing tissues than in chloroplast-containing tissues. Since GGPP synthase must also occur during chloroplast development, this gene may be a single gene that is differentially expressed, or it may be a member of a multigene family with this member expressed specifically during chromoplast development.

Table II
Genes Known to Affect Chromoplast Development in *Lycopersicon esculentum*[a]

Genes			Polyenes in Fruit										Total polyenes	Phenotype of Fruit Color
			phytoene →	phytofluene →	ζ-carotene →	neurosporene →	lycopene →	γ-carotene →	β-carotene	δ-carotene →	α-carotene →	pro-γ-carotene		
r	*t*	*at*												
+	+		■	■			■	■	■				87	red
+	−		■	■	■		■	■	■	■		■	158	tangerine
−	+		■	■			■		■				4	yellow
−	−		■	■			■		■	■			34	yellow-tangerine
		−	■	■			■		■				13	apricot
−	+	−	■	○			■		■				2	yellow-apricot
+	+	−	■	■	■		■		■	■			29	tangerine apricot
−	−	−	■	■									10	yellow-tangerine-apricot
hp	*B*	*mo$_B$* *Del*												
−			■	■			■	■	■				88	high pigment
	+		■	■			■	■	■				50	intermediate β[1]
	+	−	■	■			■	■	■				80	high β[1]
		+	■	■		■	■	■	■	■	■	■	84	delta (δ-carotene)
−		+	■	■	■	■	■	■	■	■	■	■	55	high delta (δ-carotene)
gh	*nn*													
−			■	■			○						295	ghost
	−		■	■			○	○	○					rin

Note. Wild-type genotypes are represented as +; homozygous recessives as −. Presence of a polyene is shown for substantial amounts (■) or trace amounts (○). Polyenes in the presumed normal pathway of biosynthesis of β-carotene is shown in by dark shading; unusual or aberrant carotenoids are shown by light shading. Gene symbols: *r, red; t, tangerine, at, apricot; hp, high pigment; B, beta; mo$_B$, modifier-beta; Del, Delta;* and *gh, ghost.*
[a] From Goodwin 1980.
[1] Normal fruits ($B^+B^+mo_B^+mo_B^+$) are red and lycopene comprises 90% of the pigments. Fruits with the $B^-B^-mo_B^+mo_B^+$ or $B^-B^-mo_B^-mo_B^-$ genotypes are orange and contain 50% lycopene–50% β-carotene, or 10% lycopene–90% β-carotene, respectively.

III. Isolation of Chromoplasts

Strategies for the isolation of intact, functional chromoplasts are similar to those for the isolation of chloroplasts, except that chromoplasts are typically more fragile than chloroplasts from the same plant. With one notable exception (Iwatsuki *et al.,* 1984), ripe tomatoes have not yielded intact chromoplasts. The most successful isolations have been from fruits of *C. annuum.* Camara *et al.*

Developmental stages
0 1 2 3 4 5

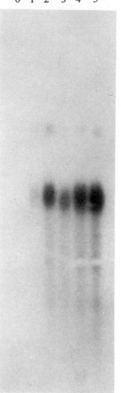

Fig. 2 Appearance of *ChrB* transcripts during chromoplast development in *Capsicum annuum* (from Newman, 1993). A northern blot of total RNA from fruits of *C. annuum* var. Albino were probed with 911H, a cDNA of ChrB. The stages of fruit development are: 1, white; 2, yellowish orange; 3, orange; 4, reddish orange; 5, red. The cDNA was kindly supplied by M. Kuntz.

(1983) described the separation of enzymatically active chromoplasts by differential centrifugation. We adapted their procedure to the general method for the isolation of plastids by isopycnic sedimentation in isosmotic density gradients (Hadjeb *et al.*, 1988; Price *et al.*, 1987).

Tissue disruption: Chromoplasts are sensitive to shear. The conditions of tissue disruption should be just sufficient to disrupt a reasonable fraction of cells but mild enough to preserve the integrity of the chromoplasts.

Temperature: If the objective is to recover biochemically active chromoplasts, solutions and equipment must be ice cold. For the isolation of proteins and nucleic acids, ordinary cold-room conditions are sufficient.

Osmotic protection: As with chloroplasts, chromoplasts require osmotic protection, as provided by 0.3 *M* sorbitol.

Table III
Genes and Enzymes of Carotenoid Biosynthesis in Higher Plants

Gene	Enzyme	Product	Location of enzyme	Reference
Ggps1	GGPP synthase	GGPP	Chromoplast stroma	Spurgeon *et al.*, 1984; Dogbo *et al.*, 1988; Kuntz *et al.*, 1992;
	Isopentenyl pyrophosphate isomerase	Dimethylallyl pyrophosphate	Chromoplast stroma	Spurgeon *et al.*, 1984; Dogbo *et al.*, 1988
Psy1	Phytoene synthase	Phytoene	Envelope	Lütke-Brinkhaus *et al.*, 1982; Dogbo *et al.*, 1988; Bartley *et al.*, 1992
	Phytoene dehydrogenase	Lycopene	Envelope	Lütke-Brinkhaus *et al.*, 1982
Pds1	Phytoene desaturase	Various	Chromoplast membrane	Bartley *et al.*, 1991; Schmidt *et al.*, 1989; Mayer *et al.*, 1990
Lcy1	Lycopene cyclase	α- and β-carotenes	Not yet isolated from plants	
Vde1	Violaxanthin de-epoxidase	5,6-dihydroxy carotenoids	Thylakoid lumen	Hager and Holocher, 1994
Zds1	ζ-Carotene desaturase	Neurosporene	Not yet isolated from plants	

Note. The designations and references for plant genes encoding enzymes of carotenoid biosynthesis are from Scolnik (1994).

Separation from other cell constituents: Chromoplasts should be separated from other constituents of the tissue brei as rapidly as possible. After the tissue is broken, the chromoplasts are pelleted and immediately resuspended in fresh medium.

Particle density: Chromoplasts are considerably less dense than chloroplasts. The banding density of chromoplasts from red fruits of *C. annuum* in isosmotic gradients of silica is approximately 1.04 compared to 1.08 for chloroplasts from green fruits. *Note.* The very low density of chromoplasts originally led Camara *et al.* (1989) to the refutable conclusion that chromoplasts would not enter Percoll gradients.

Gradient material: Silica sols are the gradient materials of choice for the isolation of membrane-bound organelles because they do not contribute to the osmotic potential (Pertoft and Laurent, 1977; Price *et al.*, 1987). Percoll (Pharmacia-LKB) contains a monomolecular layer of polyvinylpyrrolidone around each particle of silica, which shields charged oxygen on the silica from proteins in the sample. Ludox AM (DuPont) is more reactive but far less expensive. *Note.* Because of the dramatic effect of osmotic potential on the hydration of membrane-bound organelles, the banding densities of these particles are much lower in gradients of Percoll than in sucrose gradients. Exposure to high osmotic potential can irreversibly damage plastids.

Stability: Despite all precautions, chromoplasts degrade rapidly and should be used for functional assays within minutes after isolation.

A. Isolation from Fruits of *C. annuum* (Hadjeb *et al.*, 1988)

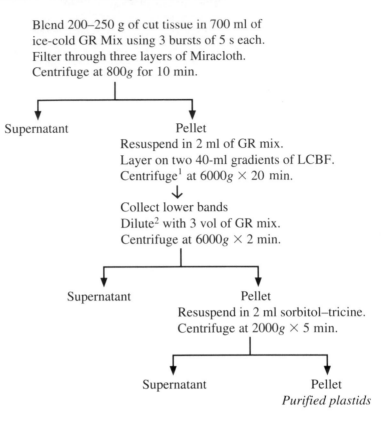

Blend 200–250 g of cut tissue in 700 ml of
ice-cold GR Mix using 3 bursts of 5 s each.
Filter through three layers of Miracloth.
Centrifuge at 800*g* for 10 min.

Supernatant   Pellet

Resuspend in 2 ml of GR mix.
Layer on two 40-ml gradients of LCBF.
Centrifuge[1] at 6000*g* × 20 min.

Collect lower bands
Dilute[2] with 3 vol of GR mix.
Centrifuge at 6000*g* × 2 min.

Supernatant   Pellet

Resuspend in 2 ml sorbitol–tricine.
Centrifuge at 2000*g* × 5 min.

Supernatant   Pellet
                           Purified plastids

B. Solutions Required

1. Gradient composition

Component	Starting solution	Limiting solution
5× GR mix	8 ml	8 ml
Glutathione	7 mg	7 mg
Isoascorbate buffer	0.4 ml	0.4 ml
LCBF (chromoplasts)	2 ml	16 ml
Deionized water	to 40 ml	to 40 ml

[1] Swinging-bucket rotor.
[2] Invert the tubes gently three times to mix.

2. 5× GR (Grind–Resuspension) Mix

Amount	Component	Concentration at 1× (mM)
4.45 g	NaP$_2$O$_7$	1
119.15 g	Hepes	50
601.2 g	Sorbitol	330
40 ml	0.5 M Na$_2$EDTA	2
10 ml	M MgCl$_2$	1
10 ml	M MnCl$_2$	1

Dissolve the pyrophosphate in 20 ml boiling water.
Dissolve balance in 1500 ml deionized water and combine.
Adjust pH to 6.8 with 6 N NaOH.
Dilute to 2 liters with deionized water.
Divide into 400-ml aliquots and store at −20°C.

3. 1× GR Mix

Amount	Component	Concentration
40 ml	5× GR mix	
2 ml	0.5 M Isoascorbate buffer	5 mM

Dilute to 200 ml with deionized water.

4. Isoascorbate Buffer

Amount	Component	Concentration
4.4 g	Isoascorbic (araboascorbic) acid	0.5 M
0.595 g	Hepes	0.05 M

Titrate to pH 7 with 6 N NaOH.
Dilute to 50 ml with deionized water.
Store frozen in aliquots of 2 ml.

5. LCBF

Amount	Component	Concentration
100 ml	Ludox AM[a]	100% (v/v)
3 g	polyethylene glycol[b]	3% (w/v)
1 g	Bovine serum albumin	1% (w/v)
1 g	Ficoll	1% (w/v)

[a] Purified according to the protocol "Ludox."
[b] Carbowax 8000, Union Carbide.

6. 5× Sorbitol Tricine

Amount	Component	Concentration at 1× (mμ)
150.3 g	Sorbitol	330
22.4 g	Tricine	50

Adjust to pH 8.4 with 6 N KOH.
Bring to a final volume of 500 ml with
  deionized water.
Heat sterilize.
When diluting to 1×, check the pH and readjust
  to 8.4.

IV. Purification of Ludox

The silica sol Ludox AM (E. I. DuPont) is an industrial chemical and contains preservatives that are toxic to many enzymes. Ludox may be purified according to the procedure of Price and Dowling (1977).

Load approximately 250 g activated charcoal, 6–14 mesh[3] in deionized water into a C26/100 chromatographic column.[4]

Wash the column in an upward direction with deionized water to remove fines.

While the column is washing, prepare 12 liters of Ludox AM (or other silica sol) as follows:

Place the Ludox in a large vessel with a magnetic stirrer and monitor the pH.

Add solid Dowex 50W-×8 or other cation-exchange resin of the styrene–sulfonic acid type with vigorous stirring until pH is about 7.[5]

Remove the resin by filtration through Miracloth or a fine screen.

After the column has been thoroughly washed, pump the deionized and neutralized silica sol upward[6] through the column at approximately 5 ml/min.

Collect the treated silica sol in 2-liter fractions and label them sequentially.

Test each fraction by the wiggle test:[7] dilute 1 ml of a suspension of *Euglena gracilis* with 9 ml of the silica sol. Observe the motility of the cells under the

---

[3] Fisher Scientific.

[4] The size of the chromatographic column may be varied. The volume of silica sol that may be safely treated is about 4 liters/100 g of charcoal. Column flow rates should be proportional to 1 ml/min/cm^2 of column cross section.

[5] Equilibration time is much slower than with liquid–liquid titrations.

[6] It is important to pump *upward* so that mixing between the column water and Ludox does not occur.

[7] Motility of *Euglena* and similar flagellates are sensitive to the toxins in Ludox. Do not combine fractions until you are sure that they are nontoxic.

microscope. Fractions that cause an increase in the numbers of nonmotile cells should be discarded or pumped through a fresh column of charcoal.

Store the purified silica sol in the dark.[8]

V. Characterization of Chromoplast Membranes

For the analysis of some components of chromoplast membranes it may be sufficient to isolate total membrane proteins (Marder *et al.,* 1986). Since chromoplast membranes normally represent a small fraction of total cell membranes, for many purposes one needs first to isolate chromoplasts, and then the membranes of the chromoplasts. We have found the procedures of Peter and Thornber (1989) for the isolation and characterization of chloroplast membranes to be almost directly applicable to chromoplasts. Once chromoplast membranes are isolated, as described below, Peter and Thornber's procedure for two-dimensional electrophoresis may be employed directly (Cervantes-Cervantes *et al.,* 1990). The groups of Beyer (Beyer *et al.,* 1985), Camara (Camara *et al.,* 1989), and Sitte (Emter *et al.,* 1990) have also described procedures for the isolation of chromoplast proteins.

A. Membrane Isolation from Purified Chromoplasts [adapted from Peter and Thornber (1989)]

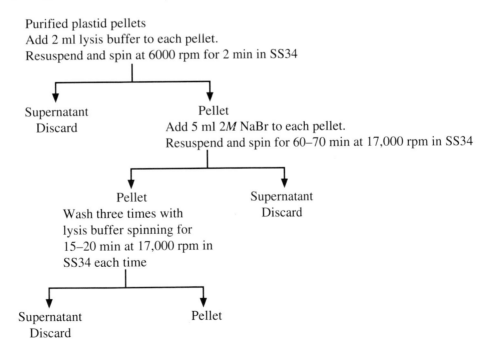

Purified plastid pellets
Add 2 ml lysis buffer to each pellet.
Resuspend and spin at 6000 rpm for 2 min in SS34

Supernatant
Discard

Pellet
Add 5 ml 2*M* NaBr to each pellet.
Resuspend and spin for 60–70 min at 17,000 rpm in SS34

Pellet
Wash three times with
lysis buffer spinning for
15–20 min at 17,000 rpm in
SS34 each time

Supernatant
Discard

Supernatant
Discard

Pellet

---

[8] The sols may be heat-sterilized prior to storage.

Determine protein concentration
in a 10- or 15-μl aliquant
by the bicinchoninic acid method.
⇓
Adjust protein concentration to
1.1 mg/ml with extraction buffer
⇓
Mix gently, divide into small
aliquants, and quick-freeze in
liquid nitrogen (or a mixture of dry ice and ethanol)
⇓
Store at $-70°C$

B. Solutions Required

1. Lysis Buffer (1 Liter)

Amount	Component	Final concentration (mM)
4.48 g	Tricine-NaOH	25
3.8 g	EDTA-Na$_2$	1
to 1 liter	Deionized water	

Adjust pH to 7.6 with 1 N NaOH. Heat sterilize.

2. 2 M NaBr (100 ml)

Amount	Component	Final concentration (M)
20.58 g	NaBr	2
to 100 ml	Deionized water	

Heat sterilize

3. Extraction Buffer (1 Liter)

Amount	Component	Final concentration
80 mg	Tris base	6.2 mM
3.6 g	Glycine	48 mM
10.0 ml	Glycerol	10% v/v
to 100 ml	Deionized water	

Adjust pH to 8.3 with 1 N NH$_4$OH, if necessary. Heat sterilize.

VI. Isolation of Chromoplast DNA

Most problems encountered in the isolation of DNA from plants are resolved once the organelles—nuclei, chloroplasts, etc.—have been separated from other tissue constituents. Similarly, pepper chromoplasts yield DNA suitable for restriction endonuclease digestion from purified preparations of the organelles. In this case, however, there is a trade-off between obtaining chromoplasts that are physiologically active and chromoplasts that will yield cuttable DNA. Silica sols, even Percoll, bind to DNA and interfere with its further manipulation. Extensive washing will overcome this problem, but it is simpler to use sucrose gradients.

A. Isolation of DNA from Chromoplasts and Chloroplasts of *C. annuum* (Gounaris *et al.*, 1986)

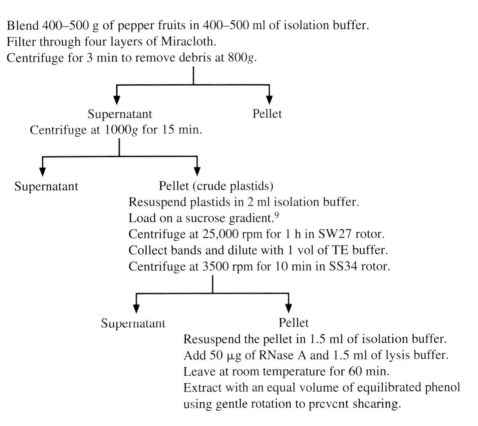

Blend 400–500 g of pepper fruits in 400–500 ml of isolation buffer.
Filter through four layers of Miracloth.
Centrifuge for 3 min to remove debris at 800*g*.

Supernatant   Pellet
Centrifuge at 1000*g* for 15 min.

Supernatant   Pellet (crude plastids)
Resuspend plastids in 2 ml isolation buffer.
Load on a sucrose gradient.[9]
Centrifuge at 25,000 rpm for 1 h in SW27 rotor.
Collect bands and dilute with 1 vol of TE buffer.
Centrifuge at 3500 rpm for 10 min in SS34 rotor.

Supernatant   Pellet
Resuspend the pellet in 1.5 ml of isolation buffer.
Add 50 μg of RNase A and 1.5 ml of lysis buffer.
Leave at room temperature for 60 min.
Extract with an equal volume of equilibrated phenol using gentle rotation to prevent shearing.

[9] Sucrose gradients yield DNA that is routinely cuttable, but it should be noted that the plastids obtained from such gradients are not intact. In addition fragments of nuclear DNA may bind nonspecifically to thylakoid membranes. For *clean* plastids, see Section III,A.

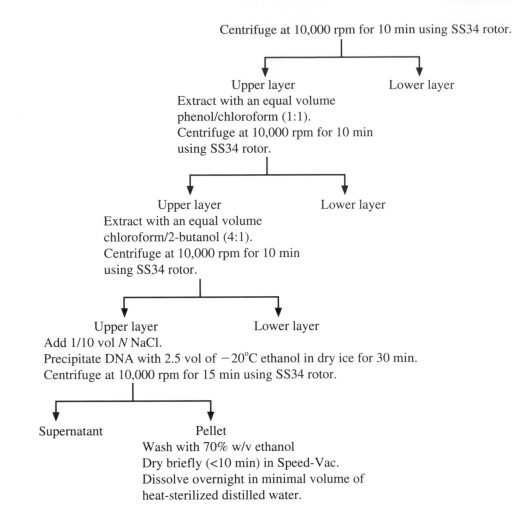

Centrifuge at 10,000 rpm for 10 min using SS34 rotor.

Upper layer   Lower layer
Extract with an equal volume
phenol/chloroform (1:1).
Centrifuge at 10,000 rpm for 10 min
using SS34 rotor.

Upper layer   Lower layer
Extract with an equal volume
chloroform/2-butanol (4:1).
Centrifuge at 10,000 rpm for 10 min
using SS34 rotor.

Upper layer   Lower layer
Add 1/10 vol N NaCl.
Precipitate DNA with 2.5 vol of $-20°C$ ethanol in dry ice for 30 min.
Centrifuge at 10,000 rpm for 15 min using SS34 rotor.

Supernatant   Pellet
Wash with 70% w/v ethanol
Dry briefly (<10 min) in Speed-Vac.
Dissolve overnight in minimal volume of
heat-sterilized distilled water.

B. Solutions Required

1. Isolation Buffer (500 ml)

Amount	Component	Final concentration
60 g	Sucrose	0.35 M
3.9 g	Trizma base	50 mM
1.8 g	EDTA	10 mM
1 ml	β-Mercaptoethanol	30 mM
0.5 g	Bovine serum albumin	0.1%

Adjust pH to 7.8

2. Gradients[10]

	Starting solution (g)	[Final]	Component	Limiting solution (g)	[Final]
For chloroplasts	9	30% (w/v)	Sucrose	24	60% (w/v)
For chromoplasts	6.5	25% (w/v)	Sucrose	19	50% (w/v)
	to 50 ml		Isolation buffer	to 50 ml	

3. Lysis Buffer

4% Na-sarkosinate in H_2O.

References

Bartley, G. E., Viitanen, P. V., Pecker, I., Chamovitz, D., Hirschberg, J., and Scolnik, P. A. (1991). Molecular cloning and expression in photosynthetic bacteria of a soybean cDNA coding for phytoene desaturase, an enzyme of the carotenoid biosynthesis pathway. *Proc. Natl. Acad. Sci. U.S.A.* **88,** 6532–6536.

Bartley, G. E., Viitanen, P. V., Bacot, K. O., and Scolnik, P. A. (1992). A tomato gene expressed during fruit ripening encodes an enzyme of the carotenoid biosynthesis pathway. *J. Biol. Chem.* **267,** 5036–5039.

Bathgate, B., Purton, M. E., Grierson, D., and Goodenough, P. W. (1985). Plastid changes during the conversion of chloroplasts to chromoplasts in ripening tomatoes. *Planta* **165,** 197–204.

Beyer, P. (1989). Carotene biosynthesis in daffodil chromoplasts: On the membrane-integral desaturation and cyclization reactions. *In* "Physiology, Biochemistry, and Genetics of Nongreen Plastids" (C. T. Boyer, J. C. Shannon, and R. C. Hardison, eds.), pp. 157–170. Rockville, Maryland: American Society of Plant Physiologists.

Beyer, P., and Kleinig, H. (1989). Molecular oxygen and the state of geometric isomerism intermediates are essential in the carotene desaturation and cyclization reactions in daffodil chromolasts. *Eur. J. Biochem.* **184,** 141–150.

Beyer, P., Weiss, G., Kleinig, H. (1985). Solubilization and reconstitution of the membrane-bound carotenogenic enzymes from daffodil chromoplasts. *Eur. J. Biochem.* **153,** 341–346.

Britton, G. (1993). Carotenoids in chloroplasts pigment-protein complexes. *In* "Pigment-Protein Complexes in Plastids. Synthesis and Assembly" (C. Sundqvist and M. Ryberg, eds.), pp. 447–483. San Diego: Academic Press.

Camara, B., Bardat, F., Dogbo, O., Brangeon, J., and Moneger, A. (1983). Terpenoid metabolism in plastids. Isolation and biochemical characteristics of *Capsicum annuum* chromoplasts. *Plant Physiol.* **70,** 1562–1563.

Camara, B., Bousquet, J., Cheniclet, C., Carde, J.-P., Kuntz, M., Evrard, J.-L., and Weil, J.-H. (1989). Enzymology of isoprenoid biosynthesis and expression of plastid and nuclear genes during chromoplast differentiation in pepper fruits (*Capsicum annuum*). *In* "Physiology, Biochemistry, and Genetics of Nongreen Plastids" (C. T. Boyer, J. C. Shannon, and R. C. Hardison, eds.), pp. 141–156. Rockville, Maryland: American Society of Plant Physiologists.

Carde, J. P., Camara, B., and Cheniclet, C. (1988). Absence of ribosomes in *Capsicum* chromoplasts. *Planta* **173,** 1–11.

[10] For a large number of separations, it might be easier to use sterile solutions of 75% w/v sucrose and a 5× stock of the sucrose/tris/EDTA components of the isolation medium. β-Mercaptoethanol and bovine serum albumin would have to be added immediately before use.

Cervantes-Cervantes, M., Hadjeb, M., Newman, L. A., and Price, C. A. (1990). ChrA is a carotenoid-binding protein in chromoplasts of *Capsicum annuum*. Plant Physiol. **92,** 1241–1243.

Covello, P. S., and Gray, M. W. (1993). On the evolution of RNA edition. *Trends Genet.* **9,** 265–268.

Deruère, J., Römer, S., d-Harlingue, A., Backhaus, R. A., Kuntz, M., Camara, B. (1994a). Fibril assembly and carotenoid overaccumulation in chromoplasts: A model for supramolecular lipoprotein structures. *Plant Cell* **6,** 199–133.

Deruére, J., Bouvier, F., Steppuhn, J., Klein, A., Camara, B., and Kuntz, M. (1994b). Structure and expression of two plant genes encoding chromoplast-specific proteins: occurence of partially spliced transcripts. *Biochem. Biophys. Res. Commun.* **199,** 1144–1150.

Dogbo, O., Laferrière, A., D'Harlingue, A., and Camara, B. (1988). Carotenoid biosynthesis: Isolation and characterization of a bifunctional enzyme catalyzing the synthesis of phytoene. *Proc. Natl. Acad. Sci. U.S.A.* **85,** 7054–7058.

Emter, O., Falk, H., and Sitte, P. (1990). Specific carotenoids and proteins as prerequisites for chromoplast tubules formation. *Protoplasma* **157,** 128–135.

Goodwin, T. W. (1980). "The Biochemistry of the Carotenoids." Second ed., Vol. I. London: Chapman and Hall.

Goodwin, T. W. (1988). "Plant Pigments." San Diego: Academic Press.

Gounaris, I., Michalowski, C. B., Bohnert, H. J., and Price, C. A. (1986). Restriction and gene maps of plastid DNA from *Capsicum annuum:* Comparison of chloroplast and chromoplast DNA. *Curr. Genet.* 7–16.

Grimm, B., Kruse, E., and Kloppstech, K. (1989). Transiently expressed early light-inducible thylakoid proteins share transmembrane domains with light-harvesting chlorophyll binding proteins. *Plant Mol. Biol.* **13,** 583–593.

Hadjeb, N. "Isolation of *ChrA,* A Chromoplast-Specific Gene from *Capsicum annuum.*" Ph.D., Rutgers University, 1993.

Hadjeb, N., Gounaris, I., Price, C. A. (1988). Chromoplast-specific proteins in *Capsicum annuum. Plant Physiol.* **88,** 42–45.

Hager, A., and Holocher, K. (1994). Localization of the xanthophyll-cycle enzyme violaxanthin de-epoxidase within the thylakoid lumen and abolition of its mobility by a (light-dependent) pH decrease. *Planta* **192,** 581–589.

Hurtado-Hernandez, H., and Smith, P. G. (1985). Inheritance of mature fruit color in *Capsicum annuum. J. Hered.* **36,** 211–213.

Iwatsuki, N., Hirai, A., and Asahi, T. (1985). A comparison of tomato fruit chloroplast and chromoplast DNAs as analyzed with restriction endonucleases. *Plant Cell Physiol.* **26,** 599–602.

Iwatsuki, N., Moriyama, R., and Asahi, T. (1984). Isolation and properties of intact chromoplasts from tomato fruits. *Plant Cell Physiol.* **25,** 763–768.

Karrer, P., and Jucker, E. (1948). *Carotenoide.* Basel: Birkhauser.

Kreuz, K., and Kleinig, H. (1984). Prenyllipid synthesis in spinach leaf cells. Compartmentation of enzymes for isopentenyldiphosphate formation. *Eur. J. Biochem.* **141,** 531–535.

Kuntz, M., Römer, S., Suire, C., Hugueney, P., Weil, J. H., Schantz, R., and Camara, B. (1992). Identification of a cDNA for the plastid-located geranylgeranyl pyyrophosphate synthase from *Capsicum annuum:* Correlative increase in enzyme activity and transcript level during fruit ripening. *Plant J.* **21,** 25–34.

Lawrence, S. D., Cline, K., and Moore, G. A. (1993). Chromoplast-targeted proteins in tomato (*Lycopersicon esculentum* Mill.) fruit. *Plant Physiol.* 102:789–794.

Lers, A., Levy, H., and Zamir, A. (1991). Co-regulation of an *elip*-like gene and β-carotene biosynthesis in the alga *Dunaliella bardawil. J. Biol. Chem.* **266,** 13698–13705.

Levy, H., Tal, T., Shaish, A., and Zamir A. (1993). Cbr, an algal homolog of plant early light-induce protein, is a putative zeoxanthin binding protein. *J. Biol. Chem.* **268,** 20892–20896.

Lütke-Brinkhaus, F., Liedvogel, B., Kreuz, K., and Kleinig, H. (1982). Phytoene synthase and phytoene dehydrogenase associated with envelope membranes from spinach chloroplasts. *Planta* **156,** 176–180.

Marder, J. B., Mattoo, A. K., and Edelman, M. (1986). Identification and characterization of the *psbA* gene product: The 32-kDa chloroplast membrane protein. *Methods Enzymol.* **118,** 384–396.

Mayer, M. P., Beyer, P., and Kleinig, H. (1990). Quinone compounds are able to replace molecular oxygen as terminal electron acceptor in phytoene desaturation in chromoplasts of *Narcissus pseudonarcissus* L. *Eur. J. Biochem.* **191,** 359–364.

Milicua, J. C. G., Juarros, J. L., De Las Rivas, J., Ibarrondo, J., and Gomez, R. (1991). Isolation of a yellow carotenoprotein from carrot. *Phytochemistry* **30,** 1535–1537.

Newman, L. A. "ChrA and ChrB: Genes and Proteins in Chromoplasts of Higher Plants." Ph.D., Rutgers University, 1993.

Newman, L. A., Hadjeb, N., and Price, C. A. (1989). Synthesis of two chromoplast-specifc proteins during fruit development in *Capsicum annuum. Plant Physiol.* **91,** 455–458.

Newman, L. A., Hadjeb, N., Vainstein, S., Smirra, I., Vishnevetsky, M., and Price, C. A. Distribution of ChrB and related proteins among chromoplasts of various plant species. In preparation.

Ngernprasirtsiri, J., Kobayashi, H., Akazawa, T. (1988). DNA methylation occurred around lowly exprcsscd genes of plastid DNA during tomato fruit development. *Plant Physiol.* **88,** 16–20.

Odland, M. L., and Porter, A. M. (1938). Inheritance of immature fruit color of peppers. *Proc. Am. Hort. Sci.* **36,** 647–657.

Oren-Shamir, M., Hadjeb, N., Newman, L. A., and Price, C. A. (1993). Occurrence of the chromoplast protein ChrA correlates with a fruit-color gene in *Capsicum annuum. Plant Mol. Biol.* **21,** 549–554.

Pertoft, H. T., and Laurent, C. (1977). Isopycnic separation of cells and cell organclles by centrifugation in modified colloidal silica gradients. *In* "Methods of Cell Separation" (N. Catsimpoolas, ed.), Vol. 1, pp. 25–65. New York: Plenum.

Peter, G., and Thornber, F. J. P. (1989). Electrophoretic procedures for fractionation of Photosystem I and II pigment-proteins of higher plants and for determination of their subunit composition. "Methods in Plant Biochemistry: Proteins and Nucleic Acids," Vol. 2. (L. J. Rogers, ed.). New York: Academic Press.

Price, C. A., Cushman, J. C., Mendiola-Morgenthaler, L. R., and Reardon, E. M. (1987). Isolation of plastids in density gradients of Percoll and other silica sols. Vol. 148. Meth. Enzymol., ed. R., Douce and L. Packer, eds.). New York: Academic Press.

Price, C. A., and Dowling, E. (1977). On the purification of the silica sol Ludox AM. *Anal. Biochem.* **82,** 243–245.

Schmidt, A., Sandmann, G., Armstrong, G. A., Hearst, J. E., and Böger, P. (1989). Immunological detection of phytoene desaturase in algae and higher plants using an antiserum raised against a bacterial fusion-gene construct. *Eur. J. Biochem.* 184:375–378.

Scolnik, P. (1994). Genes of carotenoid biosynthesis. In preparation.

Sitte, P., Falk, H., and Liedvogel, B. (1980). "Chromoplasts. Pigments in Plants" (F. C. Czygan, ed.) Gustav Fischer Verlag.

Smith, P. G. (1950). Inheritance of brown and green mature fruit color in peppers. *J. Hered.* 41:138–140.

Spurgeon, S. L., Sathyamoorthy, N., and Porter, J. W. (1984). Isopentenyl pyrophosphate isomerase and prenyltransferase from tomato fruit plastids. *Arch. Biochem. Biophys.* 230:446–454.

Thompson, J. A. (1980). Apparent identity of chromoplast and chloroplast DNA in the daffodil, *Narcissus pseudonarcissus. Z. Naturforsch.* 1101–1103.

Vainstein, A., Halevy, A. H., Smirra, I., and Vishnevetsky, M. (1994). Chromoplast biogenesis in Cucumis sativus corollas. *Plant Physiol.* **104,** 321–326.

CHAPTER 15

Methods for Isolation and Analysis of Polyribosomes

Eric Davies* and Shunnosuke Abe†

*School of Biological Sciences
University of Nebraska
Lincoln, Nebraska 68588-0118
†Laboratory of Molecular Cell Biology
College of Agriculture
Ehime University
Matsuyama 790, Japan

I. Introduction

A. Overview

Polysomes consist of two or more ribosomes traversing a strand of mRNA, translating the nucleotide sequence into the corresponding amino acid sequence (Davies and Larkins, 1980; Larkins, 1985). One reason for isolating polysomes is to resolve them on sucrose gradients to determine the extent of ribosome loading. A typical polysome profile is shown in Fig. 1. The second main reason is to provide a source of mRNA for *in vitro* translation or cDNA probing. Since the latter will be dealt with by Vayda (Chapter 25), we will concentrate on polysome isolation and separation.

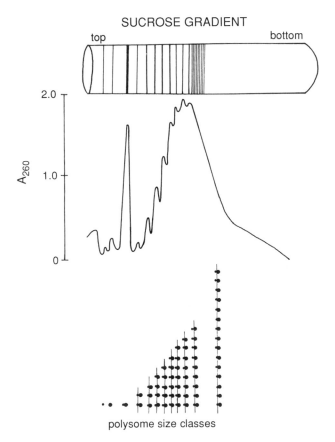

Fig. 1 Diagrammatic representation of a polyribosome profile. A mixture of polyribosomes was layered on a sucrose gradient and centrifuged. Top, the hypothetical banding achieved by the different ribosome sub-classes. Center, the actual profile of their UV absorbance. Bottom, each size class; from the left, the 40 S subunit, the 60 S subunit, the monosome, dimer, trimer, etc. Taken from Davies and Larkins (1980), with permission.

B. The Two Major Artifacts

Two major artifacts can occur during polysome isolation, degradation of polysomes by RNases and trapping of polysomes by subcellular complexes to which they bind. Simultaneous resolution of these two artifacts can be very difficult (Fig. 2). These potential artifacts arise as a consequence of typical plant cell structure: a tough wall that resists breakage, a massive vacuole occupying as much as 80% of the cell volume and often containing noxious compounds. The

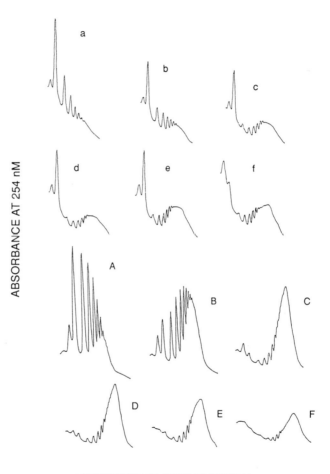

Fig. 2 Effects of Tris–HCl on polysome retention and integrity. Pea root tissue was homogenized in CSB + PTE containing various levels of Tris–HCl at pH 8.5, filtered and centrifuged for 15 min at 27,000× g to yield the supernatant (released) polysomes and the pelletted (retained) polysomes. The profiles correspond to: released polysomes, a–f; retained polysomes, A–F. Tris–HCl (mM) was added as follows: a/A, 0; b/B, 5; c/C, 50; d/D, 100; e/E, 150; f/F, 200. Taken from You et al. (1992), with permission.

cytoplasm is typically a thin layer between the vacuole and the cell wall, and forces needed to break the wall rupture the vacuole mixing its contents with those of the cytoplasm. Thus there is no such thing as a "physiological" buffer for most plant systems: the role of the extraction buffer is to counter the noxious effects of the vacuolar contents and must be quite "nonphysiological."

1. Polysome Aggregation–Disaggregation

The major factor influencing the number of ribosomes per mRNA molecule is RNase, an enzyme that is difficult to inhibit, and many polysome isolation protocols are designed solely to prevent its action (Davies and Larkins, 1980; Larkins, 1985). RNase activity initially causes conversion of large polysomes into small ones, and only at later stages do monosomes accumulate. Disaggregation can also result from ribosome run-off, and this leads to accumulation of monosomes and only a slight shift from large to smaller polysomes. These two processes can be distinguished either by visual examination of profiles (Davies and Larkins, 1974) or by computer simulation (Vassart *et al.*, 1970). In addition to disaggregation, artifacts can arise through polysome aggregation or "clumping" brought about by reagents such as bentonite, by high levels of Mg^{2+}, and perhaps through hydrophobic nascent chains (Larkins, 1985; Stankovic *et al.*, 1993). It is wise not to focus too intently on RNase degradation, otherwise artifacts arising from changes in subcellular location of polysomes may be neglected.

2. Polysome Localization

When any tissue is homogenized, components that were together *in vivo* can become separated, while those that were separate can get stuck together, but there is no easy way to tell whether either of these events has occurred. There is general acceptance that 80S ribosomes occur in the cytoplasm as free polysomes (FP) and membrane-bound polysomes (MBP). There is also abundant evidence for the existence of cytoskeleton-bound polysomes (CBP) and cytoskeleton-membrane-bound polysomes (CMBP) in both animals (Hesketh and Pryme, 1991; Pachter, 1992) and plants (Davies *et al.*, 1991; You *et al.*, 1992; Davies *et al.*, 1992; Zak *et al.*, 1995; see Chapter 16, this volume), but absolute proof of their existence remains elusive. Since much of the argument used against the existence of CBP and CMBP in plants has arisen from the use of high ionic strength polysome isolation buffers developed by us, we describe here buffers specifically formulated to maintain cytoskeleton integrity, and show how use of these buffers strongly suggests that CBP and CMBP do exist.

II. Materials

A. Chemicals

Mostly reagent grade, e.g., Sigma, Research Organics: Detergents: polyoxyethylene 10-tridecyl ether (PTE), sodium deoxycholate, Triton-X 100. RNase inhibitors; heparin, ribonucleoside vanadyl complexes. Protease inhibitors: Phenylmethyl sulfonyl fluoride (PMSF)—highly toxic. Miscellaneous: EGTA, fluorinert, Hepes, sucrose (we purify our own by passing solutions of commercial cane sugar through activated charcoal to remove RNase and UV-absorbing materials), Tris–HCl.

B. Supplies

Glass homogenizer, mortar and pestle, pestle for microfuge tube, gradient maker. Note: We make our gradients for 5-ml tubes by pipetting 2.3 ml of 60% sucrose into an ultracentrifuge tube, then carefully overlayering with 2.3 ml of 15% sucrose using a pipette with a hole in the side, stoppering the tube, and laying it on its side for 3 h to diffuse into a gradient. As many as 20 *identical* gradients can be made in about 15 min.

C. Equipment

High-speed centrifuge (e.g., Sorvall), rotors (e.g., HB4), Corex tubes; microfuge and tubes; ultracentrifuge (e.g., Beckman), rotors (e.g., SW41, SW55), and polyallomer tubes: gradient monitor and fractionator (e.g., ISCO UA5 monitor and ISCO 185 density gradient fractionator). Note: Instead of piercing the tubes and discarding them after one run, we use a device that pumps the displacing solution (sucrose, or fluorinert for heavy gradients) through the top of the gradient tube.

D. Buffers

1. Buffer A (200 mM Tris–HCl, pH 8.5, 50 mM KCl, 25 mM MgCl$_2$) is a high ionic strength buffer designed to minimize degradation by RNase (Davies *et al.,* 1972, Larkins and Davies, 1975).

2. Buffer B (50 mM Tris–HCl, 25 mM KCl, 10 mM MgCl$_2$) is used for polysome resuspension, sucrose pads, and sucrose gradients (Davies *et al.,* 1972).

3. Buffer C (5 mM Hepes, 10 mM MgOAc, 2 mM EGTA, 1 mM PMSF, pH 7.5) is a cytoskeleton-stabilizing buffer that retains polysomes on the cytoskeleton (Abe and Davies, 1991).

4. Buffer U (buffer A supplemented with 2 mM EGTA, 100 μg/ml heparin, 2% PTE, and 1% DOC) is used to solubilize virtually all polysomes from any source (Abe *et al.,* 1992). Note: Ingredients must be added in the above order, making sure that PTE is thoroughly dissolved before slowly adding DOC. If a precipitate forms before DOC is dissolved, remake the solution with somewhat more PTE (e.g., 3%) or less DOC (e.g., 0.75%). We do not know why precipitation occasionally occurs.

===== ## III. Methods

A. Isolation of Total Polysomes

This allows undegraded FP, MBP, CBP, and CMBP to be isolated together as supernatant polysomes for analyzing directly on gradients, or partly purified as pellets for layering on gradients, or for mRNA analysis by *in vitro* translation or cDNA probing. The protocol is based on buffer U, which was specifically designed to prevent RNase action and to release polysomes from all subcellular locations. All operations should be performed at less than 4°C.

1. Grind tissue in at least 5 vol ice-cold buffer U either by hand in a teflon or glass homogenizer for cell cultures, directly in a microfuge tube for small pieces of tissue such as root tips, or in a mortar and pestle for most samples. For especially tough tissue such as leaves it is best to pulverize in liquid N_2 before transferring the frozen powder to buffer U.

2. Filter through nylon cloth or Miracloth to retain wall materials. With samples homogenized in a microfuge tube, this step can be omitted.

3. Centrifuge for 5–10 min at $27,000\times g$, preferably in a swinging bucket rotor to remove debris.

4a. Layer the supernatant directly on gradients to see polysome distribution. We layer 200 μl on 15–60% sucrose gradients in buffer B using 5-ml tubes in an SW 50- or 55-type rotor and spin for 60 min at 45,000 rpm or 50 min at 55,000 rpm. Samples are scanned at 254 nm on an ISCO or similar gradient monitor.

4b. Layer about 4 ml supernatant over a 0.5- to 1-ml, 50–60% sucrose pad in buffer B and centrifuge for at least 2 h at 40,000 rpm in an SW 41 rotor, or at 50,000 rpm in an SW 50 (55) rotor. For larger (slower) rotors, times must be increased accordingly. To increase the proportion of large, actively translating polysomes and decrease yields of monosomes and small polysomes use shorter run times, and larger or denser pads. Polysomes can be pelleted more quickly in fixed angle rotors, but they are not generally as clean. The polysome pellet is then resuspended in buffer U and layered on gradients to obtain for polysome profiles (cf. Fig. 4A), or they can be resuspended in water for *in vitro* translation.

5. Analysis of polysome profiles. See Section IV; a typical profile is presented in Fig. 1.

B. Free and Membrane–Bound Polysomes

This allows the isolation of two putative polysome populations, the FP and MBP.

1. Grind tissue as above, but using buffer A instead of buffer U.

2. Filter and centrifuge at $27,000\times g$, as above. The supernatant contains the presumed FP, and is analyzed as in 4a or 4b above.

3. Resuspend the pellet in 1 to 5 vol of buffer A plus nonionic detergent and recentrifuge. Most people use Triton X-100 at up to 2%; we use 0.5% PTE (since it does not absorb at 254–260 nm). Analyze the released polysomes (MBP) as in 4a or 4b above.

C. Cytoskeleton-Bound and Other Polysomes

This allows the sequential isolation of five putative polysome populations, the FP, MBP, CBP, CMBP, and TBP the "tightly bound polysomes."

1. Grind tissue in at least 5 vol buffer C (10 vol is preferred, since buffer C resists RNase poorly). Filter and centrifuge at 27,000× g as above. The supernatant polysomes are the FP, and are analyzed as in 4a or 4b above. To prevent degradation by RNase, the samples should be converted to buffer U by adding 4 vol of 5× U and held on ice prior to further processing. Tissues may contain 20–50% of the total polysomes as FP. All other polysomes are in the pellet and are sequentially solubilized as follows.

2. Resuspend pellet in C + PTE to disrupt membranes; centrifuge for 5 min at 27,000× g to leave MBP in the supernatant. Again, convert to buffer U to maintain polysome integrity. In most tissues, we find few polysomes in this fraction. All other polysomes are still in the pellet.

3. Resuspend pellet in C plus 200 mM Tris–HCl, pH 8.5, to disrupt the polysome–cytoskeleton interaction and centrifuge for 5 min at 27,000× g to leave CBP in the supernatant. Convert to buffer U. Up to 60% of the total polysomes may be found in this fraction.

4. Resuspend pellet in C + PTE + Tris and recentrifuge to release the CMBP. Convert to buffer U. For some reason both detergent and ionic agents must be present at the same time to release these polysomes. In corn endosperm they make up 75% of the total polysomes.

5. Resuspend pellet in U and recentrifuge to release the TBP. A typical example of FP, MBP, CBP, CMBP, and TBP from peas is given in Fig. 3.

D. Simultaneous Analysis of FP, CBP, and CMBP

This is done either to visualize changes in polysomes within and between different populations on a gradient or to analyze gradient fractions for mRNA, actin, membranes (see Chapter 16 for further details).

1. Grind sample in buffer C, filter, and layer directly over a 20–80% gradient in buffer C (to maintain the integrity of the cytoskeleton).

2. Centrifuge at 50,000 rpm in an SW 50 rotor for 45 min and scan at 254 nm.

Typical examples of pea polysomes isolated in C, C + PTE, and C + PTE + DOC are shown in Fig. 4, whereas Fig. 5 shows a sample isolated in C and centrifuged on gradients for varying periods of time.

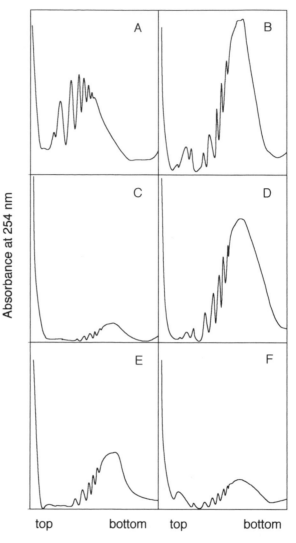

Fig. 3 Sequential extraction of FP, CBP, MBP, CMBP, and TBP. Pea stem tissue was homogenized in buffer C, filtered, and contrifuged at 27,000× g to yield FP in the supernatant. One sample was held on ice (A) or adjusted to buffer U to prevent degradation (B). The pellet was extracted sequentially in C + PTE to release MBP (C), C + Tris to release CBP (D), C + PTE + Tris to release CMBP (E), and finally buffer U to release TBP (F). The supernatants were layered on 0.5-ml pads of 60% sucrose in (B) and centrifuged for 2 h at 250,000× g. Pelletted polysomes were resuspended in buffer U, layered on 15–60% gradients in (B) and analyzed in the normal manner. From Steele and Davies, unpublished.

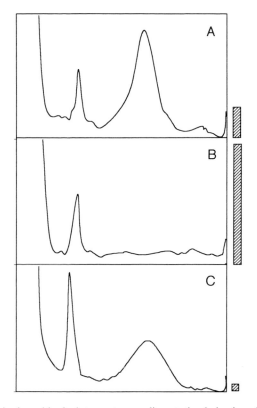

Fig. 4 Effect of nonionic and ionic detergents on sedimentation behavior of FP, CBP, and CMBP. Pea stem tissue was homogenized in buffer C (A), C + PTE (B), or C + PTE + DOC (C), filtered, layered directly on 20–80% gradients in buffer C, and centrifuged at 250,000× *g* for 1 h. In (A), three polysome populations are present; free monosomes and polysomes in the upper region; a major peak of CMBP two-thirds of the way down; CBP in the pellet (hatched bars). In (B), the major peak (CMBP) has disappeared and the amount of CBP in the pellet has increased. In (C), all the polysomes were solubilized and sedimented as typical FP. From Ito and Abe, unpublished.

IV. Results, Discussion, and Analysis

A. Analysis of Polysome Profiles

1. Percentage Polysomes (%P) and Percentage Large Polysomes (%LP)

The simplest way to determine whether a treatment has caused a shift in polysome distribution is to measure the area under the polysome region (×100) divided by the area under the entire profile to give %P (Davies *et al.*, 1972). More information can be obtained by measuring the area under the region of large polysome (LP), i.e., polysomes with 6 or more ribosomes per mRNA, divided by the area of total polysomes (*P*) to give %LP. An increase in %LP

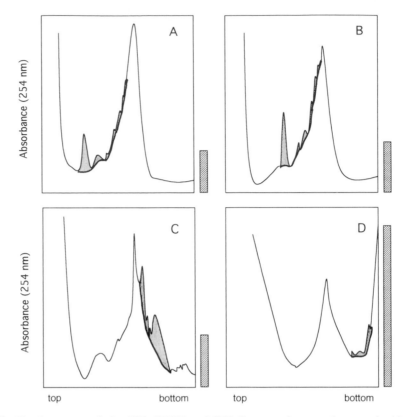

Fig. 5 Simultaneous analysis of FB, CMBP, and CBP. Pea stem tissue was homogenized in buffer C, filtered, layered directly on 20–80% gradients in C, and centrifuged at 250,000× g for 1 h (A), 4 h (B), 16 h (C), or 30 h (D). The darkened region on the gradient corresponds to FP, which sediment slowly through the gradient. The major peak corresponds to CMBP, which reach their isopicnic point by 30 min and remain there (without trapping the migrating FP). The hatched area at the bottom is pelleted material that corresponds to CBP (A–C) or CBP supplemented with newly-arrived FP (D). From Davies and Stankovic, unpublished.

indicates there has been a shift from small to large polysomes, i.e., that ribosome loading on mRNA has occurred (Davies and Larkins, 1973).

2. Quantitation of Polysomal mRNA

Polysomes are separated by rate-zonal (not equilibrium) centrifugation and sedimentation rate is based on polysome size, i.e., the number of ribosomes per mRNA. Valuable information can be obtained by measuring the area of each peak and dividing that of the 2-mer by 2, the 3-mer by 3, and so on. This gives the relative amount of polysomal mRNA in each peak and hence the relative amount in any given tissue. The method has been used to show that auxin does

(Davies and Larkins, 1973), but wounding does not (Davies and Schuster, 1981), cause an increase in polysomal mRNA at the same time as an increase in poly-somes.

3. Assaying Polysomal mRNase Activity

The above method has been modified to furnish a very sensitive assay for endolytic mRNase activity (Davies and Larkins, 1974). Since the number of interribosomal bonds is one less than the number of ribosomes in each polysome (i.e., $n - 1$), dividing the area of each peak by $n - 1$ will measure the relative number of interribosomal bonds, and a decrease in interribosomal bonds is proportional to RNase activity.

4. Interpretation of Changes in Ribosome Loading on mRNA

The extent of ribosome loading is often naively related to the rate of protein synthesis *in vivo;* i.e., the presumption is made that the more polysomes and less monosomes present in a tissue, the higher is its protein synthesizing capacity. This holds true only when translation is limited by initiation. A decrease in initiation results in the accumulation of monosomes and less ribosomes per mRNA. However, if termination is inhibited, ribosomes will accumulate on the mRNA to form larger polysomes, yet protein synthesis will slow down. Indeed, the termination inhibitor cycloheximide is often added to mammalian tissues prior to extraction to cause ribosome "pile-up," thereby protecting polysomal mRNA from RNase. Further, if elongation is inhibited, the rate of ribosome movement will slow down, as will the rate of protein synthesis, yet there will be no change in monosome:polysome ratios. We are aware of only one study providing evidence for a general inhibition of elongation: the wound-induced inhibition of protein synthesis in pea stems (Davies *et al.,* 1986), although com-puter simulations of such possibilities were developed earlier (Vassart *et al.,* 1970).

B. Subcellular Localization of Polysomes

The difficulty of obtaining undegraded polysomes still associated with the cytoskeleton is exemplified by the profiles in Fig. 2, which show that increasing levels of Tris provide increasing protection against RNase, but that greater than 50 mM causes release of polysomes from the cytoskeleton. Indeed, Tris–HCl was the only RNase inhibitor tested that allowed retention while preventing degradation, whereas heparin, ribonucleosyl–vanadyl complexes, and K^+ all caused release at levels lower than those needed for protection against RNase (You *et al.,* 1992).

The presence of different populations of polysomes in pea stems can be seen by the release of polysomes from pellets by sequential extraction with different reagents (Fig. 3). With buffers designed to maintain integrity of both the cytoskeleton and the membranes, only FP are solubilized, but these are degraded (Fig. 3A) unless the extract is converted immediately to buffer U (Fig. 3B). The pellet will then yield MBP when solubilized by nonionic detergent (Fig. 3C), CBP with high monovalent cation (Fig. 3D), CMBP with both nonionic detergent and cation (Fig. 3E) and TBP with buffer U (Fig. 3F).

These polysomes behave differently when extracts obtained in different ways are layered on gradients made up in buffer C. The huge peak present in extracts devoid of detergent (Fig. 4A) disappears when nonionic detergent is present (Fig. 4B). The disappearance of polysomes and membranes from the peak after detergent treatment accompanies an increase in membranes at the top of the gradient and an increase in cytoskeleton proteins and polysomes in the pellet, strongly suggesting that the peak contains CMBP. Extracts in DOC yield a typical "soluble" polysome profile (Fig. 4C), indicating that polysomes have been dislodged from the cytoskeleton. Detailed analyses of these profiles are given in Chapter 16. The simultaneous analysis of FP, CMBP, and CBP can be achieved by layering samples on gradients and centrifuging for various periods. The peak (CMBP) rapidly reaches its isopicnic point, the CBP pellet, whereas the FP slowly migrate through the gradient (Fig. 5).

C. Rates of Initiation, Elongation, and Termination *in Vivo*

If tissue is labeled briefly with [^{35}S]methionine, polysomes are isolated, and incorporated label is measured in each fraction, the *in vivo* protein-synthesizing capacity of FP, CBP, etc., can be estimated. If label in the released (completed) proteins is compared with that in the nascent polypeptides, estimates of relative rates of initiation, elongation, and termination can be measured. Such measurements have rarely been published with higher plant systems, although we have shown that wounding of aged pea tissue increases the label in nascent polypeptides (still in polysomes) and reduces the released protein (Davies, 1993).

D. *In Vitro* Translation with Polysomes

This topic will be dealt with by Vayda in Chapter 25 on *in vitro* translation, but we offer the following suggestions specifically for polysomes. The amount and purity of polysomes can be estimated most easily from UV absorbance at 260, 280, and 320–330 nm. "Clean" polysomes generally show 260/280 ration between 1.8 and 1.9, and contain approximately equal amounts of protein and RNA. If the ratio is 1.6 or lower, the isolation procedure should be controlled for contamination with proteins or other materials. High absorbance at 320 or 330 nm indicates the presence of poorly resuspended, turbid materials, which may be removed by brief centrifugation in a microfuge. Polysomes are best used immediately after isolation. If they are to be stored, a temperature of −85°C or lower is recommended. Even at −85°C, polysomes gradually aggregate and become increasingly pelletable in the microfuge. To lessen this, we recommend sealing under N_2 and storing at or below $-85°C$. Precipitation on storage is especially prevalent with polysomes isolated from tissues high in oxidizing capacity, and in those instances we recommend using a reducing agent such as dithiothreitol or 2-mercaptoethanol in the grinding buffer.

Acknowledgments

This work was supported by NSF Grant IBN-9310508, the UN-L Research Council, the UN-L Center for Biotechnology (to E.D.), and the Japanese Ministry of Education, Sciences and Culture, Calsonic, Inc., and BioCraft, Tokyo (to S.A.). We thank Ms. Weimin You and Bratislav Stankovic for help with experiments.

References

Abe, S., and Davies, E. (1985). Quantitative isolation of undegraded polysomes from aged pea epicotyls in the absence of artefacts. *Plant Cell Physiol.* **26,** 1499–1509.

Abe, S., and Davies, E. (1991). Isolation of F-actin from pea stems: Evidence from fluorescence microscopy. *Protoplasma* **163,** 51–61.

Abe, S., Ito, Y., and Davies, E. (1992). Co-sedimentation of actin, tubulin and membranes in the cytoskeleton fractions from peas and mouse 3T3 cells. *J. Exp. Bot.* **43,** 941–949.

Davies, E. (1993). Intercellular and intracellular signals in plants and their transduction via the membrane-cytoskeleton interface. *Seminars Cell Biol.* **4,** 139–147.

Davies, E., and Larkins, B. A. (1973). Polyribosomes from peas. II. Polyribosome metabolism during normal and hormone-induced growth. *Plant Physiol.* **52,** 339–345.

Davies, E., and Larkins, B. A. (1974). Polyribosome degradation as a sensitive assay for endolytic messenger-ribonuclease activity. *Anal. Biochem.* **61,** 155–164.

Davies, E., and Larkins, B. A. (1980). Ribosomes. *In* "The Biochemistry of Plants: A Comprehensive Treatise" (P. K. Stumpf and E. E. Conn, eds.), Vol. 1. The Plant Cell. pp. 413–435. New York: Academic Press.

Davies, E., and Schuster, A. M. (1981). Intercellular communication in plants: Evidence for a rapidly-generated, bidirectionally-transmitted wound signal. *Proc. Natl. Acad. Sci. U.S.A.* **78,** 2422–2426.

Davies, E., Larkins, B. A., and Knight, R. H. (1972). Polyribosomes from peas. An improved method for their isolation in the absence of ribonuclease inhibitors. *Plant Physiol.* **50,** 581–584.

Davies, E., Ramiah, K. V. A., and Abe, S. (1986). Wounding inhibits protein synthesis yet stimulates polysome formation in aged, excised pea epicotyls. *Plant Cell Physiol.* **27,** 1377–1386.

Davies, E., Fillingham, B. D., Ito, Y., and Abe, S. (1991). Evidence for the existence of cytoskeleton-bound polysomes in plants. *Cell Biol. Int. Rep.* **15,** 973–981.

Davies, E., Comer, E. C., Lionberger, J. M., Stankovic, B., and Abe, S. (1992). Cytoskeleton-bound polysomes in plants: Polysome-cytoskeleton-membrane interactions in corn endosperm. *Cell Biol. Int.* **17,** 331–339.

Hesketh, J. E., and Pryme, I. F. (1991). Interaction between mRNA, ribosomes and the cytoskeleton. *Biochem. J.* **277,** 1–10.

Larkins, B. A. (1985). Polyribosomes. *In* "Modern Methods of Plant Analysis" (H. F. Linskens and J. F. Jackson, eds.), New Series, Vol. 1. Cell Components, pp. 331–352. Springer-Verlag. Berlin.

Larkins, B. A., and Davies, E. (1975). Polyribosomes from peas. V. An attempt to characterize the total free and membrane-bound polysomal populations. *Plant Physiol.* **55,** 749–756.

Pachter, J. S. (1992). Association of mRNA with the cytoskeletal framework: Its role in the regulation of gene expression. *Crit. Rev. Eukaryotic Gene Expression* **2,** 1–18.

Stankovic, B., Abe, S., and Davies, E. (1993). Co-localization of polysomes, cytoskeleton, and membranes with protein bodies from corn endosperm. *Protoplasma* **177,** 66–72.

Vassart, G. M., Dumont, J. E., and Cantraine, F. R. L. (1970). Simulation of polyribosome disaggregation. *Biochem. Biophys. Acta* **224**, 1019–1026.

You, Y., Abe, S., and Davies, E. (1992). Cosedimentation of pea root polysomes with the cytoskeleton.
Cell Biol. Int. Rep. **16**, 663–673.

Zak, E. A., Karavaiko, N. N., Sokolov, O. I., Nikolaeva, M. K., and Klyachko, N. L. (1995). Identification of nonribosomal polypeptides in the polysome preparations from *Vicia faba* L. leaves. *Russ.
J. Plant Physiol.* **42**, 68–74.

CHAPTER 16

Methods for Isolation and Analysis of the Cytoskeleton

Shunnosuke Abe* and Eric Davies†

*Laboratory of Molecular Cell Biology
College of Agriculture
Ehime University
Matsuyama 790, Japan
†School of Biological Sciences
University of Nebraska
Lincoln, Nebraska 68588-0118

I. Introduction

The plant cytoskeleton (CSK) consists of microfilaments (MF) and microtubules (MT) interchanging dynamically with their corresponding monomeric proteins, actin and tubulin (Seagull, 1989) and perhaps of intermediate filaments (IF) (Menzel, 1993). Monomeric actin (globular, or G-actin) and tubulin have been isolated from many plant tissues and described in detail (Fosket, 1989). Since this review is concerned with isolation of the CSK itself, we will not deal further with solubilized monomers.

Relatively little is known about actin-binding proteins (ABP) and tubulin-binding proteins (TBP) and even less about MF-associated proteins (MFAP) and MT-associated proteins (MTAP). This is because isolation of actin as MF and tubulin as MT is a prerequisite for identifying MFBP, MTAP, and any other cellular components that might associate *in vivo* with the CSK. Much of the information concerning MFAP in animal tissues has come from studies in which MF were isolated and proteins binding to these filaments were identified. There are few such studies with higher plant tissues, primarily because methods have only recently become available for the isolation of abundant amounts of MF (Abe and Davies, 1991). Similarly, very little has been published on MTAP (Fosket, 1989). A few biochemical studies exist on ABP (Fosket, 1989), on myosin (Grolig *et al.*, 1988), on spectrin (Faraday and Spanswick, 1993), and on IF (Menzel, 1993).

A major reason for the relative lack of information on the plant CSK compared with the animal CSK is that most plant cells are surrounded by a tough cell wall, which must be ruptured to isolate cytoplasmic components (Abe and Davies, 1991). Unfortunately, this disrupts the vacuolar membrane, releasing potentially noxious compounds into the extract and thus denying the use of a truly "physiological" buffer as employed by animal workers. Instead, buffers are formulated to buffer (literally "to protect against") these noxious compounds, and are deliberately "nonphysiological." Here, we describe methods employing buffers designed to keep the CSK intact and as close to its native state as possible, and we discuss some of the findings that come from these studies.

II. Materials

A. Chemicals

Mostly reagent grade, e.g., Sigma, Research Organics. Detergents: lithium dodecyl sulfate (LDS), polyoxyethylene 10-tridecyl ether (PTE), sodium deoxycholate (DOC), sodium dodecyl sulfate (SDS), Triton-X 100, Tween 20. RNase inhibitors: heparin (Sigma, H3393). Protease inhibitors: phenylmethyl sulfonyl fluoride (PMSF)—highly toxic. Fluorescent probes (Molecular Probes, Eugene, Oregon): 3,3′dihexyloxacarbocyanine iodide ($DiOC_6$) for membranes, acridine orange (AO) for DNA, FITC–colcemide for MT, RITC–(rhodamine-) phalloidin for F-actin, thiazole orange (TO) for RNA or polysomes. Electrophoresis and Western blotting reagents: acrylamide, ammonium persulfate, bromophenol blue (BPB), dimethylformamide, methylene-bisacrylamide (MB) (Biorad), defatted milk. Antibodies (Amersham): against actin (RPN 350), α-tubulin (RPN 356),

β-tubulin (RPN 357), biotinylated anti-mouse immunogloblin, species-specific whole antibody from sheep (RPN1001), streptavidin alkaline phosphatase conjugate (RPN1234), Miscellaneous: 5-bromo-4-chloro-3-indolylphosphate p-toluidine salt (BCIP), diethanolamine, dithiocrythritol (DTE), EGTA, glycerol, glycine, Hepes, magnesium chloride, 2-mercaptoethanol (BME), methanol, nitroblue tetrazolium (NBT), sodium chloride, sucrose (we purify our own by passing cane sugar through activated charcoal to remove RNase and UV-absorbing materials), Tris–HCl.

B. Supplies

Homogenizer, mortar and pestle, Pasteur pipets, Pipetman, forceps, scissors, filter paper (Whatman 3MM), PVDF membrane (Millipore, Immobilon-P), thick plastic bag, Miracloth (Calbiochem, Inc.), Cellophane sheet for drying gel.

C. Equipment

Centrifuges and accessories: high-speed centrifuge (e.g., Sorvall), swinging bucket rotor (e.g., HB4), Corex tubes, microfuge and tubes, ultracentrifuge (e.g., Beckman), swinging bucket rotor (e.g., SW41, SW55), and polyallomer tubes; fluorescence microscope (Nikon, Zeiss); heat block or boiling water bath; power supply for electrophoresis (500 V, 500 mA) with constant voltage; DC power supply for Western blotting (100 V, 500 mA) with constant current mode; electrophoresis tank; Western blotting apparatus; reciprocal shaker.

D. Buffers Used in Cytoskeleton Isolation

1. Cytoskeleton Stabilizing Buffer (CSB)

5 mM Hepes–3.2 mM KOH, pH 7.5, 10 mM Mg(OAc)$_2$, 2 mM EGTA, pH 7.5, 1 mM PMSF*, pH 7.5. CSB can be supplimented with 0.2 M sucrose and/or 0.5% PTE.* Other proteinase inhibitors can also be used. Buffer C (for gradients) consists of 5 mM Hepes–3.2 mM KOH, 10 mM Mg(OAc)$_2$.

Note: Stock solutions of 0.5 M EGTA and 0.1 M PMSF are adjusted with NaOH to attain the correct pH and so the total monovalent cation content of CSB is about 9 mM. All these stock solutions should be refrigerated. 1M Hepes and 1M Mg(OAc)$_2$ can be stored indefinitely at $-20°$C.

Role of buffer components: The most crucial component of CSB is an appropriate concentration of Mg^{2+}. At less than 5 mM, there is a great reduction in the yield of sedimentable CSK, even in the presence of low concentrations of monovalent cations, whereas at 25–50 mM Mg^{2+}, Mg–actin paracrystals can form (Abe and Davies, 1991). Low ionic strength buffers are needed to maintain the integrity of the CSK when detergents are present. Hepes allows isolation of large fragments of CSK, whereas Tris–HCl can cause massive disruption in some tissues (You *et al.,* 1992). Even though monovalent cations reduce CSK integrity, KOH or NaOH must be used to bring the pH to 7.5. PMSF is included to prevent the action of a range of proteases, whereas EGTA reduces the activity of a Ca^{2+}-dependent protease (Abe and Davies, 1991). The nonionic detergent PTE is used to release membranes from the CSK, and we prefer it to Triton-X and other more common detergents since it does not absorb at 260 nm (Section III,A). Sucrose is included (especially when PTE is absent and thus "fragile" membranes are present) to maintain the integrity of cytoskeleton–membrane (CM) complexes (Section III,B); it should be free of RNase (to lessen nucleic acid degradation) and proteinases (to lessen CSK degradation).

2. Buffer U (to Remove Polysomes from CM Complexes)

200 mM Tris–HCl, pH 8.5, 50 mM KCl, 25 mM $MgCl_2$, 2 mM EGTA, 100 μg/ml heparin, 2% PTE, 1% DOC. Note: See Chapter 15 for details on making buffer U. We normally make it as a 5$\times$ stock. Buffer A consists of 200 mM Tris–HCl, pH 8.5, 50 mM KCl, and 25 mM $MgCl_2$.

E. Buffers Used in Cytoskeleton Analysis

1. Staining Solutions

a. RITC–(rhodamine)–phalloidin or FITC–phalloidin dissolved in methanol at 3.3 μM and diluted with CSB to 1.2 μM for use; for F-actin, 3,3'dihexyloxacarbocyanine iodide ($DiOC_6$) at 3–5 μg/ml for membranes, acridine orange (AO) 3–5 μg/ml for DNA, FITC–colcemide (1 μM) for MT, thiazole orange (TO) at 3–5 μg/ml for RNA or polysomes (Molecular Probes, Inc., Oregon). All are dissolved in CSB, except where noted.

b. Antifading agent: A saturated solution in water (about 1 M) of DTE. This should be kept frozen until use and then allowed to warm up to room temperature and shaken well to make sure it is fully saturated. This is very important; at lower concentrations it is far less effective. Use 1 drop (5–10 μl) of saturated DTE per sample on slide (usually 30–40 μl) in CSB. The final concentration of DTE is between 200 and 250 mM.

2. Electrophoresis and Western Blotting

a. Sample buffer: 10 mM Tris–HCl (pH 6.8), 1.8% LDS, 20% glycerol, 2% BME, 0.005% BPB. Note: We use LDS instead of SDS because LDS does not precipitate when samples are refrigerated.

b. Running buffer: 25 mM Tris base, 190 mM glycine, 0.1% SDS.

c. Stacking gel: 125 mM Tris–HCl (pH 6.8), 0.1% SDS, 3–5% acrylamide as AMB 37:1.

d. Separating gel: 375 mM Tris–HCl (pH 8.8), 0.1% SDS, 7.5–15% acrylamide: methylene-bisacrylamide (AMB), 37:1.

Note: Polymerize with 0.1% ammonium persulfate and 0.1% TEMED before pouring gel. Do not disturb after pouring gel. Slight distortion of the gel will cause an uneven pattern of bands, especially at high voltage. Under our electrophoretic conditions, actin and both tubulin subunits from animal sources migrate close to their M_r, i.e., actin at 41.6, α- and β-tubulin at 50 kDa while plant actin migrates to 42 kDa, α-tubulin to 46–47 kDa, and β-tubulin to 48–50 kDa. It is necessary to specify buffers and acrylamide concentration, since different mobilities of tubulin subunits will be obtained (Fosket, 1989).

e. Western blotting buffer: 25 mM Tris base, 190 mM glycine, 0.03% SDS, 20% methanol. Note: Amounts of SDS and methanol can be varied to obtain maximal transfer.

f. Substrates for alkaline phosphatase (10-ml volume): 100 mM diethanolamine–HCl (pH 9.5), 5 mM MgCl₂, 1.65 mg BCIP in 33 μl dimethylformamide, 3.3 mg NBT in 44 μl 70% dimethylformamide.

g. Others: TBS, 20 mM Tris–HCl (pH 7.6), 137 mM NaCl; TBS-T, 0.1% Tween 20 in TBS (TBS and TBS-T can be made as 10× stocks); blocking solution, 5% defatted dry milk or 1% BSA in TBS-T.

III. Methods

A. Isolation of Cytoskeleton-Enriched Fractions from Plants

The first methods for isolating relatively intact CSK in amounts sufficient for biochemical analysis employed protoplasts from carrot suspension cells (Hussey *et al.,* 1987). These pioneering methods suffer from two potential drawbacks. First, they are suitable only for tissues that yield protoplasts easily. Second, the conditions needed to release protoplasts can themselves cause substantial changes in the CSK and thus lead to artifacts (Tan and Boss, 1992). Recently, we have developed methods for the isolation of abundant amounts of relatively intact CSK from pea stems (Abe and Davies, 1991), corn endosperm (Abe *et al.,* 1992), pea roots (You *et al.,* 1992), and a wide variety of plant and animal tissues including dicots, monocots, gymnosperms, ferns, mosses, algae, cold-blooded animals, and warm-blooded animals. Examples of pelleted and stained CSK fractions from *Phyllostachys, Cycas,* and *Chara* are shown in Fig. 1 (unpublished results).

The method suitable for most plant tissues is as follows. Note: All steps are conducted at 2–4°C.

1. Grind tissue gently with a mortar and pestle in 5 to 10 vol of CSB plus 0.5% PTE.

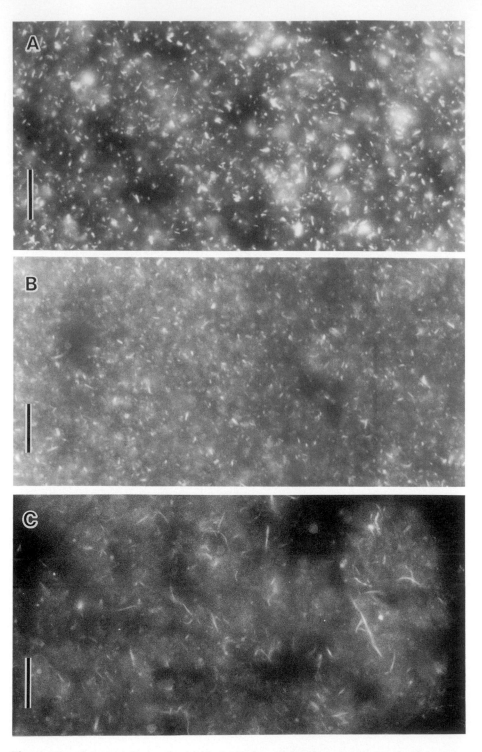

Fig. 1 Cytoskeleton pellets stained with rhodamine–phalloidin and viewed under fluorescence. Filters used; excitation 546–520 nm, emission 580 nm. (A) Young shoot of *Phyllostachys pubescens* (bamboo) (27,000×*g* pellet); (B) endosperm of *Cycas* (4000×*g* pellet); (C) internodal cells of *Chara australis* (4000×*g* pellet). Bars, 10 μm.

2. Filter (do not squeeze) the brei through two layers of Miracloth.

3. Centrifuge sequentially for 5 min at 250×g, 15 min at 4000×g, 15 min at 27,000×g, and for 2 h at >100,000×g, in a swinging bucket rotor. Save pellets at each step for analysis.

4. Analyze each pellet for the presence of CSK components. (See Section III,E). Note: with different tissues, different pellets constitute the major CSK fraction. In corn endosperm, most of the CSK and associated protein bodies are found in the 250×g pellet, in pea stems most is in the 4000×g pellet, and in pea roots most is in the 27,000×g pellet. In many tissues, very fine fragments also sediment at 100–300,000×g. The centrifugal force required to pellet CSK depends not only on the tissue, but also on the severity of grinding. Major contaminants will reflect the g forces employed to obtain the CSK fraction. For instance, in corn endosperm (the 250×g pellet), the main contaminants are protein bodies and nuclei (Abe *et al.*, 1991), whereas for leaves the contaminants are largely chloroplasts.

B. Isolation of Cytoskeleton–Membrane Complexes

This is done in essentially the same manner as above for the CSK itself, except that detergent is omitted from the grinding buffer and higher levels of monovalent cation can be included to reduce nonspecific binding, but an osmoticum (e.g., 0.2 M sucrose) should also be included (Abe *et al.*, 1992, 1994).

To isolate the CM complex:

1. Grind tissue in CM-stabilizing buffer (CMSB), i.e., CSB plus 0.2 M sucrose and 150 mM KOAc.

2. Filter (do not squeeze) through Miracloth.

3. Centrifuge at 250×g for 5 min to remove debris. Note: in corn endosperm, most of the polysomes are in this pelleted fraction (Davies *et al.*, 1993; Stankovic *et al.*, 1993).

4. Centrifuge (e.g., 15 min, 27,000×g) to pellet the CM complex.

5. Wash the pellet with water and drain on Kimwipes. This pellet contains the CM complex and associated polysomes. Most of the monosomes and soluble proteins are discarded in the supernatant.

6. Analyze for CSK proteins, membrane phospholipids, and ribosomes (see Section III,E).

When isolated in this way, the CM complex contains abundant amounts of membrane phospholipids, actin, tubulin, and streptavidin-binding proteins (SBPs) (Abe *et al.*, 1992; Ito *et al.*, 1994), polyribosomes (Chapter 15; see next section), various enzymes in the phosphatidylinositol pathway (Tan and Boss, 1992), and numerous other proteins (Ito *et al.*, 1994) including a putative ribosome–cytoskeleton binding protein (Abe *et al.*, 1995). In peas, membrane lipids in this fraction are phosphatidylcholine, phosphatidylethanolamine, and phosphatidylinositol (Ito *et al.,* 1994).

C. Identification of Cytoskeleton–Membrane–Polysome Complexes

1. Layer samples obtained above, either the Miracloth filtrate (step 2, Section III,B) or the pellet resuspended in CMSB (step 4, Section III,B) on gradients of 20–80% sucrose in buffer C and centrifuge at $250,000 \times g$ for 1 h.

2. Scan the gradient at 254 nm (for RNA), 280 nm (for protein), or both. Since the gradient is very dense, a dense chase solution such as fluorinert is needed to displace the gradient through the UV monitoring device (see Chapter 15).

3. Collect fractions and analyze for CSK proteins, membrane phospholipids, and polyribosomes (Fig. 2).

Note: The dense sucrose gradients are needed because the cytoskeleton–membrane–polysome (CMP) complex is itself a dense structure (Ito *et al.,* 1994). When samples contain additional ions (e.g., 150 mM KOAc), we recommend using the same concentration in the buffer C gradients, to avoid causing nonspecific interactions during sedimentation.

D. Release of Membranes and Polysomes from Cytoskeleton–Membrane–Polysome Complexes

When 0.5% PTE is added to a CMP complex and analyzed as above, membranes are solubilized, and the resulting cytoskeleton–polysome (CP) complex becomes sufficiently dense that it sediments to the bottom of the gradient. Addition of one-fourth volume of 5× buffer U to a similar sample results in release of polysomes, and these can be characterized by gradient separation (see Chapter 15).

Note: The CM complex may yield what looks like a typical polysome profile. However, RNase will not cause conversion of material in the peak to monosomes and small polysomes, thus showing it is not a polysome profile (Ito *et al.,* 1994).

E. Analysis of Components in the Cytoskeleton Fraction

1. Fluorescence Microscopy

a. *Microfilaments*

Rhodamine-conjugated phalloidin reliably stains the CSK pellets from a wide variety of tissues and is very easy to use (Fig. 1). Staining with antibodies against CSK proteins is more difficult and will not be described here.

i. Place a 5-μl aliquot of methanol stock of rhodamine–phalloidin or other fluorescent conjugate in a microfuge tube.

ii. Dry *in vacuo* for 10 min using a SpeedVac.

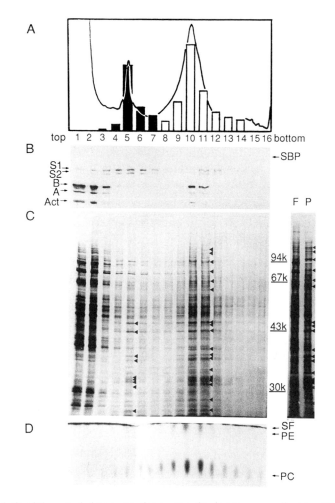

Fig. 2 Analysis of the cytoskeleton–membrane complex in sucrose gradient fractions of pea stem homogenate for ribosomes, cytoskeleton proteins, and phospholipids. Pea stem tissue was ground in CSB containing 7% sucrose and filtered through Miracloth, and the filtrate was layered onto linear 20 to 80% sucrose gradients in buffer C and centrifuged at $300,000 \times g$ for 50 min. Gradients were monitored at 254 nm (A) and 0.3 ml fractions were collected and assayed. (A) Monosomes (solid bars) and polysomes (open bars); (B) Western blot of CSK proteins, actin (Act), α-tubulin (A), β-tubulin (B), 78-kDa SBP (SBP), and smaller SBPs (S1 and S2); (C) silver-stained proteins; (D) phosphatidylethanolamine (PE), phosphatidylcholine (PC), phosphatidylinositol (PI), and solvent front (SF). In (C) F and P represent silver-stained proteins of the filtrate and the pellet ($15,000 \times g$, 15 min), respectively. Solid triangles indicate proteins that were predominant in the pellet, but were also found in the peak and in the monosome fractions. About 16 and 2% of total CSK proteins were found in the peak region (10–11) and the gradient pellet, respectively. About 20% of the total CSK proteins sedimented at $100,000 \times g$ for 1 h, and presumably were in the polymerized form (from Ito *et al.*, 1994, by permission of Oxford University Press).

Note: This dried material can be stored indefinitely desiccated at $-20°C$.

iii. Add 3 μl of a saturated solution of DTE to the tube.

iv. Add 10 μl of resuspended pellet in CSB (lacking detergent) or any other liquid sample to be examined.

v. Mix gently and stand on ice for 5 min.

vi. Place a 2- to 3-μl drop on a glass slide.

vii. Immediately place a coverslip carefully over the sample and allow it to spread. Note: Do not slide the coverslip about and do not try to squeeze bubbles out, or you will generate artificially aligned or broken structures. If you get many bubbles, discard and prepare a new sample.

viii. Seal the coverslip with nail polish.

ix. View the slide under a fluorescence microscope with the appropriate filter combination.

Notes: Use large-bored pipets (or cut off the tip of regular pipets) to prevent shearing of the CSK. If fluorescence fades rapidly, check the DTE solution. If it colored, make a fresh solution. If it is not saturated, add more solid DTE. If it is not warm enough, warm it up to 30°C and shake well. Destaining is not necessary. For observation of fluorescence from RITC conjugates, use green (546 nm excitation and 580 nm emission). For FITC conjugates, use blue (495 nm excitation and 520 nm emission). Chloroplasts are normally present in the CSK pellet from green tissues and chlorophyll fluoresces in the red region (around 640 nm) upon green excitation. To exclude chlorophyll autofluorescence, we use a special band path filter (570–590 nm).

b. Microtubules

Staining of MT with FITC–colcemide can be performed as above for rhodamine–phalloidin staining of F-actin. However, fluorescence of microtubules (tubulin) in these CSK pellets from plant sources is not very clear (Abe and Davies, 1991; Fosket, 1989). Other papers deal quite adequately with MT staining (Lloyd, 1987; Hussey *et al.*, 1987) and with MT–membrane complexes (Laporte *et al.*, 1993).

c. Membranes

Detection of membranes in isolated fractions by fluorescence can be done with $DiOC_6$ using blue excitation (Stankovic *et al.*, 1993). Detection of membrane phospholipids can be done with thin-layer chromatography using silica gel treated with 10 mM sodium acetate, and developed with a suitable solvent (Abe *et al.*, 1992). A typical example is shown in Fig. 2D.

d. Polysomes and RNA

Many polysomes are found in the CSK pellets from plants and these are relatively undegraded in tissues low in RNase such as pea stems (Davies *et al.*,

1991), but are degraded in tissues high in RNase such as pea roots (You *et al.,* 1992). Detection of polysomes (RNA) by fluorescence can be done with AO, which also stains DNA, or TO, which in our hands is more specific for RNA. Since the same filter set-up is used for TO as for $DiOC_6$, simultaneous visualization of membranes and polysomes cannot be achieved. We have used this method to show colocalization of F-actin, membranes, and RNA (polysomes) around protein bodies in corn endosperm (Stankovic *et al.,* 1993). The rRNA content of CSK fractions can be determined by GPS gradient centrifugation and ribosomal proteins can be identified by SDS–PAGE (see Chapter 15).

The easiest and most common method to assay for polysomes is to release them from the CM complex by buffer U and analyze the released polysomes on sucrose gradients:

i. Resuspend CSK pellet in Buffer U and vortex thoroughly.

ii. Add proteinase K (10 mg/ml stock in water) to 100 μg/ml.

iii. After 5 to 30 min of incubation on ice, layer a 200-μl aliquot onto a 15–60% sucrose gradient in Buffer B (50 mM Tris–HCl, pH 8.5, 25 mM KOAc, 10 mM $Mg(OAc)_2$) and process in the normal manner (see Chapter 15).

2. Electrophoresis

Gel electrophoresis can be performed using equipment from various suppliers. Generally, larger size gels are better for resolution but electrophoresis takes much longer (6–12 h). Increasing the voltage to speed up the process causes massive band distortion. A mini-gel system is convenient for quick results, since it takes only about 1 h, but band resolution is not as good. Since the plant CSK fraction contains many proteins, a larger size gel is recommended for CSK analysis.

To circumvent these difficulties, we have designed a method to run standard gels under high voltage which is almost as rapid as mini-gels, and which does not cause band distortion. With this equipment, electrophoresis is performed at 35 to 50 V/cm for 50 to 60 min. Overheating (band distortion) and band diffusion are significantly reduced by having both edges of the glass plates insulated and directly cooled by buffers. With a stacking gel over the separating gel, we recommend using half voltage until proteins pass through the stacking gel and then raising to full voltage after the samples have entered the separating gel. This has the great advantage of allowing analysis of 40–48 samples at a time (two gels with 20–24 wells each can be installed in the equipment) within just 60 min. Figure 3 shows a typical example of CSK proteins separated on such a gel.

3. Western Blotting

a. Remove gel from glass plates, and immerse in Western blotting buffer for 15 min with gentle shaking.

b. Place the equilibrated gel on five sheets of filter paper stacked on the negative electrode plate of the blotting apparatus.

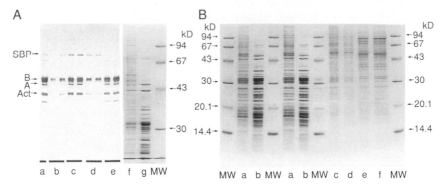

Fig. 3 High voltage SDS–PAGE of cytoskeleton fractions. (A) Various CSK fractions were electrophoresed at 350 V for 60 min in a 10% gel, transferred to PVDF membrane, and probed with antibodies to actin and α- and β-tubulin on the same "triple blot." Lanes are: a, homogenate; b, 250× *g* pellet; c, 4000× *g* pellet; d, 300,000× *g* pellet; e, 300,000× *g* supernatant. Lane f is the Brilliant blue-stained protein spectrum of the 4000× *g* pellet in the same gel; MW, molecular weight markers (94, 67, 43, 30 kDa). (B) CSK fractions and ribosomal proteins were electrophoresed in a 15% gel with 12 wells at 350 V for 60 min and stained with Brilliant blue. Lanes are: a, 4000× *g* pellet; b, ribosomes; c, membrane fraction obtained with CSB; d, membrane fraction obtained with buffer A (see Chapter 15); e, supernatant of c; f, supernatant of d; MW, molecular weight markers (94, 67, 43, 30, 20.5, and 14 kDa).

c. After the gel is positioned, stack an additional six sheets of filter paper above the positive electrode plate. These electrodes should be as flat as possible, since any slight deviation might cause uneven transfer and proteins can diffuse out from the edges.

d. Block the membrane with freshly obtained, defatted dry milk.

e. Incubate with antibodies in a sealed plastic bag with a flat weight covering its whole surface. This greatly reduces the amount of antibodies needed—only 2 or 3 ml of liquid is sufficient for blotting 2 sheets of standard size (10 × 15 cm) membrane.

f. Incubate with biotinylated secondary antibodies.

g. Incubate with streptavidin-conjugated alkaline phosphatase at 1 : 1000. Note: this will also stain SBP that have not bound the antibody; thus negative controls must be used.

h. Develop the membrane with substrates (BCIP, NBT) in diethanolamine buffer (pH 9.5) under constant shaking for several minutes.

i. Wash the membrane thoroughly with water and let it dry on a paper towel.

Note: Wash membrane after each of steps d to g.

In plant cells, the mobilities of all three major CSK proteins (actin and α- and β-tubulin) are different enough to probe on the same blot by mixing the three antibodies and using our "triple-blotting" technique (Abe *et al.,* 1992). The triple-blot has the distinct advantage of showing relative amounts of the three proteins in the same gel lane and it cuts down on the number of gels and blots needed. This method is especially convenient when analyzing multiple fractions from gradients (Ito *et al.,* 1994).

IV. Discussion

The general methods described here for CSK isolation are simple and reproducible, can be done quickly, and work with various plant sources (see Fig. 1). In tissues such as corn endosperm where the CSK can be obtained by sedimenting for 5 min at $250 \times g$, it takes less than 15 min for isolation. These methods involve simple procedures (homogenization and differential centrifugation) using equipment normally available in most laboratories and can easily be scaled up for preparative purposes. For instance, isolation of the CSK from 50 g of tissue in 250 ml of CSB can be done easily. These simple and quick methods are finding increasing usage. We have used them to isolate undegraded cytoskeleton-bound polysomes (Davies *et al.,* 1991; Abe *et al.,* 1992; You *et al.,* 1992) and cytoskeleton-membrane-bound polysomes (CMBP) (Stankovic *et al.,* 1993). Others have used these methods to identify CMBP in rice endosperm (Li *et al.,* 1993a,b) and in broad bean leaves (Zak *et al.,* 1995) and to show that a variety of kinases from the phosphatidylinositol pathway (Tan and Boss, 1992) and a PIPK activator that shows actin bundling and elongation factor activity (Yang *et al.,* 1993) are associated with the CSK in carrot cells.

Acknowledgments

This work was supported by NSF Grant IBN-9310508, the UN-L Research Council, the UN-L Center for Biotechnology (to E.D.), the Japanese Ministry of Education, Sciences and Culture, Calsonic, Inc., and BioCraft, Tokyo (to S.A.). We thank Ms. Yoko Ito, and Hiroko Doi for help with experiments.

References

Abe, S., and Davies, E. (1991). Isolaton of F-actin from pea stems: Evidence from fluorescence microscopy. *Protoplasma* **163,** 51–61.

Abe, S., Ito, Y., and Davies, E. (1992). Co-sedimentation of actin, tubulin and membranes in the cytoskeleton fractions from peas and mouse 3T3 cells. *J. Exp. Bot.* **252,** 941–949.

Abe, S., Ito, Y., and Davies, E. (1994). Association of cytoskeletal proteins in the membrane-bound polysome fraction from peas using conventional polysome isolation buffers. *Plant Physiol. Biochem.* **32,** 547–555.

Abe, S., Ito, Y., and Davies, E. (1995). Isolation of a heparin sensitive, ribosome sedimenting factor from the cytoskeleton fractions of peas and corn. *Plant Physiol. Biochem.* **33.**

Abe, S., You, W., and Davies, E. (1991). Protein bodies in corn endosperm are enclosed by and enmeshed in F-actin. *Protoplasma* **165,** 139–149.

Davies, E., Fillingham, B. D., Ito, Y., and Abe, S. (1991). Evidence for the existence of cytoskeleton-bound polysomes in plants. *Cell Biol. Int. Rep.* **15,** 973–981.

Davies, E., Comer, E. C., Lionberger, J. M., Stankovic, B., and Abe, S. (1993). Cytoskeleton-bound polysomes in plants. III. Polysome-cytoskeleton-membrane interactions in corn endosperm. *Cell Biol. Int. Rep.* **17,** 331–340.

Faraday, C. D., and Spanswick, R. M. (1993). Evidence for a membrane skeleton in higher plants: A spectrin-like polypeptide co-isolates with rice root plasma membranes. *FEBS Lett.* **318,** 313–316.

Fosket, D. E. (1989). Cytoskeletal proteins and their genes in higher plants. *In* "The Biochemistry of Plants," vol. 15, pp. 393–454, New York: Academic Press.

Grolig, F., Williamson, R. E., Parke, J., Miller, C., and Anderton, B. H. (1988). Myosin and Ca^{2+}-sensitive streaming in the alga Chara: Detection of two polypeptides reacting with a monoclonal anti-myosin and their localization in the streaming endoplasm. *Eur. J. Cell Biol.* **47,** 22–31.

Hussey, P. J., Traas, J. A., Gull, K., and Lloyd, C. W. (1987). Isolation of cytoskeletons from synchronized plant cells: The interphase microtubule array utilizes multiple tubulin isotypes. *J. Cell Sci.* **88,** 225–230.

Ito, Y., Abe, S., and Davies, E. (1994). Colocalization of cytoskeleton proteins and polysomes with a membrane fraction from peas. *J. Exp. Bot.* **45,** 253–259.

Laporte, K., Rossignol, M., and Traas, J. A. (1993). Interaction of tubulin with the plasma membrane: Tubulin is present in purified plasmalemma and behaves as an integral membrane protein. *Planta* **191,** 413–416.

Li, X., Franceschi, V. R., and Okita, T. W. (1993a). Segregation of storage protein mRNAs on the rough endoplasmic reticulum membranes of rice endosperm cells. *Cell* **72,** 869–879.

Li, X., Wu, Y., Zhang, D.-Z., Gillikan, J. W., Boston, R. S., Franceschi, V. R., and Okita, T. W. (1993b). Rice prolamine protein body biogenesis: A BiP-mediated process. *Science* **262,** 1054–1056.

Lloyd, C. W. (1987). The plant cytoskeleton: The impact of fluorescence microscopy. *Annu. Rev. Plant Physiol.* **38,** 119–199.

Menzel, D. (1993). Chasing coiled coils: Intermediate filaments in plants. *Bot. Acta* **106,** 294–300.

Seagull, R. W. (1989). The plant cytoskeleton. *Crit. Rev. Plant Sci.* **8,** 131–167.

Stankovic, B., Abe, S., and Davies, E. (1993). Co-localization of polysomes, cytoskeleton, and membranes with protein bodies from corn endosperm: Evidence from fluorescence microscopy. *Protoplasma* **177,** 66–72.

Tan, Z., and Boss, W. F. (1992). Association of phosphatidylinositol kinase, phosphatidylinositol monophosphate kinase, and diacylglycerol kinase with the cytoskeleton and F-actin fractions of carrot (*Daucus carota* L.) cells grown in suspension culture. *Plant Physiol.* **100,** 2116–2120.

Yang, W., Burkhart, W., Cavallius, J., Merrick, W. C., and Boss, W. F. (1993). Purification and characterization of a phosphatidylinositol 4-kinase activator in carrot cells. *J. Biol. Chem.* **268,** 392–398.

You, W., Abe, S., and Davies, E. (1992). Cosedimentation of pea root polysomes with the cytoskeleton. *Cell Biol. Int. Rep.* **16,** 663–673.

Zak, E. A., Karavaiko, N. N., Sokolov, O. I., Nikolaeva, M. K., and Klyachko, N. L. (1995). Identification of nonribosomal polypeptides in the polysome preparations from *Vicia faba* L. leaves. *Russ. J. Plant Physiol.* **42,** 68–74.

CHAPTER 17

Isolation and Characterization of Plasmodesmata

Bernard L. Epel, Bella Kuchuck, Guy Kotlizky, Shomrat Shurtz, Michael Erlanger, and Avital Yahalom

Botany Department
George S. Wise Faculty of Life Sciences
Tel Aviv University
Tel Aviv 69978, Israel

I. Introduction

Plasmodesmata are dynamic membrane specializations that traverse plant cell walls forming aqueous channels linking adjacent cells. The plasmodesma, like its animal counterpart, the gap junction, functions in the cytoplasmic movement of metabolites and ions, and there is strong evidence that plasmodesmata may function to provide a mechanism for intercellular signaling. In addition to their normal physiological function, plasmodesmata can be altered and exploited by viruses as conduits for viral spread from cell to cell. There are also some indications that plasmodesmata in companion cells may be specialized, allowing for the transport of proteins from the companion cells to sieve tube elements.

Although plasmodesmata were first described more than 100 years ago, detailed studies of the structure and function of plasmodesmata were not begun until the past 10–15 years. Studies of thin section by electron microscopy have led to a number of different models (for review see Robard and Lucas, 1990). A consensus model of the plasmodesma is that it is a tubular membranous structure between 20 and 100 nm in diameter and between 100 to over 1000 nm in length. The outer limit of a plasmodesma is formed by the plasmalemma, which is continuous from cell to cell. Within the interior of the plasmalemma tubular envelope runs a strand of modified endoplasmic reticulum that is apparently appressed as seen in static ER micrograph studies and has been termed by various workers as "desmotubule" or "appressed endoplasmic reticulum." Computer-enhanced image processing of TEM micrographs of plasmodesmata has enabled a clearer resolution of substructural components of plasmodesmata. According to these studies, globular proteins are embedded in both the inner and the outer leaflets of the desmotubule and in the inner surface of the plasmalemma. Such studies, however, have not forwarded our understanding of the molecular nature or physiological functions of plasmodesmata.

Until recently, little was known about the composition of plasmodesmata or about the proteins involved in the regulation of plasmodesmatal conduction. A major obstacle to a biochemical and molecular analysis of plasmodesmata has been the difficulty in the isolation of these structures. The remainder of this chapter will be devoted to describing techniques developed in the authors' laboratory for the isolation and characterization of plasmodesmata.

II. Materials

Plant material: Maize seeds (*Zea mays* L. cv Jubilee) can be obtained from Roger Bros. Co., Idaho Falls.

Nitrogen pressure bomb: Parr Instrument Co. Cell Disruption Bomb, Model 4635; Internal volume 920 ml, maximum sample size 600 ml, minimum sample size 20 ml in test tube. Manufacturer, Parr Instrument Co., 211 Fifty-third Street, Moline, IL 61265.

Protease inhibitors: Leupeptin hemisulfate (Sigma Chemical Co., St. Louis, MO, Cat. No. 12884). *Stock solution:* Dissolve 10 mg/ml H_2O; divide into aliquots of 100 μl, and store in microfuge tubes at $-20°C$.

Pepstatin A (Sigma Chemical Co., Cat. No. P 4265): *Stock solution:* Prepare 1 mM stock in dimethyl sulfoxide (DMSO) and store in 100-μl aliquots in microfuge tubes at $-20°C$.

PMSF (Sigma Chemical Co, St Louis, cat #P7626): *Stock solution:* Prepare 500 m*M* stock in DMSO. Store at 4°C.

Epiamastatin hydrochloride (Sigma Chemical Co., St. Louis, Cat. No. E 3389): Add 5 μg/ml directly to cellulase solution.

1,10-phenanthroline (Merck, Darmstadt, Germany, cat # 7225): Add 2 mg/ml directly to cellulase solution.

HM/8.5 homogenization buffer: 20 m*M* Tris–HCl, pH 8.5; 0.25 *M* sucrose; 10 m*M* EGTA; 2 m*M* EDTA.

HM/7.5 Homogenization buffer: 20 m*M* Tris–HCl, pH 7.5; 0.25 *M* sucrose; 10 m*M* EGTA; 2 m*M* EDTA.

AMM buffer: Wall hydrolysis buffer/plasmodesmatal wash buffer: 10 m*M* sodium acetate buffer, pH 5.3; 0.7 *M* mannitol; 2m*M* MgCl$_2$; 0.02% NaN$_3$.

Cellulase: Prepare a 10% solution of Cellulase TC from *Trichoderma reesi* (Serva Feinbiochemica GmbH and Co., Heidelberg, Germany, Cat. No. 16421) in AMM Buffer. Clear the solution of insoluble material by centrifugation at 16,000 × *g* for 30 min. Dilute the cleared 10% cellulase solution with 19 vol of AMM to a final concentration of 0.5% cellulase. For each milliliter of cellulase solution, add 5 μg epiamastatin, 1 μl leupeptin stock, 1 μl pepstatin stock, 2 μl PMSF (500 m*M* stock), 2 mg 1,10-phenanthroline.

2× SB: Double strength electrophoresis sample buffer: 4.4 g sodium dodecyl sulfate (SDS); 22 ml glycerol; 20 ml 0.5 *M* Tris–HCl, pH 6.8; 11 ml 2-mercaptoethanol; water to 100 ml.

Bovine serum albumin protein standards, 1.5 μg/μl BSA stock: Float 30 mg BSA on 10 ml double distilled water and let protein dissolve into water. Do not mix until protein dissolves into water. Gently stir and add 10 ml 2× SB to make 1.5 μg/μl stock solution. Dilute BSA stock to make the following protein standards.

1.0 μg/μl standard	4 ml BSA stock + 2 ml 1× SB
0.75 μg/μl standard	3 ml BSA stock + 3 ml 1× SB
0.5 μg/μl standard	2 ml BSA stock + 4 ml 1× SB
0.25 μg/μl standard	1 ml BSA stock + 5 ml 1× SB
0.125 μg/μl standard	0.5 ml BSA stock + 5.5 ml 1× SB

Coomassie brilliant blue-R-250 (CBB) solution: 1.25 g Coomassie brilliant blue R-250; 500 ml methanol; 100 ml glacial acetic acid; water to final volume of 1 liter; if necessary, filter on Whatman 1-mm filter paper.

Destain I: 500 ml methanol; 100 ml glacial acetic acid; water to final volume 1 liter.

Destain II: 70 ml methanol; 50 ml glacial acetic acid; water to final volume of 1 liter.

Electrotransfer buffer: 30 ml 500 m*M* ethanolamine titrated with 500 m*M* glycine to pH 9.5; 240 ml methanol; water to final volume of 1200 ml.

Ponceau S protein stain: 1 g Ponceau S (Ponceau S, Sigma Chemical Co, St Louis, Cat. No. P 3504); 280 ml 25% trichloroacetic acid; water to final volume of 1 liter.

TBS ×4 stock: 96.8 g Tris–HCl; 92.8 g NaCl; 1900 ml water; adjust pH to 7.4; water to final volume 2 liters.

TTBS: 250 ml TBS ×4 stock; 1 ml Tween 20 (final concentration 0.1%); water to final volume 1 liter.

Blocking solution: 250 ml TBS ×4 stock; 50 g nonfat dry milk; 10 ml 2% sodium azide; water to final volume 1 liter.

Alkaline phosphatase developing buffer: 12.2 g Tris; 5.84 g NaCl; 0.4 g $MgCl_2$; 900 ml H_2O; adjust pH to 9.5; water to final volume 1 liter.

NTB: 0.5 g *p*-nitro blue tetrazolium chloride (Sigma Cat. No. N 6876); 10 ml 70% dimethylformamide; Store at −20°C.

BCIP: 0.5 g bromo-4-chloro-3-indolyl phosphate, *p*-toluidine salt (Sigma, Cat. No. B 8503); 14.7 ml dimethylformamide; store at −20°C.

III. Methods

A. Protein Content Analysis: Marder CBB Filter Paper Method

Protein content analysis is performed by a modification of the method of Marder *et al.* (1986). This is a very convenient and simple method for determining approximate protein concentrations for membrane proteins that are extracted with SDS. With a pencil, mark a 4 by 8 grid on a 6 by 12-cm rectangular piece of Whatman 3 MM. Place filter paper on sheet of Parafilm and spot 5-μl aliquots of each BSA protein standard dissolved in SB ×1 (0.125, 0.25, 0.5, 0.75, 1.0, 1.5, 2.0 μg/μl) onto filter paper. Serially dilute isolated wall protein 1:1 and 1:3 (v/v) in SB ×1 and spot 5 μl of undiluted and diluted isolated wall protein into labeled grid on filter paper. Dry filter paper and stain for 10–20 min in a Coomassie brilliant blue-R-250 (CBB) solution. Place filter paper in Destain I for about 10 min, and then into Destain II until background stain is removed. Add small piece of polyurethane foam to absorb CBB and speed up destaining. Estimate protein concentration by comparing staining intensity of serially diluted sample with that of dilution series of standards.

B. Isolation of Plasmodesmata Embedded in Clean Cell Walls

The strategy for isolating plasmodesmata involves two major steps. The first step involves isolating a fraction enriched in plasmodesmata. Since plasmodesmata are membranous structures embedded in cell walls, isolated cell wall would constitute an enriched fraction (Yahalom *et al.* 1991; Kotlizky *et al.*, 1992). The second step involves the enzymatic digestion of the encasing cell wall to release free plasmodesmata.

1. Plant Material

Surface sterilize 1 kg maize seed with 2 liters of a 1% (final concentration) solution of sodium hypochloride (bleach) in tap water for 10 min. Rinse seeds two to three times with tap water to get rid of residual hypochloride and then imbibe in running tap water overnight.

Seeds may be planted either in moist vermiculite in plastic trays (for upper plant parts) or in rolled paper towels (for roots) and grown in the dark for 5 days at 25°C. To grow seedlings in vermiculite, plant imbibed seed in five trays (surface area of a tray is about 600 cm^2) on a 2-cm-deep layer of premoistened vermiculite and cover with about 1 cm moist vermiculite. Do not oversaturate vermiculite with water. Decant excess water. To grow seedlings in rolled paper towels: space imbibed seeds 1 cm apart on the dry paper towel about one-third distant from the top of the sheet; roll the towel tightly and fasten with masking tape. Pack rolls in bucket or other tall container that will support the towel; Add water to container to one-third height of rolled paper towel. Capillary action will keep upper two-thirds of towel wet and well aerated.

Excise 10 mm mesocotyl section adjacent to coleoptile node. Each tray should yield about 20–25 g of mesocotyl sections. Root sections are taken from a 10-mm region above the root cap. Coleoptiles and leaves can be separated by nicking the turgid coleoptile about 2 mm above coleoptile node and about 5 mm below coleoptile tip. Bend the turgid coleoptile to cause it to break at nicks. Remove coleoptile tip exposing upper region of leaf. Slit coleoptile longitudinally to base and pull leaf downward along slit until leaf is separated from coleoptile. Freeze separated sections and stored at −70°C. Use tissue only from regions that are still undergoing cell elongation or cell expansion.

2. Preparation of Mesocotyl Cell Walls

1. Immerse mesocotyl segments in liquid nitrogen in a large porcelain mortar and pulverized with pestle to a fine powder. Batches of about 25 g of tissue are recommended.

2. Transfer pulverized frozen mesocotyls to a fresh mortar chilled to 4°C containing, for each gram of tissue, 2 ml homogenization media HM/8.5. For each ml HM/8.5, add 2 μl leupeptin stock and 4 μl of stock solution of PMSF. The PMSF should always be added directly to the tissue in the homogenization medium upon commencement of grinding. Homogenization should be performed for at least 5 min.

3. In a cold room, separate the cell walls by filtration onto nylon cloth (200 mesh) and express all liquid leaving a damp cell wall mat (use plastic or latex gloves). Spread out nylon cloth on a cold glass plate and remove cell wall fraction by carefully scraping with a spatula.

4. Repeat one to three more times until about 100 g of tissue has been processed.

5. Combine and resuspend the cell wall fraction in a minimal volume of homogenization media HM/7.5 to give a thin paste (about 30–40 ml/100 g tissue), add PMSF, 2μl/ml buffer, and rehomogenize with cold mortar and pestle for

5 min. It is important that the suspension not be too dilute; otherwise one does not get a good disruption of cells. At the end of 5 min of grinding, the volume of the grinding medium is brought to 200 ml, using grinding medium fortified with 2 μl/ml PMSF.

6. Transfer homogenate to a plastic 250-ml Sorvall GSA centrifuge bottle and float bottle on ice water within the cylinder of a Paar 4635 cell disruption bomb (Parr Instrument Co., Moline, IL). Place a magnetic stirring bar in bottle, seal bomb, placed on a magnetic stirring plate, and pressurize to 10 MPa from a nitrogen bottle. After a few minutes there is generally a pressure drop as nitrogen dissolves into the sample. Repressurize to 10 MPa.

The sample should be continuously stirred under 10 MPa nitrogen for at least 20 min. It may be necessary to add additional nitrogen in order to maintain the desired pressure level.

7. After 20–30 min, discharge the cell wall fraction from the bomb into a 1-liter side arm suction flask held on ice. The sample should be released slowly and the pressure within the bomb maintained at about 8–10 MPa by adding nitrogen from time to time as the pressure drops. It is advisable to have the flask securely clamped and to seal the flask neck with parafilm to prevent splattering (gas can escape through the side arm). The disruption process occurs as the sample is decompressed as it passes from the high-pressure environment inside the bomb to atmospheric pressure.
Note: Care must be taken when the last of the liquid is expelled, as there is a sudden burst of gas and all your sample may end up painting the ceiling or the walls or giving you a shampoo.

The discharged sample is frothy and charged with nitrogen bubbles, and must be degassed before filtration. Degas sample by repeatedly applying vacuum for a few seconds and then rapidly releasing the vacuum to break the froth formed. The presence of bubbles prevents good washing and filtration of the sample.

8. Filter degassed sample on nylon cloth as described above.

9. Rehomogenized as in step 5, disrupt in the nitrogen bomb as in step 6 and 7, and filter as in step 8.

10. Repeat step 9 at least twice more.

11. Check the sample with a phase microscope to ascertain 100% cell breakage. If necessary, repeat the above homogenization, nitrogen bomb disruption, and filtration a number of times until a clean wall prep is obtained.

12. Resuspend wall fraction in 200 ml HM/7.5 and filter on nylon cloth as above.

13. Repeat step 12. Examine filtrate for presence of cellular organelles. Repeat step 12 until filtrate is clean. The purity of the plasmodesmata fraction is dependent on the purity of the cell wall fraction. By employing a highly purified cell wall fraction that is free of extraneous organelles and/or extraneous membranes, a highly enriched preparation of plasmodesmata can be obtained.

C. Isolation of Plasmodesmata by Enzymatic Hydrolysis of Walls

The procedure for the isolation of plasmodesmata is based on enzymatic hydrolysis of the walls of a clean cell wall fraction, resulting in the release of free plasmodesmata and plasmodesmatal clusters.

1. Method A: General Method

1. Suspend pelleted clean cell walls in Sorvall GSA 250-ml plastic centrifuge bottle in the 0.5% cellulase solution containing protease inhibitors (5 μg epiamastatin/ml, 1 μl leupeptin stock/ml, 1 μl pepstatin stock/ml, 4 μl PMSF/ml (500 mM stock), 2 mg/ml 1,10-phenanthroline), and 0.02% azide at ratio of 2 ml cellulase solution for each gram of fresh tissue used in preparing wall fraction. We generally use walls prepared from 100-g mesocotyls.

2. Hydrolyze cell walls by incubating suspension for 14–17 h at 37°C with shaking on a rotary shaker at 90 rpm.

3. Sediment undigested walls by centrifugation at 1500 × g for 10 min in HB-4 (Sorvall) swinging bucket rotor.

4. Decant supernatant and discard pellet which contains residual undigested wall.

5. Centrifuge supernatant at 10,000 × g for 20 min in HB-4 rotor.

6. Discard supernatant. Resuspend pellet in 4.5 ml AMM buffer in a 10 ml-tapered tube.

7. Centrifuge in Sorvall using swing bucket rotor (HB-4) at 1500 × g for 10 min.

8. Decant supernatant which contains released plasmodesmata. Discard pellet. Dispense 1.5 ml into three microfuge tubes and centrifuge supernatant in HB-4 at 10,000 × g for 20 min.

9. Decant and discard supernatant. Pellets contain isolated plasmodesmata (about 100 μg protein/microfuge tube).

2. Method B: Isolation of Wall-Free Plasmodesmata Prefixed for Electron Microscopy

In order to preserve membrane structure for electron microscopy, plasmodesmata must undergo a very mild fixation prior to digestion with cellulase.

1. Resuspend clean cell walls prepared from 100 g tissue as described above in 100 ml of HM/7.5. Add 400 μl PMSF and 1 ml glutaraldehyde (25% glutaraldehyde stock). Final glutaraldehyde concentration is 0.25%.

2. Incubate at room temperature with occasional shaking for 15 min.

3. Sediment walls by centrifugation in a HB-4 swinging bucket rotor at 2000 × g for 10 min. Discard supernatant.

4. Resuspend walls in 10 mM glycine buffer, pH 7.5, and shake for 15 min at room temperature. Note: glycine quenches any excess glutaraldehyde.

5. Sediment walls by centrifugation in HB-4 swinging bucket rotor at 2000 × *g* for 10 min. Discard supernatant.

6. Repeat steps 4 and 5 once.

7. Resuspend walls within a 250-ml Sorvall GSA centrifuge bottle in 0.5% cellulase solution containing protease inhibitors and azide at ratio of 2 ml cellulase solution for each gram of fresh tissue used in preparing wall fraction. Adjust pH if necessary to pH 5.3.

8. Hydrolyze cell walls by incubating suspension for 14–17 h at 37°C with gentle shaking in a rotary shaker at 90 rpm.

9. Sediment undigested walls by centrifugation at 1500 × *g* for 10 min in HB-4 swinging bucket rotor.

10. Decant supernatant and discard pellet which contains residual undigested wall.

11. Centrifuge supernatant at 10,000 × *g* for 20 min in the HB-4 rotor.

12. Discard supernatant. Resuspend pellet in 3 ml plasmodesmatal wash buffer and place in two microfuge tubes and centrifuge in the HB-4 rotor at 1500 × *g* for 10 min.

13. Decant supernatant into microfuge tubes; discard pellet. Centrifuge supernatant in HB-4 at 10,000 × *g* for 20 min.

14. Decant and discard supernatant. The pellet contains isolated plasmodesmata.

D. Electron Microscopy: Fixation, Dehydration, and Embedding in Epoxy

All manipulations with fixatives should be carried out in a ventilation hood.

1. Resuspend plasmodesmata isolated by Method B in microfuge tube in 1 ml of 0.2 *M* sodium cacodylate buffer, pH 7, containing 3.5% glutaraldehyde. Allow fixation to proceed for 2 h at room temperature.

2. Pellet plasmodesmata by centrifugation at 10,000 × *g* for 20 min.

3. Remove excess fixative by washing pellet three times in 0.2 *M* cocodylate buffer, pH 7. The plasmodesmata should remain in each wash for at least 5 min before pelleting.

To visualize the plasmodesmata by transmitting electron microscopy, fix plasmodesmata pellet in 1% osmium tetroxide for 1 h, dehydrated in a graded series of ethanol (5 min each, 50, 70, 90, 100%), and embed in Epon 812. Cut thin sections (70 nm) with diamond knife, place on copper grids, and contrast sections with uranyl acetate and lead citrate using standard microscopic techniques.

E. Preparation of Antibodies against Putative Plasmodesmatal-Associated Proteins

1. Protein Extraction from Clean Walls

To each of three-microfuge tubes individually containing clean cell walls isolated from 1 g mesocotyl, add 250 μl of electrophoresis sample buffer SB ×1.

Close tube, mix well, puncture a pin-hole in the cap, and boil for 3 min. Centrifuge for 5 min at full speed in microfuge. Transfer supernatant by pipette from each tube to three additional samples of wall. If necessary, add SB to bring supernatant to 250 μl. Mix, boil, and centrifuge as above. Combine supernatant from three tubes. Determine protein concentration by Marder CBB filter paper method.

2. Detection of Proteins in Gels after SDS–PAGE by Silver Staining

This method, developed by Blum *et al.* (1987), is very simple to perform. Use only double-distilled water (dd) in all solutions.

1. Fix the proteins in the gel for at least 1 h (to overnight) in about 100 ml of 50% methanol, 12% acetic acid, to which is added just before use 50 μl formaldehyde solution (37%).

2. Wash gel three times for 20 min in 50% ethanol.

3. Pretreat gel for exactly 1 min in 98 ml of a freshly prepared solution of sodium thiosulfate (20 mg $Na_2S_2O_3$–$5H_2O$ in 100 ml dd water). Attention: set aside 2 ml of the solution for use in step 7.

4. Wash gel three times for exactly 20 s in about 100 ml dd water.

5. Soak the gel for 20 min in 100 ml of fresh silver reagent (200 mg silver nitrate in 100 ml water) containing 75 μl formaldehyde (37% stock).

6. Wash the gel two times for exactly 20 s in 100 ml dd water.

7. Develop silver in 100 ml developer solution prepared by adding 2 ml of fresh sodium thiosulfate solution (see step 3) and 50 μl formaldehyde (37%) to 100 ml of 0.6 M sodium carbonate stock solution. Develop for 2 to 10 min until satisfactory staining is obtained.

8. Wash the gel two times for 2 min in 100 ml dd water.

9. Stop the silver development reaction by soaking gel for 10 min in 100 ml of 50% methanol, 12% acetic acid.

10. Wash gel in 50% methanol. Gel may be stored in dark at 4°C for a few weeks.

3. Preparation of Polyclonal Antibodies

Polyclonal antibodies are prepared according to the method of Knudsen (1985). Separate extracted proteins by preparative SDS–PAGE on a Mighty Small II 7 × 8-cm vertical slab unit (Hoefer Scientific Instruments, San Francisco, CA) employing a 1.5-mm spacer and a 6.8-mm preparative well. For separating PAP 41 and other proteins larger than 36 kDa, we routinely use 10% polyacrylamide; for PAP26 and other proteins in the range 15–36 kDa, we use 15% polyacryl-amide. Add a few grains of bromophenol blue as a tracking dye. Load 400–500 μg of protein in SB ×1 (about 300–350 μl extract) and electrophorese at 100-V constant voltage until bromophenol tracker dye leaves bottom of gel.

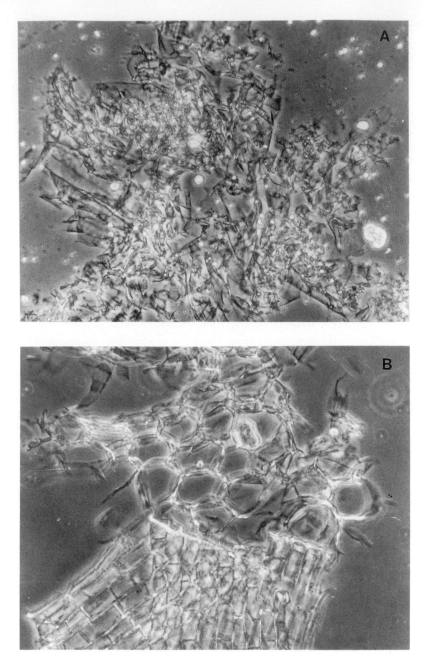

Fig. 1 Phase-contrast micrograph of cell wall matter from an isolation procedure in which (A) fresh tissue was repeatedly homogenized and the cell wall fraction separated from the homogenate by centrifugation at $600 \times g$ for 10 min; (B) liquid nitrogen-pulverized tissue homogenized, filtered through 200-μm mesh cloth, and passed through nitrogen disruption bomb as described under Methods. Note the presence in (A) of numerous starch granules that cosediment with walls isolated by centrifugation, and cytoplasmic debris adhering to cell walls. In the procedure described under Methods, the wall preparation is devoid of plastids, starch granules, and adhering cytoplasmic debris (B). From Kotlizky *et al.* (1992), with permission. Copyright Blackwell Scientific Publications Limited.

Transfer proteins to nitrocellulose paper (Schleicher & Schuell, Dassel, Germany, BA 85 cellulosenitrate, pore size 0.45 μm) by electrotransfer (Mighty Small Transphor TE 22, Hoefer Scientific Instruments, San Francisco, CA) for 1–1.5 h at 170 V in alkaline electrotransfer buffer according to Szewczyk and Kosloff (1985). Briefly stain proteins for 3–5 min on nitrocellulose with Ponceau S protein stain solution. From two nitrocellulose blots, cut out strips (2 × 6.8 mm) containing the band of interest. Note: The band should be well resolved from all other bands. Wash strips for about 30–60 s with 10 mM NaOH to destain strips and then wash in distilled water. Thoroughly dry the strips, cut them into small pieces, and add the pieces to 250 μl DMSO in order to dissolve nitrocellulose. If necessary, add additional DMSO to dissolve completely nitrocellulose. Note: The nitrocellulose must be absolutely dry before dissolving in DMSO.

Thoroughly mix dissolved nitrocellulose with 500 μl Freund's adjuvant (Difco Laboratories, Detroit, MI). Use complete Freund's adjuvant for primary injection and incomplete Freund's adjuvant for subsequent booster injections. Space booster injections 21–28 days apart.

Prior to primary injection, prepare preimmune serum: Bleed each rabbit from ear and collect 15–25 ml blood. Allow the blood to clot and separate the serum from clot by centrifugation in a clinical centrifuge at 1000 × g for 10 min. Add azide to serum to prevent bacterial contamination (final concentration 0.02%) and store in 100-μl aliquots at −70°C.

Inject protein–nitrocellulose mixture subcutaneously using a 1-ml tuberculin syringe fitted with a 25-gauge needle. Inject into 7–10 sites on the side of white female New Zealand rabbit. Give the first booster injection 21 to 28 days after primary injection and the second booster after an additional 21–28 days as above, but using incomplete Freund's adjuvant. Bleed rabbits from the ear 10 days after each boost (7–15 ml after first boost and 30–50 ml after second and later boosts), separate the serum, and immunoprobe for antibody response. Note: A good serum should work at a dilution of at least 1 : 10,000. If necessary, give a third boost after an additional 21–28 days. When good titer is obtained, divide the serum into 100-μl aliquots and freeze at −70°C. Do not thaw and refreeze antiserum repeatedly. Thawed aliquots can be stored at 4°C 1 to 2 months with little decrease in activity.

F. Immunoprobing

Extract proteins from cell wall or plasmodesmatal fraction as described above, separate by SDS–PAGE, and electrotransfer as described above. Employ a preparative comb that also has a well for molecular weight markers. Stain the electrotransferred proteins with Ponceau S protein stain, cut off the strip containing the molecular weight markers, and mark the position of the markers with a pen. Destain Ponceau using 10 mM NaOH, wash the blot with water, and block the blot for 1 h with 5% milk powder in TBS containing 0.02% azide. Cut the

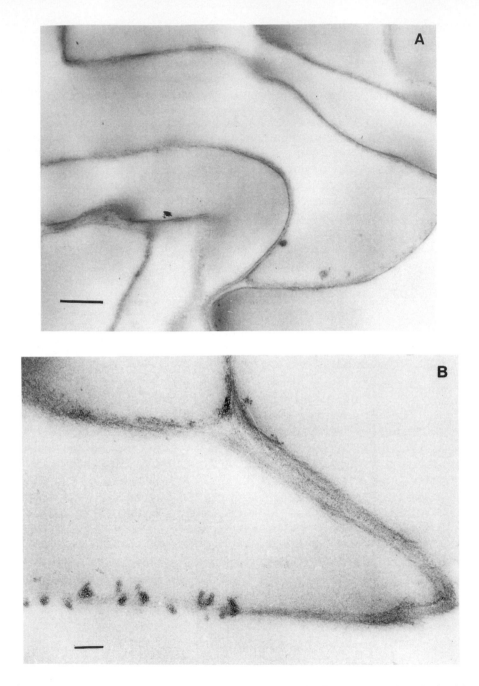

Fig. 2 Transmission electron micrograph of thin sections of cell wall preparation obtained by procedure described under Methods. (A) Low magnification showing cell wall preparation clear of particulate contamination and free of adhering plasma membrane. Bar, 500 nm. (B) Cell wall preparation free of cytoplasmic inclusions and adhering plasma membrane but showing presence of clustered plasmodesmata embedded in the cell wall. Bar, 200 nm. From Kotlizky *et al.* (1992), with permission. Copyright Blackwell Scientific Publications Limited.

blot lengthwise into 15 strips each about 0.4 mm wide. Strips can be used for titrating antibody.

Dilute serum obtained from rabbits taken 10 days after first boost 1:100, 1:200, 1:500 and after second and subsequent boosts 1:500, 1:1,000, 1:5,000, 1:10,000, 1:20,000, 1:40,000 in blocking solution. Make a similar dilution series with preimmune serum.

Incubate the antigen-containing nitrocellulose test strips overnight in each dilution of antiserum and preimmune serum with gentle rocking at room temperature.

Wash Blot three times 10 min each with TTBS. Incubate blot for 1 h in goat anti-rabbit antibody coupled to alkaline phosphatase (Jackson ImmunoResearch Laboratories, Inc., West Grove, PA) at 1:4000 dilution in blocking solution.

Wash Blot three times, 10 min each, with TTBS. Visualize antibody binding by incubating blot in 20 ml alkaline phosphatase developing buffer to which was added 100 μl NBT and 100 μl BCIP.

IV. Critical Aspects of the Procedure

In order to obtain a clean plasmodesmatal preparation, it is essential to start with a cell wall fraction that is totally free of any intact cells, trapped subcellular cytoplasmic organelles, adhering plasma membrane, and adsorbed cytoplasmic proteins. This is not an easy task, and no one procedure can be relied on to fulfill the requirements in all cases.

The selection of the biological material is of utmost importance in the successful isolation of a wall-free plasmodesmatal fraction. We chose to work with the elongation zone of the mesocotyl, the first internode of young etiolated maize seedlings. This section consists mainly of young cells having only primary cell walls and that are free of chloroplasts. Cells with secondary cell walls are more difficult to break. The presence of intact cells will contaminate the preparation with nonplasmodesmatal membranes and make the identification of plasmodesmatal proteins very difficult.

Isolation of plasmodesmata from leaves presents special problems. The highly differentiated leaf cells are dedicated to photosynthesis and are loaded with membranes associated with the photosynthetic apparatus. Thus, even a small amount of contamination by a few unbroken cells will result in the masking of the contribution of plasmodesmata. In green corn leaves, the vascular bundle sheath cells are very resistant to breakage. Using the above unmodified procedure, the wall fraction was light green. Microscopic examination showed the presence of intact vascular bundle sheath cells. Similarly, guard cells may be resistant to breakage. Thus, leaf tissue may require many additional grindings and passage through the nitrogen bomb in order to ensure complete breakage of all cells.

Note: Extraction of the cell wall fraction with acetone may remove pigments, but the chloroplast proteins still remain wall-associated and are not extracted. Similarly, extraction of the wall pellet with Triton-100 extracts many membrane proteins and will result in removal of chlorophyll; however, not all chloroplast proteins or other membrane proteins are extracted by this method. Thus, the goal of 100% cell breakage remains essential.

Amyloplasts, chloroplasts, and other heavy organelles will sediment even at very low centrifugal forces. Thus, it is very difficult to separate these heavy organelles from the cell wall by differential sedimentation. We find filtration absolutely essential in order to remove these contaminants.

In preparing the cellulase solution, it is also absolutely essential to clear the enzyme solution of any insoluble matter. The presence of insoluble material in the enzyme preparation that sediments at the centrifugal force used to sediment the isolated plasmodesmata will contaminate the plasmodesmata.

Commercial cellulases contain in addition to polysaccharidases, proteases and, possibly, other hydrolytic enzymes. These proteases and hydrolases may result in damage to the plasmodesmata during the prolonged incubation period required to hydrolyze the cell wall and release the plasmodesmata. Cellulase TC from *T. reesi* purchased from Serva was found to cause minimal proteolytic damage to the plasmodesmata. The following cellulases were found to cause significant proteolytic damage during prolonged hydrolysis of cell walls: *Trichoderma viride* cellulase concentrate (Miles Kali-chemie GmbH & Co., Hanover); Onozuka R-10 Cellulase from *T. viride* (Yakult Honshe Co., Tokyo); Cellulysin cellulase from *T. viride* (Calbiochem, San Diego); Driselase (Kyowa Hakko Kogyo Co., Tokyo); Pectinase Rohament PS from *Aspergillus niger* (Serva Feinbiochemica GmbH & Co., Heidelberg).

During incubation of cell walls in cellulase solution, care must be taken in agitating the solution. The suspension must not be overagitated, since the fragile plasmodesmata, upon release, will lyse due to shear.

V. Results

A comparison of cell wall preparations prepared using centrifugation realtive to those prepared by filtration is shown in Fig. 1. As a result of centrifugation, starch granules cosediment with cell walls. Figure 1A illustrates a preparation isolated by filtration but without repeated passage through the nitrogen bomb. Walls prepared using the nitrogen disruption bomb followed by filtration through 200-μm mesh nylon show very little evidence of cytoplasmic contamination and no presence of starch granules (Fig. 1B). A transmitting electron micrograph of thin sections of a wall preparation obtained by the above procedure is given in Fig. 2. At low magnification (Fig. 2A), the section of wall material appears free of cytoplasmic inclusions and adhering plasma membrane. The only membranous

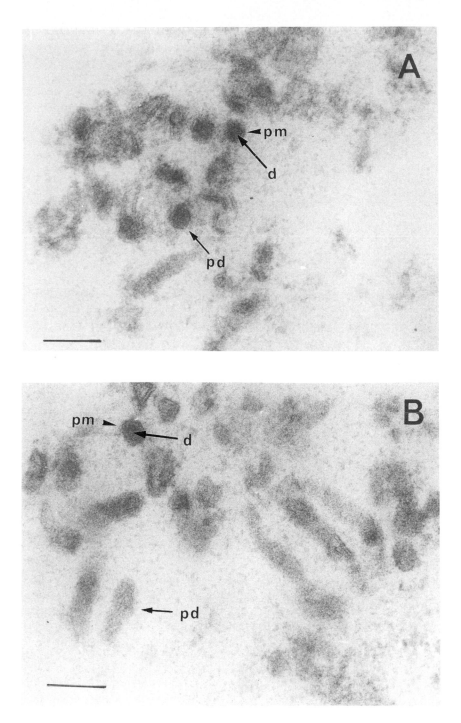

Fig. 3 Transmission electron micrographs of thin sections of isolated plasmodesmata. Clean cell walls were prepared and digested as described under Methods. Following digestion, the residual cell walls were sedimented at $600 \times g$ for 10 min and the supernatant was centrifuged at $10,000 \times g$ for 20 min. The cell wall fraction and the 10-kg pellet were fixed, embedded, and cut as previously described (Kotlizky *et al.,* 1992). d, desmotubule; pm, plasmalemma; pd, plasmodesmata. Bar, 100 nm.

structures evident are clusters of plasmodesmata embedded in the cell wall (Fig. 2B).

Figure 4 (lane A) illustrates the proteins associated with a clean cell wall fraction, as visualized using silver staining. These proteins we functionally term wall-associated proteins (WAPs). Figure 4 (lane B) illustrates proteins associated with a plasmodesmatal fraction isolated from the mesocotyl of etiolated maize seedlings. These we term plasmosdesmata-associated proteins (PAPs).

In Fig. 3 is shown a field of isolated cell wall free plasmodesmata obtained after 17 h enzymatic digestion of a clean cell wall fraction. Many plasmodesmata appear as clusters and are apparently held together by cellulase-resistant wall material.

VI. Conclusions and Perspectives

The techniques described in this chapter have contributed to a first characterization of plasmodesmatal proteins associated with the mesocotyl of etiolated maize seedlings. Work in press from the authors' laboratory indicates that there

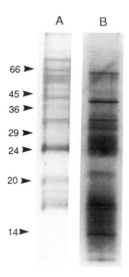

Fig. 4 SDS–PAGE analysis of proteins stained with silver extracted from (A) a clean cell wall fraction and (B) the $10,000 \times g$ pelleted fractions obtained by differential centrifugation following cellulase digestion of isolated clean maize mesocotyl cell walls. Clean maize mesocotyl cell walls were suspended, per gram fresh weight tissue, in 2 ml of reaction buffer containing 0.5% cellulase TC and incubated for 14 h at 37°C with shaking. Residual undigested walls were sedimented at $600 \times g$ for 10 min, and the supernatant was fractionated by differential centrifugation at $10,000 \times g$ for 20 min and $90,000 \times g$ for 1 h. The pelleted fractions were washed and boiled in sample buffer for 5 min and the extracted proteins (approx. 7 μg/lane) were separated by SDS–PAGE (15% polyacrylamide) and silver stained as described under Methods.

are differences in organ- and tissue-specific expression of PAPs. Plasmodesmata from different organs and tissues must be isolated and organ- and tissue-specific PAPs identified, and their genes cloned and sequenced. Affinity-purified antibody reagents to each PAP can be used to map immunocytologically different tissues and organs for cell specific expression of each PAP. These antibody probes together with nucleic acid probes can be employed to examine the effect of viral infection on the composition of the plasmodesmata and the effect of the virus on the level of PAP expression.

Acknowledgments

We thank Jacob Delarea for excellent technical aid in producing the electron micrographs of isolated plasmodesmata. This research was supported by grants from The Fund for Basic Research administered by the Israeli Academy of Sciences and Humanities (Project No. 361/88), the United States–Israel Binational Agricultural Research and Development Fund (Project No. US-1384-87), the Ministry of Science and Technology, Israel, and the Gesellschaft fuer Biotechnologische Forschung (GBF), Braunschweig, Germany, and the Karse-Epel Fund for Botanical Research at Tel Aviv University.

References

Blum, H., Beier, H., and Gross, H. J. (1987). Improved silver staining of plant proteins, RNA and DNA in polyacrylamide gels. *Electrophoresis* **8,** 93–99.

Knudsen, K. A. (1985). Proteins transferred to nitrocellulose as immunogens. *Anal. Biochem.* **147,** 285–288.

Kotlizky, G., Shurtz, S., Yahalom, A., Malik, Z., Traub, O., and Epel, B. L. (1992). An improved procedure for the isolation of plasmodesmata embedded in clean maize cell walls. *Plant J.* **2,** 623–630.

Marder, J. B., Mattoo, A. K., and Edelman, M. (1986). Identification and characterization of the psbA gene product: The 32-kDa chloroplast membrane protein. *Methods Enzymol.* **118,** 384–396.

Robards, A. W., and Lucas, W. J. (1990). Plasmodesmata. *Annu. Rev. Plant Physiol.* **41,** 369–419.

Szewczyk, B., and Kozloff, L. M. (1985). A method for the efficient blotting of strongly basic proteins from sodium dodecyl phosphate polyacryamide gels to nitrocellulose. *Anal. Biochem.* **150,** 403–407.

Yahalom, A., Warmbrodt, R. D., Laird, D. W., Traub, O., Revel, J. P., Willecke, K., and Epel, B. L. (1991). Maize mesocotyl plasmodesmata proteins cross-react with connexin gap junction protein antibodies. *Plant Cell* **3:**407–417.

CHAPTER 18

Characterization and Isolation of the Chloroplast Protein Import Machinery

Karin Waegemann and Jürgen Soll

Botanisches Institut
24098 Kiel, Germany

I. Introduction

Most chloroplast proteins are nuclear-encoded, synthesized in the cytoplasm, and subsequently imported into the organelle. The characterization of the different steps and components that are involved in this process has been a major research topic over the last few years. Recent reviews, which emphasize theoretical aspects of protein import, include those by Keegstra *et al.* (1989), de Boer and Weisbeek (1991), and Soll and Alefsen (1993). This review is aimed to provide a practical introduction to chloroplast import field, dealing with the *in vitro* synthesis of precursor proteins, as well as with the way in which the protein passes across the two envelope membranes into the organelle (for another practical review on this topic see Perry *et al.*, 1991).

II. Isolation of Intact Chloroplasts

Most import experiments are done with chloroplasts isolated from spinach or pea tissues. It is best to use young plant material because chloroplasts isolated from developing leaves import precursor proteins most efficiently (Dahlin and Cline, 1991).

Harvest 50–100 g leaves of 8- to 10-day-old peas. It is useful to collect the tissue at the end of the dark period to avoid accumulation of starch inside the chloroplasts. Homogenize the leaves together with 300 ml of ice-cold isolation buffer (330 mM sorbitol, 20 mM Mops, 13 mM Tris, 0.1% BSA, 3 mM MgCl$_2$, pH 7.6) in a Waring blender with several short (3–5 s) bursts. The homogenate is then filtered through four layers of cheesecloth and one layer of nylon gaze (30 μm) and centrifuged for 1 min at 1500 $\times$ g (Sorvall, SS34 rotor). Remove the supernatant and resuspend each pellet carefully in about 0.5 ml of ice-cold HMS buffer (330 mM sorbitol, 50 mM Hepes/KOH, pH 7.6, 3 mM MgCl$_2$). Overlay the resuspended crude chloroplasts onto two silica sol gradients (Percoll) prepared in advance [7 ml 80% (v/v) Percoll in 330 mM sorbitol, 50 mM Hepes/KOH, pH 7.6, and 12 ml 40% (v/v) Percoll in 330 mM sorbitol, 50 mM Hepes/KOH, pH 7.6] and centrifuge in a swinging bucket rotor (Sorvall, HB-4 rotor) for 5 min at 8000 $\times$ g. After centrifugation two bands are obtained. The upper band, which contains broken chloroplasts and thylakoids, is discarded. The lower band, which represents the intact chloroplasts, is carefully collected and transferred to a clean tube. Dilute the suspension with at least 4 vol of ice-cold HMS buffer and centrifuge for 1 min at 1500 $\times$ g (Sorvall, SS34 rotor). Resuspend the chloroplast pellet and wash again with about 20 ml HMS buffer. Recover intact chloroplasts by centrifugation for 45 s at 1500 $\times$ g. Resuspend the final pellet in a small volume (0.5–1 ml) HMS buffer so that the final chlorophyll concentration is 2–4 mg of chlorophyll per milliliter. Keep the isolated chloroplast suspension on ice and in the dark until use.

III. Synthesis of Precursor Proteins

Most import experiments are done with precursor proteins generated in an *in vitro* translation system. There are two commonly used translation systems (1) the rabbit reticulocyte system; (2) the wheat germ system. Transcription–translation of a cloned gene yields generally a radiochemically pure precursor protein, which can be imported into chloroplast readily, i.e., without further purification. The disadvantage is that only very small quantities of protein are synthesized, although these are of high specific activity.

The second method of choice for the production of precursor proteins is the expression in bacteria. This system needs some experience, because the radioactive protein has to be purified from the bacterial lysate before use. Proteins expressed in bacteria are often synthesized as insoluble inclusion bodies, which make the purification rather simple, but a detergent or a denaturant is then needed (urea, guanidiumHCl, or HCl) to solubilize the protein before use. Furthermore it is often difficult to import these kind of proteins into the organelle, because some (Waegemann *et al.*, 1990) but not all (Pilon *et al.*, 1990) preproteins require the presence of additional protein factors to attain import competence. The main advantage of this system is that large amounts of precursor proteins can be obtained by this method.

A. *In Vitro* Transcription

The gene for the precursor protein has to be cloned into a vector containing a suitable promotor. In our hands, the best results were obtained with the SP6 RNA-polymerase, but T3 and T7 can also be used. The plasmid should be linearized with a suitable restriction enzyme and subsequently extracted with phenol/chloroform. After precipitation with 1/10 vol 4 M LiCl and 1 vol propan-2-ol the DNA pellet is resuspended in H_2O. Since a eukaryotic translation system will be used, *in vitro* capping of the DNA with P^1-5'-(7-methyl)-guanosine-P^3-5'-guanosine (m^7 GpppG) is recommended (Krieg and Melton, 1987). The transcription is performed in a final volume of 50 μl in polymerase buffer (supplied together with the polymerase) supplemented with 100 U RNase inhibitor, 10 mM DTT, 25 μg BSA, 0.5 mM m^7 GpppG, 0.5 mM each ATP, CTP, and UTP, 2–3 μg linearized plasmid, and 30 U SP6 RNA-polymerase (alternatively 100 U T_3- or T_7-polymerase can be used). After initiation for 15 min at 37°C, RNA synthesis is completed by adding 1.2 mM GTP and continuing the incubation for 2 h at 37°C. The synthesized mRNA can be used in the translation assay without further purification.

B. *In Vitro* Translation

We generally use rabbit reticulocyte lysate and [^{35}S] methionine for translation. A standard translation assay with a final volume of 100 μl consists of 20 mM Hepes/KOH, pH 7.6, 10 mM creatine phosphate, 12 μg creatine phosphate kinase, 50 μM of each amino acid with the exception of methionine, 50–180 mM K-acetate, 33% v/v rabbit reticulocyte lysate, 260 μCi [^{35}S]methionine, and 10 μl transcript. The K$^+$ concentration has to be optimized for each batch of RNA and reticulocyte lysate, respectively. Incubate for 60–90 min at 30°C. It is important to control the temperature carefully. Temperatures higher than 30°C are not useful, because the synthesized precursor protein will be folded to a higher degree and will therefore lose import competence (Soll *et al.,* 1992). The translation product can be stored in small aliquots in liquid N_2 for several weeks. The method for translation in a wheatgerm system is described in detail by Perry *et al.* (1991).

C. Expression in Bacteria

The gene of the desired precursor protein has to be cloned into a suitable vector. Vectors and bacterial strains for the expression of proteins in bacteria are available from several suppliers. It is possible to include any radioactive amino acid in the growing culture. In our lab we use [^{35}S] sulfate. Growing of the cells is performed according to Sabohl and Ochoa (1974). Briefly, an overnight culture is grown in a medium with a sulfur concentration (0.2 m*M*) slightly higher than that required for growth. The next day, the cells are resuspended in fresh medium and after another incubation for 1 h the expression of the precursor protein is induced with IPTG. At the same time [^{35}S]sulfate (0.5 mCi/ml) is added to the culture. The duration of the induction has to be optimized for each expressed protein. Proteins expressed in bacteria are often synthesized as insoluble inclusion bodies, which can be isolated rather quickly (Paulsen *et al.,* 1990). The bacteria are harvested by centrifugation (7000 × *g;* 5 min) and are resuspended in cold lysis buffer [50 m*M* Tris/HCl, pH 8, 25% (w/v) sucrose, 1 m*M* EDTA]. Lysozyme is then added to a final concentration of 2 mg/ml and the mixture is incubated for 30 min on ice. After addition of $MgCl_2$ (0.5 m*M*) and $MnCl_2$ (0.05 m*M*) the culture is treated with DNase I (40 µg/ml) for another 30 min on ice. Add 2 vol of cold detergent buffer [20 m*M* Tris/HCl, pH 7.5, 200 m*M* NaCl, 1% (w/v) deoxycholate, 1% (w/v) Nonidet P-40, 2 m*M* EDTA 10 m*M* 2-mercaptoethanol] and mix very carefully and extensively. After centrifugation (7000 × *g,* 10 min) the supernatent is removed and the pellet is resuspended in cold Triton buffer [20 m*M* Tris/HCl, pH 7.5; 0.5% (v/v) Triton-X-100, 1 m*M* EDTA, 10 m*M* 2-mercaptoethanol]. Mix well and centrifuge as above. The pellet is washed at least once more with Triton buffer until it appears white. Wash once again with cold Tris buffer (50 m*M* Tris/HCl, pH 7.6; 1 m*M* EDTA, 10 m*M* DTT). The collected inclusion bodies are finally resuspended in Tris buffer and stored as a suspension in small aliquots in liquid N_2. The inclusion bodies have to be solubilized with 8 *M* urea, 6 *M* guanidine–HCl, or 10 m*M* HCl prior to the import experiment.

IV. Import into Chloroplasts

The import process can be divided into several steps: (1) Binding of the precursor to surface-exposed receptor proteins on the organelle; (2) Translocation across the two envelope membranes, utilizing the chloroplast import machineries; (3) Processing of the precursor by the stroma localized processing peptidase

(SPP); (4) Assembly into holoenzyme complexes in the stroma, or further sorting to the thylakoids.

The first two steps can be separated in the *in vitro* experiment by adding different amounts of ATP and by controlling temperature. The complete translocation of precursor protein destined for the stromal or thylakoid compartment requires the hydrolysis of ATP (1–3 mM), which can be added externally or can be synthesized by the organelle itself by the photosynthetic light reactions. Furthermore, proper transport occurs efficiently only at adequate temperatures (20–25°C). A standard import assay in our lab consists of chloroplasts equivalent to 15 μg chlorophyll in a final volume of 100 μl. First prepare the import buffer (330 mM sorbitol, 50 mM Hepes/KOH, pH 7.6, 3 mM MgSO$_4$, 10 mM methionine, 20 mM K-gluconate, 10 mM NaHCo$_3$, 2% BSA, and 3 mM ATP). If a radioactive amino acid other than methionine is used for the preparation of precursor proteins, this amino acid should be added unlabeled to the import reaction to dilute the radioactive one. Finally the translocation product (0.5–5 μl depending on the quality of the translated precursor) is added to the import buffer, and the reaction is started by addition of intact chloroplasts. It is best to perform the transport reaction at a temperature of 25°C with gentle agitation. The incubation time has to be established for each precursor. In general, transport is linear over a range of 5–15 min. After the import reaction has been completed, the vials are set on ice and the import reaction is transferred on top of preprepared cold 300-μl 40% Percoll cushions (in 330 mM sorbitol, 50 mM Hepes/KOH, pH 7.6) and centrifuged for 5 min at 3300 × g. The supernatant, which contains broken chloroplasts and nonbound precursor, is removed. The pellet, which consists of the intact organelles with imported and bound precursor, is washed twice with 100 μl of HMS buffer. Chloroplasts are collected by centrifugation for 1 min at 800 × g. Either the pellet can be prepared for electrophoresis, i.e., solubilized in 50 μl of SDS sample buffer and boiled for 3 min (Laemmli, 1970), or the organelles can be treated with protease (thermolysin) to remove external bound precursors (Cline *et al.*, 1984). The chloroplasts are resuspended in 100 μl of HMS buffer supplemented with 0.5 mM CaCl$_2$. Thermolysin (0.1 mg/mg chlorophyll) is added and the plastids are incubated for 20 min on ice. The digestion is terminated by 10 mM EDTA. The chloroplasts are then reisolated (800 × g; 1 min) and washed once in 100 μl of HES buffer (330 mM sorbitol; 50 mM Hepes/KOH, pH 7.6, 5 mM EDTA). The import reactions are analyzed by SDS–PAGE and fluorography (Bonner and Laskey, 1974). An example of the different steps in pSSU translocation into isolated chloroplasts is shown in Fig. 1. pSSU is imported into the organelle under standard import conditions (3 mM ATP; 25°C, 10 min) and is processed to the mature form inside (lane 1, 2). Because of the very efficient import of pSSU, only low amounts of bound precursor can be detected. The bound precursor is sensitive to treatment of the organelles with thermolysin, whereas the mature form is protected inside the chloroplast from protease action (lane 2).

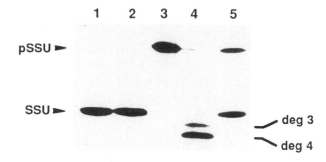

Fig. 1 Translocation of pSSU into isolated intact chloroplasts. pSSU is imported into the organelle under standard import conditions (3 mM ATP; 25°C, 10 min) (lane 1). The processed, mature SSU is protease protected in the organelle (lane 2). In the presence of low ATP (100 μM) the precursor is not completely imported but appears to be bound to the chloroplasts (lane 3). Digestion with thermolysin shows that pSSU is already inserted into the envelope membranes yielding the translocation intermediates deg 3 and 4 (lane 4). If reisolated, chloroplasts with bound pSSU are supplemented with higher ATP (3 mM) and warmed to 25°C the precursor completes its way into the organelle and is processed to the mature form (lane 5).

Proteins destined for the outer chloroplast envelope follow a different pathway. These proteins (e.g., OEP 7; OM 14), which have no cleavable transit sequences, do not require protease-sensitive receptors on the organellar surface and they do not require ATP for incorporation. It seems that they insert more or less spontaneously into the outer membrane (Salomon *et al.*, 1990; Li *et al.*, 1991; Soll *et al.*, 1992).

V. Binding to Chloroplasts

For the characterization of the binding process of a precursor protein to the chloroplast surface, it is necessary to separate binding from complete translocation. In principle there are two ways to achieve this: (1) The import experiment is incubated on ice instead of at 25°C, because low temperatures block full translocation but do not affect binding (Friedman and Keegstra, 1989); (2) The binding experiment is done at low ATP concentration (<100 μM) (Olsen *et al.*, 1989). For this method it is useful to keep the isolated chloroplasts 30 min in the dark on ice prior to the experiment to lower the internal ATP level. Furthermore, the import reaction must be performed in the dark to prevent new ATP generation by photophosphorylation. Often it is also necessary to separate the precursor protein from the ATP-generating system (creatinphosphat, creatin phosphat kinase) in the translation mixture. This can be done by spinning the translation product through a Sephadex G-25 column (Perry *et al.*, 1991). Another method used in our laboratory is to precipitate the precursor with $(NH_4)_2SO_4$ from the translation mixture. This is done by adding $(NH_4)_2SO_4$ to a final concentration of 52% (w/v) to the translation product. Incubate for 30 min on ice and centrifuge for 10 min at 15,000 × g. The pellet is dissolved in a small volume of water or buffer and can be added directly to the binding assay. The problem with both methods is that the precursor proteins partly lose their import competence during the pretreatments, probably due to a change in their conformations.

With some precursor proteins (e.g., pSSU), the protein will not be found at the receptor under the conditions used above [i.e., low ATP ($<100\ \mu M$)], but the precursor is already inserted into the two envelope membranes. If the "binding" reaction is post-treated with thermolysin, translocation intermediates that are partially protected against the protease can be detected (Fig. 1, lanes 3 and 4) (Waegemann and Soll, 1991). It is also possible to perform the transport reaction in two separate steps: the first involves binding of the precursor to the chloroplast under conditions of low ATP ($<100\ \mu M$) and/or low temperature (4°C). After reisolation of the organelles, the bound and inserted precursor can be chased into the chloroplasts by adding higher concentrations of ATP (1–3 mM) and by shifting to higher temperature (25°C) (Fig. 1, lane 5). Bound and partially translocated precursors (e.g., deg 3, deg 4) can also be chased into the organelle, indicating that they are on the proper route into the organelle.

VI. Criteria for the Specifity of Binding and Transport

If the import characteristics for a new precursor protein are to be established, control experiments should be carried out to examine whether the protein follows a specific translocation pathway. Some aspects are common to most precursor proteins that are directed to the inner envelope, stroma, or thylakoids: (1) Interaction with the chloroplast surface depends on a cleavable targeting signal; (2) Binding occurs to protease-sensitive receptor proteins on the organellar surface; (3) Translocation depends on the hydrolysis of ATP. For proteins destined for the outer envelope membrane the above-described criteria can be different, e.g., OEP 7 and OM 14 (see above).

A. Receptor Dependency

To examine whether a protein uses a surface exposed polypeptide component for the import, it is common to pretreat chloroplasts with protease. Thermolysin has been shown to digest proteins that are attached to the organellar surface (Cline et al., 1984). To investigate receptors, intact chloroplasts are prepared as described in Section II. The protease treatment is performed using 0.75 mg thermolysin/mg chlorophyll in HMS buffer supplemented with 0.5 mM CaCl$_2$ and incubating for 30 min on ice. The reaction is terminated by adding EDTA to a final concentration of 10 mM. The chloroplasts are then reisolated using another small Percoll gradient containing 5 mM EDTA. The chloroplasts are washed once in HES buffer, once in HMS buffer, and finally resuspended in a small volume of HMS buffer. The import experiment is then performed in comparison with undigested chloroplasts under standard conditions.

If there is a receptor dependency for the precursor protein under evaluation, a further experiment can be done to test whether the new protein uses the same receptor as other precursors that have already been examined (e.g., pSSU, pLHCP, pPC, pFd). For these studies competition experiments with two precursors are performed. It is best to use an overexpressed protein as the second precursor. The import experiments are run under standard conditions, but the second precursor is added in increasing amounts. In this case the import reaction must be initiated by adding the organelles. If two proteins use identical receptors, the overexpressed precursor protein should block the import of the other.

B. ATP Dependency

Several groups have shown that the hydrolysis of ATP is required for the import of most proteins into chloroplasts. To investigate the energy requirement for a new precursor the chloroplasts should be isolated from plants kept in the dark for 12 h. Endogenous ATP should be depleted by preincubation for 30 min on ice prior to the import experiment. In general, it is sufficient to perform an import experiment in the dark without adding ATP, because the ATP from the translation mixture should not be sufficient to complete transport. However, (*a*) if higher amounts of translation products are added to the assay (>3–5% v/v), or (b) if the incubation time is longer than 5 min, or (*c*) if too many chloroplasts (>10–15 μg chl) are used, the translation product should be pretreated with apyrase (Sigma, Grade VIII), an ATP-hydrolyzing enzyme. The pretreatment of the precursor is performed with 10 units/ml for 15 min at 25°C. If ATP is to be readded to the apyrase-treated precursor to regain translocation, the preincubation should be done with 2 U apyrase/ml.

C. Correct Localization and Integration of the Imported Protein

For the characterization of proper protein import, the right localization of the translocated protein must be established. For this purpose, chloroplasts must be subfractionated into the major compartments (envelope membranes, stroma, thylakoid membranes, and thylakoid lumen), as described in detail by Perry *et al.* (1991). For proteins destined to a membrane compartment the correct integration in the bilayer should be examined. This can be done by digestion of the isolated membranes after completion of the import experiment with protease. The degradation pattern of the radioactive imported protein must be identical to that of the protein *in situ*.

Isolated membranes are extracted with salt (e.g., 0.5 *M* NaCl) or high pH (e.g., 0.1 *N* NaOH) to determine whether the imported protein is a peripheral or an integral membrane protein. Membranes are treated with NaOH or NaCl for 30 min on ice and reisolated by centrifugation at 250,000 × *g* for 20 min. The supernatant, which contains peripheral associated proteins, is precipitated with 10% trichloroacetic acid. Both fractions are further analyzed by SDS–PAGE.

VII. Import of Overexpressed Precursor Proteins

Protein import into chloroplasts and other organelles depends on a certain unfolded or loosely folded conformation of the precursor (della Cioppa and Kishore, 1988; Eilers and Schatz, 1986). Proteins synthesized in an *in vitro* translation assay under appropriate conditions will retain such an import-competent conformation, because both systems contain components (e.g., hsp 70) that retard the folding (Zimmermann *et al.,* 1988). If an overexpressed, purified precursor is used that has to be denatured and unfolded in urea prior to the import experiment, it depends on the protein (hydrophobicity, size, etc.) used and on the time, temperature, and volume of the assay whether it will be imported into the organelle. For example, the overexpressed precursor of ferredoxin (pFd) or pSSU, which is unfolded in 6 M urea and diluted into a standard import assay, is effectively imported (Pilon *et al.,* 1990; Waegemann, unpublished). The overexpressed precursor of the light-harvesting chlorophyll a/b binding protein (pLHCP), however, seems to require soluble factors present in leaf extract or reticulocyte lysate for translocation after denaturation in urea (Waegemann *et al.,* 1990).

The two methods that our laboratory uses to import overexpressed precursor proteins are described below:

1. The overexpressed and purified precursor of the small subunit of Rubisco (pSSU) can be translocated without soluble factors. The insoluble inclusion bodies that contain the precursor are harvested by centrifugation at 10,000 $\times$ g for 5 min and dissolved in a small volume of 8 M urea. After another centrifugation (10,000 $\times$ g; 2 min) the supernatant, which contains the dissolved unfolded pSSU, is diluted immediately into the import assay. The final urea concentration should be below 200 mM.

2. The overexpressed and purified pLHCP, which is also synthesized as inclusion bodies, has to be dialyzed with soluble proteins from pea leaves (LE) prior to the import experiment. Briefly, inclusion bodies are centrifuged and denatured in 8 M as above and then diluted to a final urea concentration of 400 mM into HMS buffer containing LE (780 μg ml^{-1}). Subsequently the assay (50 μl) is immediately placed on a dialysis membrane filter (Millipore, "VS," 0.025 μm) and dialyzed against 40 ml of HMS buffer for 30 min at 4°C. The import experiment is then performed under standard conditions.

VIII. Translocation into Isolated Outer Chloroplast Membranes

It is also possible to use the isolated outer envelope membrane of chloroplasts for binding and insertion experiments with precursor proteins. We have established the isolated outer membrane as an enriched translocation system that can partially translocate precursors (e.g., pSSU) in a specific (ATP-, transit-sequence, and receptor-dependent) manner. For the use of this model system highly purified envelope membranes must be prepared. This can be done from pea chloroplasts by the method of Keegstra and Youssif (1986). The outer membranes are further purified through a linear sucrose gradient (0.6–1.2 M sucrose in 10 mM NaP$_i$, pH 7.4) at 125,000 × g for 16 h. Outer membranes are recovered from the gradient, diluted with 4 vol of 10 mM NaP$_i$, pH 7.4, and pelleted by centrifugation (125,000 × g, 1 h). The final pellet, which represents highly purified chloroplasts outer envelopes, is resuspended in 10 mM NaP$_i$, pH 7.6, and stored in small aliquots in liquid N$_2$. Another prerequisite is the correct, i.e., right-side-out, orientation of the membrane vesicles (Waegemann *et al.,* 1992).

The binding experiment is performed with 10 μg of purified outer membranes, which are washed once in HM buffer (25 mM Hepes/KOH, pH 7.6; 5 mM MgCl$_2$). The membrane pellet is resuspended in a final volume of 20 μl in HM buffer, supplemented with 1% BSA and 100 μM ATP with translation product (1–2 μl). Incubate for 10 min at 25°C and reisolate the membranes with bound precursor by centrifugation (250,000 × g; 25 min) through a 500-μl sucrose cushion (0.2 M sucrose in HM buffer). The pellet is washed once in HM buffer and then prepared for gel electrophoresis or post-treated with thermolysin. The membrane pellet is resuspended in 25 μl HM buffer (+0.5 mM CaCl$_2$) for the protease treatment, then thermolysin is added to a final concentration of 1 μg thermolysin/10 μg envelope protein and incubated for 90 s at 25°C. The reaction is terminated by EDTA (final concentration 10 mM).

A binding experiment of pSSU to isolated outer chloroplast envelope is shown in Fig. 2, lane 1. A post-treatment of the membranes with thermolysin results in protease-protected translocation intermediates deg 1 and deg 2 (lane 2), indicating a partial insertion of the precursor into the bilayer. The removal of surface-exposed polypeptides, e.g., receptor protein from envelope membrane vesicles,

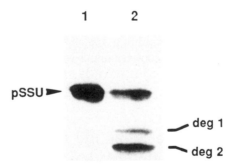

Fig. 2 Binding of pSSU to isolated outer chloroplast membranes. pSSU binds to isolated outer envelope membranes in the presence of 100 μM ATP (lane 1). Post-treatment of the membranes with thermolysin results in the occurrence of partially protease protected translocation intermediates deg 1 and 2, which are partially inserted into the outer envelope (lane 2).

is done in a manner identical to that previously described, prior to precursor binding experiments.

IX. Isolation of a Functionally Active Import Complex from Chloroplast Membranes

A further enrichment and purification of the translocation machinery in comparison to the intact organelle and the isolated outer envelope membranes can be achieved by solubilization of the envelope membranes and centrifugation through a sucrose density gradient. The partially purified membrane complex is a good tool for the identification of constituents of the machinery. In our laboratory, we have pioneered the development of a method to isolate a functionally active translocation complex from isolated outer chloroplast membranes (Soll and Waegemann, 1992). For the isolation of the import complex, 100–125 μg isolated outer envelope membranes are resuspended in 100 μl HM buffer supplemented with 1% BSA and 100 μM ATP; 10–15 μl of pSSU translation product is bound to the membranes by incubation for 10 min at 25°C. The membranes with bound precursor are reisolated by centrifugation (250,000 $\times$ g; 25 min) through a sucrose cushion, 0.2 M sucrose in ME buffer (25 mM MOPS/KOH, pH 7.2, 1 mM EDTA), and then resuspended in 200 μl ME buffer. Solubilization with 0.5% w/v digitonin is done for 10 min at room temperature, according to Kiebler *et al.* (1990). The suspension is then layered on top of a linear 5-ml sucrose gradient (5–20% sucrose in ME buffer supplemented with 0.2% digitonin and 0.05% Triton-x-100). After a 4-h centrifugation at 330,000 $\times$ g in a swinging-bucket rotor (TST 55) the gradient is manually fractionated from the top into 200-μl fractions. An aliquot of each fraction is counted in the scintillation counter. The distribution pattern of radioactivity in the fractionated gradient shows the association of pSSU with components of the membranes, which results in a shift of radioactivity to higher density. If translation product without the addition of membranes is loaded on top of a sucrose gradient, the labeled precursor protein remains on the top of the gradient. By incubating the envelope membranes with SSU instead of pSSU, no shift of the radioactive polypeptide to higher density in the gradient occurs, indicating that the interaction depends on the transit sequence. Furthermore, the detection of the import complex shows the same characteristics as the intact organelle or the isolated outer membrane system, namely the dependency on ATP and on protease-sensitive components. If the isolated complex is treated with thermolysin, identical translocation intermediates (i.e., deg 1 and 2) appear. The functional activity of the isolated complex is demonstrated in the following way: envelope membranes are first solubilized with digitonin, spun through a linear sucrose gradient, and fractionated as above. The fractions that contained the labeled precursor, i.e., the import complex in the experiment described above (in generally fractions 14–19), are collected, diluted at least twofold in ME buffer, and incubated with 10 μl pSSU translation product in the presence of 500 μM ATP for 10 min at 25°C. The complex is then reisolated by spinning overnight at 250,000 × g through a second 9-ml linear 5–20% sucrose gradient (final volume 12 ml). The gradient is fractionated (400-μl fractions) from the top. In the presence of pSSU, but not SSU, a shift of the labeled preprotein to higher density in the gradient (fraction 17–26) is detected. The interaction requires ATP and protease-sensitive components, indicating the presence of a receptor polypeptide in this complex. Protease treatment yields again translocation intermediate deg 1 and 2, demonstrating the specificity of the interaction of the precursor with the isolated membrane complex (Soll and Waegemann, 1992) and its functional activity.

References

Bonner, W. M., and Laskey, R. A. (1974). A film detection method for tritium-labeled proteins and nucleic acids in polyacryl amidgels. *Eur. J. Biochem.* **46,** 83–88.

Cline, K., Werner-Washburne, M., Andrews, J., and Keegstra, K. (1984). Thermolysin is suitable protease for probing the surface of intact pea chloroplasts. *Plant Physiol.* **75,** 675–678.

Dahlin, C., and Cline, K. (1991). Developmental regulation of the plastid protein import apparatus. *Plant Cell* **3,** 1131–1140.

de Boer, A. D., and Weisbeek, P. J. (1991). Chloroplast protein topogenesis: Import, sorting and assembly. *Biochim. Biophys. Acta* **1071,** 221–253.

della-Cioppa, G., and Kishore, G. M. (1988). Import of a precursor protein into chloroplasts is inhibited by the herbicide glyphosate. *EMBO J.* **7,** 1299–1305.

Eilers, M., and Schatz, G. (1986). Binding of a specific ligand inhibits import of a purified precursor protein into mitochondria. *Nature* **322,** 228–232.

Friedman, A. L., and Keegstra, K. (1989). Chloroplast protein import. Quantitative analysis of precursor binding. *Plant Physiol.* **89,** 993–999.

Keegstra, K., and Youssif, A. E. (1986). Isolation and characterization of chloroplast envelope membranes. *Methods Enzymol.* **118,** 316–325.

Keegstra, K., Olsen, C. J., and Theg, S. M. (1989). Chloroplastic precursors and their transport across the envelope membranes. *Annu. Rev. Plant. Physiol. Plant Mol. Biol.* **40,** 471–501.

Kiebler, M., Pfaller, R., Söllner, T., Griffith, G., Horstmann, H., Pfanner, N., and Neupert, W. (1990). Identification of a mitochondrial receptor complex required for recognition and membrane insertion of precursor proteins. *Nature* **348,** 610–616.

Krieg, P. A., and Melton, D. A. (1987). *In vitro* RNA synthesis with SP6 RNA polymerase. *Methods Enzymol.* **155,** 397–415.

Laemmli, U. K. (1970). Cleavage of structural proteins during the assembly of the head of bacteriophage T4. *Nature (London)* **227,** 680–685.

Li, H.-M., Moore, T., and Keegstra, K. (1991). Targeting of proteins to the outer envelope membrane uses a different pathway than transport into chloroplasts. *Plant Cell* **3,** 709–717.

Olsen, L. J., Theg, S. M., Selman, B. R., and Keegstra, K. (1989). ATP is required for the binding of precursor proteins to chloroplasts. *J. Biol. Chem.* **264,** 6724–6729.

Paulsen, H., Rümmler, U., and Rüdiger, W. (1990). Reconstitution of pigment-containing complexes from light-harvesting chlorophyll a/b-binding protein overexpressed in *Excherichia coli. Planta* **181,** 201–211.

Perry, S. E., Li, H.-M., and Keegstra, K. (1991). *In vitro* reconstitution of protein transport into chloroplasts. *Methods Cell Biol.* **34,** 327–344.

Pilon, M., de Boer, D. A., Knols, S. L., Koppelman, M. H. G. M., van der Graaf, R. M., de Kruijff, and Weisbeek, P. J. (1990). Expression in *Escherichia coli* and purification of a translocation-competent precursor of the chloroplast protein ferredoxin. *J. Biol. Chem.* **265,** 3358–3361.

Sabohl, S., and Ochoa, S. (1974). Preparation of radioactive initiation factor 3. *Methods Enzymol.* **30,** 39–44.

Salomon, M., Fischer, K., Flügge, U.-I., and Soll, J. (1990). Sequence analysis and protein import studies of an outer chloroplast envelope polypeptide. *Proc. Natl. Acad. Sci. U.S.A.* **87,** 5778–5782.

Soll, J., Alefsen, H., Böckler, B., Kerber, B., Salomon, M., and Waegemann, K. (1992). Comparison of two different translocation mechanisms into chloroplasts. *In* "Membrane Biogenesis and Protein Targeting" (W. Neupert and R. Lill, eds.), pp. 299–306. Amsterdam/New York: Elsevier.

Soll, J., and Waegemann, K. (1992). A functionally active protein import complex from chloroplasts. *Plant J.* **2,** 253–256.

Soll, J., and Alefsen, H. (1993). The protein import apparatus of chloroplasts. *Physiol. Plant.* **87,** 433–440.

Waegemann, K., Paulsen, H., and Soll, J. (1990). Translocation of proteins into isolated chloroplasts requires cytosolic factors to obtain import competence. *FEBS Lett.* **261,** 89–92.

Waegemann, K., and Soll, J. (1991). Characterization of the protein import apparatus in isolated outer envelopes of chloroplasts. *Plant J.* **1,** 149–158.

Waegemann, K., Eichacker, S., and Soll, J. (1992). Outer envelope membranes from chloroplasts are isolated as right-side-out vesicles. *Planta* **187,** 89–94.

Zimmermann, R., Sagstetter, M., Lewis, M. J., and Pelham, H. R. B. (1988). Seventy-kilodalton heat shock proteins and an additional component from reticulocyte lysate stimulate import of M13 procoat protein into microsomes. *EMBO J.* **7,** 2875–2880.

CHAPTER 19

Macromolecular Movement into Mitochondria

Elzbieta Glaser, Carina Knorpp, Marie Hugosson, and Erik von Stedingk

Department of Biochemistry
Arrhenius Laboratories for Natural Sciences
Stockholm University
S-106 91 Stockholm, Sweden

I. Introduction

A. General Aspects of the Intracellular Protein Transport

Genetic information in plants is localized in three subcellular compartments, the nucleus, mitochondria, and chloroplasts, hence the need for coordination of expression between the nuclear and the organellar genomes (Grivell, 1989). The great majority of plant organellar macromolecules are nuclear-encoded (Hartl *et al.,* 1989). The coding capacity of the mitochondrial and chloroplastic genomes is very limited, including 13–15 proteins encoded by the mitochondrial DNA and ca. 60 proteins encoded by the chloroplastic DNA. Since their nuclear-encoded proteins are synthesized in the cytosol, the biogenesis of all cellular organelles requires specific intracellular protein trafficking and import systems. Many of the nuclear-encoded proteins are synthesized as precursor proteins containing N-terminal extensions called signal peptides, presequences, or transit peptides, which function as organellar sorting and targeting signals (Baker and Schatz, 1991). Most of the proteins destined to mitochondria, chloroplasts, and ER contain N-terminal signals, which are proteolytically cleaved off after the completed import (Dalbey and von Heijne, 1992; Waters *et al.,* 1991). Noncleavable C-terminal and N-terminal targeting signals for transport into peroxisomes and internal signals for transport to the nucleus (Newmeyer, 1993) have also been characterized.

B. The Mitochondrial Protein Transport Process

More than 95% of the mitochondrial protein complement is synthesized outside the organelle and has to be imported. The mitochondrial protein trafficking and import process is complex and includes several events: 1. Interaction of the signal peptides of the newly synthesized protein with cytosolic factors, which facilitates recognition, prevents missorting, and confers an import competent conformation; 2. Recognition of the signal or the signal/cytosolic factor complex on the organellar membrane; 3. Translocation of the protein through the organellar membrane(s); 4. Processing of the imported polypeptide inside the organelle; 5. Intraorganellar transport of the imported protein; and 6. Assembly of the mature protein into the final functional conformation in a process, presumed often to involve organellar chaperones. Besides the proteins, some tRNAs also have to be imported into plant mitochondria. The mitochondrial protein import system is both general and specific in its properties; i.e., it recognizes all the mitochondrial precursors, yet discriminates against precursors destined to other organelles. The overall mitochondrial protein import process is very similar to that of chloroplast protein import. Taking into consideration the great excess of chloroplasts in comparison to mitochondria in photosynthetic cells, it is obvious that the sorting pressure in the plant cell is very high. Most of the mitochondrial and chloroplast proteins are synthesized with N-terminal presequences. Structural analysis of the presequences has revealed that the secondary structure of the mitochondrial and chloroplast presequences is different. The mitochondrial presequences contain an amphiphilic α-helix stretch at the N-terminus, whereas the chloroplastic presequences have an uncharged N-terminal region and an amphiphilic β-strand at the C-terminus of the presequence (von Heijne *et al.,* 1989). Chloroplast precursors destined for the thylakoid lumen also contain an added domain resembling a bacterial signal sequence. It seems that these structural differences are responsible for the organellar recognition.

The general characteristics of plant and fungal mitochondrial protein import systems have been found to be analogous (Whelan *et al.*, 1988; Chaumont *et al.*, 1990; Whelan *et al.*, 1990). However, in contrast to the results obtained in sorting experiments using fungal mitochondria (Hurt *et al.*, 1986; Pfaller *et al.*, 1989), studies in plants using isolated spinach leaf mitochondria and chloroplasts showed strict organellar specificity (Whelan *et al.*, 1990), in agreement with *in vivo* investigations using transgenic tobacco (Boutry *et al.*, 1987). Recent studies have demonstrated that, in contrast to the situation observed for yeast and mammalian cells, the plant mitochondrial processing activity is membrane-bound (Eriksson and Glaser, 1992) and an integral part of the bc_1 complex of the respiratory chain (Braun *et al.*, 1992; Eriksson *et al.*, 1993; Emmermann *et al.*, 1993; Eriksson *et al.*, 1994). *Neurospora* seems to represent an intermediate stage in the evolutionary pathway, with one of the subunits of the mitochondrial processing peptidase in the matrix, and the other existing both in the matrix and as a core 1 subunit of the bc_1 complex. This finding implies that the bc_1 complex in plants is bifunctional and is involved both in respiration and in protein processing. This opens several intriguing questions concerning interdependence of respiration, protein processing, and transport of proteins through the mitochondrial membrane.

The aim of the present review is to give a description of the basic principles and methodology used for the study of protein movement into mitochondria. We will present a combination of biochemical isolation and characterization procedures with techniques of molecular biology for studies of intracellular protein transport in plants with emphasis on mitochondrial protein import and sorting mechanisms of the nuclear-encoded polypeptides between mitochondria and chloroplasts.

II. Materials and Methods

A. Preparation of Import–Competent Plant Mitochondria

Although the first studies of *in vitro* mitochondrial protein import in fungi date from the middle 1970s, the first reports of *in vitro* plant mitochondrial protein import appeared only recently (Whelan *et al.*, 1988; Whelan *et al.*, 1990). This delay was in most part due to technical difficulties associated with the preparation of import-competent plant mitochondria. In this chapter we will describe procedures for preparation of import-competent mitochondria and methods for the *in vitro* synthesis of the precursor proteins used in *in vitro* import studies. The highest efficiency of protein import is achieved with high-quality, tightly coupled mitochondria. The methods described here are effective on both small and large scales in yielding preparations of mitochondria that are protein import-competent.

1. Spinach Leaf and Root Mitochondria

Spinach (*Spinacia oleracea,* cv. *medania*) was grown under constant temperature (18–22°C) and light conditions (10/14-h light/dark cycles) hydroponically for 6 weeks. Spinach leaf and root mitochondria were isolated as described in Fig. 1. The method is based on the previously described procedure from our laboratory (Whelan *et al.,* 1990) used for *in vitro* protein import into spinach leaf mitochondria. The preparation procedure, shown in Fig. 1, is simple and relatively fast, consisting of a differential centrifugation step of leaf or root homogenate followed by density gradient centrifugation on a discontinuous gradient of Percoll. Two hundred grams of tissue and 300 ml of 0.3 M sucrose, 50 mM Mops, 5 mM $MgCl_2$, 2 mM EDTA, 1% (w/v) BSA, 1% (w/v) PVP-40, pH 7.8 (medium A) were blended for 5 s in a Waring blender and filtered through a 60-μm nylon cloth. The filtrate was centrifuged for 2.5 min at 5000 $\times$ g. The supernatant was centrifuged at 20,000 $\times$ g for 4.5 min. The pellet of "crude" mitochondria was resuspended in 0.3 M sucrose, 10 mM Mops, 1 mM EDTA, 0.2% (w/v) BSA, pH 7.2 (medium B), homogenized, and loaded on six Percoll gradients. The gradients were centrifuged for 40 min at 11,400 $\times$ g. The mitochondria were collected as the pale band on top of the 60% cushion. The purified mitochondria were diluted 5–10 times in 0.3 M sucrose, 25 mM Mops, 5 mM $MgCl_2$, pH 7.2 (medium C), and centrifuged for 5 min at 11,400 $\times$ g, and then pelleted in a microfuge at 12,000 rpm for 5 min. Protein content was measured according to Bio-Rad (Bradford, 1976). The yield was 6.3 mg (leaf) and 5.2 mg (root) mitochondrial protein per 200 g of tissue. The respiratory activities (in nmol O_2/min/mg protein, in the presence of ADP) of leaf and root mitochondria were 20.4 and 70.4, respectively, and, with NADH as a substrate were 35 and 86, respectively. These values were comparable to those found for spinach leaf mitochondria prepared using a large-scale procedure (Hamasur *et al.,* 1990). Latency of the cytochrome c oxidase was 95 and 78% for leaf and root mitochondria, respectively.

2. Large-Scale Spinach Leaf Mitochondria

Depetiolated spinach leaves (3 kg, as 0.5-kg portions in 700 ml of medium) were homogenized for 2–3 s at high speed, in a 4-liter Waring blender. The medium comprised 0.3 M sucrose, 25 mM Mops–KOH, pH 7.8, 5 mM $MgCl_2$ 4 mM cysteine, 0.6% (w/v) polyvinylpirrolidone (PVP), and 0.2% (w/v) BSA (Hamasur *et al.,* 1990). The slurry was filtered through four layers of nylon net (60 μm) and the residue was homogenized and filtered again. Chloroplasts, starch, and cell wall fragments were sedimented at 6500 $\times$ g for 10 min in an HR-4L rotor (RC-3 centrifuge) using 1-liter bottles. The supernatant containing

200 g of tissue and 300 ml medium A were blended for 6 sec in Waring blender and filtered through a 60 µm nylon cloth. The filtrate was centrifuged for 2.5 min at 5000 g.

The supernatant was centrifuged at 20 000 g for 4.5 min.

The pellet of "crude" mitochondria was resuspended in medium B, homogenized and loaded on Percoll gradients. The gradients were centrifuged for 40 min at 11400 g.

Leaf mitochondria   Root mitochondria

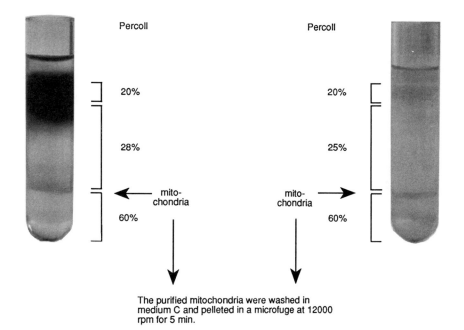

Percoll   Percoll

20%   20%

28%   25%

← mito-   mito- →
chondria   chondria

60%   60%

The purified mitochondria were washed in medium C and pelleted in a microfuge at 12000 rpm for 5 min.

Fig. 1   Preparation method for import-competent spinach root and leaf mitochondria.

the mitochondria was recentrifuged at $9000 \times g$ for 15 min using a JA-14 rotor. The mitochondrial precipitate was suspended in 50 ml of Percoll medium, composed of 0.25 M sucrose, 10 mM Mops–KOH, pH 7.2, 1mM EDTA, 0.1% (w/v) defatted BSA, 32% Percoll. The suspension of crude mitochondria was applied directly to the above Percoll medium (2.5 ml crude mitochondria to 30 ml to Percoll medium) and centrifuged at $19,000 \times g$ for 45 min. After centrifugation, the mitochondrial band, positioned close to the bottom of the tube, was collected using a bent Pasteur pipette. It was diluted 10 times with wash medium containing 0.3 M sucrose, 5 mM Mops–KOH (pH 7.2) and collected at $9000 \times g$ for 15 min. The pellets were suspended in wash medium at a protein concentration of 10–15 mg/ml. All steps were performed at 4°C.

3. Potato Tuber Mitochondria

Potato tubers (1 kg) were homogenized in a Moulinex juice centrifuge in 400 ml of medium containing 0.6 M mannitol, 10 mM EDTA, 40 mM Mops–KOH, pH 7.3, 8 mM cysteine, and 0.4% (w/v) BSA, with continuous adjustment of the pH in the range 7–7.5. The slurry was then either filtered through four layers of nylon net (60 μm) or left to settle in the cold for 5–10 min until most of the starch had sedimented, before the slurry was decanted. Starch and cell wall fragments were sedimented at $4000 \times g$ for 5 min in a Beckman centrifuge in 250-ml bottles using a JA-14 rotor. The supernatant containing the mitochondria was recentrifuged at $10,000 \times g$, for 10 min in the JA-14 rotor. The mitochondrial pellet was suspended in Percoll medium, composed of 0.4 M mannitol, 5 mM Mops–KOH, pH 7.2, 0.1% (w/v) defatted BSA, 32% Percoll. The suspension of crude mitochondria was applied directly to the above Percoll medium (2.5 ml crude mitochondria to 30 ml of Percoll medium) and centrifuged at $18,000 \times g$ for 45 min in JA-20 rotor. After centrifugation the mitochondrial band, positioned close to the bottom of the tube, was collected. The mitochondria were diluted 10 times with wash medium containing 0.4 M mannitol, 5 mM Mops–KOH, pH 7.2, 0.1% defatted BSA and collected at $9000 \times g$ for 15 min. The pellets were suspended in wash medium at a protein concentration of 20–30 mg/ml. All steps were performed at 4°C.

B. *In Vitro* Transcription and Translation of Precursor Proteins

Precursor proteins were routinely prepared from full-lenth cDNAs cloned in expression vectors. The plasmid was linearized using a suitable restriction enzyme that cuts downstream of the protein; e.g., the *Nicotiana* $F_1\beta$ precursor in a pTZ 18U plasmid was linearized using *Hin*dIII (1.5 units restriction enzyme per μg DNA for 60 min at 37°C). The linearization can be controlled by running an 0.8% agarose gel electrophoresis. The DNA was precipitated with ethanol and taken up to a concentration of 1 mg/ml. Transcription was performed with SP6 or T7 RNA polymerase. For the *Nicotiana* $F_1\beta$ precursor protein, 1.5 μg of DNA was incubated with 60 units of RNAsin (Boehringer Mannheim), 25 units of T7 RNA polymerase (Boehringer Manheim), and 0.1 unit of m75Gppp(5')G (Pharmacia) in a total volume of 50 μl. Note that RNAsin and RNA polymerases are very temperature-sensitive. The reaction mixture was preincubated for 2 min before the addition of 50 mM GTP. The reaction was carried out for 20 min at 37°C. Prior to the translation of the transcript, buffer was exchanged with water using a Sephadex G-50 spin-column according to Maniatis. Translation was routinely performed using rabbit reticulocyte lysate (Amersham) and [^{35}S]methionine (Amersham); typically 10–15 μl of the transcript was incubated with 2–3 μl [^{35}S]methionine and 50 μl rabbit reticulocyte lysate at 30°C for 1 h.

We have also had very good experience with a coupled transcription/translation system (SDS-PROMEGA TNT), which has the advantage of being considerably quicker and it abolished the need to store RNA. Routinely, for the *Nicotiana* F$_1\beta$ precursor, 60 units of RNAsin, 50 units of T7 polymerase and 1 μg of circular DNA in the buffers supplied in the kit plus 4 μl [^{35}S]methionine were incubated at 30°C for 90 min. Prior to use for import, ribosomes were removed by centrifugation at 120,000 $\times$ g for 60 min in a Beckman TLA 100.2 rotor.

C. *In Vitro* Protein Import

Routine protein import experiments (Whelan *et al.,* 1990) were carried out by incubation, at 23°C for 15 min with continuous shaking, of the *in vitro*-synthesized precursor proteins (approximately 4–6 μl, 20,000–30,000 cpm) with mitochondria (200 μg protein) in a final volume of 200 μl of a solution comprising 0.25 M mannitol, 50 mM KCl, 10 mM Mops, 2 mM ATP, 10 μM ADP, 1 mM MnCl$_2$, 1 mM methionine, 5 mM K$_3$PO4, and 1% BSA, pH 7.4, supplemented with 1 mM glycine or 1 mM succinate as respiratory substrates for spinach leaf and root mitochondria, respectively (Knorpp *et al.,* 1994). With potato mitochondria, 1 mM succinate or 2 mM NADH was used as a respiratory substrates. After protein import was completed, the samples were divided into 100-μl aliquots and placed on ice. One aliquot was incubated with proteinase K (PK, 2 μg/ml) for 30 min. After the incubation, phenylmethylsulfonyl fluoride (PMSF) was added to 2 mM. The treated mitochondria were reisolated in a microfuge at 12,000 rpm for 5 min, lysed in sample buffer, and analyzed by 12% SDS–PAGE (Laemmli, 1970). The gels were subsequently fixed and impregnated with Amersham amplifier and autoradiographed (see also Section III,B for a discussion of temperature dependence and kinetics of protein import).

D. Binding of Precursor

Binding of precursor to the mitochondrial outer membrane can be separated from the translocation process. The protein translocation process can be inhibited by different methods. One of the possibilities is to lower the temperature to 0°C by placing the import reaction mixture on ice. Another possibility is to inhibit the formation of a membrane potential, which is required for the translocation. This can be achieved by addition of ionophores, for example, valinomycin at 10 μM. Inhibition of translocation can also be achieved by depletion of matrix ATP, which is required by the mitochondrial chaperones in order to complete the translocation. Depletion of the matrix ATP can be achieved by not adding ATP to the import reaction, in combination with treatment of mitochondria with apyrase, 12.5 U/ml, (Pfanner and Neupert, 1986) and addition of oligomycin (10 μM) to the reaction mixture. Nonspecifically bound precursor can be removed by centrifugation through a cushion of, e.g., Percoll, sucrose, or silicon oil.

E. *In Vitro* Protein Processing

For measuring precursor protein processing activity, the isolated mitochondria were diluted with 0.3 M sucrose, 10 mM Mops, pH 7.5, to a protein concentration of 1 mg/ml and sonicated in the presence of 30 mM MgCl$_2$ and 1 mM PMSF using a Branson Sonifer (equipped with a microtip, setting 3) three times for 30 s at 4°C (Eriksson and Glaser, 1992). The membrane fraction was separated from the matrix by centrifugation at 105,000 $\times$ g for 45 min. The reaction for processing contained 1–2 ml of [^{35}S]methionine labeled translation product (ca. 10,000–15,000 cpm) in 0.5% w/v Triton X-100, 15 mM Tris–HCl, pH 8, 4 mM MnCl$_2$, 2 mM PMSF, and membranes/matrix (5–15 μg protein). The reaction was carried out at room temperature for 30 min. The reaction was stopped by addition of double-strength SDS–PAGE sample buffer.

III. Results and Discussion of Critical Aspects of the Procedures

A. Technical Advice

In the above sections, we have presented methods for preparation of import-competent mitochondria and precursor proteins as well as the protein import and processing recipies. In this section we highlight critical points of the procedures:

1. For preparation of import-competent mitochondria it is necessary to use plant tissue freshly taken from the plant (most critical when using root tissue). All solutions, glass, and plastic-ware used during the preparation should be prechilled on ice. The preparation procedure must be performed without delay. The purified mitochondria should be kept on ice. In order to obtain good efficiency of import, the import reaction should be started as soon as possible after the completed organelle preparation. If it is difficult to obtain coupled mitochondria, include 1 mM substrate in all buffers during the preparation. The mitochondria (in 20% glycerol) can be frozen directly after the preparation in liquid nitrogen; however, this reduces the efficiency of import by several-fold.

2. During transcription and translation of the precursor, all solutions and materials must be nuclease free. The translation product should not be stored at temperatures higher than $-80°C$. All reagents used must be stored according to manufacturers' recommendations. Store reticulocyte lysate in small aliquots, so it will not be repeatedly thawed and refrozen.

3. Protein import efficiency increases when the import reaction is exposed to agitation. The proteinase K should be freshly solubilized or frozen in aliquots and used once. The concentration of proteinase K should be adjusted for each different mitochondrial type since mitochondria differ in their sensitivity to proteinase K. After the proteinase K treatment, PMSF should be added and the mitochondria reisolated and resuspended in electrophoresis sample buffer without delay. Alternatively, the electrophoresis sample buffer can be warmed up to 90°C prior to addition to proteinase K-treated mitochondria. Precautions should also be taken to use the correct time and temperature, as described in the following discussion.

A routine import experiment (Fig. 2) demonstrates *in vitro* import of the *Neurospora* $F_1\beta$ precursor into leaf mitochondria. The 56-kDa precursor of *Neurospora* $F_1\beta$ is imported and cleaved to the mature 52-kDa form by the leaf mitochondrial import machinery. The mature form of the protein was protease-protected inside the mitochondria. The import efficiencies were generally 10–20%; i.e., 10–20% of the added precursor was imported and processed.

The following controls are required to be certain that the observed phenomenon is a specific mitochondrial protein import: 1. Inhibition of import in the presence of an ionophore (valinomycin) or an uncoupler (FCCP) to show that

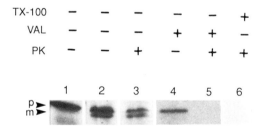

Fig. 2 Import of the *Neurospora* $F_1\beta$ precursor into spinach leaf mitochondria. Lane 1, Import mixture without mitochondria showing precursor (p) with a relative molecular mass of 56 kDa. This represents 35% present in each import reaction. Lane 2, import mixture with mitochondria showing precursor and mature (m) 54-kDa forms. Lane 3, as lane 2 with proteinase K (PK) added showing that both precursor and mature forms are insensitive to externally added protease. Hence both forms are imported products. Lane 4, import mixture with mitochondria in the presence of valinomycin and oligomycin. Precursor is bound to mitochondria but no mature form is generated. Lane 5, as lane 4 treated with PK showing that no protein has been imported into the mitochondria. Lane 6, import mixture with mitochondria. Mitochondria are lysed by the addition of Triton X-100 to 1% prior to PK treatment, indicating that protease insensitivity is due to a location inside intact mitochondria (From Whelan *et al.,* 1990. Reprinted by permission of Kluwer Academic Publishers.

the protein import is membrane potential dependent; 2. Pretreatment of the mitochondria with 1% Triton X-100 prior to proteinase K treatment in order to show that with the disrupted mitochondrial membrane both the precursor and the mature form of the imported protein are fully accessible to proteinase K. Figure 2 shows both these controls.

B. Temperature Dependence and Kinetics of *in Vitro* Import

The efficiency of import into mitochondria is temperature and time-dependent. We have reported two temperature optima for the import of the *Nicotiana* $F_1\beta$ precursor into spinach leaf mitochondria (Knorpp *et al.*, 1992) (Fig. 3), indicating that the process is controlled by more than one event with different temperature optima. The first optimum is about 15°C lower than the temperature optimum for mitochondria from other sources (Schleyer and Neupert, 1985; Sztul *et al.*, 1988). This probably reflects the different optimal growth temperatures for these different organisms. Thus, we have found that one characteristic feature of plant mitochondria is their capacity to import proteins at rather low temperatures. This is followed by an increase of import at higher temperatures, whereas at temperatures of 25°C and above a drastic decrease of import was observed (Fig. 3). Elevated temperatures would be suspected to give rates of import that would result in a higher amount of the imported protein. However, we have investigated the kinetics at these different temperature optima (15 and 25°C) and have concluded that there exists a temperature-triggered degradation pathway inside the mitochondrion (Fig. 4). This results in a rapid increase of imported precursor up to about 10 min at higher temperatures followed by a decrease of imported

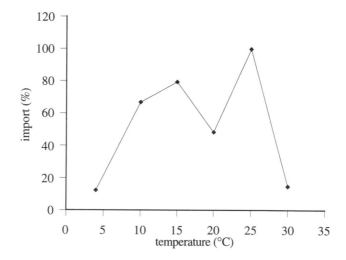

Fig. 3 Temperature dependence of *in vitro* import of *Nicotiana* $F_1\beta$ precursor protein into spinach leaf mitochondria. Import was carried out for 30 min.

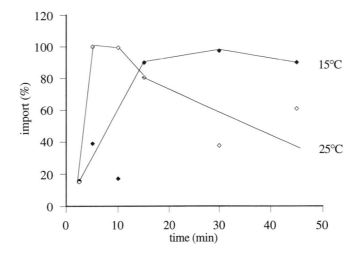

Fig. 4 Kinetics of *in vitro* import of *Nicotiana* $F_1\beta$ precursor protein into spinach leaf mitochondria at different temperatures.

protein found inside the mitochondrion. Thus a lower net amount of imported precursor is found inside the mitochondrion after 30 min at higher temperatures compared to lower (Knorpp *et al.,* 1995).

IV. Conclusions and Perspectives

In studies of *in vitro* protein sorting between mitochondria and chloroplasts, homologous organelle system was used; i.e., mitochondria and chloroplasts prepared from spinach leaves were used as well as mitochondrial and chloroplastic precursor proteins (Whelan *et al.,* 1990). These studies show a very high organellar specificity of precursor proteins in the plant cell and are in contrast to *in vitro* import experiments with heterologous systems. It will be of interest to extend these studies and use homologous precursor proteins and a homologous cytosol preparation.

Analysis of the secondary structures of the mitochondrial and chloroplast presequences shows structural differences, which may account for sorting (von Heijne *et al.,* 1989). However, some of the presequences coupled to the reporter proteins cannot differentiate between organelles (Chaumont and Boutry, 1994). The role of the mature portion of the precursor in the import process is still an open question. It will be also valuable to investigate the involvement of the molecular chaperones and other cytosolic factors as well as the role of organellar phospholipids in the import and sorting processes. Also of high interest for future studies are degradation events, which are crucial for understanding the fate of the imported precursor. It would be of interest to investigate whether the observed mitochondrial protein degradation is related to the amount of the mitochondrial chaperones Hsp 60, Hsp 70 (von Stedingk and Glaser, 1995), and Hsp 10 and whether the association of the imported protein with mitochondrial chaperones facilitates or decreases degradation.

The processing enzyme in plant mitochondria has been shown to be a part of the bc$_1$ complex of the respiratory chain (Braun *et al.*, 1992; Eriksson *et al.*, 1993). Integration of the processing activity into the bc$_1$ complex opens intriguing questions concerning regulation of the bioenergenetic events in mitochondria. What is the topology of the complex in comparison to import sites? Is import of the precursors dependent on processing? Does the respiratory function of the bc$_1$ complex influence import and processing? The physiological and bioenergenetic significance of the association of the respiratory function and the processing activity and the regulation and mutual interdependence of the translocation of the precursor, processing, and electron transfer in plants are important questions for future investigations.

References

Baker, K. P., and Schatz, G. (1991). Mitochondrial essential for viability mediate protein import into yeast mitochondria. *Nature* **349,** 205–208.

Boutry, M., Nagy, F., Poulsen, C., Aoyagi, K., and Chua, N.-H. (1987). Targeting of bacterial chloramphenicol acetyltransferase to mitochondria in transgenic plants. *Nature* **328,** 340–342.

Bradford, M. M. (1976). A rapid and sensitive method for the quantitation of microgram quantities of protein utilizing the principle of protein-dye binding. *Annu. Biochem.* **72,** 248–254.

Braun, H.-P., Emmermann, M., Kruft, V., and Schmitz, U. K. (1992). The general mitochondrial processing peptidas from potato is an integral part of cytochrome c reductase of the respiration chain. *EMBO J.* **11,** 3219–3227.

Chaumont, F., and Boutry, M. (1994). Protein import into plant mitochondria. *In* "Advances in Cellular and Molecular Biology of Plants" (C.S. Levings, III and I. K. Vasil, ed.), vol. 2: Molecular Biology of the Mitochondria, pp. 207–235. Kluwer Acad. Publ., Amsterdam.

Chaumont, F., O'Riordan, V., and Boutry, M. (1990). Protein transport into mitochondria is conserved between plant and yeast species. *J. Biol. Chem.* **265,** 16856–16862.

Dalbey, R. E., and von Heijne, G. (1992). Signal peptidases in procaryotes and eukaryotes—A new protease family. *Trends Biochem. Sci.* **18,** 474–478.

Emmermann, M., Braun, H.-P., Arretz, M., and Schmitz, U. K. (1993). Characterization of the bifunctional cytochrome *c* reductase-processing peptidase compex from potato mitochondria. *J. Biol. Chem.* **268,** 18936–18942.

Eriksson, A., and Glaser, E. (1992). Mitochondrial processing proteinase: A general processing proteinase of spinach leaf mitochondrial is a membrane-bound enzyme. *Biochim. Biophys. Acta* **1140,** 208–214.

Eriksson, A., Sjöling, S., and Glaser, E. (1993). A general processing protease of spinach leaf mitochondria is a associated with the bc$_1$ complex of the respiratory chain. *In* "Plant Mitochondria" (A. Brennicken and U. Kuch, ed.) pp. 233–241. Berlin; Germany: VHS Verlagsgesellscaft.

Eriksson, A., Sjöling, S., and Glaser, E. (1994). The ubiquinol cytochorome c oxidasereductase complex of spinach leaf mitochondria is involved in both respiration and protein processing. *Biochim. Biophys. Acta* **1186,** 221–231.

Grivell, L. A. (1989). Nucleo-mitochondrial interaction in yeast mitochondrial biogenesis. *Eur. J. Biochem.* **182,** 477–493.

Hamasur, B., Birgersson, U., Eriksson, A., and Glaser, E. (1990). Large scale purification of spinach mitochondria—Isolation and immunological studies of the F$_1$-ATPase. *Physiol. Plant.* **78,** 367–373.

Hartl, F. U., Pfanner, N., Nicholson, D. W., and Neupert, W. (1989). Mitochondrial protein import. *Biochim. Biophys. Acta* **988,** 1–45.

Hurt, E. C., Soltanifar, N., Goldschmidt-Clermont, M., Rochaix, J.-D., and Schatz, G. (1986). The cleavable pre-sequence of an import chloroplast protein directs attached polypeptides into yeast mitochondria. *EMBO J.* **5,** 1343–1350.

Knorpp, C., Hugosson, M., and Glaser, E. (1992). Temperature dependence and kinetics of plant mitochondrial protein import. *In* "Molecular, Biochemical and Physiological Aspects of Plant Respiration" (H. Lambers and L. van der Plas, eds.), pp. 361–365. The Hague: SPB Academic Publishing.

Knorpp, C., Hugosson, M., Sjöling, S., Eriksson, A., and Glaser, E. (1994). Tissue-specific differences of the mitochondrial protein import machinery. In vitro import, processing and degradation of the F$_1\beta$ subunit of the ATP synthase in spinach leaf and root mitochondria. *Plant Mol. Biol.* **26,** 571–579.

Knorpp, C., Sziqyarto, C., and Glaser, E. (1995). Evidence for a novel ATP-dependent membrane-associated protease in spinach leaf mitochondria. *Biochem. J.,* in press.

Laemmli, U. K. (1970). Cleavage of structural proteins during the assembly of the head of bacteriophage T4. *Nature* **227,** 680–685.

Newmeyer, D. D. (1993). The nuclear pore complex and nucleocytoplasmic transport. *Curr. Opinion Cell Biol.* **5,** 395–407.

Pfaller, R., Pfanner, N., and Neupert, W. (1989). Mitochondrial protein import: Bypass of proteinaceous surface receptors can occur with low specificity and efficiency. *J. Biol. Chem.* **264,** 34–39.

Pfanner, N., and Neupert, W. (1986). Transport of F$_1$-ATPase subunit β into mitochondria depends on both a membrane potential and nucleoside triphosphates. *FEBS Lett.* **209,** 152–156.

Schleyer, M., and Neupert, W. (1985). Transport of proteins into mitochondria: Translocation intermediates spanning contact sites between outer and inner membranes. *Cell* **43,** 339–350.

Sztul, E. A., Chu, T. W., Strauss, A. W., and Rosenberg, L. E. (1988). Import of the malate dehydrogenase precursor by mitochondria. Cleavage within leader peptide by matrix protease leads to formation of intermediate-sizied form. *J. Biol. Chem.* **263,** 12085–12091.

von Heijne, G., Steppuhn, J., and Herrmann, R. (1989). Domain structure of mitochondrial and chloroplast targeting peptides. *Eur. J. Biochem.* **180,** 535–545.

Von Stedingk, E., and Glaser, E. (1995). The molecular chaperone Mhsp 72 is partially associated with the inner mitochondrial membrane both in normal and heat stressed *Spinacia oleracea.* (1995). *Biochem. Int.* **35,** 1307–1314.

Waters, M. G., Griff, I. C., and Rothman, J. E. (1991). Proteins involved in vesicular transport and membrane fusion. Curr. Opin. *Cell Biol.* **3,** 615–620.

Whelan, J., Dolan, L., and Harmey, M. A. (1988). Import of precursor proteins into *Vibica faba* mitochondria. *FEBS Lett.* **236,** 217–220.

Whelan, J., Knorpp, C., and Glaser, E. (1990). Sorting of precursor proteins between isolated spinach leaf mitochondria and chloroplasts. *Plant Mol. Biol.* **14,** 977–982.

CHAPTER 20

Targeting of Proteins to the Nucleus

James C. Carrington

Department of Biology
Texas A&M University
College Station, Texas 77843

I. Introduction

Nuclear processes are involved in all aspects of plant cell growth and development. Understanding how proteins and nucleic acids are routed into and out of the nucleus, therefore, is of central importance to understanding how plant cells are regulated.

The nucleus is a complex organelle enclosed by a double-membrane system through which macromolecular traffic must flow (Forbes, 1992). Protein translocation into the nucleus, unlike other organelles, does not require traversal of the lipid bilayers. Rather, translocation involves active passage through nuclear pore structures spanning the envelope. The pores are composed of two stacked coaxial rings, each composed of eight subassemblies arranged symmetrically around a central plug. Although the diffusion exclusion limit of the pore is relatively high, permitting passage of macromolecules of up to 60 kDa, transport of physiologically relevant proteins is dependent on cytosolic and pore-associated receptors that interact with one or more nuclear localization signals (NLSs) within the translocated protein (Silver, 1991). Most NLSs fall into one of three classes. The SV40 T-antigen-like signal (e.g., PPKKKRKV) is composed of a series of basic residues in a single cluster (Kalderon *et al.,* 1984). The bipartite class of NLSs (e.g., KRPAATKKAGQAKKKKLD from nucleoplasmin) contains two short, basic clusters separated by variable spacing of around 10 residues (Robbins *et al.,* 1991). The MATα2-like NLSs (e.g., NKIPIKD) contain a hydrophobic sequence flanked by basic residues (Hall *et al.,* 1984), although this class appears to be the least represented among nuclear proteins. Despite the existence of discrete classes of NLSs, a strict consensus sequence does not emerge when comparing different signals within a class. Importantly, the general features of nuclear transport of animal, fungal, and plant proteins are similar, that is, it is a signal-dependent, receptor-mediated process (Raikhel, 1992).

This chapter focuses on methods involved in analysis of nuclear translocation properties of proteins in plant cells. I will emphasize those methods that can be adapted to plant molecular biology laboratories lacking extensive cell biology experience.

II. Scope and Application

The techniques presented here employ a gene fusion strategy using a reporter system based on ß-glucuronidase (GUS), which has a number of features ideally suited for use in plant nuclear targeting studies (Jefferson, 1987). These include the availability of convenient histochemical and fluorometric assays to conduct *in situ* and quantitative biochemical analyses, the lack of appreciable endogenous GUS activity in most plants, and the high stability of GUS in plant cells and extracts. Given its relatively large size, wild-type GUS does not diffuse into the nucleus, but rather behaves like a cytosolic protein. Fusion of a sequence containing a functional NLS results in translocation of GUS from the cytosol to the nucleus (Restrepo *et al.,* 1990). Importantly, GUS retains high activity when heterologous polypeptides are fused to the N- or C-terminus, thereby allowing flexibility in the design of gene fusions. Provided that the gene encoding a protein of interest is in hand, there are numerous applications of these methods, including mapping of NLSs and analysis of regulatory mechanisms controlling nuclear transport.

The application of these methods depends on the ability to introduce DNA transiently or stably into plant cells. Protoplasts from leaves or cell cultures can be transfected with a plasmid containing an expression cassette composed of gene regulatory elements and the GUS fusion sequence. Nuclear transport of the GUS fusion protein can then be assessed using microscopic or fractionation techniques (Carrington *et al.,* 1991). Although not discussed here, transient assays can also be performed in intact tissues using biolistic particle bombardment (Varagona *et al.*, 1992). Both of these transient assays permit rapid analysis of nuclear transport and are particularly useful when large numbers of fusion proteins are being tested. Alternatively, the GUS fusion expression cassette can be integrated into the plant genome stably, providing a convenient and unlimited source of cells with which to conduct translocation experiments.

III. Methods

A. Expression Vectors

A set of vectors has been developed for introduction of gene fusion into plant cells. The base vector is pRTL2 (Carrington *et al.*, 1990), which contains a modified 35S transcriptional promoter from cauliflower mosaic virus, the 5′ nontranslated region (NTR) sequence from tobacco etch virus (TEV), a multiple cloning site, and the 35S transcriptional terminator/polyadenylation sequence (Fig. 1A). The promoter contains a tandem duplication of the sequence between −90 and −418, which contains enhancer-like control elements. The TEV NTR constitutes the initial transcribed sequence and has been shown to enhance translation efficiency in a cap-independent manner (Carrington and Freed, 1990). This vector also contains a short region of the TEV open reading frame, although this DNA is removed by insertion of heterologous sequences. Insertions in pRTL2 are most often conducted using restriction or PCR fragments containing an *Nco*I site, CCATGG (start codon underlined), at the 5′ end of the open reading frame. Plasmids derived from pRTL2 are useful for transient expression of genes in transfected protoplasts or tissues.

For nuclear transport studies, at least two control plasmids should be analyzed in addition to the experimental constructs. First, pRTL2-GUS, which encodes a nonfused GUS protein, should serve as a cytosolic control. Second, a plasmid that encodes a GUS fusion protein that localizes to the nucleus should serve as a positive translocation control. We use pRTL2-GUS/NIa, which encodes a fusion protein containing the TEV NIa protein linked to the C-terminus of GUS. In fact, the pRTL2-GUS/NIa serves as a useful vector for construction of GUS gene fusions (Fig. 1B). Heterologous DNA fragments can be fused to the GUS sequence by first removing the NIa sequence by cleavage with *Bgl*II plus *Bam*HI, or *Bgl*II plus *Xba*I, and then inserting the new DNA as a *Bgl*II–*Bam*HI or *Bgl*II–*Xba*I fragment. If gene fusions encoding the heterologous protein fused to the N-terminus of GUS are desired, we use pRTL2-NIa/GUS as the base vector (Fig. 1C). The NIa sequence is removed as an *Nco*I–*Bgl*II restriction fragment, and foreign sequences are then inserted as *Nco*I–*Bgl*II fragments.

For stable transformation of plants, the expression cassette is removed from the pRTL2-based plasmid and subcloned into the binary vector pGA482 (or one of its derivatives; An, 1987). This vector can be introduced into *Agrobacterium tumefaciens* strain LBA4404 for standard plant transformation.

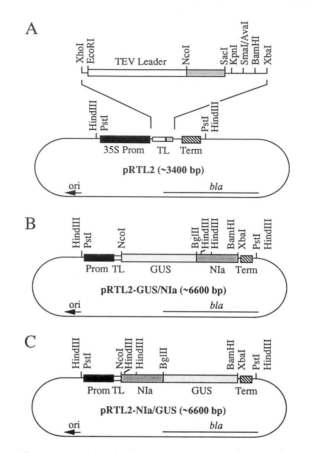

Fig. 1 Diagrammatic representation of plant expression vectors. The positions of relevant restriction endonuclease sites, 35S transcriptional promoter (Prom), 35S transcriptional terminator (Term), TEV 5′ nontranslated leader sequence (TL), plasmid origin of replication (ori), and β-lactamase gene (*bla*) are shown. (A) pRTL2. The shaded region immediately after the leader sequence represents the initial coding sequence of the TEV genome. The *Nco*I site (CCATGG) after the leader sequence contains the translation initiation codon. (B) pRTL2-GUS/NIa. The positions of GUS and NIa coding sequences are shown. (C) pRTL2-NIa/GUS.

B. Nuclear Transport in Protoplasts

1. Preparation of Protoplasts from Leaf Tissue

Leaf protoplasts are obtained from *Nicotiana tabacum* cv "Xanthi nc." It is important to use unblemished leaves from healthy plants (12 to 15 inches tall), and to use leaves that are relatively large but not yet fully expanded. The following is a general method for protoplasts isolation, and can be used in conjunction with transfection mediated by electroporation or polyethylene glycol.

a. Preparation of protoplast enzyme solution.

First, prepare 10× Macro Salts:

1.5 g monosodium phosphate ($NaH_2PO_4 \cdot 2H_2O$)

9.0 g calcium chloride ($CaCl_2 \cdot 2H_2O$)

25 g ammonium nitrate (NH_4NO_3)

1.34 g ammonium sulfate [$(NH_4)_2SO_4$]

2.5 g magnesium sulfate ($MgSO_4 \cdot 7H_2O$)

Combine each component and dissolve in 1 liter deionized H_2O. Filter-sterilize and store at 4°C.

Next, prepare the protoplast enzyme solution:

1.5% cellulase RS (Yakult Honsha, Tokyo)

0.1% Driselase (Sigma)

0.5% Macerozyme R-10 (Yakult Honsha)

1× macro salts

0.55 M mannitol

pH 5.9 with acetic acid or potassium hydroxide

Combine each component, dissolve thoroughly in deionized H_2O, and filter-sterilize. Aliquot in 10-ml volumes and store frozen at −20°C.

b. Surface-sterilize the leaves in 1% bleach for 4 min. Rinse with sterile 0.55 M mannitol (Sigma Chemical Co.).

c. Dust the underside of the leaves lightly with carborundum, and gently abrade with a cotton-tip applicator.

d. Remove the mid-vein, and "score" the underside by pressing with the edge of a razor blade at 0.5-cm intervals. The leaves should have visible marks at the razor blade pressure points.

e. Place the leaves in petri dishes containing protoplast enzyme solution. Use 10 ml for every 1 g of tissue. Incubate in the dark at room temperature or 28°C for 12–16 h without shaking.

f. Filter the released protoplasts through 350- and 80-μm gauze and layer onto 10 ml of 0.55 M sucrose in a 50-ml conical Falcon tube. Spin at 45 × g for 7 min and remove the dark green, live protoplast layer from the top of the sucrose with a wide-bore pipette. The dead cells and incompletely digested tissue fragments will form a pellet.

g. Wash the protoplasts with 0.55 M mannitol by two sedimentation cycles at 100 × g for 4 min. Resuspend cells in a solution appropriate for transfection.

2. Transfection of Protoplasts

There are several reliable methods to introduce DNA into protoplasts. Electroporation and polyethylene glycol-mediated transfection are used most commonly.

Both work well with tobacco protoplasts, although we find that electroporation results in greater cell death. Extensive details for each method, as well as protoplast culturing procedures, are available elsewhere (Carrington and Freed, 1990; Negrutiu *et al.*, 1987).

3. Analysis of Nuclear Transport by Fractionation

Analysis of nuclear transport of the GUS fusion protein by organelle fractionation should be conducted 24 h after transfection. Survival of protoplasts decreases upon prolonged incubation, and the regenerating cell wall presents a problem for efficient lysis. Due to the high sedimentation coefficient of nuclei, it should be relatively straightforward to isolate nuclei from nonnuclear cell constituents. In practice, however, quantitative recovery of intact nuclei is often difficult, and the importance of fractionating control samples in parallel with experimental samples cannot be overemphasized. These should include cells transfected with pRTL2-GUS (cytosolic control), pRTL2-GUS/NIa (nuclear control), and a mock-transfected control.

The following protocol employs physical disruption of protoplasts and limited treatment with Triton X-100. Although nuclei are relatively resistant to this nonionic detergent, the fractionation should be carried out as quickly as possible.

a. Harvest protoplasts by sedimentation at $100 \times g$ for 5 min.

b. Resuspend protoplasts in nuclear isolation buffer (2.5×10^5 protoplasts/250 μl) consisting of 0.25 M sucrose, 20 mM Tris–HCl, 1.5 mM MgCl$_2$, 0.14 M NaCl, 10 mM ß-mercaptoethanol, 5% (vol/vol) glycerol, 0.3% (v/v) Triton X-100, pH 6.8.

c. Rapidly draw the protoplast suspension 5 to 10 times through a 25-gauge needle attached to a 1-ml syringe. Be consistent with the number of passages through the needle for all samples. The efficiency of lysis, and the integrity of the nuclei, can be checked by mixing 5 μl of lysate with 5 μl of Azure C [1% (w/v) in 250 mM sucrose]. Intact nuclei should stain light blue or purple.

d. Layer 125 μl lysate onto a 15/50% Percoll (Pharmacia) step gradient in a 1.5-ml microcentrifuge tube. Use nuclear isolation buffer to prepare the Percoll dilutions. Each Percoll step should consist of 100 μl.

e. Spin the tubes in a swinging bucket rotor at $2000 \times g$ for 15 min at 10°C. Intact nuclei will collect at the interface between the Percoll layers, while the cytosolic/lysed organelle fraction will remain on top of the 15% layer.

f. Withdraw the cytosolic fraction, and then the nuclear fraction, with a micropipette, and normalize the volumes by addition of nuclear isolation buffer to the fraction with lesser volume.

g. Add 1/10 vol of 1% (w/v) sodium lauryl sarcosine to each fraction to dissolve membranes. The fractions are now ready to assay for GUS activity by the fluorometric method.

4. Fluorometric GUS Assay

a. Prepare at least two 10-fold serial dilutions of both fractions from each sample using GUS assay buffer [40 mM monosodium phosphate, 10 mM EDTA, 0.1% (v/v) Triton X-100, 0.1% (w/v) sodium lauryl sarcosine, 0.07% (v/v) ß-mercaptoethanol, pH 7].

b. To 25 μl of each undiluted and diluted fraction, add 25 μl of freshly prepared substrate solution, consisting of 2 mM 4-methylumbelliferyl glucuronide (Jersey Lab Supply) in GUS assay buffer.

c. Incubate at 37°C for 30 min. Add 1.5 ml of stop buffer composed of five parts GUS assay buffer and four parts 0.2 M sodium carbonate.

d. Read samples using a mini-fluorometer (Model TKO-100, Hoefer Scientific Instruments). Use the diluted and undiluted fractions from the mock-transfected protoplasts to set the mini-fluorometer baseline. See Jefferson (1987) for calculation of GUS units.

5. Analysis of Nuclear Transport by *in Situ* Histochemical Assay

Visualization of GUS fusion proteins in the nucleus is achieved by reacting transfected protoplasts with 5-bromo-4-chloro-3-indolyl-ß-D-glucuronic acid (X-gluc).

a. At 24 h post-transfection, replace the protoplast culture medium with fresh medium that also contains 1.2 mM X-gluc, 0.5 mM potassium ferricyanide, 0.5 mM potassium ferrocyanide, 10 mM EDTA, 5 μg/ml DAPI. Incubate at room temperature.

b. Monitor the color development in the culture dish with an inverted microscope, or remove samples periodically to a clean slide to view with a compound microscope.

c. Cells should be photographed using brightfield or DIC optics. Confirm the location of the nucleus by fluorescence microscopy, which will detect the DAPI stain. Cytosolic localization of the GUS fusion protein is detected as a nearly uniform blue or indigo color throughout the protoplast, whereas nuclear localization is detected by colocalization of color with the DAPI stain and by confinement to a discrete round or ovular region of the cell.

C. Nuclear Transport in Transgenic Plants

Transgenic plants expressing GUS fusion protein genes should be screened for GUS activity, and the fusion protein should be detected by an immunoblot assay using anti-GUS serum or an antiserum specific for the polypeptide fused to GUS. Localization of the fusion protein *in situ* is carried out using a rapid epidermal strip histochemical assay (Restrepo *et al.*, 1990). The fusion protein can also be localized using mirotome sections cut from embedded transgenic tissue. Alternatively, GUS activity can be assayed in nuclear and cytosolic fractions from protoplasts prepared from the transgenic plants. Again, control plants that express nonfused GUS and a nuclear GUS fusion protein should be analyzed in parallel with experimental plants.

1. Analysis of Nuclear Transport in Epidermal Leaf Strips

The epidermis is useful for qualitative *in situ* localization of GUS fusion proteins because a single layer of cells can be isolated, reacted with X-gluc, and spread on a microscope slide. The most useful epidermal cells are those located on the underside of the mid-vein and petiole, as they remain intact better than epidermal cells in interveinal areas. Also, the mid-vein/petiole epidermal cells are more regular in shape.

a. Excise a leaf, including the petiole, from the transgenic plants. Small plants (>10 cm tall) work best.

b. Using small forceps with flat edges near the tip, grasp the epidermis on the underside of the mid-vein near the base of the leaf. Gently pull away the epidermis in a strip, removing cells in the direction of the petiole. After the initial strip is placed in X-gluc solution (step c), remove a mid-vein epidermal strip in the opposite direction. At least two strips can be taken from each leaf.

c. Place epidermal strips in X-gluc solution containing 1.2 mM X-gluc, 20 mM potassium phosphate, 0.5 mM potassium ferricyanide, 0.5 mM potassium ferrocyanide, 10 mM EDTA, 5 μg/ml DAPI, pH 7, making certain that the interior side of the strip is in contact with the solution. Multiple samples are processed easily using a 96-well microtiter dish. Incubate at room temperature.

d. As color develops, transfer epidermal strips to a drop of Aquamount (Polysciences, Inc.) mounting medium on a clean microscope slide and apply coverslip.

e. View the strips using DIC and fluorescence microscopy.

IV. Results and Critical Aspects of the Procedures

A. Activity of GUS Fusion Proteins

ß-Glucuronidase retains activity when heterologous proteins are fused to the N- or C-terminus, although the activity in some cases may be only a small fraction of that exhibited by nonfused GUS. We have found that addition of protein sequences to the C-terminus of GUS generally results in fusion proteins with enzymatic activities higher than those resulting when the same sequence is fused to the N-terminus (Restrepo *et al.,* 1990). Importantly, NLSs within fusion polypeptide sequences are active when present at either the N- or the C-terminus of GUS.

B. Localization of Nuclear Proteins in Protoplasts

1. Nuclear and Cytosolic Fractionation

Typically, recovery of nuclei after the Percoll gradient step is around 50% of the theoretical maximum. The other 50% may be lost due to incomplete cell lysis, nuclear lysis, incomplete recovery from the Percoll interface, or sedimentation through the interface.

In the experiment shown in Table I, protoplasts were transfected with pRTL2-derived plasmids encoding GUS, GUS/NIa fusion protein, GUS/NIa 1-76 in which only the initial 76 amino acid residues of NIa were fused to GUS, and GUS/NIaΔ43-46 in which amino acid residues 43–46 were lacking from the GUS/NIa fusion protein. The GUS and GUS/NIa samples served as the cytosolic- and nuclear-localized controls, respectively. The GUS/NIa 1–76 protein contained a segment of NIa harboring the NLS. The GUS/NIaΔ43-46 protein contained a deletion within the NLS. Most (98%) of the activity from cells expressing nonfused GUS was in the cytosolic fraction. Normally, between 95 and 99.5% of the GUS activity will fractionate with the cytosol. In contrast, nearly one-half of the activity from protoplasts expressing GUS/NIa or GUS/NIa 1-76 was detected in the nuclear fraction. Variation in nuclear transport efficiency of GUS/NIa in different experiments was common. In some experiments, as little as 10% of the GUS/NIa activity fractionated with the nuclei. Only 9% of the activity from cells expressing the NLS-defective GUS/NIaΔ43-46 fusion protein fractionated with the nuclei.

Table I
GUS Fusion Protein Activity in Nuclear and Cytosolic Fractions From Protoplasts

Protein	% Nuclear[a]	% Cytosolic[b]	Nuclear transport efficiency (%)[c]
GUS	2	98	0
GUS/NIa	40	60	100
GUS/NIa 1-76	52	48	132
GUS/NIaΔ43-46	9	91	18

[a] The percentage of activity in the nuclear fraction was calculated as nuclear-associated activity/total activity. Total activity was the sum of activity in the nuclear and cytosolic fractions.

[b] The percentage of activity in the cytosolic fraction was calculated as cytosolic-associated activity/total activity.

[c] Nuclear transport efficiency is a relative measure of nuclear-associated activity using GUS and GUS/NIa samples as the 0 and 100% efficiency standards, respectively. The ratio was calculated for each sample as follows:

$$\frac{\text{sample nuclear activity} - \text{GUS nuclear activity}}{\text{GUS/NIa nuclear activity} - \text{GUS nuclear activity}}$$

When analyzing GUS fusion proteins containing defects in a putative NLS, it is useful to express nuclear transport as a percentage of nuclear transport measured with the corresponding GUS fusion protein containing an active NLS (Table I) (Carrington *et al.,* 1991). The NLS-active construct serves as the 100% nuclear transport efficiency control, whereas the nonfused GUS control serves as the 0% efficiency control. The nuclear transport efficiency of an experimental sample can only be calculated using the controls analyzed in parallel within the same experiment. However, the calculated nuclear transport efficiency values can be compared across experiments using different protoplast preparations, assuming the same 100 and 0% controls are used in all experiments.

A potential problem arises if a cytoplasmic GUS fusion protein has an affinity for the nuclear envelope or the endoplasmic reticulum. It is difficult to rid nuclei of peripherally associated proteins using the lysis and crude fractionation techniques presented; purity of nuclei is sacrificed for high yield. Fusion proteins that associate peripherally with the nucleus will be indistinguishable from genuine nuclear proteins. Confirmation that a protein is localized within the nucleus can be achieved using one of the qualitative *in situ* assays.

2. Histochemical Localization

Protoplasts subjected to the transfection procedure in the absence of DNA exhibit no histochemical reaction after incubation with X-gluc (Color Plate 3A). Protoplasts expressing nonfused GUS show a general, non-localized blue color (Color Plate 3B). In contrast, cells expressing GUS/NIa exhibit a blue histochemical reaction in the nucleus (Color Plate 3C). The nucleolus often stains more intensely than the nucleoplasm, but the basis for this is not known. Either NIa directs accumulation partially in the nucleolus and partially in the nucleoplasm, or the hydrolyzed X-gluc product has an affinity for the nucleolus. We have seen this histochemical staining pattern with most of the nuclear-associated fusion proteins that we have examined (Li and Carrington, 1993; Restrepo *et al.,* 1990).

The most critical aspect of this procedure is the length of time of the X-gluc reaction. Depending on the health of the cells and the efficiency of transfection, a colorimetric reaction may be seen in as little as 10 min. Cells should be viewed frequently and photomicrographs taken at various times after addition to X-gluc. We have observed consistently that the hydrolyzed X-gluc product in the cytoplasm is cytolytic, whereas X-gluc product in the nucleus has no cytolytic effect. In other words, prolonged incubation of protoplasts expressing nonfused GUS or a cytoplasmic GUS fusion protein with X-gluc results in cell lysis. Like the fractionation method described above, variation in the degree of nuclear accumulation of an NLS-containing fusion protein is sometimes observed, demanding that the assay be repeated several times with different batches of protoplasts. In addition, the intensity of staining in nuclear and nonnuclear regions of protoplasts can be measured from photographic negatives by a colorimetric quantification method (Howard *et al.,* 1992).

C. Localization of Nuclear Proteins in Transgenic Plants

Epidermal leaf strips taken from nontransgenic plants fail to react with X-gluc, and thus appear clear (Color Plate 3D). In the presence of DAPI, which binds DNA, nuclei are clearly visualized by UV fluorescence microscopy. Cells from leaf strips of GUS-expressing transgenic plants display a uniform, nonlocalized histochemical reaction (Color Plate 3E). Nuclei from these cells are also easily visualized with DAPI staining. Cells from leaf strips of GUS/NIa-expressing transgenic plants show GUS activity predominantly in the nucleus (Color Plate 3F). Like the transfected protoplasts stained with X-gluc, the nucleolus often accumulates significantly more hydrolyzed product than the surrounding nucleoplasm. A minor complication with this assay, however, is that DAPI fluorescence is quenched when nuclei are stained heavily with the X-gluc product (Color Plate 3F). Nuclei with lower levels of hydrolyzed X-gluc show weak fluorescence.

Multiple transgenic plants expressing any given GUS fusion protein should be analyzed. Extreme care should be applied to minimize tissue damage while peeling the epidermis from the leaf. Excess tissue damage will result in disruption of nuclei and release of nuclear proteins into the cytoplasm. Care should also be taken to avoid peeling multiple layers of cells from the mid-vein, although it is nearly impossible to remove only a single layer of cells throughout the entire strip.

V. Conclusions

These methods have been useful to detect and map NLS in NIa and other proteins. The quantitative fractionation method using transfected protoplasts and the qualitative *in situ* localization techniques provide complementary approaches to address issues relating to nuclear transport of proteins in plants.

Acknowledgments

The author sincerely appreciates several co-workers, including Deon Freed, Xiao Hua Li, María Restrepo-Hartwig, and Ruth Haldeman-Cahill, who made substantial contributions to the development and application of these methods. Research on the control of nuclear translocation of NIa is funded by the National Institutes of Health (AI27832).

References

An, G. (1987). Binary Ti vectors for plant transformation and promoter analysis. *Methods Enzymol.* **153,** 293–305.

Carrington, J. C., and Freed, D. D. (1990). Cap-independent enhancement of translation by a plant potyvirus 5' nontranslated region. *J. Virol.* **64,** 1590–1597.

Carrington, J. C., Freed, D. D., and Leinicke, A. J. (1991). Bipartite signal sequence mediates nuclear translocation of the plant potyviral NIa protein. *Plant Cell* **3,** 953–962.

Carrington, J. C., Freed, D. D., and Oh, C.-S. (1990). Expression of potyviral polyproteins in transgenic plants reveals three proteolytic activities required for complete processing. *EMBO J.* **9,** 1347–1353.

Forbes, D. J. (1992). Structure and function of the nuclear pore complex. *Annu. Rev. Cell Biol.* **8,** 495–527.

Hall, N. N., Hereford, L, and Herskowitz, I. (1984). Targeting of E. coli ß-galactosidase to the nucleus in yeast. *Cell* **36,** 1057–1065.

Howard, E. A., Zupan, J. R., Citovsky, V., and Zambryski, P. C. (1992). The VirD2 protein of *Agrobacterium tumefaciens* contains a C-terminal bipartite nuclear localization signal: Implications for nuclear uptake of DNA in plant cells. *Cell* **68,** 109–118.

Jefferson, R. A. (1987). Assaying chimeric genes in plants: The GUS gene fusion system. *Plant Mol. Biol. Rep.* **5,** 387–405.

Kalderon, D., Richardson, W. D., Markham, A. F., and Smith, A. E. (1984). Sequence requirements for nuclear location of simian virus 40 large-T antigen. *Nature* **311,** 33–38.

Li, X. H., and Carrington, J. C. (1993). Nuclear transport of tobacco etch potyviral RNA-dependent RNA polymerase is highly sensitive to sequence alterations. *Virology* **193,** 951–958.

Negrutiu, I., Shillito, R., Potrykus, I., Biasini, G., and Sala, F. (1987). Hybrid genes in the analysis of transformation conditions. 1. Setting up a simple method for direct gene transfer in plant protoplasts. *Plant Mol. Biol.* **8,** 363–373.

Raikhel, N. V. (1992). Nuclear targeting in plants. *Plant Physiol.* **100,** 1627–1632.

Restrepo, M. A., Freed, D. D., and Carrington, J. C. (1990). Nuclear transport of plant potyviral proteins. *Plant Cell* **2,** 987–998.

Robbins, J., Dilworth, S. M., Laskey, R. A., and Dingwall, C. (1991). Two interdependent basic domains in nucleoplasmin nuclear targeting sequence: Identification of a class of bipartite nuclear targeting sequence. *Cell* **64,** 6 15–623.

Silver, P. A. (1991). How proteins enter the nucleus. *Cell* **64,** 489–497.

Varagona, M. J., Schmidt, R. J. and Raikhel, N. V. (1992). Nuclear localization signal(s) required for nuclear targeting of the maize regulatory protein Opaque-2. *Plant Cell* **4,** 1213–1227.

CHAPTER 21

Import into the Endoplasmic Reticulum

Aldo Ceriotti,* Emanuela Pedrazzini,* Marcella De Silvestris,† and Alessandro Vitale*

*Istituto Biosintesi Vegetali
Consiglio Nazionale delle Ricerche
Via Bassini 15
20133 Milano, Italy
†Dipartimento di Farmacologia
Università di Milano
Via Vanvitelli 32
20129 Milano, Italy

I. Introduction

Proteins destined to the endoplasmic reticulum (ER), the Golgi complex, the vacuole, and the plasma membrane are first imported into the ER before being targeted to their final destination (or retained in the ER itself as permanent residents). Import into the ER is therefore a common step in the synthesis of a large set of proteins destined to various compartments of the cell (Vitale *et al.,* 1993). The mechanism by which efficient segregation is obtained has been studied mainly using mammalian and yeast systems (for reviews, see Rapoport, 1991; Sanders and Schekman, 1992). The synthesis of proteins that are imported into the ER begins on cytosolic ribosomes. Generally, soon after translation has started the growing polypeptide chain and the ribosome are targeted to the ER membrane and translocation (or insertion into the ER membrane) occurs cotranslationally. Targeting is achieved by a specific signal (the signal peptide) that is located at the N-terminus in the case of soluble proteins. This signal is recognized by a ribonucleoprotein particle (signal recognition particle, SRP) (Walter and Blobel, 1981) that mediates the targeting of the nascent-chain and the ribosome to the ER membrane. Interaction with the membrane involves an ER-located receptor, the SRP receptor or docking protein (Gilmore *et al.,* 1982; Meyer *et al.,* 1982). After targeting has been achieved, the signal peptide is released from the SRP and the polypeptide is transferred across the membrane, most likely through a proteinaceous channel that contains, as a key component, the yeast protein Sec61p or its mammalian homolog (Görlich *et al.,* 1992; Müsch *et al.,* 1992; Sanders *et al.,* 1992). In the case of soluble proteins the signal peptide is normally cotranslationally removed by an ER-located protease (the signal peptidase complex (Shelness and Blobel, 1990)), whereas membrane proteins often have an uncleaved signal that also performs the function of a membrane-spanning segment. Proteins belonging to the heat shock family have also been implicated in the process of protein import into the ER (Zimmermann *et al.,* 1988; Sanders *et al.,* 1992; Miernyk *et al.,* 1992; Brodsky *et al.,* 1993; Nicchita and Blobel, 1993). The SRP-dependent mechanism of protein import into the ER appears to be ubiquitous but alternative pathways may exist (Rapoport, 1991; Sanders and Schekman, 1992).

Available information indicates a high degree of conservation of signals and translocation machinery between the animal and the plant kingdoms (Bassüner *et al.,* 1984; Prehn *et al.,* 1987; Campos *et al.,* 1988), although plant and mammalian components are not always fully interchangeable (Prehn *et al.,* 1987; Miernyk and Shatters, 1992).

In this chapter we will describe methods that can help in determining whether a protein is targeted to the ER of plant cells and to characterize some of the early processing steps that occur in this compartment, namely signal peptide cleavage and glycosylation. This is normally achieved using a combination of *in vivo* and *in vitro* approaches, the feasibility of which will depend on factors such as the level of expression of the protein, the tissue where it is expressed, and the availability of antibodies and DNA clones. The methods described here should also be helpful for the study of the translocation and processing apparatus of the plant cell.

II. Methods

A. *In Vitro* Assays for Translocation into the ER

Natural and synthetic mRNAs can be translated *in vitro* in the absence or in the presence of microsomes isolated from various sources. Analysis of the translation products can provide information about the presence of an ER targeting signal, the site of signal peptide cleavage, the cotranslational addition of asparagine-linked oligosaccharide side chains (N-glycosylation), and the topology with respect to the ER membrane. Although such studies can be performed using mRNA isolated from plant tissue, the availability of a cDNA clone coding for the protein of interest will greatly facilitate the task. In this case, the coding sequence is subcloned downstream from a phage promoter and synthetic mRNA is produced *in vitro* using the corresponding phage-encoded polymerase. Three phage polymerases (SP6, T7, and T3) are currently used for *in vitro* transcription. Although artifacts, such as initiation at internal ATGs or premature termination, may complicate the pattern of the translation products as analyzed by SDS–PAGE, one major product is often obtained when this mRNA is translated *in vitro,* eliminating the need for immunoselection of the protein of interest.

1. *In Vitro* Transcription

a. *Preparation of the DNA template*

A variety of suitable transcription vectors are commercially available; these include the Bluescript series (Stratagene), and the pGEM and pSP series (Promega). We routinely employ pSP64T (Krieg and Melton, 1984) as a cloning vector. Among other features, this vector provides a 5′ untranslated sequence that can replace the 5′ untranslated sequence of the original transcript especially when the presence of out-of-frame ATGs and poly(G)stretches (derived from cDNA cloning) might negatively influence translation efficiency (Galili *et al.,* 1986). For transcription the plasmid is linearized with an enzyme that cuts the polylinker sequence 3′ to the inserted coding sequence, leaving a blunt end or a 5′ overhang. Obviously, such a site should be absent from the coding region that is to be transcribed. Aberrant transcription initiation can occur if a 3′ overhang is produced (Schenborn and Mierendorf, 1985). After linearization, the reaction is made 0.3 M in sodium acetate, extracted once with phenol–chloroform, once with chloroform, and once with ether. Two volumes of ethanol are added. The tube is incubated for 10 min on ice and then spun for 30 min at room temperature in a microfuge (15,000 × g). The pellet is washed twice with 80% ethanol to remove salt (phage polymerases work better in low salt buffers), and the DNA is resuspended in water at a concentration of 1 mg/ml.

b. *Transcription*

We typically set up 50-μl transcription reactions. This can yield sufficient mRNA for more than 100 *in vitro* translations. Capped or uncapped mRNA can be produced. Capping is achieved by simply including the cap analog m⁷G(5′)ppp(5′)G. Although capping can enhance translation in reticulocyte lysates, we find uncapped transcript to be fully adequate for general *in vitro* studies. A small amount of radiolabeled UTP can be included to quantitate mRNA synthesis. All solutions are prepared taking care to avoid RNase contamination.

Solutions. 10× transcription buffer (400 mM Tris–Cl, pH 7.5, 60 mM MgCl$_2$, 20 mM spermidine); 100 mM DTT; 10 mM each of ATP, CTP, GTP, UTP dissolved in water, pH 7.5; RNase inhibitor (RNasin) 40 U/μl; 10 mM m⁷G(5′)ppp(5′)G, sodium (cap analog, Pharmacia Cat. No. 27-4635), dissolved in water; acetylated BSA (1 mg/ml); SP6, T7, or T3 polymerase, normally supplied at 40 U/μl; 7 M NH$_4$OAc, dissolved in sterile water and filtered through a 0.22-μm filter.

Reaction. The reaction is set up at room temperature to avoid spermidine-induced precipitation of the DNA template. Mix in the following order:

Linearized plasmid (1 mg/ml)	5 μl
Sterile water	to a final volume of 50 μl
10× transcription buffer	5 μl
BSA 1 mg/ml	5 μl
RNasin 40 U/μl	1 μl
DTT 100 mM	5 μl
rNTPs mix, 10 mM each	2.5 μl
10 mM m^{7}G(5′)ppp(5′)G (optional)	2.5 μl
[α-^{32}P]UTP 37 kBq/μl	1 μl
RNA polymerase	50 U

Components are carefully mixed by pipetting up and down and the reaction is then incubated for 1 h at 40°C (SP6) or 37°C (T3 and T7). At the end of the incubation period 2 μl of the reaction is diluted to 10 μl with water, and 4 × 2-μl aliquots of the diluted transcription mix are spotted on DE81 paper (Whatman). Two aliquots are counted directly, whereas the other two are washed five times (2 min for each wash) in 50 mM NaHPO$_4$, once with ethanol and once with acetone, and then dried and counted. One hundred percent incorporation corresponds to about 34 μg of RNA.

The rest of the transcription reaction is diluted to 100 μl with H$_2$O, made 0.7 M in NH$_4$OAc, and extracted once with phenol–chloroform, once with chloroform, and once with ether. Two and a half volumes of ethanol are added and, after overnight storage at −20°C, the RNA is collected by centrifugation (30 minutes at 15,000 × g in a microfuge). The pellet is washed with 80% ethanol, dried, and resuspended in water to give a final concentration of 100 μg/ml.

2. Use of Microsomal Membranes in *in Vitro* Translation Systems

The microsome preparation most widely used to study import into the ER is derived from canine pancreas. Plant preproteins are efficiently imported into canine microsomes, and available information indicates that the mechanism of protein import and the specificity of processing events such as signal peptide removal and N-glycosylation have been conserved through evolution. However, exceptions might exist. Data obtained with canine microsomes can be complemented by *in vivo* data, or by the use of *in vitro* systems supplied with plant microsomes. A few different plant microsomal preparations, competent in protein translocation, have been described (Higgins and Spencer, 1981; Prehn *et al.*, 1987; Hattori *et al.*, 1987; Campos *et al.*, 1988; Osteryoung *et al.*, 1992). As an example of such systems we will describe the preparation of translocating microsomes from developing *Phaseolus vulgaris* cotyledons and their use in a wheat germ cell-free translation system. When this system is programmed with synthetic mRNA coding for phaseolin (the major storage protein in bean seeds) translocation, proteolytic processing and glycosylation of the *in vitro* synthesized protein are observed. Although the efficiency of these processes is lower than the one observed in the system composed by reticulocyte lysate and canine membranes, the plant-derived system allows the same kind of analysis addressed using the mammalian system.

a. Preparation of Microsomes

Microsomes are prepared from developing *P. vulgaris* cotyledons basically as described by Higgins and Spencer (Higgins and Spencer, 1981). A nuclease treatment is included to eliminate the background due to endogenous mRNA.

Solutions:

Buffer A: 50 mM triethanolamine-HOAc (TEA-HOAc), pH 7.5, 50 mM KOAc, pH 7.5, 5 mM Mg(OAc)$_2$, 2 mM DTT, 0.25 M sucrose.

Buffer B: 25 mM TEA-HOAc, pH 7.5, 4 mM DTT, 0.25 M sucrose.

Buffer C: 100 mM TEA-HOAc, pH 7.5, 20 mM EDTA.

Buffer D: 25 mM TEA-HOAc, pH 7.5, 25 mM KOAc, pH 7.5, 2 mM Mg(OAc)$_2$, 4 mM DTT, 0.5 M sucrose.

Buffer E: 25 mM TEA-HOAc, pH 7.5, 1 mM DTT, 0.25 M sucrose.

50 mM CaCl$_2$.

100 mM EGTA, pH 8.3, with NaOH.

Nuclease S 7 from *Staphylococcus aureus* (micrococcal nuclease, Boehringer Cat. No. 107 921) is prepared in water at a concentration of 2000 U/ml and stored at $-80°C$.

Buffers A, B, C, D, and E are filtered through a 0.45-μm filter. DTT is added from a 1 M stock solution in water (stored at $-80°C$) just before use. Whenever possible, equipment (glassware, mortars, homogenizers, etc) is treated at 200°C for 2 h to destroy nucleases.

Procedure:

All procedures are performed at 4°C, preferably in a cold room. Green, midmaturation *P. vulgaris* cotyledons (cultivar Greensleeves, length ~1 cm) are collected and either immediately used or frozen on dry ice and stored at $-80°C$. Twenty grams of cotyledons is homogenized in a mortar using 100 ml of buffer A. The buffer is added progressively during homogenization. If frozen material is used, it is first ground to a fine powder in a mortar that has been cooled at 80°C. The homogenate is sequentially spun for 10 min at 1000 $\times$ g_{max} and then for 10 min at 10,000 $\times$ g_{max} in a HB 6 rotor (Sorvall). The final supernatant is then spun at 93,000 $\times$ g_{av} for 90 min in a 50.1 Ti rotor (Beckman). Supernatant is removed by aspiration and the pellet is carefully resuspended in 20 ml of buffer B. The first milliliter of buffer is added in 50-μl aliquots, while the pellet is resuspended with a glass rod. The resuspended membranes are homogenized using a Dounce homogenizer (loose-fitting pestle) and then diluted with an equal volume of buffer C. After a 10-min incubation the sample is placed on a cushion of buffer D (10 ml of resuspended membranes on 7 ml of buffer D in a polycarbonate bottle for rotor 50.1 Ti) and spun at $93,000 \times g_{av}$ for 60 min. After complete removal of supernatant and cushion, the pellets are resuspended in minimal volume of buffer E (e.g., 1 ml) and gently homogenized with a loose-fitting pestle in a Dounce homogenizer. A_{280} is determined after dilution of a 10-μl aliquot to 1 ml with 1% SDS. The microsome preparation is diluted with buffer E to give a final concentration of 50 A_{280} units/ml, split in small aliquots, frozen in liquid nitrogen, and stored at $-80°C$.

Just before use, membranes are quickly thawed in a 30°C water bath and immediately transferred onto ice. The sample is made 2 mM in $CaCl_2$, and micrococcal nuclease is added to a final concentration of 40 U/ml. Nuclease treatment is for 10 min at 23°C. The nuclease is then inactivated by adding EGTA to a final concentration of 4 mM.

The preparation and use of canine pancreatic microsomes has been described in detail (Walter and Blobel, 1983). These can also be purchased from commercial suppliers, including Boehringer, Amersham, and Promega.

b. In Vitro Translation

We perform *in vitro* translations using a wheat germ extract (WG) in combination with bean microsomes, or a reticulocyte lysate (RL) in combination with canine microsomes. The preparation of WG (Anderson *et al.*, 1983) and RL (Jackson and Hunt, 1983) has been described in detail; commercial preparations are available. We normally set up reactions in a minimal volume of 12.5 μl containing 1 μl of RNA and up to 2 μl of microsomes. The amount of RNA to be added will depend on whether natural total RNA, a poly(A)$^+$ preparation, or an *in vitro* transcript is used. A final concentration of 100–200 μg/ml in the case of total RNA, and of 5–20 μg/ml in the case of a poly(A)$^+$ preparation is often found to give good stimulation of translation. In the case of capped *in vitro* transcripts, we normally add 20–50 ng to a 12.5-μl reaction. Uncapped transcript are used at a fivefold higher concentration. However, in all cases, a range of concentrations should be tested to determine optimal conditions.

Translation reactions are set up on ice by first preparing a master mix containing all components except for microsomes and RNA. The mix is then split into aliquots, to which microsomes and RNA are added, in this order. It is important to mix well all components, since the translation mix is very viscous. The amount of microsomal membranes to be added should be determined empirically. We suggest setting up on ice a 25-μl reaction containing 4 μl of bean microsome preparation (50 A_{280} units/ml) and 2 μl of RNA solution. Half of this reaction is then serially diluted into three 12.5-μl reactions where microsomes have been replaced by 2 μl of water. Canine microsomes should be titrated in the range 0.25–2 eq. (as defined by Walter and Blobel, 1983) for each 12.5-μl reaction.

Translation is typically performed for 1 h at 25°C in the case of WG, and at 30°C in the case of RL. At the end of the incubation, the tubes are transferred on ice, RNase A (stock solution 10 mg/ml, dissolved in water and stored at 20°C) is added to a final concentration of 50 μg/ml, and the tubes are incubated for 10 min at room temperature. RNase treatment abolishes further incorporation, and degrades labeled *in vitro* transcripts and tRNAs charged with radioactive amino acids, either of which might otherwise interfere with the detection of the radiolabeled protein after SDS–PAGE and fluorography.

Proper controls should be included in any assay. Reactions in which microsomal membranes or RNA are excluded are essential. When a new RNA preparation is used, it is wise to include a reaction containing a suitable RNA that has been previously found to be efficiently translated.

Figure 1 shows a comparison of the translation products obtained in the plant and mammalian systems. A synthetic transcript coding for β-type phaseolin (β-PHSL) was translated in the WG or in the RL system, either in the absence or

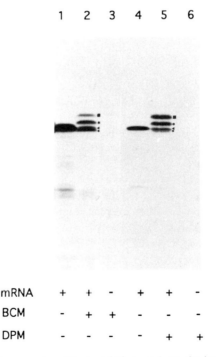

	1	2	3	4	5	6
mRNA	+	+	-	+	+	-
BCM	-	+	+	-	-	-
DPM	-	-	-	-	+	+

Fig. 1 SDS–PAGE and fluorography of the translation products obtained in a WG (lanes 1 to 3) or RL (lanes 4 to 6) translation system supplemented with a synthetic transcript coding for β-type PHSL (lanes 1,2,4, and 5) in the absence (lanes 1 and 4) or in the presence of microsomal membranes obtained either from bean cotyledons (lanes 2 and 3) or dog pancreas (lanes 5 and 6). Arrowhead, nontranslocated precursor; dot, translocated product from which the signal peptide has been removed; asterisk, singly glycosylated PHSL; square, fully glycosylated PHSL. BCM, bean cotyledon microsomes; DPM, dog pancreas microsomes.

in the presence of microsomal membranes isolated from bean cotyledons or dog pancreas. In bean cotyledonary cells, β-PHSL is segregated into the ER lumen where signal cleavage occurs and the polypeptide is modified by the addition of either one or two N-linked oligosaccharide chains (Bollini *et al.*, 1983). In the absence of added membranes one single major polypeptide is synthesized both in WG and in RL. Addition of membranes at the beginning of translation results in the appearance of three additional polypeptides. The faster migrating one is generated by the removal of the signal peptide, whereas the other two represent chains that, in addition to signal peptide removal, have been subjected to glycosylation at one or two sites. The ratio between unprocessed, signal sequence-minus and glycosylated polypeptides is different in the two systems, the mammalian one being generally more efficient. However, the same qualitative picture emerges from the two different translation–translocation systems.

3. Analysis of Translation Products

Translation products are typically analyzed by SDS–PAGE either directly or after various manipulations that are necessary to confirm the preliminary evidence obtained by simply comparing the molecular weight of the proteins synthesized in the absence and in the presence of microsomes. We normally run gels that are 1.5 mm thick and 16 cm long, according to the general methods of Laemmli (1970), with an acrylamide : bisacrylamide ratio of 200 : 1. The translation mixture is a very concentrated protein solution. It is therefore important to dilute it enough in sample buffer so that SDS is not limiting. We routinely load 5 μl of translation per well after dilution in 25 μl of sample buffer (containing 1% SDS).

In general, the presence of a cleavable signal peptide is indicated by the appearance of a slightly faster migrating band when the translation is supplemented with microsomes. Signal peptides are variable in length and sequence (von Heijne, 1985). Therefore, the effect of signal peptide cleavage on electrophoretic mobility can differ from case to case. Obviously, the detection of a change in electrophoretic mobility becomes more difficult as the apparent molecular weight of the protein of interest increases. Definitive proof that signal cleavage has occurred can be obtained by Edman degradation analysis.

Protease protection analysis will indicate whether complete segregation into the microsomes has occurred. Treatment with high pH buffers will convert the closed microsomal vesicles into sheets, causing the release of soluble and peripheral proteins (Fujiki *et al.*, 1982). After high-pH treatment, the membranes can be reisolated to distinguish between membrane-integrated and peripheral proteins. Finally, addition of oligosaccharide side chains is normally suggested by the presence of one or more bands that migrate more slowly than the product obtained in the absence of microsomes. However, the effects of signal processing and addition of glycan chains may compensate each other. Removal of the oligosaccharide chains by enzymatic treatment will resolve the situation. These techniques are described below.

a. N-terminal Sequence Analysis

For protein sequence analysis, the *in vitro* transcript is translated in the absence and in the presence of microsomal membranes using a labeled amino acid that will allow to distinguish unambiguously the primary translation products from the processed protein. We recommend parallel experiments using two different amino acids. As an example, one of the authors has used [³H]leucine and [³H]proline for this purpose (Fabbrini *et al.*, 1991). The putative signal peptide cleavage site can be predicted following rules that have been found to be generally applicable also to plant secretory proteins (Watson, 1984; von Heijne, 1986; Folz and Gordon, 1987; Miernyk and Nelsen, 1989). After translation, the mixture that has been supplemented with microsomes can be treated with protease to obtain complete degradation of any nontranslocated proteins (Fabbrini *et al.*, 1991). The translation products are then separated by SDS–PAGE and electroblotted on polyvinylidene difluoride membranes (Problott, Applied Biosystems) at 300 mA for 50 min using 10 mM 3-(cyclohexylamino)-1-propanesulfonic acid, pH 11, 10% methanol as transfer buffer. If necessary, a translation mixture labeled with [³⁵S]methionine is also loaded on the gel to provide a marker that will allow detection and precise localization of the desired translation product after autoradiography of the filter. The part of the filter where the band has been transferred is excised and placed in the cartridge box of the sequencer, beneath a polybrene-preconditioned trifluoroacetic acid-treated glass-fiber filter (Matsudaira, 1987). Automated Edman degradation can be performed on a Gas-Phase Sequencer Model 470A (Applied Biosystems) using the manufacturer program ATZ470-1. The radioactivity in each fraction cycle is determined by scintillation counting.

b. Protease Protection

Fully segregated products will be protected by the microsomal membrane from the attack of added proteases, whereas nontranslocated products will be degraded. Treatment with proteases can therefore be used to demonstrate that translocation has occurred or to prepare samples for Edman degradation that will be devoid of untranslocated (i.e., unprocessed) radiolabeled protein. In the case of membrane-spanning proteins part of the amino acidic sequence will be normally accessible to exogenous proteases. Therefore protease treatment can be expected to produce one or more protected fragments that can be identified by the use of domain specific antibodies (Wessel *et al.*, 1991).

Various proteases have been used in this kind of assay. We use proteinase K, a nonspecific serine protease, at a final concentration of 500 μg/ml. The reaction also contains 1 mM CaCl$_2$ as a membrane stabilizer. Proteinase K is dissolved at a concentration of 5 mg/ml in 50 mM Tris–HCl, pH 8, 1 mM CaCl$_2$ and preincubated for 15 min at 37°C to digest any contaminating lipases. This solution can be stored at −20°C.

A 50-μl translation reaction is terminated by the addition of 2.5 μl of a 1 mg/ml solution of RNase A. After a 10-min incubation at room temperature, the translation mix is made 1 mM in CaCl$_2$, and three 16-μl aliquots are processed as follows. One aliquot is supplemented with 2 μl of water and 2 μl of 50 mM Tris–HCl, pH 8, 1 mM in CaCl$_2$. A second aliquot is supplemented with 2 μl of water and 2 μl of proteinase K solution (5 mg/ml). To the third aliquot 2 μl of 10% Triton X-100 and 2 μl of proteinase K solution (5 mg/ml) are added. After careful mixing, the tubes are incubated for 1 h on ice. Proteinase K is then inactivated by the addition of 2 μl of 100 mM phenylmethanesulfonyl fluoride (PMSF) dissolved in ethanol (stable for at least 1 year when stored at −20°C). After a 10-min incubation on ice the samples are ready to be employed for SDS–PAGE analysis. Membranes will be disrupted in the reaction supplemented with Triton X-100, exposing all translation products to the action of the protease. This will show that the protected products are not inherently resistant to proteolytic attack.

c. Enzymatic Deglycosylation of Translation Products

A common modification occurring on proteins that are inserted into the ER is the addition of preformed oligosaccharide chains to asparagine residues within the context Asn–X–Ser/Thr. These chains can be removed from *in vitro*-synthesized polypeptides by treatment with the enzyme endo-β-N-acetyl-glucosaminidase H (endoH). The enzyme will cleave the added oligosaccharide side chain leaving a single N-acetylglucosamine linked to the asparagine residue. This will result in a decrease in apparent molecular mass (about 2 kDa) that can be easily detected on SDS–PAGE. In some cases the oligasaccharide chain(s) can be removed using the folded protein as the substrate, but in other cases complete denaturation is required. When more than one chain is added to a single polypeptide, partial digestion will allow determination of their number (Gallagher *et al.,* 1988).

To obtain complete deglycosylation, 10 μl of translation mix is diluted with 40 μl of 100 mM Tris–HCl, pH 8, 1% SDS, 1% β-mercaptoethanol and heated for 4 min in a boiling water bath. The sample is then diluted with 450 μl of 150 mM Na-citrate, pH 5.5, and split into two aliquots, to one of which 10 mU of endoH is added. The other half of the sample serves as the control. The reaction is then allowed to proceed for 16 h at 37°C. The incubation is terminated by addition of an equal volume of cold 30% TCA, and the tubes are incubated for 1 h on ice. The protein is recovered by centrifugation for 30 min at 10,000 × g and the pellet is washed twice with acetone, dried, and resuspended in SDS–PAGE sample buffer.

d. High pH Treatment of Microsomal Vesicles

Translation mix (5 μl) is thoroughly mixed with 45 μl of 1.8 M sucrose dissolved in 0.11 M Na$_2$CO$_3$ and placed at the bottom of a centrifuge tube for a TLA 100 rotor (Beckman). The solution is then overlaid with 140 μl of 1.2 M sucrose dissolved in 0.1 M Na$_2$CO$_3$, and with 50 μl of 0.25 M sucrose in 0.1 M Na$_2$CO$_3$. After centrifugation for 30 min at 95,000 rpm at 4°C in an Optima TL centrifuge, the top 140 μl, containing membranes and membrane-associated proteins, is recovered. The sample is diluted with 130 μl of water, 30 μg of hemoglobin is added as carrier, and the sample is TCA-precipitated and analyzed as described for endoH-treated samples.

B. *In Vivo* Assays for Translocation into the ER

1. Background

This assay is based on the fact that the density of ER-derived microsomes, and therefore the behavior of microsomes on isopicnic sucrose gradients, is influenced by the presence of polysomes bound to the cytosolic face of the ER membrane, which are engaged in the synthesis of polypeptides to be translocated into the ER (Quail, 1979). These polysomes remain attached to the ER membrane if tissue homogenization is performed in the presence of Mg^{2+} and are instead released in the presence of EDTA, a Mg^{2+} chelating agent, the ion being necessary for the association of the small and large ribosomal subunits. The Mg^{2+}-dependent change of density of microsomes is known as "magnesium-induced shift." Because the density of the other organelles, compartments and vesicles of the cell is not Mg^{2+}-dependent, magnesium-induced shift has been widely used on material from several plant tissues to identify the ER as the compartment of synthesis of protein destined to distal locations of the secretory pathway (see for example Bollini *et al.*, 1982) as well as to identify activities and proteins located in the ER as permanent residents, including protein disulfide-isomerase (Roden *et al.*, 1982), glycoprotein arabinosyl transferase (Andreae *et al.*, 1988), and BiP (D'Amico *et al.*, 1992).

The *in vivo* insertion of cloned proteins into the ER can therefore be verified by the expression of the recombinant protein in transgenic plants and a magnesium-induced shift assay. Here we give a protocol for performing this assay on small (3–6 cm long) leaves from axenic tobacco plants.

2. Assay

Two dark green leaves (3–6 cm long) are cut from the plant and weighed (approximately 0.3 g each). Each leaf is individually homogenized in an ice-cold mortar with 6–8 weights of ice-cold homogenization buffer (100 mM Tris–Cl, pH 7.8, 10 mM KCl) containing 12% (w/w) sucrose and either 1 mM EDTA or 2 mM MgCl$_2$. The homogenates are centrifuged at 1000 $\times$ g for 10 min at 4°C. About 1 ml of each supernatant is loaded on an 11-ml linear 16–55% (w/w) sucrose gradient made on top of a 0.5-ml cushion of 55% sucrose. The gradients are made in homogenization buffer containing either 1 mM EDTA or 2 mM MgCl$_2$.

After centrifugation at 35,000 rpm for 2 h at 4°C in a Beckman SW40 rotor, fractions of about 600 μl are collected. In these conditions, the ER will distribute around 1.14 g/cm³ in 1 mM EDTA and 1.17 g/cm³ in 2 mM MgCl$_2$, as indicated by SDS–PAGE and Western blot analysis of the gradient fractions using antiserum against the ER resident BiP (Fig. 2).

Leaves of transgenic plants expressing a recombinant protein can be labeled with radioactive amino acids before homogenization, using the protocol given below, and the position of the recombinant protein within the gradient can be determined by immunoprecipitation followed by SDS–PAGE and fluorography. If the recombinant protein is inserted into the ER and is destined to distal locations of the secretory pathway, it will show a magnesium-induced shift only after pulse-labeling; conversely, if it is ER-resident, it will cofractionate with the ER also after a chase. The magnesium-induced shift of a recombinant ER resident can also be detected by Western blot analysis, without the need of labeling, provided that the level of expression is sufficiently high.

A protease protection assay will help establish the topology of the protein with respect to the membrane. The assay is performed using a microsomal fraction purified on discontinuous sucrose gradients (D'Amico *et al.*, 1992). It will usually be necessary, after protease treatment, to immunoprecipitate the radiolabeled protein before SDS–PAGE and fluorography. A method for immunoprecipitation is described elsewhere in this volume (Chapter 24).

3. Pulse–Chase Labeling of Tobacco Leaves

Each leaf is labeled individually by immersing the petiole into 300 μl of MS salts containing 5.55 MBq/ml of Tran ³⁵S-Label (ICN), in a 50-ml Falcon tube.

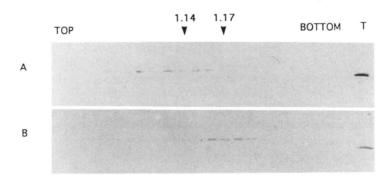

Fig. 2 Magnesium-induced shift in the distribution of BiP polypeptides. Homogenates obtained from tobacco leaves were fractionated as described in the text in the presence of either 1 mM EDTA (A) or 2 mM MgCl$_2$ (B). The proteins in each fraction were separated by SDS–PAGE and electroblotted on nitrocellulose membrane. The membrane was then developed using an anti-tobacco BiP antibody (Denecke *et al.*, 1991) and the ECL protein detection system (Amersham). T, total homogenate.

To facilitate the uptake of labeled medium, the tube is subjected to vacuum for 1 min. The tube is then left open during 4 h of pulse-labeling at room temperature. Three hours after the beginning of labeling, the radioactive incubation medium is discarded and substituted with 300 μl of MS salts without radioactive amino acids. For the chase, the medium is substituted with 300 μl of MS salts supplemented with 10 mM methionine and 50 mM cysteine. During the chase, the lid is placed on top of the tube and is loosely screwed on. We do not notice any obvious signs of leaf stress during a 16-h chase.

It should be noted that this protocol includes a 4-h period of pulse-labeling. This ensures very efficient protein labeling. It also ensures a high probability that a relevant proportion of radiolabeled recombinant protein is still present in the ER at the end of the pulse, even in cases where the protein is destined to more distal locations of the secretory pathway. It has been shown that different proteins transported along the secretory pathway have different kinetics of transport between the various membranous compartments (Gomez and Chrispeels, 1993). Therefore, the length of pulse-labeling optimal to allow magnesium-induced shift of the newly synthesized and ER-inserted proteins should be empirically determined for each case.

III. Concluding Remarks

In this chapter we have described a series of methods employed to determine whether a protein is imported into the ER and to characterize some of the early processing events that occur in this compartment. Although *in vitro* studies are extremely useful, they should be complemented, whenever possible, by data obtained *in vivo*. A combination of these approaches will be necessary to improve our understanding of protein import into the plant ER, and to determine the degree of conservation of this process throughout evolution.

Acknowledgments

We thank Roberto Bollini, Nica Borgese, Serena Fabbrini, and Barbara Valsasina for helpful suggestions and for critical reading of the manuscript. This work was partially supported by CNR Grants RAISA (Subproject No. 2, Paper No. 1953) and Biotecnologic e Biostrumentazione.

References

Anderson, C. W., Straus, J. W., and Dudock, B. S. (1983). *In* "Methods in Enzymology" (R. Wu, L. Grossman, and K. Moldave, eds.), Vol. 101, pp. 635–644. New York: Academic Press.
Andreae, M., Lang, W. C., Barg, C., and Robinson, D. G. (1988). *Plant Sci.* **56,** 205–212.
Bassüner, R., Wobus, U., and Rapoport, T. A. (1984). *FEBS Lett.* **166,** 314–320.
Bollini, R., Van der Wilden, W., and Chrispeels, M. J. (1982). *Physiol. Plant.* **55,** 82–92.
Bollini, R., Vitale, A., and Chrispeels, M. J. (1983). *J. Cell Biol.* **96,** 999–1007.
Brodsky, J. L., Hamamoto, S., Feldheim, D., and Schekman, R. (1993). *J. Cell Biol.* **120,** 95–102.

Campos, N., Palau, J., Torrent, M., and Ludevid, D. (1988). *J. Biol. Chem.* **263,** 9646–9650.

D'Amico, L., Valsasina, B., Daminati, M. G., Fabbrini, M. S., Nitti, G., Bollini, R., Ceriotti, A., and Vitale, A. (1992). *Plant J.* **2,** 443–455.

Denecke, J., Goldman, M. H. S., Demolder, J., Seurinck, J., and Botterman, J. (1991). *Plant Cell* **3,** 1025–1035.

Fabbrini, M. S., Valsasina, B., Nitti, G., Benatti, L., and Vitale, A. (1991). *FEBS Lett.* **286,** 91–94.

Folz, R. J., and Gordon, J. I. (1987). *Biochem. Biophys. Res. Commun.* **146,** 870–877.

Fujiki, Y., Hubbard, A. L., Fowler, S., and Lazarow, P. B. (1982). *J. Cell Biol.* **93,** 97–102.

Galili, G., Kawata, E. E., Cuellar, R. E., Smith, L. D., and Larkins, B. A. (1986). *Nucleic Acids Res.* **14,** 1511–1524.

Gallagher, P., Henneberry, J., Wilson, I., Sambrook, J., and Gething, M.-J. (1988). *J. Cell Biol.* **107,** 2059–2073.

Gilmore, R., Walter, P., and Blobel, G. (1982). *J. Cell Biol.* **95,** 470–477.

Gomez, L., and Chrispeels, M. J. (1993). *Plant Cell* **5,** 1113–1124.

Görlich, D., Prehn, S., Hartmann, E., Kalies, K.-U., and Rapoport, T. A. (1992). *Cell* **71,** 489–503.

Hattori, T., Ichihara, S., and Nakamura, K. (1987). *Eur. J. Biochem.* **166,** 533–538.

Higgins, T. J. V., and Spencer, D. (1981). *Plant Physiol.* **67,** 205–211.

Jackson, R. J., and Hunt, T. (1983). *In* "Methods in Enzymology" (S. Fleischer and B. Fleischer, eds.), Vol. 96, pp. 50–74. New York: Academic Press.

Krieg, P. A., and Melton, D. A. (1984). *Nucleic Acids Res.* **12,** 7057–7070.

Laemmli, U. K. (1970). *Nature* **227,** 680–685.

Matsudaira, P. (1987). *J. Biol. Chem.* **262,** 10035–10038.

Meyer, D. I., Krause, E., and Dobberstein, B. (1982). *Nature* **297,** 647–650.

Miernyk, J. A., Duck, N. B., Shatters, R. G. J., and Folk, W. R. (1992). *Plant Cell* **4,** 821–829.

Miernyk, J. A., and Nelsen, T. C. (1989). *Plant Physiol.* **89 Suppl.,** 62.

Miernyk, J. A., and Shatters, R. G. (1992). *Plant Physiol.* **99 Suppl.,** 44.

Müsch, A., Wiedmann, M., and Rapoport, T. A. (1992). *Cell* **69,** 343–352.

Nicchitta, C. V., and Blobel, G. (1993). *Cell* **73,** 989–998.

Osteryoung, K. W., Sticher, L., Jones, R. L., and Bennett, A. B. (1992). *Plant Physiol.* **99,** 378–382.

Prehn, S., Wiedmann, M., Rapoport, T. A., and Zwieb, C. (1987). *EMBO J.* **6,** 2093–2097.

Quail, P. H. (1979). *Ann. Rev. Plant Physiol.* **30,** 425–484.

Rapoport, T. A. (1991). *FASEB J.* **5,** 2792–2798.

Roden, L. T., Miflin, B. J., and Freedman, R. B. (1982). *FEBS Lett.* **138,** 121–124.

Sanders, S. L., and Schekman, R. (1992). *J. Biol. Chem.* **267,** 13791–13794.

Sanders, S. L., Whitfield, K. M., Vogel, J. P., Rose, M. D., and Schekman, R. W. (1992). *Cell* **69,** 353–365.

Schenborn, E. T., and Mierendorf, R. C. (1985). *Nucleic Acids Res.* **13,** 6223–6236.

Shelness, G. S., and Blobel, G. (1990). *J. Biol. Chem.* **265,** 9512–9519.

Vitale, A., Ceriotti, A., and Denecke, J. (1993). *J. Exp. Bot.* **44,** 1417–1444.

von Heijne, G. (1985). *J. Mol. Biol.* **184,** 99–105.

von Heijne, G. (1986). *Nucleic Acids Res.* **14,** 4683–4690.

Walter, P., and Blobel, G. (1981). *J. Cell Biol.* **91,** 557–561.

Walter, P., and Blobel, G. (1983). *In* "Methods in Enzymology" (S. Fleisher and B. Fleisher, eds.), Vol. 96, pp. 84–93. New York: Academic Press.

Watson, M. E. E. (1984). *Nucleic Acids Res.* **12,** 5145–5164.

Wessel, H. P., Beltzer, J. P., and Spiess, M. (1991). *In* "Methods in Cell Biology" (A. M. Tartakoff, ed.), Vol. 34, pp. 287–302. San Diego: Academic Press.

Zimmermann, R., Sagstetter, M., Lewis, M. J., and Pelham, H. R. B. (1988). *EMBO J.* **7,** 2875–2880.

CHAPTER 22

Protein–Protein Interactions within the Endoplasmic Reticulum

Jeffrey W. Gillikin,* **Elizabeth P. B. Fontes,†** **and Rebecca S. Boston***

*Department of Botany
North Carolina State University
Raleigh, North Carolina 27695-7612
†BIOAGRO-Sector de Biologia Molecular de Plantas
Universidade Federal de Viçosa
Viçosa, MG, Brasil 36570,000

I. Introduction

Translocation of proteins into the ER[1] is the initial transport step for proteins with ER, Golgi, lysosomal, vacuolar, or extracellular destinations. The process of protein translocation into the ER involves a complex series of events and interactions among numerous proteins, including the nascent polypeptide, signal

---

[1] Abbreviations used: AMP-PNP, 5'-adenylylimidodiphosphate; ATP, adenosine, 5'-triphosphate; ATP-γ-S, adenosine 5'-O-(3-thiotriphosphate); BCIP, 5-bromo-4-chloro-3-indolylphosphate *p*-toluidine salt; BiP, binding protein; BSA, bovine serum albumin; DAF, days after flowering; DAP, days after pollination; De*-B30, Defective endosperm-B30; ECL, enhanced chemiluminescence;

recognition particle, signal sequence receptor, signal peptidase, and several ER-resident proteins. The protein transport process in plant cells has been the focus of numerous studies (for review see Chrispeels, 1991; Bednarek and Raikhel, 1992; and Vitale *et al.,* 1993), yet surprisingly little is known about the protein–protein interactions that occur at each step of the secretory process. Recent work in yeast, animal, and plant systems has implicated ER-resident proteins in the folding, assembly, and modification of proteins translocated into the ER (for review see Gething and Sambrook, 1992).

Most seed storage proteins are transported through the ER en route to specialized storage organelles. It is likely that nascent storage proteins encounter ER-resident proteins such as BiP, grp94, PDI, and PPIase prior to attainment of proper tertiary and/or quaternary structure (reviewed by Gething and Sambrook, 1992). In this chapter, we will focus on identification of protein–protein interactions within the ER and ER-derived protein bodies. The methods described here were optimized for analysis of interactions among the ER-resident protein, BiP, and storage proteins of maize and soybean. With slight modification, however, they should also be adaptable for studying protein–protein interactions involving any protein for which antibodies are available.

Proteins such as BiP, which facilitate protein folding through the recognition and stabilization of nonnative polypeptide intermediates, have been termed molecular chaperones (Ellis, 1987; Gething and Sambrook, 1992). BiP, the ER-resident member of the stress-70 chaperones, shares homology with hsp 70 family members and is involved in many vital functions including protein translocation, protein folding, and assembly of oligomeric complexes (for review see Gething and Sambrook, 1992). Binding of BiP to target polypeptides is noncovalent and easily disrupted by ATP. Apparently, ATP hydrolysis is important in the dissociation event because nonhydrolyzable ATP analogs are ineffective in dissociating polypeptides. Recently, evidence from *in vitro* experiments with bacterial and cytosolic hsp 70s indicates that K^+ may also be involved in facilitating the dissociation of protein–hsp 70 complexes after the binding of Mg–ATP (Palleros *et al.,* 1993).

Currently, a number of methods are available for detecting BiP–protein interactions. These methods generally require that the proteins of interest be soluble in aqueous solution and the procedures be performed under nondenaturing conditions. The most commonly used techniques for detecting BiP–protein interactions include indirect immunoprecipitation (Bole *et al.,* 1986; D'Amico *et al.,*

EGTA, ethylene glycol-*O-O'*-bis(2-aminoethyl)-*N,N,N',N'*-tetraacetic acid; ER, endoplasmic reticulum; grp94, 94-kDa glucose-regulated protein; Hepes, *N*-[2-hydroxyethyl]piperazine-*N'*-[2-ethanesulfonic acid]; HRP, horseradish peroxidase; hsc70, 70-kDa heat shock cognate protein; hsp70, 70-kDa heat shock protein; IgG, G class immunoglobulin; MMSB, modified microfilament stabilization buffer; NBT, nitroblue tetrazolium; PDI, protein disulfide isomerase; PMSF, phenylmethylsulfonyl fluoride; PPIase, peptidyl prolyl *cis-trans* isomerase; SDS–PAGE, sodium dodecyl sulfate–polyacrylamide gel electrophoresis; TBS, Tris-buffered saline; TBST, Tris-buffered saline supplemented with 0.1% (v/v) Tween 20; Tris, tris(hydroxymethyl) aminomethane.

1992; Vitale *et al.*, 1995), native gel electrophoresis (Freiden *et al.*, 1992), cosedimentation of proteins through a sucrose gradient (Li *et al.*, 1993), and chemical or photocross-linking followed by immunoprecipitation (Sanders *et al.*, 1992; Müsch *et al.*, 1992). Although these techniques can be used for the detection of many protein–protein complexes, BiP-mediated interactions can be easily distinguished from nonspecific or specific, nonchaperone-mediated interactions by their disruption in the presence of ATP. In the presence of nonhydrolyzable ATP analogs, such as AMP-PNP and ATP-γ-S, these interactions are stable. Thus, sensitivity of BiP–protein binary complexes to ATP can be used as a diagnostic feature of BiP-mediated interactions.

II. Interaction of BiP with Storage Proteins in *Glycine max*

A. Principle of the Assay

Indirect immunoprecipitation of proteins has been extensively used to analyze protein–protein interactions during the ordered assembly of multimeric complexes as well as to assess any biochemical activities that may be associated with these interactions. This technique is based on the coimmunoprecipitation of an antigen and any associated polypeptides. Successful detection of the interaction between two polypeptides requires a relatively stable protein–protein complex, steric availability of recognizable epitopes upon protein association, and no cross-reactivity between antibodies and associated proteins. If neither antibody can cross-react with the reciprocal protein, the capacity of both proteins to be immunoprecipitated by either antibody must result from the formation of a stable complex between the proteins of interest.

In the particular case of the association between BiP and plant storage proteins, the interaction is expected to be transient. As a result, during protein trafficking to protein bodies, only a small fraction of the storage proteins are likely to be associated with BiP at any given time. Such a situation would lead to a relatively low proportion of BiP-associated proteins and a high background of free (unbound) proteins. Despite these difficulties, we have used coimmunoprecipitation as a qualitative assay to study the involvement of a putative soybean BiP in the biosynthetic transport pathway of the storage protein ß-conglycinin, in immature seeds.

B. Immunoprecipitation of ß-Conglycinin with Anti–BiP and Anti–ß-Conglycinin Antisera

1. Materials and Equipment

a. Plant Material

Glycine max (soybean) seeds from the Brazilian variety (UFV-5 harvested 25 DAF.

b. Chemicals

Reagent grade ATP, glucose, HCl, $MgCl_2$, PMSF, NaCl, Triton X-100, and Tris; protein–A Sepharose (Sigma Chemical Co); hexokinase, lyophilized powder (Merck); CNBr-activated Sepharose 4B (Pharmacia); acrylamide, ammonium persulfate, bis-acrylamide, BCIP, NBT, and TEMED (GIBCO BRL).

c. Antibodies

Rabbit anti-maize BiP antiserum either affinity-purified by selection with a ß-galactosidase soybean–BiP fusion protein (clone-selected antibody) or coupled to protein-A Sepharose (see Fontes *et al.,* 1991; Harlow and Lane, 1988, for description of antibody preparation, coupling, and purification); chicken anti-ß-conglycinin antiserum used without purification or after coupling to CNBr-activated Sepharose (see Leal, 1991; coupling was performed according to manufacturer's instructions); mouse antiserum against an unrelated protein coupled to protein–A Sepharose; alkaline phosphatase-conjugated goat anti-rabbit IgG and rabbit anti-chicken IgG.

d. Equipment

Microcentrifuge and tubes, mortar, pestle, protein electrophoresis and electrotransfer unit.

2. Methods

1. Using an ice-cold mortar and pestle, grind three to five fresh soybean seeds harvested 25 DAF in extraction buffer [50 m*M* Tris–HCl, pH 7.5, at 25°C, 100 m*M* NaCl, 0.5% (v/v) Triton X-100, 1 m*M* PMSF, 2 U/ml hexokinase, 5 m*M* glucose] at a ratio of 1 g of tissue/5 ml of extraction buffer.

2. Incubate the protein extract at 0°C for 10 min.

3. Remove cellular debris by centrifugation at $13,000 \times g$ for 10 min at 4°C.

4. Transfer supernatant to a new centrifuge tube and place on ice.

5. Preclear the supernatant by incubation with 0.1 ml of protein–A Sepharose [50% slurry in TBS ($1 \times$ TBS = 20 m*M* Tris–HCl, pH 7.5, at 25°C, and 140 m*M* NaCl)] for 30 min at 4°C.

6. Pellet the protein–A Sepharose beads by centrifugation at $13,000 \times g$ for 10 min.

7. Prepare duplicate 0.5-ml aliquots of supernate for each antibody to be used for immunoprecipitation. To each tube of the first pair, add 0.1 ml anti-maize BiP antiserum linked to protein A–Sepharose (50% slurry in TBS). To each tube of the second pair, add 0.1 ml anti-ß-conglycinin antiserum linked to CNBr–Sepharose 4B (50% slurry in TBS).

8. To one tube of each pair, add ATP to 2 m*M*.

9. Add 60 μl of 10× TBS to each tube and incubate at 4°C for 4 h under gentle agitation.

10. Isolate the immune complexes by centrifugation at 8000 × g for 5 min.

11. Wash the Sepharose beads 7× with 1 ml of 50 mM Tris–HCl, pH 7.5, at 25°C, 0.5 M NaCl, 1% (v/v) Triton X-100, and 1 mM PMSF.

12. Resuspend the immune complexes with 40 μl of SDS sample buffer (Laemmli, 1970).

13. Dissociate immune complexes by boiling the sample for 5 min.

14. Separate the polypeptides by SDS–PAGE through a 10% polyacrylamide gel. The identity of the immunoprecipitated proteins can be confirmed by immunoblotting using either anti-ß-conglycinin antiserum or anti-BiP antiserum. Develop the immunoblots by mixing 33 μl BCIP and 44 μl NBT together in 10 ml of 0.1 M Tris–HCl, pH 9.5, at 25°C, 0.1 M NaCl, and 50 mM MgCl$_2$.

3. Results and Discussion

The ability to assay protein–protein interactions by indirect immunoprecipitation of proteins represents a rapid method for identification of protein–protein complexes. Here we describe an immunological procedure to assay for BiP-storage protein association in whole-cell extracts from soybean cotyledons. This method is simple, reliable, nonradioactive, and straightforward to interpret based on the diagnostic features of dissociation by ATP and resistance to high salt described above.

BiP and ß-conglycinin were coimmunoprecipitated by antibodies against either protein, but not by antibodies against an unrelated control protein (Figs. 1A and 1B). Comparison of the amount of α and α' subunits immunoprecipitated directly with anti-ß-conglycinin antiserum and indirectly with anti-BiP antiserum revealed that only a small fraction of ß-conglycinin subunits was associated with BiP (Fig. 1B). Direct immunoprecipitation of α and α' subunits from 0.5 μl extract (Fig. 1B) was as efficient as indirect immunoprecipitation of the same polypeptides from 0.5 ml extract using anti-BiP antiserum (Fig. 1B). Therefore, the concentration of free ß-conglycinin was at least 1000-fold higher than BiP-associated ß-conglycinin. This ratio is not surprising since the interaction of BiP with normal proteins is expected to be very transient, whereas interactions with assembly-deficient phaseolin mutants has been observed to be much higher (Pedrazzini et al., 1994). Thus, the efficiency of BiP dissociation from ß-conglycinin subunits and the subsequent assembly of the released subunits in their trimeric forms may be very high.

For the detection of BiP–ß-conglycinin association, it is necessary to maximize the amount of newly synthesized ß-conglycinin in the extract. This is most easily accomplished by choosing a stage of cotyledon development where ß-conglycinin gene transcription is maximized (25 DAF), but large amounts of storage protein have not yet accumulated in protein vacuoles (Walling et al., 1986). Even under optimal conditions, immunoprecipitation must be performed in the presence of large amounts of storage proteins that contribute dramatically to increases in

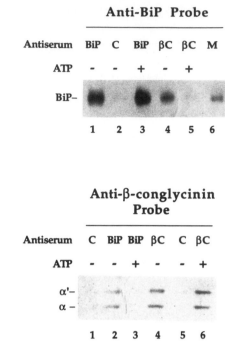

Fig. 1 Indirect immunoprecipitation of BiP and soybean storage proteins. Immunoprecipitation reactions were carried out as described in Section II,B,2 using rabbit anti-maize BiP antiserum cross-linked to protein A–Sepharose (BiP), chicken anti-ß-conglycinin antiserum linked to CNBr–Sepharose 4B (ßC), or control antiserum cross-linked to protein–A Sepharose (C). (A) Immunoblot of the immunoprecipitated products probed with clone-selected anti-BiP antiserum and goat anti-rabbit IgG secondary antibody conjugated to alkaline phosphatase. Initial immunoprecipitations were carried out with anti-BiP antiserum minus ATP (1 ml extract), lane 1; unrelated antiserum minus ATP (1 ml extract), lane 2; anti-BiP antiserum plus ATP (1 ml extract), lane 3; anti-ß-conglycinin minus ATP (1 ml extract), lane 4; anti-ß-conglycinin plus ATP (1 ml extract), lane 5. Lane 6 contains a sample of the initial protein extract (50 μl). The position of BiP is shown to the left of the blot. (B) Immunoblot of immunoprecipitated products probed with chicken anti-ß-conglycinin antiserum and rabbit anti-chicken IgG secondary antibody conjugated to alkaline phosphatase. Immunoprecipitations were performed with control antiserum minus ATP (500 μl extract), lane 1; anti-BiP antiserum minus ATP (500 μl extract), lane 2; anti-BiP antiserum plus ATP (500 μl extract), lane 3; anti-ß-conglycinin antiserum minus ATP (0.5 μl extract), lane 4; unrelated antiserum (0.5 μl extract), lane 5; anti-ß-conglycinin antiserum plus ATP (0.5 μl extract), lane 6. The position of α and α' subunits are shown to the left of the blot.

nonspecific immunoprecipitation. Using the procedure described here, we typically have protein concentrations of 50 mg/ml after the first centrifugation step. A second drawback is the tendency of highly concentrated proteins to aggregate and become trapped by the immunobeads, thereby confusing the results. Yet a third problem is the presence of endogenous ATP, which must be depleted to preserve the integrity of BiP–protein complexes. We have found that three steps are essential for optimal detection of signal above background: (1) grinding should be performed in the presence of glucose and hexokinase to rapidly remove free ATP from the system, (2) whole-cell extracts must be incubated with protein–A Sepharose beads prior to the immunoprecipitation, and (3) the washing steps should be carried out at high concentrations of nonionic detergent and salt. For globulin soybean storage proteins, the extensive washes with a high salt buffer contribute not only to reduction of nonspecific binding but also, and primarily, to promotion of complete solubilization of the ß-conglycinin subunits. We have found that these conditions reduce the nonspecific binding below the limits of detection, as judged by the failure of an unrelated antibody to immunoprecipitate either BiP or ß-conglycinin subunits from whole-cell protein extracts (Figs. 1A and 1B).

III. Interaction of BiP with Prolamine Protein Bodies

The preservation of BiP–protein interactions and the solubility of both soybean globulins and BiP in high salt solutions allow for straightforward analysis of complexes. In contrast, the insolubility of many cereal storage proteins in aqueous solutions requires that an alternative approach be developed. In this section we describe a method we have developed for studying the interactions of BiP within protein bodies in maize. In addition to the limitations imposed by the insolubility of the prolamine fraction in aqueous solutions, the number of available techniques that can be used to characterize BiP–prolamine interactions is further reduced by the lack of the reactive amino acid lysine, which is commonly used in chemical cross-linking experiments. Due to these limitations, BiP–prolamine interactions are not amenable to characterization by the techniques described for soybean proteins. Therefore, we will describe two methods that we routinely use in our laboratory for the isolation of protein bodies as well as a method for the detection of BiP–protein body interactions.

A. Isolation of Protein Bodies

1. Cytoskeletal Preparation

Cytoskeletal preparations from endosperm tissue can be isolated by modification of a procedure previously described by Abe *et al.* (1991). These modifications allow for the isolation of a corn (*Zea mays* L.) endosperm protein body fraction that (1) is devoid of starch, (2) maintains cytoskeletal integrity, and (3) maintains membrane integrity as judged by insensitivity of proteinase K. All procedures are carried out at 4°C.

a. Materials and Equipment
Plant material:
Endosperm regulatory mutant De*-B30 from the maize inbred line B37. Characterization of the mutant maize line has been described previously by

F. Salamini and colleagues (Salamini *et al.,* 1979), who kindly provided us with seed stocks.

Chemicals:

Reagent grade EGTA, glucose, Hepes, KOH, MgOAc, and sucrose; hexokinase, lyophilized powder (Worthington, Catalog No. LS02512)

Equipment and Supplies:

Centrifuge tubes (15 ml), Miracloth (Calbiochem), mortar, pestle, preparative centrifuge, and swinging bucket rotor.

b. Methods

1. Dissect endosperm from five immature kernels harvested 20 DAP.

2. Gently grind the endosperm tissue in 10 ml of MMSB [10 mM Hepes–KOH, pH 7.5, 5 mM MgOAc, 2 mM EGTA, and 7.2% (w/v) sucrose] supplemented with 5 mM glucose and 5 units/ml hexokinase in a mortar and pestle.

3. Filter the homogenate through Miracloth and collect the filtrate into a beaker.

4. Layer the filtrate onto a 2.5-ml sucrose pad (10 mM Hepes–KOH, pH 7.5, 5 mM MgOAc, 2 mM EGTA, and 2 M sucrose) in a 15-ml Corex tube.

5. Centrifuge the filtrate at 4100 × g in a swinging bucket rotor for 10 min.

6. Aspirate off 8 ml from the upper portion of the tube without disturbing the cytoskeletal fraction located at the sucrose interface.

7. Gently remove the protein bodies from the interface to a clean beaker and dilute to 10 ml with MMSB.

8. Layer the protein bodies over a second sucrose pad and repeat steps 5 and 6.

9. Remove the protein bodies from the interface and transfer to 1.5-ml microcentrifuge tubes for further analysis.

2. Purification of Protein Bodies

The isolation of protein bodies was performed essentially as previously described by Larkins and Hurkman (1978). This method can also be used for preparation of ER fractions. All procedures are carried out at 4°C.

a. Materials and Equipment

Plant material and equipment are described in Section III,A,1,a, with the exception of the following:

Microcentrifuge and tubes, ultracentrifuge and tubes, and swinging bucket rotor

Chemicals:

Reagent grade glucose, HCl, KCl, MgCl$_2$, sucrose, and Tris; hexokinase, lyophilized powder (Worthington, Catalog No. LS02512).

b. Methods

1. Grind endosperm from five kernels harvested 20 DAP in extraction buffer [10 mM Tris–HCl, pH 8.5, at 25°C, 10 mM KCl, 5 mM MgCl$_2$, 7.2% (w/v) sucrose, 10 mM glucose, and 5 units/ml hexokinase] at a ratio of 1 mg endosperm/2 μl extraction buffer in a mortar and pestle.

2. Centrifuge the homogenate at 80 × g for 5 min to remove starch and other cellular debris.

3. Load the supernatant onto discontinuous sucrose gradients composed of 2 ml of 2 M sucrose, 4 ml of 1.5 M sucrose, 4 ml of 1 M sucrose, and 2 ml of 0.5 M sucrose each in 10 mM Tris–HCl, pH 8.5, at 25°C, 10 mM KCl, and 5 mM MgCl$_2$.

4. Centrifuge at 80,000 × g for 30 min.

5. Aspirate off the upper 9 ml of the gradient.

6. Carefully remove the protein bodies from the 1.5/2 M sucrose interface and transfer to 1.5-ml microcentrifuge tubes for further analysis.

3. Results and Discussion

Prolamine protein bodies can be isolated from maize by a number of techniques. We have chosen to present two isolation techniques that yield protein bodies of different purity. The first method we describe results in a crude cytoskeletal fraction containing the protein bodies and the second method provides a purified preparation of protein bodies. The protein bodies recovered by each method are similar with respect to membrane integrity as judged by insensitivity to proteinase K digestion in the absence of detergent (data not shown), but differ in the amount of time and equipment required for preparation.

The first isolation method is a modification of a method previously described (Abe *et al.,* 1991). The original method employed a low-ionic-strength buffer that helped maintain cytoskeletal integrity as determined by BODIPY 503/512 Phallacidin staining (data not shown), but did not provide the osmoticum necessary to maintain the ER/protein body membranes. We have experienced difficulty in recovering protein bodies after centrifugation at 30 × g during the wash steps. Moreover, use of higher centrifugal forces (660 × g) to pellet the protein bodies was detrimental to maintaining protein body integrity. To compensate for the absence of osmoticum, sucrose has been included in the MMSB at a final concentration of 7.2% (w/v). We have found that the integrity of both the cytoskeleton and the protein bodies is maintained if we collect them on a 2 M sucrose pad. The use of the sucrose pad also provides an efficient means of removing the starch granules, which are common contaminants of the original preparation reported by Abe *et al.* (1991). It is imperative that any modification to the protocol avoid the incorporation of cytoskeleton-disrupting agents such as Tris–HCl, KCl, KI, and (NH$_4$)$_2$SO$_4$. Likewise, lyophilized hexokinase powder is preferred over hexokinase in 3 M (NH$_4$)$_2$SO$_4$ since this is a cytoskeleton-disrupting agent.

The second isolation method we used was previously described by Larkins and Hurkman (1978). If the protein bodies are destined for use in the ATP release assay, the isolation protocol can be used as described. However, if once foresees using the protein bodies in NHS-ester cross-linking experiments, Tris–HCl should be replaced by buffers such as Hepes–KOH, bicarbonate/carbonate, borate, or phosphate to prevent interference with the cross-linking reaction.

Because both methods result in preparations of intact protein bodies, the method of choice is dependent upon the availability of equipment and the fate of the protein bodies after isolation. The cytoskeletal preparation offers several advantages over purification of protein bodies on a discontinuous gradient. It is rapid and requires less preparation time and no ultracentrifugation steps. In addition, it yields protein bodies that are still attached to ER and thus provides a system for studying interactions within both the ER and the protein bodies. Its usefulness is limited, however, by the likelihood of contamination by other cytosolic stress-70 proteins that could interfere if one is looking for events specific for protein bodies. These limitations can be overcome by choosing the discontinuous sucrose gradient procedure. Protein bodies isolated from the discontinuous gradient lack contamination by cytosolic hsc70, which has been shown to be involved in the *in vitro* translocation of proteins into microsomes (Miernyk *et al.,* 1992). Furthermore, because microsomes are separated from the protein bodies, fractions at the 1/1.5 M sucrose interface can be removed for study of BiP–protein interactions specific to the ER.

B. ATP-Dependent Release of BiP from Immature Protein Bodies

1. Materials and Equipment

a. Antibodies

Rabbit anti-maize BiP antiserum (previously described by Fontes *et al.,* 1991). HRP-conjugated goat anti-rabbit IgG (H + L) (Bio-Rad).

b. Chemicals

ATP and AMP-PNP (Sigma Chemical Co., Catalog Nos. A9187 and A2647, respectively); ECL Western blotting detection system (Amersham Life Sciences); X-OMAT film for autoradiography (Kodak); nitrocellulose (Schleicher & Schuell); acrylamide (Amresco); bis-acrylamide, ammonium persulfate, and TEMED (Bio-Rad); Reagent grade EGTA, Hepes, MgOAc, KOH, sucrose, Triton X-100, and Tween 20.

c. Equipment and Supplies

Microcentrifuge, tubes, protein electrophoresis and electrotransfer unit

2. Methods

The procedure for the ATP release assay is shown schematically in Fig. 2. A detailed step-by-step protocol is described below.

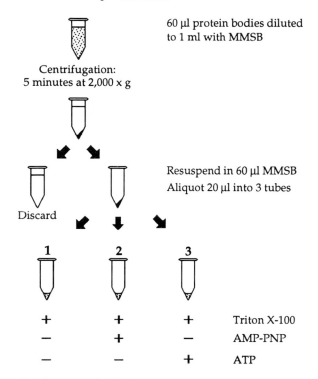

60 µl protein bodies diluted
to 1 ml with MMSB

Centrifugation:
5 minutes at 2,000 x g

Resuspend in 60 µl MMSB
Aliquot 20 µl into 3 tubes

Discard

1   **2**   **3**

+   +   +   Triton X-100

–   +   –   AMP-PNP

–   –   +   ATP

Incubate samples for 30 minutes on ice
Centrifugation: 15,800 x g for 5 minutes
Transfer supernate to a fresh microcentrifuge tube

Analyze samples by SDS-PAGE
and immunoblotting

P   S   P   S   P   S

Fig. 2 Schematic diagram of the *in vitro* assay for ATP-dependent release of BiP from prolamine protein bodies from immature maize kernels.

1. Aliquot 60 µl of protein bodies (prepared by either method described in Sections III,A,1 or III,A,2) into a 1.5-ml microcentrifuge tube.

2. Dilute the sample to 1 ml with MMSB.

3. Centrifuge the protein bodies at 2000 × g for 5 min.

4. Carefully discard the supernate.

5. Resuspend the pellet in 60 µl MMSB.

6. Aliquot 20 µl of resuspended protein bodies into three microcentrifuge tubes.

7. Add 80 μl of 1.25 mM Hepes, pH 7, 0.625% (v/v) Triton X-100 to the first tube.

8. Add 80 μl of 1.25 mM AMP–PNP, 1.25 mM Hepes, pH 7, and 0.625% (v/v) Triton X-100 to the second tube.

9. Add 80 μl of 1.25 mM ATP, 1.25 mM Hepes, pH 7, and 0.625% (v/v) Triton X-100 to the third tube.

10. Incubate tubes on ice for 30 min.

11. Centrifuge samples 15,800 $\times$ g for 5 min.

12. Remove supernates (90 μl) to new 1.5-ml microcentrifuge tubes.

13. Fractionate samples through a 10% SDS–polyacrylamide gel (Laemmli, 1970).

14. Transfer proteins to nitrocellulose using a semidry electrophoretic transfer cell and the buffer system described by Bjerrum and Schafer-Nielsen (1986).

15. The filter should be blocked in 1$\times$ TBST supplemented with 1% (v/v) Tween 20 for 1 h at room temperature. We have found no need for additional blocking agents such as BSA and nonfat dry milk.

16. Incubate the filter with primary antibody for 1 h. We routinely use a 1:10,000 dilution of rabbit anti-BiP antiserum in 1$\times$ TBST, but the dilution should be determined for each individual antiserum.

17. Wash filter 3$\times$ for 10 min each wash in 1$\times$ TBST.

18. Incubate the filter with secondary antibody for 1 h at room temperature. We use a 1:10,000 dilution of HRP-conjugated goat anti-rabbit IgG.

19. Wash filter 3$\times$ for 10 min each wash in 1$\times$ TBST.

20. Develop immunoblot with detection reagents provided with ECL Western blotting kit according to manufacturer's instructions.

3. Results and Discussion

In this section, we describe a method that can be used to detect the interaction of BiP with prolamine protein bodies from cereal crops. Protein bodies prepared by either of the techniques described above can be used equally well to detect the interaction of BiP with protein bodies. The data derived from these experiments can be analyzed by either immunoblot as described above or ELISA (not shown). If one chooses an ELISA for data analysis, care must be taken to remove the Triton X-100 from the supernate since it can interfere with the binding of proteins to the ELISA plates.

Treatment of protein bodies with ATP (1 mM) in the presence of Triton X-100 resulted in the dissociation of BiP from the prolamine protein bodies (Fig. 3). If Triton X-100 was omitted from the assay, the membrane was presumably left intact and, even in the presence of ATP, BiP was not released from the protein bodies (data not shown). The detection of the low levels of BiP in the crude preparation after treatment with Triton X-100 alone could be explained

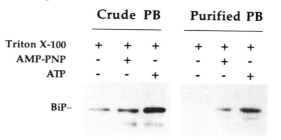

	Crude PB			Purified PB		
Triton X-100	+	+	+	+	+	+
AMP-PNP	–	+	–	–	+	–
ATP	–	–	+	–	–	+
BiP–						

Fig. 3 Effect of ATP and a nonhydrolyzable ATP analog on the release of BiP from prolamine protein bodies from immature maize kernels. Protein bodies were isolated as a cytoskeletal fraction (crude) and a discontinuous sucrose gradient fraction (purified). Protein bodies were treated with 0.5% (v/v) Triton X-100 or 0.5% (v/v) Triton X-100 and 1 mM AMP–PNP or 1 mM ATP at 4°C for 30 min. Following incubation, reactions were subjected to SDS–PAGE and immunoblotting with anti-maize BiP antiserum. Position of purified BiP is shown at the left of the blot.

either as the release of soluble BiP from the ER fraction of the crude preparation or as contamination of the preparation with hsc70. Treatment of purified protein bodies with Triton X-100 alone did not promote the release of immunoreactive polypeptides in the absence of ATP or AMP-PNP (Fig. 3). A low but detectable level of BiP was released from protein bodies treated with AMP-PNP (Fig. 3). A similar result was obtained with rice prolamine protein bodies where the amount of BiP released with ATP-γ-S was 25 to 30% of that observed with ATP (Li *et al.,* 1993). We have observed that Li$^+$, the AMP–PNP counterion, can dissociate a small amount of BiP from the protein bodies. Such dissociations could probably be avoided by using Mg-AMP-PNP.

IV. Conclusions

Defining the role of ER-resident proteins in the process of protein folding and assembly is important for understanding the protein transport process in plant cells. The methodologies presented here have been described primarily as they apply to molecular chaperone participation in plant storage protein translocation and packaging under study in our laboratories. Many of the same approaches can be applied to investigation of other protein–protein interactions in the ER. The techniques can be readily modified to accommodate differences in protein solubility as well as differences in the ratio of bound to free protein. Furthermore, it is possible to use both crude and pure fractions of protein bodies without jeopardizing the specificity of the interactions. Thus, these techniques may be helpful in identifying chaperone–protein interactions. In turn, it should be possible to exploit such findings to maximize the efficiency of heterologous protein production in transgenic plants.

Acknowledgments

We thank Fan Zhang for invaluable photographic assistance and the members of the Boston lab group for helpful comments and discussion. We also thank Dr. Maurilio A. Moreira, who kindly provided the anti-ß-conglycinin antiserum, Dr. Ralph Dewey in whose lab the soybean work was initiated, and Dr. Alessandro Vitale for communicating results prior to publication. This work was supported by Grant 91-37304-5842 from the U.S. Department of Agriculture (R.S.B) the North Carolina Agricultural Research Service (R.S.B.), and Grant 64.92.0005.00 from the Brazilian government (PADCT/FINEP) (E.P.B.F.). J.W.G. is the recipient of a postdoctoral fellowship from the North Carolina Biotechnology Center.

References

Abe, S., You, W., and Davies, E. (1991). Protein bodies in corn endosperm are enclosed by and enmeshed in F-actin. *Protoplasma* **165,** 139–149.

Bednarek, S. Y., and Raikhel, N. V. (1992). Intracellular trafficking of secretory proteins. *Plant Mol. Biol.* **20,** 133–150.

Bjerrum, O. J., and Schafer-Nielsen, C. (1986). Buffer systems and transfer parameters for semidry electroblotting with a horizontal apparatus. *In* "Electrophoresis '86" (M. J. Dunn, ed.), pp. 315–327. Deerfield Beach: VCH Publishers.

Bole, D. G., Hendershot, L. M., and Kearney, J. F. (1986). Posttranslational association of immunoglobulin heavy chain binding protein with nascent heavy chains in nonsecreting and secreting hybridomas. *J. Cell Biol.* **102,** 1558–1566.

Chrispeels, M. J. (1991). Sorting of proteins in the secretory system. *Annu. Rev. Plant Physiol. Plant Mol. Biol.* **42,** 21–53.

D'Amico, L., Valsasina, B., Daminati, M. G., Fabbrini, M. S., Nitti, G., Bollini, R., Ceriotti A., and Vitale, A. (1992). Bean homologs of the mammalian glucose-regulated proteins: Induction by tunicamycin and interaction with newly synthesized seed storage proteins in the endoplasmic reticulum. *Plant J.* **2,** 443–455.

Ellis, J. (1987). Proteins as molecular chaperones. *Nature (London)* **328,** 378–379.

Fontes, E. B. P., Shank, B. B., Wrobel, R. L., Moose, S. P., OBrian, G. R., Wurtzel, E. T., and Boston, R. S. (1991). Characterization of an immunoglobulin binding protein homolog in the maize *floury*-2 endosperm mutant. *Plant Cell* **3,** 483–496.

Freiden, P. J., Gaut, J. R., and Hendershot, L. M. (1992). Interconversion of three differentially modified and assembled forms of BiP. *EMBO J.* **11,** 63–70.

Gething, M. J., and Sambrook, J. (1992). Protein folding in the cell. *Nature (London)* **355,** 33–44.

Harlow, E., and Lane, D., eds. (1988). "Antibodies: A Laboratory Manual." Cold Spring Harbor, NY: Cold Spring Harbor Laboratory.

Laemmli, U. K. (1970). Cleavage of structural proteins during the assembly of the head of bacteriophage T4. *Nature (London)* **227,** 680–685.

Larkins, B. A., and Hurkman, W. J. (1978). Synthesis and deposition of zein in protein bodies of maize endosperm. *Plant Physiol.* **62,** 256–263.

Leal, M. C. (1991). Quantificacao da gloculina 11S e 7S e adsorcao de hexANAL pelas proteinas de soja. Tese de MS, UFV, Viçosa, Brasil. 60 p.

Li, X., Wu, Y., Zhang, D. Z., Gillikin, J. W., Boston, R. S., Franceschi, V. R., and Okita, T. W. (1993). Rice prolamine protein body biogenesis: A BiP-mediated process. *Science* **262,** 1054–1056.

Miernyk, J. A., Duck, N. B., Shatters, R. G., Jr., and Folk, W. R. (1992). The 70-kilodalton heat shock cognate can act as a molecular chaperone during the membrane translocation of a plant secretory protein precursor. *Plant Cell* **4,** 821–829.

Müsch, A., Wiedmann, M., and Rapoport, T. A. (1992). Yeast Sec proteins interact with polypeptides traversing the endoplasmic reticulum membrane. *Cell* **69,** 343–352.

Palleros, D. R., Reid, K. L., Shi, L., Welch, W. J., and Fink, A. L. (1993). ATP-induced protein-hsp70 complex dissociation requires K⁺ but not ATP hydrolysis. *Nature (London)* **365,** 664–666.

Pedrazzini, E., Giovinazzo, G., Bollini, R., Ceriotti, A., and Vitale, A. (1994). Binding of BiP to assembly-defective protein in plant cells. *Plant J.* **5,** 103–110.

Salamini, F., DiFonzo, N., Gentinetta, E., and Soave, C. (1979). A dominant mutation interfering with protein accumulation in maize seeds. *In* "Seed Protein Improvement in Cereals and Grain Legumes." Proceedings of Food and Agriculture Organization. International Atomic Energy Agency Symposium, pp. 97–108. International Atomic Energy Agency; Vienna, Austria.

Sanders, S. L., Whitfield, K. M., Vogel, J. P., Rose, M. D., and Schekman, R. W. (1992). Sec61p and BiP directly facilitate polypeptide translocation into the ER. *Cell* **69,** 353–365.

Vitale, A., Ceriotti, A., and Denecke, J. (1993). The role of the endoplasmic reticulum in protein synthesis, modification and intracellular transport. *J. Exp. Bot.* **44,** 1417–1444.

Vitale, A., Bielli, A., and Ceriotti, A. (1995). The binding protein associates with monomeric phaseolin. *Plant Physiol.* **107,** 1411–1418.

Walling, L., Drew, G. N., and Goldberg, R. B. (1986). Transcriptional and post-transcriptional regulation of soybean seed protein mRNA levels. *Proc. Natl. Acad. Sci. U.S.A.* **83,** 2123–2127.

CHAPTER 23

Assaying Proteins for Molecular Chaperone Activity

Garrett J. Lee

Department of Biochemistry
University of Arizona
Tucson, Arizona 85721

I. Introduction

Molecular chaperones play critical roles in protein folding, assembly, and translocation (Gething and Sambrook, 1992). For example, chaperone-mediated protein folding is necessary during and after *de novo* protein synthesis or as a repair mechanism following protein denaturation. Additionally, molecular chaperones maintain other proteins in translocationally competent, unfolded conformations and also function during subsequent folding. *In vitro*, many experiments have confirmed Anfinsen's theory (Anfinsen *et al.*, 1961) that the primary sequence of a protein contains the information required to direct proper protein folding. However, the folding of purified proteins *in vitro* rarely occurs at the efficiencies observed *in vivo*. Recent evidence has suggested that molecular chaperones are probably the cellular factors responsible for this discrepancy (Hartl *et al.*, 1994). The cellular requirement for chaperones is thought to be due to the fact that adverse conditions can exist in cells that jeopardize productive protein folding. These conditions include the high cellular concentration of folding intermediates that are prone to aggregation or misfolding, or can result from the presence of environmental stresses such as heat shock, which promote protein unfolding.

Several properties define molecular chaperone activity. These are the ability (1) to recognize and bind unfolded proteins; (2) to suppress aggregation during protein unfolding and folding; (3) to influence the yield of folding; and (4) to carry out properties 2 and 3 at approximately stoichiometric levels (Jakob and Buchner, 1994). In almost every case reported, molecular chaperones increase folding yields of other proteins by suppressing their aggregation rather than accelerating the folding reaction. Thus, molecular chaperones are not likely to alter the mechanism by which their substrates fold, but instead, allow spontaneous folding to occur while reducing the flux through nonproductive pathways. It is believed that molecular chaperones accomplish this task by masking interactive surfaces that are exposed in unfolded proteins.

The vigorous study of molecular chaperones began in plants with the indication that the chloroplast-localized rubisco subunit binding protein (also termed Cpn60[1]), an HSP60 homolog related to the bacterial GroEL protein, might be essential for proper folding and assembly of rubisco (Barraclough and Ellis, 1980). Since that time, several highly conserved classes of molecular chaperones, most of them HSPs and constitutively expressed HSP homologs, have been identified in many divergent organisms. Such proteins with demonstrated *in vitro* molecular chaperone activity include the HSP90, HSP70, HSP60, TCP-1, and small HSP classes of proteins (Hendrick and Hartl, 1993; Jakob and Buchner, 1994). Of these five classes of proteins, the HSP70, HSP60, and TCP-1 classes have been shown to require ATP for the release of substrates stably bound to the chaperones (Hendrick and Hartl, 1993).

HSP90, HSP70, HSP60, TCP-1, and small HSP homologs have all been identified in plants, and in many cases, within multiple cellular compartments (Vierling, 1991; Mummert *et al.,* 1993). However, with the exception of the chloroplast-localized Cpn60, plant homologs of the above five classes of proteins have not been thoroughly characterized with regard to chaperone activity. Research in our laboratory has focused on the function of plant small HSPs, which appear to posses chaperone activity *in vitro* (Figs. 1–3) (Lee *et al.,* 1995). Still, the majority of chaperone research has focused on proteins from bacterial, yeast, or mammalian cells.

This chapter describes three general purpose assays for *in vitro* assessment of molecular chaperone activity utilizing citrate synthase (CS) as the protein substrate. CS has been widely used as a substrate because nonnative forms of the enzyme are recognized by a number of divergent chaperones (Buchner *et al.,*

[1] Abbreviations: Cpn, chaperonin; acetyl-CoA, acetyl coenzyme A; CS, citrate synthase; Hepes, 4-(2-hydroxyethyl)-1-piperazine ethanesulfonic acid; HSP, heat shock protein; Tris, tris(hydroxymethyl)-amino-methane.

1991; Wiech *et al.*, 1992; Zhi *et al.*, 1992; Jakob *et al.*, 1993). These assays test the ability of a potential chaperone to refold chemically denatured CS, to prevent thermal inactivation of CS, and to prevent thermal aggregation (insolubilization) of CS. Other enzymes such as dimeric rubisco (Goloubinoff *et al.*, 1989), firefly luciferase (Frydman *et al.*, 1992; Schröeder *et al.*, 1993), porcine malate dehydrogenase (Hartman *et al.*, 1993), and bovine muscle lactate dehydrogenase (Lee *et al.*, 1995) have proven to be useful substrates and can also be evaluated using the guidelines presented here. All these enzymes have simple and sensitive activity assays and are good chaperone substrates due to the tendency of their nonnative states to aggregate in the absence of chaperones.

II. Materials and Equipment

Pig heart citrate synthase, approximately 8 mg/ml, supplied as an ammonium sulfate suspension (Sigma C-3260)

Tris

Hepes

Guanidine hydrochloride, ultrapure

Acetyl CoA, sodium salt (Sigma A-2056)

Oxalacetic acid (Sigma O-4126)

Spectrophotometer

Quartz cuvettes

Water bath

III. General Considerations

The following assays utilize citrate synthase, a homodimer of 50-kDa subunits, at a concentration of 150 nM monomers. This concentration maximizes assembly from the unfolded state but minimizes aggregation; both processes are competitive with respect to one another and are highly dependent on the concentration of unfolded molecules (Buchner *et al.*, 1991). Since chaperones often display maximal activity at a chaperone protomer-to-substrate ratio of 1:1, the potential chaperone should be initially evaluated at approximately 150 nM.

Because of the abundance of known molecular chaperones in most cells, a reasonable enrichment of the protein to be tested should be sought. If chaperone activity of the protein preparation is not initially detected, other factors that may stimulate chaperone activity should be considered. These may include other protein components, nucleotide triphosphates, or cations. For example, Cpn60 requires the cochaperone Cpn10, MgATP, and potassium ions for proper folding of substrates (Viitanen *et al.*, 1990). Substrate specificity should also be considered, and lead the investigator to evaluate potentially more physiologically relevant substrates. Once chaperone activity is detected, an equivalent weight of a negative control protein should be substituted for the chaperone in order to verify that the observed effects are specific for the protein being tested. Lysozyme, bovine serum albumin, and IgG have also been widely used. Preferably, the control protein should have a molecular weight and isoelectric point similar to those of the chaperone being evaluated.

A. Assay 1: Refolding of Chemically Denatured Citrate Synthase

1. Principle

In this assay, CS is dissociated and unfolded in 6 M guanidine hydrochloride. Dithiothreitol is included during the denaturation step to reduce the disulfide bonds present in the native protein. Denaturant is rapidly removed from CS by 100-fold dilution into refolding buffer supplemented with chaperone. For comparison, the refolding reaction is carried out in the absence of chaperone. Refolding is allowed to proceed for defined periods of time, and aliquots are removed and assayed for CS enzymatic activity. The reappearance of CS activity is indicative of CS reaching its native conformation. Reactivation yields in the presence of the potential chaperone are then compared to unassisted yields. Unassisted reactivation is low and is typically less than 20% of the activity of an equivalent amount of nondenatured CS. An example of CS refolding is shown in Fig. 1, in which PsHSP18.1, a small HSP from pea (DeRocher *et al.,* 1991), enhances CS folding.

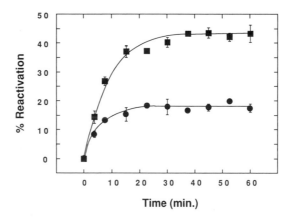

Fig. 1 PsHSP18.1 (dodecamer of 18.1-kDa subunits) enhances the refolding of chemically denatured CS. CS monomers (15 μM) were denatured in 6 M guanidine hydrochloride, then diluted 100-fold into solutions lacking smHSPs (•), or supplemented with 150 nM PsHSP18.1 (■). At the times indicated, CS activity was determined. Reactivation is expressed as a percentage relative to the activity of an equivalent amount of nondenatured CS. Substitution of PsHSP18.1 with 32 μg/ml catalase (equivalent weight of 150 nM HSP18.1) resulted in 16% CS reactivation after 60 min (not shown), indicating that the enhancement was specific to PsHSP18.1.

2. Procedure

1. Determine the concentration of the commercial CS stock solution by measuring A_{280} of an appropriately diluted sample. For CS, $A_{280}(0.1\%) = 1.78$ for a 1-cm path length (Singh et al., 1970).

2. Denature CS in 6 M guanidine hydrochloride in a 0.5-ml microfuge tube by combining the following in the following order. Vortex after each addition and allow to stand at least 1.5 h at 25°C to denature CS.

57.4 mg	guanidine hydrochloride (will displace 48 μl of solution when dissolved)
variable H$_2$O	(add an appropriate volume to bring the final volume to 100 μl)
10 μl	1 M Tris–HCl, pH 8
4 μl	50 mM dithiothreitol
75 μg CS	(volume will vary depending on concentration of CS stock solution)

Total volume 100 μl

3. Renature CS at 25°C in 1.5-ml microfuge tubes by combining the following in the following order.

125.0 μl	200 mM Tris–HCl, pH 8
122.5 μl	H$_2$O and chaperone buffer only, or appropriate volume of H$_2$O plus potential chaperone or control protein
2.5 μl	denatured CS (from step 2 above)

Total volume 250 μl

3. Important Considerations

Consistent pipetting and addition of denatured CS are critical. To avoid high local concentrations of unfolded CS, denatured CS should be pipetted directly into rapidly vortexing solution. Continue vortexing 3 s after addition. For a zero time point for refolding, immediately withdraw 20 μl of the refolding solution after addition of CS, and assay for CS activity as described below. No CS activity should be present. Thereafter, assay 20-μl aliquots for CS activity over time for up to 60 min.

The optimal substrate to chaperone stoichiometry should also be investigated by holding the substrate concentration and refolding time constant, but varying the amount of added chaperone.

4. CS Activity Assay

This assay measures CS utilization of acetyl CoA by monitoring the breakage of the thioester bond of acetyl CoA, which absorbs at 233 nm (Ochoa, 1957). Prepare a solution of 425 μM oxalacetic acid in 50 mM Tris–HCl, pH 8. Zero a spectrophotometer set at 233 nm against this solution. In a quartz cuvette, combine 470 μl of the oxalacetic acid/Tris–HCl solution with 10 μl of a 7.5 mM acetyl-CoA solution. Initiate the reaction by adding a 20-μl aliquot withdrawn from the refolding reaction. Vortex and measure A_{233} every 5 s for 1 min. Plot A_{233} versus time and determine the negative slope of the linear portion of the plot, which is proportional to the initial rate of the reaction.

For comparison, assay the activity of an equivalent amount of CS not treated with guanidine hydrochloride and express reactivated CS activities as percentages relative to this activity.

B. Assay 2: Protection against Citrate Synthase Thermal Inactivation

1. Principle

Since many HSPs have been shown to possess molecular chaperone activity, this and the following assay evaluate chaperone activity at elevated temperatures. In this assay, CS is incubated in the presence or absence of the potential chaperone at 38°C. This temperature corresponds to a moderate heat stress temperature for plants. After the high temperature incubation, samples are moved to 22°C, a temperature at which CS can refold if protected by a chaperone during the high-temperature incubation. Throughout the high- and low-temperature incubations, aliquots are removed and measured for residual CS activity. Typically after 60 min at 38°C in the absence of chaperone, CS loses almost all activity in an irreversible manner. In contrast, protection by a chaperone allows CS to refold at the permissive temperature of 22°C. An example is shown in Fig. 2, in which PsHSP18.1 facilitates CS refolding.

2. Procedure

In a 13 × 100 mm borosilicate glass tube, combine the following at room temperature:

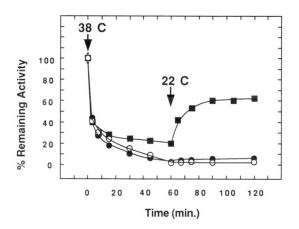

Fig. 2 PsHSP18.1 prevents irreversible thermal inactivation of CS at 38°C. CS monomers (150 nM) were incubated at 38°C in the absence of (•) or presence of 32 μg/ml catalase (○), or 150 nM HSP18.1 (■). Where indicated, samples were shifted to 22°C. CS enzymatic activity was determined at the times indicated.

250 μl   100 mM Hepes–KOH, pH 8

3.75 μg   CS (volume will vary depending on concentration of CS stock solution)

Adjust the volume to 500 μl with H_2O and chaperone buffer only, or H_2O plus potential chaperone or control protein

Prior to heating at 38°C, remove a 20-μl aliquot for the zero time point and assay CS activity at room temperature as described above. Cover the sample with parafilm and place in a 38°C water bath. Over time, for up to 60 min, remove 20-μl aliquots and measure CS activity. After removing the aliquot at the 60-min time point, place the sample in a 22°C bath and continue periodic measurement of CS activity for an additional 60 min.

Express remaining CS activity as a percentage relative to the initial activity present at time zero.

C. Assay 3: Protection against Citrate Synthase Thermal Aggregation

1. Principle

Light scattering due to protein aggregation (insolubilization) can be detected by spectrophotometric measurement of the apparent absorbance of a solution at 320 nm (Jaenicke and Rudolph, 1989). In the case of CS, detectable aggregation begins at a temperature of 45°C, but may be suppressed by the presence of a chaperone. For example, the presence of increasing amounts of PsHSP18.1 suppresses CS aggregation as seen in Fig. 3.

2. Procedure

In a spectrophotometric cuvette, combine the following at room temperature, mix, and cover:

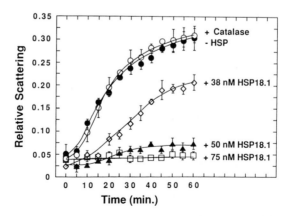

Fig. 3 PsHSP18.1 protects CS from thermal aggregation. CS monomers (150 nM) were incubated at 45°C in the absence or presence of increasing amounts of HSP18.1 as shown. Where indicated, catalase was added at a concentration of 32 μg/ml in the absence of PsHSP18.1. At the times indicated, samples were monitored for their apparent absorbance at 320 nm, which is indicative of light scattering due to CS aggregation. Relative scattering is expressed in arbitrary units.

500 μl   100 mM Hepes–KOH, pH 7.5
7.5 μg   CS (volume will vary depending on concentration of CS stock solution)

Adjust volume to 1 ml with H$_2$O and chaperone buffer only, or H$_2$O plus potential chaperone or control protein.

Zero a spectrophotometer set at 320 nm against a blank containing 50 mM Hepes–KOH, pH 7.5, then measure A_{320} of the sample for the zero time point. Place the blank and the sample in a 45°C water bath. Every 5 min, remove the blank from the water bath and zero the spectrophotometer, then measure A_{320} of the sample. Place the blank and the sample back in the water bath immediately after reading. Plot relative scattering (arbitrary units proportional to apparent absorbance) versus time.

General comments: The potential chaperone should be evaluated alone for light scattering over the assay period to verify that the protein itself does not become insoluble at 45°C. Use of a water bath can be eliminated if the spectrophotometer is equipped with a temperature control unit.

Although the chaperone being tested may facilitate reactivation of CS at 22°C following incubation at 38°C (Assay 2), reactivation may not be observed after initial incubation at a higher temperature such as 45°C. For example, reactivation of CS occurs in the presence of PsHSP18.1 following 38°C treatment (Fig. 2). In contrast, after 60 min at 45°C, PsHSP18.1 cannot facilitate CS reactivation after a similar shift to 22°C (not shown). These results suggest that higher temperatures can irreversibly inactivate CS even in the presence of a chaperone that prevents CS insolubilization.

It should be noted that some molecular chaperones possess disaggregating capabilities. For example, DnaK and GroEL, the *Escherichia coli* HSP70 and HSP60 homologs, respectively, have been shown to disaggregate and subsequently reactivate heat-induced RNA polymerase aggregates (Ziemienowicz *et al.,* 1993). For a given experiment, the inactivation profile will vary depending on the protein substrate being used, the incubation temperature, and the functional properties of the chaperone. The alteration of such variables may be useful for further chaperone characterization.

VII. Conclusions

Because of the fundamental nature of protein folding, molecular chaperones have emerged as a class of proteins that have a profound impact on many cellular processes. The study of chaperones has revolutionized the understanding of protein folding pathways in cells. In recent years the number of known molecular chaperones has grown considerably and new chaperones will certainly continue to be identified. Several assays that test for the properties most consistent with molecular chaperone activity have been presented here. Although the protein being tested may demonstrate more specific protein–protein interactions *in vivo,* the methods presented here should be useful as an initial diagnostic tool that may later be modified accordingly.

Acknowledgment

I thank Dr. Elizabeth Vierling for valuable discussions and in whose lab these methods were developed and used to characterize small HSPs from pea.

References

Anfinsen, C. B., Haber, E., Sela, M., and White, F. H. (1961). The kinetics of formation of native ribonuclease during oxidation of the reduced polypeptide chain. *Proc. Natl. Acad. Sci. U.S.A.* **47,** 1309–1314.

Barraclough, R., and Ellis, R. J. (1980). Protein synthesis in chloroplasts. IV: Assembly of newly-synthesized large subunits of ribulose bisphosphate carboxylase in isolated pea chloroplasts. *Biochim. Biophys. Acta* **608,** 19–31.

Buchner, J. Schmidt, M., Fuchs, M., Jaenicke, R., Rudolph, R., Schmid, F. X., and Kiefhaber, T. (1991). GroE facilitates refolding of citrate synthase by suppressing aggregation. *Biochemistry* **30,** 1586–1591.

DeRocher, A. E., Helm, K. W., Lauzon, L. M., and Vierling, E. (1991). Expression of a conserved family of cytoplasmic low molecular weight heat shock proteins during heat stress and recovery. *Plant Physiol.* **96,** 1038–1047.

Gething, M.-J., and Sambrook, J. (1992). Protein folding in the cell. *Nature* **355,** 33–45.

Goloubinoff, P., Christeller, J. T., Gatenby, A. A., and Lorimer, G. H. (1989). Reconstitution of active ribulose bisphosphate carboxylase from an unfolded state depends on two chaperonin proteins and Mg-ATP. *Nature* **342,** 884–889.

Hartl, F.-U., Hlodan, R., and Langer, T. (1994). Molecular chaperones in protein folding: The art of avoiding sticky situations. *Trends Biochem. Sci.* **19,** 20–25.

Hartman, D. J., Surin, B. P., Dixon, N. E., Hoogenraad, N. J., and Høj, P. B. (1993). Substoichiometric amounts of the molecular chaperones GroEL and GroES prevent thermal denaturation and aggregation of mammalian mitochondrial malate dehydrogenase *in vitro. Proc. Natl. Acad. Sci. U.S.A.* **90,** 2276–2280.

Hendrick, J. R., and Hartl, F.-U. (1993). Molecular chaperone functions of heat-shock proteins. *Annu. Rev. Biochem.* **62,** 349–384.

Jaenicke, R., and Rudolph, R. (1989). Folding proteins. *In* "Protein Structure. A Practical Approach" (T. Creighton, ed.), pp. 191–222. Oxford, England: IRL Press.

Jakob, U., Gaestel, M., Engel, K., and Buchner, J. (1993). Small heat shock proteins are molecular chaperones. *J. Biol. Chem.* **268,** 1517–1520.

Jakob, U., and Buchner, J. (1994). Assisting spontaneity: The role of Hsp90 and small Hsps as molecular chaperones. *Trends Biochem. Sci.* **19,** 205–211.

Lee, G. J., Pokala, N., and Vierling, E. (1995). Structure and in vitro molecular chaperone activity of small heat shock proteins from pea. *J. Biol. Chem.,* in press.

Mummert, E., Grimm, R., Speth, V., Eckerskorn, C., Schlitz, E., Gatenby, A. A., and Schäfer, E. (1993). A TCP1-related molecular chaperone from plants refolds phytochrome to its photoreversible form. *Nature* **363,** 644–648.

Ochoa, S. (1957). Crystallin condensing enzyme. *Biochem. Prep.* **5,** 19–30.

Schröeder, H., Langer, T., Hartl, F.-U., and Bukau, B. (1993). DnaK, DnaJ and GrpE form a cellular chaperone machinery capable of repairing heat-induced protein damage. *EMBO J.* **12,** 4137–4144.

Singh, M., Brooks, G. C., and Srere, P. A. (1970). Subunit structure and chemical characteristics of pig heart citrate synthase. *J. Biol. Chem.* **245,** 4636–4637.

Vierling, E. (1991). The roles of heat shock proteins in plants. *Annu. Rev. Plant Physiol. Plant Mol. Biol.* **42,** 579–620.

Viitanen, P. V., Lubben, T. H., Reed, J., Goloubinoff, P., O'Keefe, D. P., and Lorimer, G. H. (1990). Chaperonin-facilitated refolding of ribulose bisphosphate carboxylase and ATP hydrolysis by chaperonin 60 (GroEL) are K$^+$ dependent. *Biochemistry* **29,** 5665–5671.

Wiech, H., Buchner, J., Zimmermann, R., and Jakob, U. (1992). Hsp90 chaperones protein folding *in vitro. Nature* **358,** 169–170.

Zhi, W., Landry, S. J., Gierasch, L. M., and Srere, P. A. (1992). Renaturation of citrate synthase: Influence of denaturant and folding assistants. *Protein Sci.* **1,** 522–529.

Ziemienowicz, A., Skowyra, K., Zeilstra,-Ryalls, J., Fayet, O, Georgeopoulos, C., and Zylicz, M. (1993). Both the *Escherichia coli* chaperone systems, GroEL/GroES and DnaK/DnaJ/GrpE can reactivate heat-treated RNA polymerase. *J. Biol. Chem.* **268,** 25425–25431.

CHAPTER 24

The Use of Protoplasts to Study Protein Synthesis and Transport by the Plant Endomembrane System

Jürgen Denecke[*] and Alessandro Vitale[†]

[*]Department of Molecular Genetics, Uppsala Genetic Center
Swedish University of Agricultural Sciences
Box 7010
S-75007 Uppsala, Sweden
[†]Istituto Biosintesi Vegetali
Consiglio Nazionale delle Ricerche
via Bassini 15
20133 Milano, Italy

I. Introduction

The endomembrane system comprises a series of compartments, the major of which are the endoplasmic reticulum (ER), the Golgi complex, and the vacuole.

These structures are interconnected by vesicular traffic. This network of compartments is involved in the synthesis of a large number of molecules. Some of these fulfill a structural role, such as within the primary cell wall, whereas others have more specialized rôles. Well-studied examples are the secreted hydrolases that are produced by the aleurone layer in germinating seeds, or the storage proteins accumulated in vacuoles during seed development, or defense-related proteins produced in response to pathogens and directed either to the vacuole or to the cell surface. Protein synthesis by the endomembrane system occurs on membrane-bound polysomes and translocation of the nascent polypeptides generally occurs cotranslationally. From the ER, vesicular traffic transports proteins to the Golgi complex, and from here to the vacuole or the apoplast. Polypeptide folding and modifications such as glycosylation, hydroxylation, formation of disulfide bridges, oligomerization, and proteolytic cleavage occur either co- or post-translationally in specific compartments along the transport pathway. Hence, the endomembrane system must be considered a complex machinery devoted to the synthesis, modification, assembly, and transport of proteins to various locations within or outside the cell. This machinery includes chaperones that are involved in the translocation, folding, and assembly of proteins, enzymes with protein modifying functions, and receptors that direct the traffic of molecules through the endomembrane system. The functional analysis of this machinery and the identification of signals in the structure of transported proteins have become major research topics in cell biology. Such studies have taken great advantage from the development of techniques for the transient expression of recombinant genes in plant protoplasts. The main contribution of this methodology is that processes can be studied *in vivo* without losing the possibility to manipulate events by the use of drugs, culture conditions, and mutational analysis of proteins.

II. Transient Expression of Genes in Plant Protoplasts

A. Applications and Limitations of Transient Expression Methods Using Protoplasts

Transient expression techniques using protoplasts have been used to study gene expression in monocotyledonous as well as dicotyledonous plant species. Plant protoplasts can be derived from a vast array of tissues, and plasmid-borne genes can be introduced either by exposure to transient electric fields (Fromm *et al.,* 1987) or by treatment with membrane destabilizing agents (Shillito *et al.,* 1985; Werr and Lörz, 1986). Gene products are detected rapidly, typically from hours up to several days after DNA transfer. Since the majority of the introduced DNA remains extrachromosomal, gene expression is transient in nature.

For the identification of regulatory sequences in promoter regions, reporter enzymes such as neomycin phosphotransferase II (NptII, Reiss *et al.,* 1984), chloramphenicol acetyl transferase (CAT, Gorman *et al.,* 1982), phosphinothricin acetyl transferase (PAT, Thompson *et al.,* 1987), β-glucuronidase (GUS, Jefferson *et al.,* 1987, or firefly luciferase (De Wet *et al.,* 1987) have been used as an indirect way to estimate the transcription rate of a given promoter. mRNA levels of transiently introduced genes can also be measured directly. Experiments are fast since plant regeneration is not required, and results are reproducible due to an absence of position effects acting on the extrachromosomal gene. To compare promoters, to perform deletion analysis, and to study regulation in function of hormones and physical parameters such as temperature, internal marker genes on the same plasmid are required to normalize transfection efficiency. The latter is dependent on the topology, the size, and the purity of the introduced plasmid.

In contrast to the advantages offered by protoplasts, there are some drawbacks. In many cases of biological relevance, protoplasts cannot be used to study regulatory regions of promoters. In some cases, for example, the barley aleurone layer, regulatory factors typical for this specific tissue are well represented in protoplasts prepared from the tissue, and regulated gene expression can be demonstrated (Jacobsen and Beach, 1985). *In situ* GUS detection experiments on immobilized protoplasts in some cases indicate that, regardless of the transfection method, a small percentage of the protoplast population can account for the majority of GUS activity of the whole cell suspension, and that GUS activities in these cells in some cases exceed well over an order of magnitude those levels seen in cells of transgenic plants (Denecke *et al.*, 1989). The presence of high copy numbers of a certain gene may also result in titration of transcription factors, and the measured gene activity may therefore not represent the situation *in vivo*.

A number of basic processes are conserved between cells of different tissues of a plant. In these cases, transient expression techniques in plant protoplasts have now become the model system of choice. Processes that can be studied using this methodology include the maturation of transcripts, translation initiation on *in vitro* synthesized transcripts, protein synthesis and protein turnover, post-translational processing, protein folding, -oligomerization, and interactions with chaperones, and also the transport of proteins to different destinations. In terms of the functional analysis of the plant endomembrane system, protoplasts have been used to demonstrate that protein secretion is independent of active cis-acting sorting information and occurs by default (Denecke *et al.*, 1990), to identify and characterize vacuolar transport signals (Dombrowski *et al.*, 1993; Holwerda *et al.*, 1992), and to analyze ER retention signals (Denecke *et al.*, 1992; Denecke *et al.*, 1993). It has also been shown that the high copy number of genes in transfected protoplasts does not saturate the transport machinery in the endomembrane system (Denecke *et al.*, 1990; Denecke *et al.*, 1992), indicating that results obtained with such expression systems have biological relevance. Recently, transient expression has been used to monitor the interaction of a structurally defective protein with the ER chaperone BiP (Pedrazzini *et al.*, 1993).

In the following sections, we provide an overview of methods based on transient expression in plant protoplasts for the study of protein synthesis and transport by the endomembrane system.

B. Electroporation Protocol

To study protein synthesis and transport in plant cells, large numbers of cells are typically used, and this requires a fast and efficient transfection method. We consider electroporation as the method of choice, since it comprises only one step and can be scaled up easily. However, if an electroporation device is not available, protoplast treatments with polyethylene glycol and calcium can be used to transfect DNA. This method is more elaborate and less reproducible, but can yield transfection efficiencies similar to that of electroporation; all down-stream methods, which are described in subsequent sections, are independent of the method of transfection. The given protocol is optimized for tobacco leaf protoplasts, but has also been successfully used with potato and *Arabidopsis* leaf protoplasts.

1. Preparation of Electroporation-Competent Protoplasts

All manipulations are done in a sterile laminar flow bench at approximately 22°C. Leaves of axenically grown tobacco plants are used. These plants are grown on "half MS medium" (50% MS salts, 2% sucrose) in large jars (10-cm-diameter, 10–15 cm high) in 16 h light and 8 h darkness regime to obtain optimal leaf material. Dark green leaves (4–5 cm) of 4- to 6-week-old plants are cut at the lower epidermal surface using a scalpel every 1–2 mm, without cutting through the whole leaf. The leaves are then carefully transferred onto the surface of 7 ml TEX medium (B5 salts, 250 mg/liter NH_4NO_3, 750 mg/liter $CaCl_2$ $2H_2O$, 500 mg/liter MES, 0.4 M sucrose, brought to pH 5.7 with KOH) containing 0.4% cellulase (Onozuka R10) and 0.2% Macerozyme (Onozuka R10), contained in a standard disposable plastic petri dish. The leaves are placed on the liquid surface with the upper side up, and without wetting the upper side, so that the complete surface of the liquid is covered, but without the leaves overlapping. The plates are incubated for 6–16 h at 22°C. The faster the digest is finished, the higher the yield of viable protoplasts with high capacity to synthesize proteins. Each digestion plate yields about 5×10^6 cells. Using this technique, it is routinely possible to prepare 10^8 protoplasts.

The digestion mix is filtered through a sterile 100-μm mesh nylon filter, washed with an equal volume of electroporation buffer (80 mM KCl, 4 mM $CaCl_2$, 2.4 g Hepes, 0.4 M sucrose, brought to pH 7.2 with KOH), and aliquots of 50 ml of digestion filtrate are centrifuged in Falcon tubes for 10 min at 80× g in a swing out rotor at room temperature. Living protoplasts will float, whereas dead cells and debris will stay in suspension or form a pellet. The underlying solution and the pellet are removed using a long sterile Pasteur pipet and a peristaltic pump. When penetrating the floating cell layer with the pipet, it is necessary to form a "window" by pushing the cells from the center to the sides. The cells are resuspended in 25 ml of electroporation buffer and spun again and the underlying solution and the (much smaller) pellet are removed. The washing procedure is repeated twice and the protoplasts are resuspended to obtain $2–5 \times 10^6$ cells/ml. If large numbers of protoplasts are prepared, it is possible to pool the suspensions of two Falcon tubes after the first washing step to reduce the number of tubes to be handled. Incubation of electroporation-competent protoplasts in electroporation buffer for more than 1 h prior to electroporation reduces the viability of the cells and should be avoided. We recommend electroporation only of freshly prepared protoplasts.

2. Electroporation Procedure

Samples of 600 μl of protoplast suspension are pipetted in sterile, 1.5-ml plastic cuvettes. Ten to fifty micrograms of DNA in 100 μl of electroporation buffer is added to the protoplast suspension and mixed by gentle shaking of the cuvette. Within 5 min, the sample is electroporated. We use a homemade electroporation device connected to an ordinary power supply, in combination with a homemade stainless-steel plate electrode that fits into 1.5-ml cuvettes (i.e., Kartel 1938 P5 microcuvettes). The internal electrode distance is 3 mm, the range of capacitance is from 100 to 1000 μF, and the maximal voltage to be applied from the power supply is 300 V. For a volume of 700 μl using the above-described electroporation buffer, the optimal conditions for electroporation are 540 μF/200 volts, so the initial electric field during the pulse is 660 V/cm. Using this setup, the optimal capacitance is directly proportional to the volume of cell suspensions. The discharge is done over a period of 3 s (to guarantee total discharge), and recharging of the capacitor is done at least 3 s before a new electroporation can be carried out. Electroporation is done at room temperature (22°C). Between two electroporations with different plasmids, the electrode is rinsed in distilled water to remove the sucrose medium and the DNA, dipped in 90% ethanol and briefly flamed, and dipped in electroporation buffer (to quickly cool down the electrode). There are a number of commercially available electroporation devices that can be used for this purpose. If plate electrodes that have a 3-mm separation are used, then it is possible to use the same electroporation conditions as described above. Between electroporations with identical DNAs, it is sufficient to dip the electrode in electroporation buffer to remove membranes of dead cells that adhere to the electrode surface. After electroporation, suspensions are left in the cuvettes on the bench for 15 min without shaking. The cells are diluted 10 times with TEX medium and incubated in Petri dishes at the appropriate temperature in darkness. Gene products can be detected within 30 min and up to several days after electroporation. With this method it is possible to generate a large number of cells that will transiently synthesize a gene product. Pools from different cuvettes can be made after electroporation, and the homogeneous suspension can be divided into identical portions to be incubated under different conditions, for different times, in the presence or absence of drugs, etc.

III. Evaluation of *in Vivo* Protein Folding in the ER by Monitoring Interaction with BiP

A. Theoretical Considerations

For proteins synthesized by the rough ER, translation, translocation, and folding occur simultaneously, and these processes are catalyzed by a number of proteins located in the cytosol, the ER membrane, and the ER lumen. The lumenal binding protein (BiP) is one of the most abundant component residents of the ER and is involved as a chaperone in the processes of translocation, folding, and assembly of newly synthesized secretory proteins in the ER (Gething and Sambrook, 1992). *In vitro,* BiP binds peptides containing a high proportion of amino acids with aliphatic side chains. This observation suggests that BiP might bind to such hydrophobic regions during the process of folding and assembly of newly synthesized polypeptides *in vivo,* thus preventing permanent misfolding and irreversible protein aggregation. This model for the action of BiP in the ER implies that most, if not all, polypeptides that fold in the ER lumen will interact with BiP initially. The stability of this interaction is protein-specific and most probably depends on how fast and efficiently the individual protein acquires its correct conformation. Once this has been achieved, sequences that have affinity to BiP are not likely to be exposed on the protein surface, and therefore their possible interaction with BiP is abolished.

Monitoring the *in vivo* interaction of BiP with recombinant proteins synthesized by the ER provides a simple and efficient assay to evaluate whether a protein acquires the correct conformation in order to be transported out of the ER. This can therefore be useful as a quality control step in the production of heterologous proteins in plants, as well as in fundamental research on the *in vivo* activity of BiP. Moreover, this assay can be helpful for the analysis of protein sorting signals. One powerful approach to study sorting signals is the production of mutant proteins followed by the study of their *in vivo* destiny. However, the effects of mutations are not always predictable, and can result in the production of polypeptides that are highly unstable or that are incompetent for intracellular transport out of the ER (Hoffman *et al.,* 1988; Holwerda *et al.,* 1992; Saalbach *et al.,* 1991). Studies on mammalian cells have indicated that this abnormal behavior can be the consequence of an inability to acquire proper folding or to assemble correctly (Hurtley and Helenius, 1989). Such defective polypeptides are often found in stable association with BiP before their degradation occurs (Accili *et al.,* 1992; Hurtley *et al.,* 1989; Knittler and Haas, 1992; Suzuki *et al.,* 1991). It has been shown recently that a mutated, assembly-defective, and most probably transport-incompetent, plant storage protein is stably associated with BiP in transiently transformed tobacco protoplasts (Pedrazzini *et al.,* 1993). Therefore, the stable binding of a recombinant, mutated protein to BiP can be considered an indication that the mutant is severely affected in its conformation. In this case, such a defect, rather than a mutation in a protein sorting determinant, might explain its inability to reach the final destination.

B. Criteria for BiP Binding to Malfolded Proteins

Statements about specific interaction between BiP and a given secretory protein and conclusions about structural defects of a protein must consider the following criteria. (*a*) Both proteins must be coselected using antiserum raised against one or other of the proteins. (*b*) *In vitro* treatment with ATP must result in the release of the protein that has been coselected. The latter is based on observations indicating that the release of BiP from the bound ligand requires ATP (Munro and Pelham, 1986). (*c*) The proportion of the mutant protein bound to BiP and the *in vivo* stability of this complex must be significantly higher than that for the wild-type form of the same protein (Accili *et al.*, 1992; Machamer *et al.*, 1990; Pedrazzini *et al.*, 1993). These criteria can be easily tested in a transient expression system based on tobacco protoplasts. Antiserum either against the recombinant protein or against BiP can be used. In the first approach, immunoprecipitation will recover 100% of the foreign protein, but only the portion of total cellular BiP that is bound to the produced protein. The opposite will occur in the second approach. To compare the interaction of wild-type and mutated forms of a protein with BiP, we suggest using the first approach. If a polypeptide of 78 kDa is coselected, then ATP-mediated release should be performed, followed by immunoprecipitation of the released material with anti-BiP serum (Denecke *et al.*, 1991), to confirm the identity of the 78-kDa polypeptide. The opposite approach, to select for BiP first and then to monitor the coprecipitation of the recombinant protein, does not always provide clear-cut data. BiP is promiscuous in its binding activity, and therefore a variety of endogenous newly synthesized proteins are likely to be transiently associated with it. The recombinant foreign protein might not be immediately and easily identifiable in the pattern of the BiP-associated polypeptides, which will depend on the length of the labeling period and on metabolic conditions of the protoplasts. However, the approach gives information on the percentage of the recombinant protein bound to BiP, and may be undertaken after information from the first approach has been obtained.

The protocol given below is based on results obtained with a mutant, assembly-defective form of phaseolin transiently synthesized in tobacco protoplasts (Pedrazzini *et al.*, 1993). The number of protoplasts and labeling times will have to be optimized for each experiment, but the figures can serve as a starting point.

C. A Transient BiP-Binding Assay in Tobacco Protoplasts

1. Protoplast Labeling and Homogenization

Protoplasts (10^5) are labeled for 3–5 h with Trans^{35}S-label (4.44 MBq/ml, ICN Biomedicals) and/or L-[4,5^{3}H]leucine (4.44 MBq/ml, Amersham International) in a total volume of 100 μl TEX medium. The optimal timepoint after gene transfer for the labeling must be empirically determined for each protein. We recommend 10–15 h after gene transfer for proteins such as phaseolin, PAT, or GUS. After *in vivo* labeling of the transfected protoplasts, cells are recovered by flotation at $100\times g$ for 5 min in a 1.5-ml Eppendorf tube and removal of the medium. A compact pellet of intact cells can be obtained by adding 1 ml of 250 mM NaCl and centrifuging at $100\times g$ for 2 min. Removal of the liquid with a fine pasteur pipet connected to a water suction pump will permit washing away the sucrose-containing medium as well as most of the free label. At this stage, the material can be stored at −80°C. Protoplasts are resuspended and lysed in 4 vol of freshly made protoplast homogenization buffer (200 mM Tris–Cl, pH 8, 300 mM NaCl 2% Triton X-100, 1 mM EDTA, and 2 mM PMSF), supplemented with a 1 : 1000 dilution of antiprotease mix (10 TI units/ml aprotinin and 2 mg/ml each of antipain, chymostatin, leupeptin, pepstatin, and bestatin in 50% DMSO, stored in aliquots at −20°C). The homogenate is then centrifuged at $7000\times g$ for 5 min at 4°C in a minifuge and the insoluble pellet is discarded. This protoplast homogenate can be stored at −80°C.

2. Immunoprecipitation

All manipulations are performed on ice or at 4°C, using ice-cold buffers. A fraction of the protoplast homogenate (up to 300 μl) is brought to 1 ml with NET–gel buffer (50 mM Tris–Cl pH 7.5, 150 mM NaCl, 1 mM EDTA, 0.1% Nonidet P-40, 0.02% sodium azide, supplemented with 0.025% gelatin from autoclaved 2% stocks kept at 4°C). The inclusion of gelatin in the buffer is absolutely required to reduce non-specific co-precipitation of contaminants. The sample is centrifuged for 4 min at $12,000\times g$ in a minifuge and the supernatant is transferred into a new Eppendorf tube and is centrifuged again in the same way, to eliminate any precipitating material. The supernatant of the second precipitation is incubated with an appropriate dilution of antiserum for 1–2 h on ice. One hundred to one hundred fifty microliters of a 10% (volume hydrated resin/volume buffer) suspension of protein A–Sepharose in NET buffer (as NET–gel, but lacking gelatin) is added, and the sample is incubated for 1–2 h at 4°C with gentle agitation. The beads are pelleted by centrifugation for 2 min at $12,000\times g$ in a minifuge. The supernatant is discarded and the beads are resuspended in 1 ml of NET–gel buffer and centrifuged again. This washing step is repeated three times. Twenty to thirty microliters of 2× SDS–PAGE loading buffer is added to the beads and the suspension incubated 5 min at 90°C. The sample is centrifuged for 2 min in a minifuge at room temperature, and the supernatant is collected with an Hamilton syringe and loaded on SDS–PAGE. The gel is treated with PPO, dried, and exposed to fluorography.

3. ATP-Dependent Disruption of BiP–Protein Complexes

The *in vitro* release of BiP from the bound protein can be achieved by treating the complex with ATP (Munro and Pelham, 1986). This treatment does not interfere with the binding of the antibody to its antigen and can therefore be performed on immunoprecipitated material. After immunoprecipitation and washing with NET–gel, the beads are resuspended in 0.5–1 ml of freshly made BiP release buffer (20 mM Tris–HCl, pH 7.5, 150 mM NaCl, 0.1% Triton X-100, 6 mM Mg Cl$_2$, 3 mM ATP), incubated on ice for 30 min, and centrifuged for 2 min at 12,000× g in a minifuge at 4°C. The presence of MgCl$_2$ and the absence of EDTA favor the ATP-dependent release of BiP from the bound protein (Suzuki et $al.$, 1991). The BiP-release step is repeated twice before the beads are treated for SDS–PAGE analysis as described above. The washing steps abolish the association between BiP and and its ligands. Therefore, depending on the antiserum used for the immunoprecipitation, BiP or the bound protein(s) will be recovered in the BiP washing buffer and can be reimmunoprecipitated using specific antiserum.

D. Quantitative Considerations

BiP is an abundant protein with a relatively long half-life. When anti-BiP serum is used to immunoprecipitate total BiP from homogenate corresponding to 10^5 tobacco leaf protoplasts that have been pulse-labeled as described above, a 1-day exposure of the gel treated with PPO is sufficient to obtain a clearly detectable 78-kDa band on the fluorogram. The detection limit of the binding assay is determined by the percentage of total radiolabeled BiP, which is associated with the immunoprecipitated protein, at a given time point of labeling. For wild-type proteins the binding appears to be very transient and, for example, we are unable to detect any association between BiP and wild-type phaseolin in protoplasts pulse-labeled for 3 h. However, 5–10% of total, labeled BiP is coselected with the assembly defective phaseolin mutant after 5 h of pulse labeling. Values around or higher than 10% of the total labeled BiP have also been reported as percentages of mammalian BiP coselected with defective proteins (Knittler and Haas, 1992).

IV. Protein Transport through the Endomembrane System

Transport of soluble proteins from the ER via the secretory pathway is believed to occur independently of active sorting information by a process of "bulk flow," whereas sorting signals are required for other locations such as the vacuole or the ER (reviewed by Chrispeels, 1991). However, the efficiency of secretion is highly dependent on the protein, and may be influenced by its interactions with chaperones, by nonspecific interactions with stationary components of the endomembrane system, by protein stability in the endomembrane system, and by secondary modifications. Therefore, it cannot be assumed that a protein will be secreted efficiently simply due to the absence of sorting signals. Transient expression analysis in protoplasts offers a simple and reproducible model system to measure the rate of secretion for a given protein *in vivo* and also to study the effect of genetically engineered modifications on protein transport and stability.

A. Comparison of Secretion Rates

To study secretion quantitatively, plant protoplasts are the ideal model system since the cells do not divide during the course of the experiment. Plant protoplasts will not stick to surfaces as mammalian cells do and can be kept in homogeneous suspensions. The absence of the cell wall allows the cells to secrete proteins efficiently, and it is possible to separate the culture medium without contaminating protoplasts by flotation of the protoplasts and recovery of the medium below the cells. Similarly, protoplasts can be separated from the medium without losing or breaking too many cells. Recovery of the protein in both fractions has to be quantitative and should permit a calculation of intra- and extracellular protein concentrations in the original cell suspension. In transient expression assays, a gene that encodes a cytosolic protein is used to control for cell mortality and leakage of intercellular proteins due to the separation method. This internal marker should be located on the same plasmid as the gene that codes for the secreted protein. The internal marker can be used to set up electroporation conditions to obtain equal electroporation efficiencies. Even though no saturation effects have been seen in such transport experiments (Denecke *et al.,* 1990), it cannot be ruled out that the number of gene products can influence the rate of transport. In analyzing protein sorting signals, saturation of receptors can be a potential problem and normalization of the protein synthesis rate is essential to obtain meaningful results.

The most current method for measuring the rate of protein secretion *in vivo* involves pulse-labeling of the cells, followed by various periods of chase in nonradioactive medium. During this time, the proportions of the specific radiolabeled protein within and outside the cells is determined through immunoprecipitation. This method, which has been used in a large number of different pro- and eukaryotic cell systems, allows a direct estimation of the half-life of a protein within the cell. In plant cells, this method is not always adequate, since *in vivo* labeling times, even using isolated protoplasts, are generally an order of magnitude longer than those for other eukaryotic cells (for example, mammalian cells), exceeding several times the half-life of a rapidly secreted experimental molecule. Moreover, *in vivo* labeling and subsequent handling involve many steps. The procedure needs to be repeated several times including internal standards for immunoprecipitation. Labeled molecules can only be measured by semiquantitative scanning densitometry of autoradiograms including a dilution series of the labeled protein run in the same gel. For many proteins, labeling procedures do not allow sensitive detection of the immunoprecipitated proteins.

An alternative approach is to use transient expression of genes in protoplasts, followed by direct measurement of the proteins by sensitive and quantitative methods. This approach is particularly useful if the proteins can be detected enzymatically, and if enzyme activity reflects protein quantity. The intra- and extracellular concentration of a protein can thus be followed as a function of time, and the profiles of the accumulation curves provide comparative information about the transport rates. Thus, the intracellular level of a rapidly secreted protein will reach an equilibrium between synthesis and export early after electroporation, whereas the opposite is true for a protein whose secretion is slowed due to an ER retention signal. The "secretion index" (Denecke et al., 1990), defined as the ratio between extra- and intracellular protein quantity, is directly proportional to the rate of protein secretion and can be measured over time after electroporation. This allows a comparison of the rate of protein secretion between proteins relative to a standard. However, secretion indices do not provide direct information about intracellular protein half-life. The protocol given below is optimized for tobacco protoplasts, but may be adapted easily for protoplasts of other sources.

Electroporation of competent cells is done as described above. Kinetic measurements require pooling of cells from several electroporations and division into identical aliquots after 10-fold dilution in TEX buffer. The sensitivity of detection can be increased by removing most of the electroporation buffer below the floating cells 30 min after the electroporation. At this time, most of the living cells that have received plasmids will have reached the surface. Thus, cells can be concentrated into a small volume of TEX buffer. Aliquots of 5 ml are incubated in small petri dishes and samples are taken at different time points. Cells are separated from the medium by gently floating the protoplasts at $80\times$ g in a swing-out centrifuge (10 min). The medium below the floating cells is removed manually using an extrafine pasteur pipet. Not more than 2 ml of the medium is recovered and kept on ice. The pellet (of dead cells) and 2 ml of remaining medium are removed and the remaining suspension is resuspended in 15 ml of 250 mM NaCl. In this medium, the protoplasts will sediment. This step will concentrate the protoplasts in a small volume, and at the same time the culture medium will be washed away. Care should be taken to avoid losing viable cells after flotation and during concentration. The culture medium is concentrated 10-fold using centricon 10 membranes (Amicon), with the exact concentration factor being determined by measurement. The cells are usually resuspended in 500 μl of the appropriate protein extraction buffer. Since the cells represent the total population of the 5 ml of cell suspension, the extract is 10-fold concentrated compared to the cell suspension. By analyzing equal volumes of cell extracts and concentrated culture medium, the secretion index can be determined. It is then possible to plot the secretion index as a function of time, thereby providing a comparison of the transport rates of two proteins.

B. Analysis of Protein Sorting Signals

For the functional analysis of protein sorting signals, a model system is required to study the transport of recombinant proteins in cells that do not normally produce the protein of interest. This can be done by introducing chimeric genes into cells from a different plant species or a different tissue, so that the chosen detection method will only show the recombinant protein. Alternatively, epitope tags can be included to study proteins that are ubiquitously present; in this case, it is critical that control experiments demonstrate that the added tag does not interfere with the function or transport of the protein.

The goal of a mutational analysis is the assignment of domains in the primary structure of a protein that participate or are required in the formation of a particular structure that is recognized by a sorting receptor. Unfortunately, the main problem with this approach is that sorting signals are not always specified by conserved short linear sequences, but may comprise higher order structures involving several domains of the protein (Holwerda et al., 1992; Saalbach et al., 1991; von Schaewen et al., 1993). Moreover, mutations might result in misfolding and instability of the protein, and it is important to distinguish such mutations from mutations that only affect the functionality of the sorting signal. However, sorting signals are required to deviate proteins from the default pathway, which leads to secretion in the case of soluble proteins. Measurement of increased secretion rates is therefore an appropriate way to detect alterations in the protein domain that is responsible for the formation of such sorting signals. Misfolding is almost completely excluded in this type of assay, since it typically leads to inefficient secretion, which may not be measurable. Since the intracellular location of the protein under study is known, it is not required to dissect the intracellular location in each assay.

An alternative approach is the introduction of protein domains into passenger molecules that are normally secreted. Besides demonstrating the requirement of a certain domain in protein sorting, this approach permits analysis of a domain that is sufficient to direct the protein to a certain location. In this case, measurement of reduced secretion rates will give information about the successful introduction of sequences that reconstitute a sorting signal. However, reduced secretion rates will also be observed if the chimeric protein is misfolded and retained in the ER due to interaction with BiP. Therefore, this approach requires an analysis of the intracellular location of the chimeric protein. If vacuolar sorting signals are analyzed, it is necessary to show that the protein accumulates in the vacuole, rather than in the ER. Vacuoles can be isolated from protoplasts, and a BiP antiserum can be used to control for the absence of ER in these preparations. We recommend the inclusion of a BiP-binding assay, as described in Section III, in combination with each transport assay. As a negative control, the unmodified passenger protein is used, whereas a well-described assembly defective protein such as that described by Pedrazzini et al. (1993) can be used as a positive control for BiP association. Another, more serious, problem is the fact that probably most of the sorting signals cannot be transplanted to another protein without impeding their function or altering the structure of the rest of the protein. Therefore, the results obtained by transplantation of protein domains from one protein to another must be treated with caution. A well-described passenger protein that can be used for this approach is the enzyme phosphinothricin acetyl transferase (PAT). It has been demonstrated that PAT hybrid proteins containing C-terminal fusions remain enzymatically active (Botterman *et al.*, 1991). PAT fusions have been used to characterize C-terminal ER retention signals of soluble proteins (Denecke *et al.*, 1992; Denecke *et al.*, 1993), C-terminal vacuolar targeting signals (Denecke, unpublished results), and sorting determinants required for targeting membrane proteins to the tonoplast membrane (Höfte *et al.*, 1992). The PAT C-terminus is most likely an extended structure and addition of peptides and even transmembrane domains to the C-terminus of this protein do not have a negative effect on its enzymatic activity (Denecke, unpublished results). PAT is secreted from tobacco protoplasts if inserted into the ER lumen (Denecke *et al.*, 1990). It offers an ideal model system to study transport signals, since it does not contain competing targeting information. Another excellent passenger protein (PHALB), based on a cytosolic seed albumin fused to a signal peptide, has been shown to be more efficiently secreted than PAT (Hunt and Chrispeels, 1991). The most obvious difference between PHALB and PAT is the fact that PHALB contains consensus sites for N-linked glycosylation that are recognized by the ER machinery for glycosylation. Whether glycans are the reason for the high secretion rate of PHALB, compared to PAT or NPTII, is unknown, but these proteins may open the way to the study of glycan modification as a function of intracellular location.

Acknowledgements

A.V. thanks all of the past and present members of the group in Milan for their contribution to the development of some of the techniques described in this work. J.D. is indebted to Kjell Ove Holmström for help with the construction of the electroporation aparatus Denström 2001 and Bruno Denecke for the design and the construction of the plate electrode. Funding was provided in part by the Progetti Finalizzati "Biotecnologie e Biostrumentazione," "RAISA" (Subproject 2, Paper 1901) of the Consiglo Nazionale delle Ricerche, and the Swedish Natural Science Research Council.

References

Accili, D., Kadowaki, T., Kadowaki, H., Mosthaf, L., Ullrich, A., and Taylor, S. I. (1992). Immunoglobulin heavy chain-binding protein binds to misfolded mutant insulin receptors with mutations in the extracellular domain. *J. Biol. Chem.* **267**, 586–590.

Botterman, J., Gosselé, V., Thoen, C., and Lauwereys, M. (1991). Characterization of phosphinothricin acetyl transferase and C-terminal enzymatically active fusion proteins. *Gene* **102**, 33–37.

Chrispeels, M. J. (1991). *Annu. Rev. Plant Physiol. Plant Mol. Biol.* **42**, 21–53.

D'Amico, L., Valsasina, B., Daminati, M. G., Fabbrini, M. S., Nitti, G., Bollini, R., Ceriotti, A., and Vitale, A. (1992). Bean homologs of the mammalian glucose regulated proteins: Induction by tunicamycin and interaction with newly synthesized storage proteins in the endoplasmic reticulum. *Plant J.* **2**, 443–455.

Denecke, J., Gosselé, V., Botterman, J., and Cornelissen, M. (1989). Quantitative analysis of transiently expressed genes in plant cells. *Methods Mol. Cell Biol.* **1**, 19–27.

Denecke, J., Botterman, J., and Deblaere, R. (1990). Protein secretion in plants can occur via a default pathway. *The Plant Cell* **2**, 51–59.

Denecke, J., Souza Goldman, M. H., Demolder, J., Seurinck, J., and Botterman, J. (1991). The tobacco luminal binding protein is encoded by a multigene family. *The Plant Cell* **3**, 1025–1035.

Denecke, J., De Rycke, R., and Botterman, J. (1992). Plant- and mammalian sorting signals for protein retention in the endoplasmic reticulum contain a conserved epitope. *EMBO J.* **11**, 2345–2355.

Denecke, J., Ek, B., Caspers, M., Sinjorgo, K. M. C., and Palva, E. T. (1993). Analysis of sorting signals responsible for the accumulation of soluble reticuloplasmins in the plant endoplasmic reticulum. *J. Exp. Bot.* **44,** S., 213–221.

De Wet, J. R., Wood, J. V., De Luca, M., Helinski, D. R., and Subramani, S. (1987). The firefly luciferase gene: Structure and expression in mammalian cells. *Mol. Cell. Biol.* **7,** 725–737.

Dombrowski, J. E., Schroeder, M. R., Bednerek, S. V., and Raikhel, N. V. (1993). Determination of the functional elements within the vacuolar targeting signal of barley lectin. *Plant Cell* **5,** 587–596.

Fromm, M., Callis, J. L. P., and Walbot, V. (1987). Electroporation of DNA and RNA in plant protoplasts. *In* "Methods in Enzymology," Vol. 153, pp 351–366. New York: Academic Press.

Gething, M.-J., and Sambrook, J. (1992). Protein folding in the cell. *Nature* **355,** 33–45.

Gorman, C. M., Moffat, L. F., and Howard, B. H. (1982). Recombinant genomes which express chloramphenicol acetyl transferase in mammalian cells. *Mol. Cell. Biol.* **2,** 1044–1051.

Hoffman, L. M., Donaldson, D. D., and Herman, E. M. (1988). A modified storage protein is synthesized, processed, and degraded in the seeds of transgenic plants. *Plant Mol. Biol.* **11,** 717–729.

Höfte *et al.* (1992). *The Plant Cell* **4,** 995–1004.

Holwerda, B. C., Padgett, H. S., and Rogers, J. C. (1992). Proaleurain vacuolar targeting is mediated by short contiguous peptide interactions. *Plant Cell* **4,** 307–318.

Hunt, D. C., and Chrispeels, M. J. (1991). The signal peptide of a vacuolar protein is necessary and sufficient for the efficient secretion of a cytosolic protein. *Plant Physiol.* **96,** 18–25.

Hurtley, S. M., Bole, D. G., Hoover-Litty, H., Helenius, A., and Copeland, C. S. (1989). Interactions of misfolded influenza virus hemagglutinin with binding protein (Bip). *J. Cell Biol.* **108,** 2117–2126.

Hurtley, S. M., and Helenius, A. (1989). Protein oligomerization in the endoplasmic reticulum. *Annu. Rev. Cell Biol.* **5,** 277–307.

Jacobsen, J., and Beach, L. (1985). Control transcription of alpha amylase and rRNA genes in barley aleurone protoplasts by gibberilline and abscisic acid. *Nature* **316,** 275–277.

Jefferson, R. A., Kavanagh, T. A., Michael, W. B. (1987). GUS fusions: β-Glucuronidase as a sensitive and versatile gene fusion marker in higher plants. *EMBO J.* **6,** 3901–3907.

Knittler, M. R., and Haas, I. G. (1992). Interaction of BiP with newly synthesized immunoglobulin light chain molecules: Cycles of sequential binding and release. *EMBO J.* **11,** 1573–1581.

Machamer, C. E., Doms, R. W., Bole, D. G., Helenius, A., and Rose, J. K. (1990). Heavy chain binding protein recognizes incompletely disulfide-bonded forms of vesicular stomatitis virus G protein. *J. Biol. Chem.* **265,** 6879–6883.

Munro, S., and Pelham, H. R. B. (1986). An Hsp70-like protein in the ER: Identity with the 78 kd glucose-regulated protein and immunoglobulin heavy chain binding protein. *Cell* **46,** 291–300.

Pedrazzini, E., Giovinazzo, G., Bollini, R., Ceriotti, A., and Vitale, A. (1993). Binding of BiP to assembly-defective protein in plant cells. *Plant J.,* **5,** 103–110.

Reiss, B., Sprengel, R., Will, H., and Schaller, H. (1984). Protein fusions with the kanamycin resistance gene from Tn5. *Gene* **30,** 211–218.

Saalbach, G., Jung, R., Kunze, G., Saalbach, I., Adler, K., and Muntz, K. (1991). Different legumin protein domains act as vacuolar targeting signals. *Plant Cell* **3,** 695–708.

Shillito, R. D., Saul, M. W., Paszkowski, J., Muller, M., and Potrykus, I. (1985). High efficient direct gene transfer to plants. *Bio/Technol.* **3,** 1099–1103.

Suzuki, C. K., Bonifacino, J. S., Lin, A. Y., Davis, M. M., and Klausner, R. D. (1991). Regulating the retention of T-Cell receptor a chain variants within the endoplasmic reticulum: Ca^{2+}-dependent association with BiP. *J. Cell Biol.* **114,** 189–205.

Thompson, C. J., Movva, N. R., Tizard, R., Crameri, R., Davies, J. E., Lauwereys, M., and Botterman, J. (1987). Characterization of the herbicide-resistance gene bar from Streptomyces hygroscopicus. *EMBO J.* **6,** 2519–2523.

von Schaewen, A., and Chrispeels, M. J. (1993). Identification of vacuolar sorting information in phytohemagglutinin, an unprocessed vacuolar protein. *J. Exp. Bot.* **44,** S. 339–342.

Werr, W., and Lörz, H. (1986). Transient gene expression in a gramineae cell line: A rapid procedure for studying plant promotors. *Mol. Gen. Genet.* **202,** 471–475.

CHAPTER 25

Assessment of Translational Regulation by Run-off Translation of Polysomes *In Vitro*

Michael E. Vayda

Department of Biochemistry, Microbiology, and Molecular Biology
University of Maine
Orono, Maine 04469-5735

I. Introduction

The regulation of translational processes is an important element in the fine tuning of plant gene expression. Broad changes in gene expression are evident as changes in mRNA steady-state levels and transcriptional activity, which can be detected by RNA gel blot hybridization and nuclear run-on assays, respectively. However, not all mRNAs present in the cell are translated; polypeptides that are produced *in vivo* in the presence of labeled amino acids often are only a subset of those that can be synthesized by translation of total cellular mRNA *in vitro* using a cell-free extract. This discrepancy can be explained in part by the selective recruitment of specific mRNAs by the translational machinery. That is, only a subset of the total cellular mRNA population is associated with polyribosome ("polysome") complexes. Further, not all polysome-associated mRNAs are translated *in vivo;* some are arrested. Translational arrest has been observed following acute changes in the environment, such as the onset of hypoxia (Crosby and Vayda, 1991), the light/dark transition (Berry *et al.,* 1988), or upon Ca^{2+} influx following wounding (Davies, 1993). These observations indicate that assessment of translational activity must be made in order to fully understand plant gene regulation.

Run-off translation of polysomes *in vitro* is a method by which to assess translational competence. The method differentiates by function, not simply by the presence of mRNA species in cells or their association with polysomes. Polysomes are isolated from cells and then resuspended in an extract containing all of the factors necessary to perform the elongation reactions of translation. In this way, the synthesis of polypeptides that was initiated *in vivo* is completed *in vitro* in the presence of radioactive amino acid tracers. The result provides an idea of what mRNAs were in the process of being translated *in vivo* at the time of isolation, and also provides an estimate of the fraction of polysome complexes that were translationally active at that time. In a variation of the procedure, one can assess the capability of the extract performing the elongation reactions. This chapter describes the basic method of polysome run-off assay using polysomes isolated after varied stress conditions and a complete heterologous extract (rabbit reticulocyte lysate).

Polysomes are isolated from other cellular components by sedimentation through sucrose (as described by Chapter 15, this volume). The polysome fraction contains ribosomes, mRNA and associated proteins, but translation is suspended in the elongation process. Translation can resume *in vitro* upon add-back of soluble factors which include: elongation factor EF-1α (presents amino acyl tRNAs), EF-1$\beta\tau$ (recycles EF -1α), EF-2 (effects ribosome translocation), GTP, amino acyl tRNA synthetases, ATP, amino acids, and tRNAs. These components are present in the complete extract. Alternatively, the elongation factors can be isolated from the tissue of interest by selective ammonium sulfate precipitation (the 40–70% cut) (Lax *et al.,* 1986; Webster *et al.,* 1991). Reinitiation of ribosomes on mRNA can be achieved if initiation factors are present in a complete extract (Ramaiah and Davies, 1985). However, to simplify interpretation of results and ensure that only run-off products are produced, initiation reactions are blocked by addition of analogs of the mRNA 5′ cap such as m^7GMP (200 μg/ml) (Berry *et al.,* 1990; Crosby and Vayda, 1991). Translation is monitored by incorporation of one or more radioactive amino acid tracers. The most common choices are [^{35}S]methionine or [^{14}C]leucine. The reader should be aware that the choice of label introduces a bias into the experiment, and it is best to use a mix of labeled amino acids, although this makes the experiment considerably more expensive. Protein synthesis resulting from organellar ribosomes is blocked by the addition of chloramphenicol (25 μg/ml). Conversely, if the objective is to monitor organellar polysome activity, polysome run-off is achieved using an *E. coli* extract (Berry *et al.,* 1988; 1990) and synthesis by cytoplasmic ribosomes is inhibited using cycloheximide (50 μg/ml) and aurin tricarboxylic acid (ATA, 50 μg/ml). Thus, the method can be varied to monitor the translational competence of polysomes from different sources in a reference extract, or the same polysome preparation in translational extracts from different sources.

II. Materials

A. Equipment

Beckman J2-21 preparative centrifuge
Beckman Rotor JA-14 rotor (chilled to 4°C)
Beckman 70.1 fixed angle rotor (chilled to 4°C)
Beckman 13-ml polyallomar quick seal centrifuge tubes (product No. 342413)
Markson Miracle Mill (Model MC-170)
Autoclaved Oak Ridge tubes (Nalgene No. 3119-0050)
Autoclaved spatulas
Polyvinyl gloves
Heating block set to 20°C
p200 and p20 pipettemen with autoclaved tips
Ice bucket and ice
Eppendorf microfuge (Model 5415)
Autoclaved 1.5-ml microfuge tubes
16-gauge needle
Razor blade
Kimwipes
Beckman spectrophotometer (Model DU-40)
Quartz cuvette (VWR Cat No. 58016-469)
Sidearm Erlenmeyer flask
Glass fiber filters (2.4-cm diam.) (Schleicher & Schuell Grade 31, Order No. 37-06710)
Schleicher & Schuell filter holder (Model GV 025/0)
Whatman 3MM paper
Forceps
Bench paper (Fischer No. 14-206-30)
Scintillation counter
SDS–PAGE apparatus and power pack
Hoefer gel drier (Model SE-1160)
Kodak XAR/5 X-ray film
Film cassette

B. Reagents

Liquid nitrogen in Dewar flask

Diethylpyrocarbonate (DEP) (Sigma No. D-5158)

β-Mercaptoethanol (store at 4°C) (Sigma No. M-6250)

Polysome buffer: 200 mM Tris–HCl, pH 9, 400 mM KCl, 60 mM MgOAc, 50 mM EGTA, 250 mM sucrose, and 0.01% Triton X-100

Sucrose cushion: 1.5 M sucrose, 40 mM Tris–HCl, pH 9, 100 mM KCl, 30 mM MgOAc, 5 mM EGTA, 7 mM β-mercaptoethanol

Dupont-NEN Translation Kit (Catalog No. NEK-001)

Kit contents:

    Rabbit reticulocyte lysate

    [35S]Methionine (sp act 1000 Ci/mmol)

    1 M KOAc

    50 mM MgOAc

    Sterile water

    NEN translation cocktail (amino acids and buffer)

4 mM 7mGMP (Sigma No. M-5012, dissolved in water)

500 μg/ml chloramphenicol (Sigma No. C-0378, dissolved in ethanol)

10% Trichloroacetic acid (TCA) (Sigma No. T-4885, dissolve 10 g TCA in dH_2O to final volume of 100 ml

Stop solution: 0.5 M Tris–HCl, pH 6.5, 1% SDS, 0.05 mg/ml bromphenol blue, 1 mg/ml RNase A

95% ethanol

BioSafe II (Research Products, Inc.)

1 M sodium salicylate (Sigma No. S-3007)

C. Precautions

Solutions are DEP-treated, tubes and tips are autoclaved, and polyvinyl gloves are worn at all times to minimize introduction of RNAses that degrade polysomes. Solutions containing Tris cannot be DEP-treated; therefore these solutions should be made using DEP-treated water in autoclaved bottles. Liquid nitrogen (−156°C) can cuase burns to skin and fingers should never be immersed in it. Nitrogen expands greatly in volume when warmed to a gas and, thus, should never be placed in a tightly closed container. DEP is highly reactive; care should be taken not to contact skin. DEP spontaneously decomposes to CO_2 and H_2O upon autoclaving. The translation kit should be maintained at −70°C. The lysate is very labile and particularly sensitive to freeze/thaw cycles. Lysate should be dispensed into aliquots upon first use and kept at −70°C until seconds before addition to the reaction mix. Radioactive tracers: 35S emits moderate to weak beta particles, thus even a single sheet of paper is effective as shielding. Although the external hazard from ^{35}S, ^{14}C, and ^{3}H is minimal, special care should be taken because these nuclides are not as easily detected as strong beta emitters (such as ^{32}P) and pose a hazard if internalized. Thus, polyvinyl gloves should be worn whenever handled, protective plastic-backed bench paper should be layed on surfaces prior to beginning the procedure, and all tips, tubes, and gloves that come in contact with radioactivity should be disposed of as radioactive waste. Care should be taken to monitor the area when the experiment is completed to prevent inadvertent spread of contamination.

III. Methods

1. Weigh out 5 g of fresh tissue (Note: amount of material needed will vary with tissue type).

2. Add an excess of liquid nitrogen to tissue in weigh boat to quick-freeze. Keep submerged in nitrogen for about 3 min to ensure freezing throughout tissue.

3. Aliquot 13 ml of cold (4°C) polysome buffer into an autoclaved Oak Ridge tube. Keep tube on ice.

4. Add 14 μl β-mercaptoethanol (to a final concentration of 15 mM).

5. Transfer frozen tissue to a Miracle Mill tissue grinder, and grind with shaking for 30 s.

6. Quickly transfer the fine powder to the Oak Ridge tube using a sterile spatula. It is crucial to work quickly so that the tissue does not thaw until it is in the polysome buffer.

7. Cap tube and mix for 5 min at room temperature (until tissue thaws).

8. Centrifuge in Beckman JA-14 rotor at 15,000 $\times g$ for 15 min at 4°C.

9. During the spin, prepare the sucrose cushion: add 3 ml of cold sucrose cushion into each of two 13-ml Beckman polyallomer quick-seal tubes. (approximate volume of the homogenate is 16 ml). Keep tubes on ice.

10. Remove supernatant from step 8 with pasteur pipet and layer atop sucrose cushion. Add polysome buffer to fill tube to neck (if necessary). Seal tubes.

11. Centrifuge in Beckman 70.1 fixed-angle rotor for 4.5 h at 45 kRPM at 4°C.

12. Poke a hole through top of tube with a 16-gauge needle and remove 1 to 2 ml of supernatant into a sterile Oak Ridge tube. Cut off the top of tube with a razor blade and pour off remaining supernatant into Oak Ridge tube. Invert tube with the polysome pellet on a Kimwipe to remove last vestiges of supernatant.

13. From the translation kit, mix 170 μl sterile water, 20 μl KOAc, 5 μl MgOAc, and 55 μl translation cocktail in a sterile microfuge tube. Keep on ice.

14. Resuspend each polysome pellet in 50 μl of the translation mix prepared in step 13. Carefully pipette up and down ~20 times to resuspend completely. Combine the resuspended pellets in a sterile microfuge tube. Keep on ice.

15. Remove 10 μl of the resuspended polysomes and mix with 990 μl of dH$_2$O in a cuvette. Measure the absorbance at 260 nm using spectrophotometer. Estimate concentration of polysomes (1 A_{260} unit = 40 μg/ml)

16. Prepare translation reaction mix for each sample by adding together in a sterile microfuge tube: 5.5 μl cocktail, 2 μl KOAc, 0.5 μl MgOAc, and 5 μl [^{35}S]methionine (25 μCi/reaction).

17. Then add to reaction mix (step 16) 1 μL m^{7}GMP, 1 μL chloramphenicol, 10 μg resuspended polysomes (approximately 1–3 μl, step 15). Add sterile water to adjust volume to 18 μl.

18. Prepare a duplicate reaction mix omitting the m^{7}GMP to serve as a control for reinitiation.

19. Initiate the reaction. Rapidly thaw reticulocyte lysate (stored until now at −70°C or on bench top in liquid nitrogen) by rolling between gloved fingers. Add 12 μl of the rabbit reticulocyte lysate to each reaction. Quickly mix by pipetting or *gently* tapping the tube. Microfuge for 5 s, then incubate at 20°C for 30 min. Reaction is terminated by addition of 6 μl stop solution. Incubate at room temperature for 5 min after mixing. Sample can be stored frozen at −20°C. Omit stop solution if immunoprecipitation (Harlow and Lane, 1988) of translation products is to be done.

20. Measurement of translational activity: Remove 2-μl aliquots (in triplicate) from each reaction and add to separate tubes each containing 100 μl of 10% TCA (chilled to 4°C). Mix and keep on ice for at least 5 min to precipitate polymers. Assemble filtration apparatus and attach to a vacuum line. Recover precipitated material on a glass fiber filter by vacuum filtration. With vacuum on, rinse filter with 1 ml 10% TCA and then 1 ml 95% ethanol. Remove glass fiber filter with forceps and air-dry on a sheet of Whatman 3 MM paper. Place filter in scintillation vial with 3 ml of scintillation fluid (BioSafe II) and count radioactivity recovered.

21. Resolution of translation products: Labeled polypeptide products are resolved by 15% SDS–PAGE (Laemmli, 1970; Vayda and Schaeffer, 1988). Gel is then soaked in 1 M Na salicylate for 30 min, placed on Whatman 3MM paper, and dried using a Hoefer vacuum gel drier. The dried gel is then exposed to Kodak X-AR 5 film in a cassette at −70°C.

Optional: RNA can be isolated from the supernatant and polysome pellet fractions (after step 12) by standard phenol/chloroform extraction procedures (Sambrook *et al.*, 1989; Butler *et al.*, 1990; Crosby and Vayda, 1991). Supernatant MUST be diluted at least 2:1 with DEP-treated sterile water prior to extraction; failure to dilute supernatant will result in precipitation of sucrose. Pellet should be resuspended in 1 ml of DEP-treated water for RNA isolation. For normalization of the pellet and supernatant fractions, 20 μl of a 2 ng/μl stock solution of single-stranded M13 phage DNA (40 ng total) is added to each fraction. Note: the total RNA in each fraction will differ dramatically; this internal standard is necessary for comparison of RNA in supernatant and pellet fractions. Add equal volumes of precipitated RNA onto a test gel or slot blot and hybridize to radiolabeled M13 DNA. Take densitometer tracing of light exposures, and adjust volume of RNA such that the M13 signal is the same in all lanes. mRNA can then be assessed by gel blot hybridization (Sambrook *et al.,* 1989; Butler *et al.,* 1990). It is prudent to always include an extra set of samples for probing with M13 as a hybridization control.

IV. Critical Aspects of the Procedure

Polysomes are sensitive to RNases, temperature, and salt concentrations and dissociate with time. Thus, one should take care to keep materials on ice, minimize the time between steps, and prevent introduction of nucleases. It is a reasonable precaution to prepare redundant run-off reactions with varying amounts of polysomes, for example, 0.5, 1, 2, and 3 μl of resuspended polysomes. Control reactions with and without m^{7}GMP and containing purified RNA as template should be run in parallel to verify lysate activity and inhibition of initiation reactions by m^{7}GMP. It is important to add RNase A to the stop buffer to degrade [^{35}S]methionyl–tRNA that would otherwise precipitate in 10% TCA and confuse interpretation of the incorporation results. We have performed the run-off reactions at both 20 and 37°C. Although somewhat higher incorporation is sometimes observed at 37°C, translation products tend to appear as less distinct bands in SDS–PAGE fluorography.

V. Results and Discussion

Table I shows the incorporation of [^{35}S]methionine during run-off translation of polysomes isolated from nonstressed mature potato tubers (Nonstressed), the

Table I
Translational Activity of Potato Tuber Polysomes by Run-off *In Vitro*

	Incorporation of [^{35}S]-methionine[a]	
Template	TCA precipitable −m^{7}GMP	CPM/μl reaction +m^{7}GMP
Total RNA (1 μg)	24340	605
No RNA	580	496
Nonstressed polysomes (10 μg)	6836	6090
2 h wounded polysomes (10 μg)	9844	11626
2 h wounded 4 h hypoxic polysomes (10 μg)	846	828

[a] Incorporation is the average of TCA precipitation performed in triplicate.

same tubers 2 h after wounding (2 h Wounded), and 2 h after wounding and a subsequent 4-h period of hypoxic incubation (2 h Wounded/4 h Hypoxic). Figure 1 shows the results obtained by SDS–PAGE and fluorography of the polypeptide products of these translation reactions. These analyses reveal that wounded tuber polysomes are engaged in the synthesis of numerous polypeptides that are distinct from the few prominent polypeptides synthesized by nonstressed tuber polysomes. The 40-kDa species of nonstressed tubers is most likely patatin, the major tuber storage protein. Although polysomes from the "Wounded and Hypoxic" tubers produce the same polypeptide species as wounded aerobic tubers, the amount of product synthesized from the same amount of polysomal material is significantly less than that of wounded aerobic tuber polysomes. This result indicates that polysomes from hypoxic tissues are somehow inefficient or blocked in translation. These observations are supported by the analysis of polysome-associated mRNA species shown in Fig. 2. Upon wounding, patatin (PAT) mRNA is released from ribosomes as phenylalanine ammonia-lyase (PAL) mRNA accumulates in the cytoplasm (Butler *et al.,* 1990) and becomes polysome-

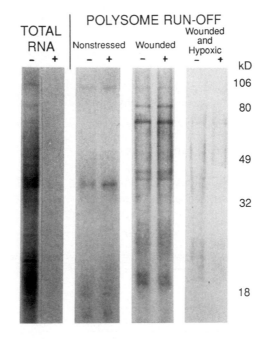

Fig. 1 Translation products resulting from polysome run-off *in vitro.* Translation reactions used as template either 2 μg deproteinized total tuber RNA (TOTAL RNA) or 10 μg of resuspended polysomes (POLYSOME RUN-OFF) isolated from mature, nonstressed potato tubers (Nonstressed), the same tuber 2 h after wounding with a pipette tip (Wounded), or the same 2 h wounded tuber after subsequent 4h incubation in a chamber flushed with argon (approximately 1–2% O_2) (Wounded and Hypoxic). − and +, absence and presence, respectively, of m^7GMP in the reaction mix. Five microliters each terminated reaction was resolved by SDS–PAGE and fluorography.

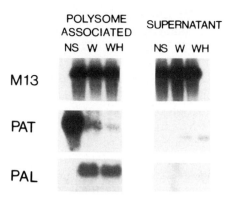

Fig. 2 RNA gel blot hybridization analysis of polysome-associated mRNA species. RNA was isolated from the resuspended polysome pellet and diluted supernatant reactions of a nonstressed tuber (NS), that tuber 2 h after wounding (W), or after 2 h of wounding and a subsequent 4 h hypoxic incubation (WH). Equivalent aliquots of RNA (relative to the M13 added) were resolved by electrophoresis through a 1.4% formaldehyde–agarose gel (Sambrook *et al.*, 1989) and blotted to nylon (Hybond-N, Amersham). Blots were probed by hybridization to ^{32}P-labeled M13 (M13) DNA, patatin (PAT) cDNA, or phenylalanine ammonia-lyase (PAL) cDNA.

associated (Fig. 2; Crosby and Vayda, 1991). However, wound-response mRNAs such as PAL remain bound to polysomes when tubers are transferred to hypoxic conditions (Fig. 2; Crosby and Vayda, 1991), although they are only very poorly translated in the run-off assay (Fig. 1) and are not translated *in vivo* (Butler *et al.*, 1990; Vayda and Schaeffer, 1988). Thus, changes in potato tuber gene expression during stress conditions is explained in part by two different translational regulatory mechanisms: wounding causes a change in mRNA species that are selected by ribosomes, and hypoxia arrests translation in the elongation process.

The nature of the translational arrest during hypoxia has been investigated in maize root tips by Webster *et al.* (1991), using a variation of the polysome run-off procedure. Rather than use a heterologous "complete" extract, they prepared extracts from root tips that were competent to perform elongation and termination reactions of translation *in vitro*. These extracts are prepared by selective precipitation with ammonium sulfate (the 40–70% fraction)(Webster *et al.*, 1991; Lax *et al.*, 1986). Polysomes isolated from aerobic maize root tips were run-off in extracts prepared from either aerobic root tips or hypoxic root tips. Because the template for run-off translation is the same, such an approach assesses differences in the "soluble component" extracts that contain translation elongation factors. Using this approach, Webster *et al.* (1991) demonstrated that extracts of hypoxic maize root tips only poorly support run-off translation of competent polysomes. Further, decreasing the pH of an aerobic corn root extract *in vitro* reduces translational competence to a low level comparable to the hypoxic root extract. We have determined by immunoblot analysis of polysome-associated polypeptides that elongation factor EF-1α accumulated aberrantly in the polysome fraction of potato tubers during hypoxia and upon acidification of the plant cell extract (Vayda *et al.,* 1995). Thus, translational arrest during hypoxia appears to result from both the inactivation of soluble elongation factors in the cytoplasm and the failure of EF-1α to release from ribosomes after amino acyl-tRNA presentation.

The analysis of specific products synthesized during run-off translation can be achieved by antibody adsorption and immunoprecipitation (Harlow and Lane, 1988) after completion of the run-off reaction (Berry *et al.,* 1988; 1990). Using this technique, Berry *et al.* (1990) demonstrated that the translation of ribulose-1,5-bisphosphate carboxylase small and large subunit polypeptides (SSU and LSU, respectively) occurs only in light-grown plants. Both SSU and LSU mRNAs are present in dark-grown plants but are not selected for translation by ribosomes. SSU and LSU mRNA are translated by cytoplasmic and chloroplastic ribosomes, respectively, as can be demonstrated by run-off translation using either rabbit reticulocyte or *E. coli* lysates (Berry *et al.,* 1988; 1990). Further, both SSU and LSU mRNA remain bound to polysomes when light-grown plants are shifted to the dark, although translation of both species is arrested (Berry *et al.,* 1988). These observations indicate that both cytoplasmic ribosomes and chloroplastic ribosomes may be regulated by similar mechanisms and in a coordinate fashion.

Polysome run-off assay can also be used to explore the ability of ribosomes to reinitiate *in vitro*. Ramaiah and Davies (1985) used wheat germ extracts depleted of ribosomes to achieve run-off translation of polysomes isolated from nonwounded or wounded pea epicotyls. Polysomes from wounded epicotyls exhibited greater translational ability in the wheat germ extract *in vitro*. However, this activity was sensitive to the addition of the initiation inhibitors ATA and pactamycin (Ramaiah and Davies, 1985). These observations indicate that ribosomes that dissociate from wounded epicotyl polysomes are more efficient at reinitiation reactions than are polysomal ribosomes from nonwounded tissue, or than monomeric ribosomes from either tissue. Translational competence may also be mediated by association of polysomes with the cytoskeleton (see Chapter 16, this volume) and by Ca^{2+}-mediated phosphorylation of elongation factors (reviewed by Davies, 1993).

VI. Conclusions

Run-off translation of polysomes is a useful technique for assessing the relative translational activity of different tissues, or the same tissue after some environmental change. The technique can be used to identify the complement of major polypeptides being synthesized in a tissue under given conditions, or can be used to determine whether specific gene products are in the process of being translated. Modification of the technique can also be used to provide information on the mechanisms by which translation is regulated. Such information is important for understanding the fine-tune control of plant gene expression.

References

Berry, J. O., Carr, J. P., and Klessig, D. F. (1988). mRNAs encoding ribulose-1,5-bisphosphate carboxylase remain bound to polysomes but are not translated in amaranth seedlings transferred to darkness. *Proc. Natl. Acad. Sci. U.S.A.* **85,** 4190–4194.

Berry, J. O., Breiding, D. E., and Klessig, D. R. (1990). Light-mediated control of translational initiation of ribulose-1,5-bisphosphate carboxylase in amaranth cotyledons. *Plant Cell* **2,** 795–803.

Butler, W., Cook, L., and Vayda, M. E. (1990). Hypoxic stress inhibits multiple aspects of the potato tuber wound response. *Plant Physiol.* **93,** 265–270.

Crosby, J. S., and Vayda, M. E. (1991). Stress-induced translational control in potato tubers may be mediated by polysome-associated proteins. *Plant Cell* **3,** 1013–1023.

Davies, E. (1993). Intercellular and intracellular signals and their transduction via the plasma membrane-cytoskeleton interface. *Seminars Cell Biol.* **4,** 139–147.

Harlow, E., and Lane, D. (1988). "Antibodies: A Laboratory Manual," pp. 421–470. Cold Spring Harbor Laboratory Press, Cold Spring Harbor, NY.

Lax, S. R., Lauer, S. J., Browning, K. S., and Ravel, J. M. (1986). Purification and properties of protein synthesis initiation and elongation factors from wheat germ. *Methods Enzymol.* **118,** 109–128.

Laemmli, U. K. (1970). Cleavage of structural proteins during the assembly of the head of bacteriophage T4. *Nature* **227,** 680–685.

Ramaiah, K. V. A., and Davies, E. (1985). Wounding of aged pea epicotyls enhances the reinitiating ability of isolated ribosomes. *Plant Cell Physiol.* **26,** 1223–1231.

Sambrook, J., Fritsch, E. F., and Maniatis, T. (1989). "Molecular Cloning: A Laboratory Manual" Cold Spring Harbor, NY: Cold Spring Harbor Laboratory Press.

Vayda, M. E., and Schaeffer, H. J. (1988). Hypoxic stress inhibits the appearance of wound-response proteins in potato tubers. *Plant Physiol.* **88,** 805–809.

Vayda, M. E., Shewmaker, C. K., and Morelli, J. K. (1995). Translational arrest in hypoxic potato tubers is correlated with the abberant association of elongation factor EF-1α with polysomes. *Plant Mol. Biol.,* in press.

Webster, C., Kim, C.-Y., and Roberts, J. K. M. (1991). Elongation and termination reactions of protein synthesis on maize root tip polysomes studied in a homologous cell-free system. *Plant Physiol.* **96,** 418–425.

PART II

Molecular Methods for Analysis of Cell Function

CHAPTER 26

Electroporation of Plant Protoplasts and Tissues

George W. Bates

Department of Biological Science
Florida State University
Tallahassee, Florida 32306

I. Introduction

Electroporation utilizes short, high-voltage electric shocks to make cells permeable to exogenous molecules. Although direct physical evidence is still lacking, it is widely believed that pores form transiently in the cell membrane in response to these shocks and that the uptake of exogenous molecules occurs through these pores. (Chang *et al.,* 1992). Electroporation is used to induce the cellular uptake of reporter dyes, enzyme substrates, nucleic acids, and even antibodies and viruses (see Chang *et al.,* 1992, for the range of applications of electroporation).

In plants, electroporation is used primarily to stimulate the uptake of plasmids for stable and transient genetic transformation. Because the cell wall is a major barrier to the diffusion of macromolecules, protoplasts are normally utilized in plant electroporation. However, recent work indicates that plasmid DNAs can be introduced into walled plant cells by electroporation (Dekeyser *et al.,* 1990; Klöti *et al.,* 1993). This application may prove particularly valuable for the genetic transformation of difficult to culture cereal species. Included in this chapter are protocols for electroporation of protoplasts, selection of stable, kanamycin-resistant transformants of tobacco, and electroporation of intact tissues.

II. Materials

A. Electroporation Instruments and Background on Electrical Conditions

Most of the instruments currently used for electroporation are capacitive-discharge instruments; a capacitor bank is charged by a high-voltage power supply and then discharged into a chamber containing the experimental material. The DC pulses used in electroporation can be either square waves or exponentially decaying capacitive discharges, but capacitive discharges are usually used in order to reduce the cost of the instrument. Although the duration and voltage of a square-wave pulse is easily defined, some convention must be used to describe an exponential decay. Early electroporation literature did not always provide clear descriptions of the pulses used. However, the accepted convention now is to give the voltage as that at the start of the pulse, and the pulse length is described by its RC time constant (the time required for the voltage to fall to 1/e or 37% of its initial value). Electroporation is driven by electric field strength, rather than voltage per se. Thus, the spacing between the electrodes in the electroporation chamber is important. The industry standard for electroporation of eukaryotic cells is a chamber with an interelectrode spacing of 0.4 cm. Thus, a 300-V pulse results in a field strength within the chamber of 750 V/cm. The voltage at which pore formation occurs increases with decreasing cell size (Tsong, 1989). For plant cells, pulses in the range 500–1000 V/cm are effective in electroporation, but for bacteria much higher voltage pulses must be used. (The higher field strength required for electroporation of bacteria are achieved, in part, by use of chambers with a 0.1-cm gap. These narrow chambers hold sample volumes that are too small for efficient work with plant protoplasts.) Effective electroporation also depends on the duration of the electric pulse. Pulses in the range 1–10 ms (milliseconds) are generally used in work with protoplasts; somewhat longer pulses are used in tissue electroporation. To a certain extent, increasing the pulse length can compensate for lower voltages, and vice versa. Electroporation instruments can be homemade or purchased commercially. Homemade equipment is inexpensive, but difficulties are encountered in the construction of a reliable switch for discharging the capacitors and in making workable electroporation chambers. A great deal of time can be spent in the construction, testing, and maintenance of a homemade instrument. I have also worked with electroporators from BRL (Cell-Porator, Electroporator System I, $2000, Life Technologies, Inc., Gaithersburg, MD) and Bio-Rad (Gene Pulser, complete system, $4600, Bio-Rad Laboratories, Inc., Hercules, CA). Both are well-designed, reliable, effective, and safe instruments and offer the valuable convenience of presterilized, disposable electroporation chambers. Chassey *et al.* (1992) provide a nearly complete list of commercial suppliers, Fromm *et al.* (1987) describe a homemade instrument. In choosing an instrument, look for one that allows both the voltage and the pulse length to be varied, and that will deliver 1- to 100-ms-long pulses of 750 V/cm (or greater) into a 0.5-ml sample containing 150 mM salts. Some of the commercial instruments are designed only for electroporation of bacteria and will provide very high voltage (1000–2000 V), but short (1- to 100-μs) pulses, and cannot deliver pulses into highly conductive (salty) media. These instruments are not effective for plant electroporation. Most commercial instruments are, however, designed for use in a variety of applications, but special attachments may have to be purchased for plant electroporation. For example, Bio-Rad's Gene Pulser has a basic pulse control unit that delivers high-voltage, short pulses. A second Capacitance Extender unit must also be purchased to obtain the long pulses used in plant electroporation.

B. Protoplasts

Protoplasts may be isolated from any source and by conventional procedures.

C. Plant Tissues

For rye leaf bases, surface-sterilized rye seeds were grown in Magenta boxes (Sigma Chemical Co., St. Louis, MO) on solid MS salts and vitamins + 0.3% gelrite, for 7 days in the dark. Leaf-base sections 1 mm long were cut from 2 to 4 mm above the apical meristem. Rye inflorescences, 1–2 mm long were isolated from field-grown plants. Immature embryos, approximately 1 mm long, were isolated from field- and incubator-grown plants of rye, wheat, and *Pennisetum*. The embryos were precultured 1–3 days on solid MS + 2 mg/l 2,4-D.

D. DNA

Plasmid DNA purified on CsCl density gradients is excellent for electroporation, but crude plasmid preparations (alkaline lysates) can also be used. The plasmid preparation should be free of RNA. If RNase is used in plasmid isolation, but sure to phenol-extract and ethanol-precipitate the DNA before use. Plasmid preparations can be stored in sterile water or TE at 4°C. Accurate quantitation of the plasmid DNA is important. CsCl-purified plasmids can be quantified spectrophotometrically by a 260/280 reading, but this may not be accurate for crude plasmid preparations because of the presence of trace amounts of phenol or other contaminants. Crude plasmid preparations can be quantified fluorometrically using Hoechst Dye 33258 (Labarca and Paigen, 1980).

E. Protoplast Electroporation Media

HBS (Hepes-buffered saline): 150 mM KCl, 4 mM CaCl$_2$, 10 mM Hepes (pH 7.2), and mannitol. The amount of mannitol must be adjusted to match the tonicity of the protoplasts. For carrot suspension cell protoplasts, we use 0.11 M mannitol, whereas 0.21 M mannitol is required to stabilize tobacco mesophyll protoplasts. HBS can be autoclaved and stored at 4°C. DNA should be added immediately before use, and the medium should be resterilized by filtration (use a low-binding, 0.2-mm cellulose acetate syringe filter, such as Nalgene No. 190-2520) after addition of the DNA.

F. Reagents and Media for Selection of Stable Transformants of Tobacco

1. Culture medium for tobacco mesophyll protoplasts, K3G: K3 salts, vitamins, and hormones (Nagy and Maliga, 1976; Chapter 29, this volume) containing 0.4 M glucose as both the carbon source and the osmotic stabilizer.

2. Tobacco callus medium, CM: Murashige and Skoog salts and vitamins plus 100 mg/l inositol, 3% sucrose, 1 mg/l banzyladenine, and 1 mg/l α-naphthaleneacetic acid (plus 0.8% agar if the medium is to be solid).

3. SeaPlaque agarose (FMC Corp.).

4. Kanamycin sulfate: stock solution, 100 mg/ml dissolved in water. Filter-sterilize and store frozen.

G. Tissue Electroporation Media

EP; 10% glucose + 10 mM Hepes + 4 mM CaCl$_2$ + 0.2 mM spermidine (pH 7.2). Autoclave without the spermidine, and then add the spermidine from a filter-sterilized stock. EP + NaCl + DNA; same as EP but with the addition of 150 mM NaCl and 100 μg/ml plasmid DNA. The medium should be resterilized by filtration after addition of the DNA.

III. Methods

All manipulations in the procedures below must be carried out under sterile conditions in a laminar-flow hood.

A. Protoplast Electroporation

1. Protoplast preparation. Protoplasts prepared by any standard method should be acceptable for electroporation (see, for example, Chapter 29 this volume).

2. After washing the protoplasts free of enzymes, pellet them by centrifugation and gently resuspend them in 10 ml of HBS + mannitol. Mix the protoplast suspension with a plastic transfer pipet until it is homogeneous, and then count the protoplasts using a hemacytometer. Place aliquots of 1×10^6 protoplasts into individual centrifuge tubes.

3. Pellet the aliquots of protoplasts and resuspend them in 0.5 ml HBS + mannitol + DNA. Transfer the 0.5-ml samples to electroporation chambers. (The cell density for electroporation should be 2×10^6 protoplasts/ml). Save the centrifuge tubes for reuse after electroporation.

4. Let the protoplasts stand for 5 min. Then resuspend them by agitating the electroporation chamber, and apply a single 10-ms (325 μF) pulse of 300 V (750 V/cm). Let the protoplasts stand for 3 to 5 min.

5. Use a transfer pipet to remove the protoplasts from the electroporation chamber, transfer them back into a centrifuge tube, and add 5 ml of culture medium. Rinse the chamber with 0.5 ml of HBS and combine the rinse with the sample.

6. Pellet the protoplasts, resuspend them in 2 ml of culture medium, and transfer to a petri dish for culture.

1. Critical Aspects of the Procedure:

1. Controls should include protoplasts mixed with DNA and cultured without being electroporated, and protoplasts electroporated in the absence of DNA.

2. It is important not to leave the protoplasts more than about 30 min in HBS. To make the timing of the operation efficient, electroporate four samples of protoplasts at a time. So, at Step 3, spin down four tubes of protoplasts, resuspend them in DNA, and transfer them to electroporation chambers, while leaving the other tubes of protoplasts to stand in HBS + mannitol without DNA. Then, begin centrifuging a second set of four samples and, at the same time, place 5 ml of culture medium into the centrifuge tubes from which the first batch of samples came. Electroporate the first batch of samples, and while they are standing after electroporation, resuspend the second batch of samples in DNA. We usually use each electroporation chamber for two or three samples before disposing of it, so the second batch of samples can be introduced into the chambers as soon as the first batch of samples has been transferred back to the centrifuge tubes. Step 7 is done after all the samples have been electroporated. In this way 12 to 16 samples can be electroporated in 30 to 40 min. If more than 16 samples are being electroporated, split the protoplasts into two batches after Step 1, and hold half

the protoplasts in protoplast wash or culture medium while the first group of protoplast samples are electroporated.

3. High-quality protoplast preparations are essential. Electroporation conditions that give optimal stable or transient transformation can kill 50% of the protoplasts. In poor-quality preparations, electric shock-induced cell death can exceed 90–95%, and the remaining protoplasts die due to the low density of viable cells in the culture. Because the shocks kill many cells, it is also useful to culture electroporated protoplasts at higher than normal cell densities (2.5×10^5/ml). Protoplast viability can be determined using the exclusion dye Evan's blue (0.1%) or by staining with fluorescein diacetate. Viability should not be measured until 12 or 24 after electroporation, as damaged protoplasts frequently take hours to die.

4. Optimal electric-field strengths and pulse lengths can vary for different cell types and may be species dependent as well. Find the optimal conditions for your experimental system by varying both the pulse length and the voltage around the settings mentioned above (10 ms, 750 V/cm). In general suspension-cell protoplasts can withstand greater shocks than mesophyll protoplasts.

5. Chilling the protoplasts on ice, either before or after electroporation, is generally unnecessary. However, when using very long (40-ms or longer) or high-voltage (1000 V/cm) pulses, significant heating of the medium can occur that will reduce cell viability. Under these conditions the samples should be kept on ice.

B. Selection of Kanamycin-Resistant, Stable Transformants following Electroporation of Tobacco Protoplasts

The following procedure can be used efficiently to recover stable transformants following the electroporation of tobacco protoplasts in the presence of a plasmid containing a functional neomycin phosphotransferase gene (such as pMON200, Rogers *et al.,* 1986). This procedure utilizes the agarose-bead culture technique of Shillito *et al.* (1983), and can probably be adapted for use with protoplasts of any species that is readily cultured. In more difficult to culture species, selection of stable transformants may be facilitated by use of a feeder layer or nurse culture (for example, see Fromm *et al.,* 1986).

The general strategy in the procedure below is to immobilize the microcolonies (in agarose) early in their growth so that individual transformants can be identified. At the same time, the osmotic strength of the medium is progressively reduced, the protoplast culture medium is replaced with a callus culture medium, and selection for transformants is carried out by addition of kanamycin.

1. Linearize the plasmid by digestion with a restriction enzyme. Do not choose an enzyme that cuts within the chimeric NPTII gene, and try to select an enzyme that cuts the plasmid at only one site. It is convenient to cut a large amount of plasmid (several hundred micrograms or more depending on the number of

experiments to be done), and then repurify the plasmid by phenol extraction and ethanol precipitation. Take the linearized plasmid up in water and adjust its concentration to about 1 μg/μl.

2. Electroporate the protoplasts and culture them in K3G in 60 × 15-mm petri plates at a cell density of 2 × 10^5 protoplasts/ml. Controls should include samples incubated in DNA but not electroporated and protoplasts electroporated in the absence of DNA.

3. At the end of the first week of culture, when the protoplasts have grown into microcolonies, they are transferred to larger dishes and mixed 1:1 with agar-containing medium. Prepare, in advance, CM medium + 0.23 M mannitol + 2.4% SeaPlaque agarose; sterilize by autoclaving. On the day of the transfer, melt this medium on a hot plate and let it stand to cool. Scrape adhering protoplast microcolonies off of the petri dish with a plastic micropipet tip and transfer 2.5 ml of cells to a new 60 × 15-mm petri dish, leaving the remaining 2.5 ml of cells in the original dish. When the CM + agarose has cooled to just above its gelling temperature, add 2.5 ml to each petri dish of microcolonies. Mix thoroughly by swirling and harden the media by refrigeration of the plates for 15 min. Culture at 27°C.

4. At the end of the second week, divide the solidified cultures into wedges using a weighing spatula, and transfer the wedges to large petri plates (100 × 15 mm). To each dish, add 5 ml of liquid CM + 0.13 M mannitol + 100 μg/ml kanamycin. (CM + 0.13 M mannitol can be prepared in advance, autoclaved, and stored at 4 or −20°C. Kanamycin should be added to this medium just before use from a frozen, filter-sterilized stock of 100 mg/ml). Culture at 27°C.

5. At the end of the third week, add 5 ml CM + 100 μg/ml kanamycin to each dish. CM can be prepared in advance and the kanamycin added just before use. This version of CM contains no mannitol.

6. At weekly intervals thereafter, remove 5 ml of liquid medium from each dish and replace it with 5 ml of fresh CM + 100 μg/ml kanamycin.

7. Kanamycin-resistant calli should be large enough to be transferred to solid medium 4 to 6 weeks after electroporation. These calli can be regenerated by standard procedures, but to ensure that the material is transformed, keep kanamycin in the culture medium until the regenerated plants have formed roots.

1. Critical Aspects of the Procedure:

1. The timing of the media changes listed below must not be adhered to rigidly. Adjustments should be made, particularly in the first 2 to 3 weeks after electroporation, depending on how fast the colonies are growing. The schedule above assumes a healthy, rapidly growing culture. If the protoplasts are dividing slowly, the cultures should be diluted more slowly. It is particularly important to delay the first dilution (Step 3) in slow-growing samples. Diluting the protoplasts too fast results in the death of many or all the growing colonies. If the

culture is diluted too fast the colonies will show extensive browning by 24 h after the dilution was made.

2. Use of mannitol as the osmoticum in the callus medium (CM) results in much better growth of the colonies than the use of glucose.

3. Selection can be started at the end of the first week, but kanamycin-resistant colonies, large enough for transfer to solid media, are obtained more rapidly if selection is delayed until the end of Week 2. This selection is still very clean; no colonies should be recovered in control (unshocked) samples cultured in the presence of kanamycin.

C. Electroporation of Tissues

This procedure is adapted from that of Dekeyser *et al.* (1990).

1. Place 20–40 tissue sections or immature embryos in 1 ml of EP medium in a 30×10-mm petri dish. Force the tissues under the liquid with fine tweezers. Incubate for 3 h at room temperature, and agitate periodically. Replace the EP medium at the end of each hour.

2. Replace the medium in the petri plate with 1 ml of EP + 100 μg/ml plasmid DNA, and incubate for 1 h.

3. Using tweezers, transfer the tissues to an electroporation chamber, then add 0.5 ml EP + 150 mM NaCl + 100 μg/ml DNA. Make sure the tissues are in the liquid at the bottom of the chamber.

4. Chill the chamber on ice for 15 min and apply a single pulse of 750 V/cm, 1180 μF.

5. Incubate the chamber on ice for an additional 10 min.

6. Using 2 ml of liquid culture medium, rinse the tissues out of the chamber and into a petri dish. Use tweezers to transfer the tissues to solid media for culture.

1. Critical Aspects of the Procedure

The pulses used for tissue electroporation are longer than those used with protoplasts. Under the conditions described above, an 1180 μF capacitor will deliver a 30-ms pulse. Other groups are using pulses as long as 100 ms. Obtaining bacteria-free plant tissues is a major problem, particularly when working with field-grown or incubator-grown plants. Preculturing the tissues for 24 to 48 h on a solid callus medium facilitates identification of contaminated tissue sections and embryos prior to electroporation. Culturing the sections on a solid, rather than a liquid, medium after electroporation also helps limit the spread of contamination.

IV. Results and Discussion

Transient expression assays of electroporated protoplasts indicate that expression of the introduced gene is maximal 24 to 48 h after electroporation (Bates et al., 1990). As seen in Fig. 1, optimum expression is obtained for pulses that kill about 50% of the protoplasts. In many published electroporation protocols, the protoplasts are chilled on ice either before or after electroporation, or both. However, transient expression of a CAT gene construct in protoplasts derived from a carrot suspension culture was the same for samples electroporated at room temperature and on ice (data not shown). In addition, holding the samples on ice after electroporation for 2, 5, 10, or 15 min in the presence of HBS + DNA, before diluting them with culture medium, did not increase expression. These results suggest that plasmid uptake through the electric field-induced pores in the membrane is electrophoretic and not diffusional.

Both transient expression and stable transformation increase with increasing DNA concentration (Fig. 2). Linear and supercoiled plasmids are equally effective in transient expression (Bates *et al.,* 1990). However, linear DNA is about 10-fold more effective than supercoiled DNA for stable transformation. Addition of carrier DNA (80 μg/ml sonicated salmon sperm DNA) gives an additional 2-fold increase in stable transformation.

The efficiency of stable transformation can be calculated as either a relative or an absolute transformation efficiency. Absolute transformation efficiency is number of kanamycin-resistant calli recovered, divided by the number of protoplasts that were electroporated. If the electroporated sample is split just before selection, then half of each sample can be grown in the absence of selection, and half grown in selection. Relative transformation efficiency is then the number of kanamycin-resistant calli divided by the total number of calli obtained. For tobacco mesophyll protoplasts, absolute transformation efficiencies are on the order of 0.1% or less, and relative transformation efficiencies are about 10-fold higher.

In experiments on tissue electroporation of cereals, the plasmid used contained the GUS gene and the BASTA gene, both of which were under the control of a strong monocot promoter. After electroporation the samples were split. Some

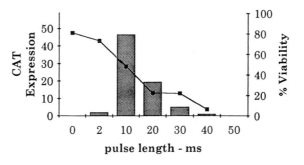

Fig. 1 Effect of pulse length on transient CAT expression. Protoplasts in from a carrot suspension culture were electroporated (750 V/cm) in the presence of a plasmid containing the CAT gene under the control of the CaMV35S promoter. Protoplast viability (the line graphs) was assessed after 24 h, and CAT activity (bar graphs) after 48 h.

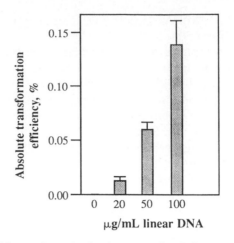

Fig. 2 Efficiency of stable transformation by electroporation. Tobacco mesophyll protoplasts were electroporated (750 V/cm, 2-ms pulse) in varying amounts of linearized plasmid containing the kanamycin gene under the control of the nopaline synthase promoter. Transformants were recovered by selection on kanamycin. Transformation efficiency is expressed as the number of transformed calli obtained, divided by the total number of protoplasts electroporated, times 100. At 20 μg/ml, supercoiled plasmid gave a transformation efficiency of 0.001%, which is more than 10-fold less than linear DNA.

of the tissues (or embryos) were assayed for GUS expression (4–7 days after electroporation) by the GUS histochemical assay (Jefferson, 1987), and the remaining tissues were cultured in the presence of BASTA. In the GUS assay, 1 to 10 (usually 1 or 2) individual, blue-staining cells were observed in about 20% of the tissue sections or embryos. These cells are intensely blue, and the blue dye is localized inside the cells and not on the cell surface. Tissues electroporated in the absence of DNA never had blue spots. However, no stable BASTA-resistant calli were obtained. In addition, increasing the pulse length, voltage, and DNA concentration and treating the tissues or embryos briefly with pectinases or cellulases did not increase the number of GUS-expressing cells per sample. GUS expression was species-dependent, the best results were obtained with immature embryos and influorescences of rye, rye leaf bases, and wheat embryos also gave an occasional positive response, and no blue-staining cells were ever observed on *Pennisetum* embryos.

V. Conclusions and Perspectives

Protoplast electroporation is an established technique that can be used efficiently for both transient expression assays and the production of stable transformants. However, electroporation of intact, walled cells is a technique still in development. Transient GUS expression indicates that DNA can be delivered to walled cells, but its efficiency is still too low for efficient stable transformation. It is hoped that the protocol provided here will assist other groups in exploring tissue electroporation.

Acknowledgment

The work on tissue electroporation was done in collaboration with Dr. Indra Vasil and members of his laboratory. I especially thank Dr. Vimla Vasil, Dr. Ana Castillo, and Mr. Mark Taylor, without whose generous help these experiments would not have been possible.

References

Bates, G. W., Carle, S. A., and Piastuch, W. C. (1990). Linear DNA introduced into carrot protoplasts by electroporation undergoes ligation and recircularization. *Plant Mol. Biol.* **14,** 899–908.

Chang, D. C., Chassy, B. M., Saunders, J. A., and Sowers, A. E., eds. (1992). "Guide to Electroporation and Electrofusion." San Diego: Academic Press.

Chassey, B. M., Saunders, J. A., and Sowers, A. E. (1992). Pulse generators for electrofusion and electroporation. *In* "Guide to Electroporation and Electrofusion" (D. C. Chang, B. M. Chassey, J. A. Saunders, and A. E. Sowers, eds.), pp. 555–569. San Diego: Academic Press.

Dekeyser, R. A., Claes, B., De Rycke, R. M. U., Habets, M. E., van Montague, M. C., and Caplan, A. B. (1990). Transient gene expression in intact and organized rice tissues. *Plant Cell* **7,** 591–602.

Fromm, M., Callis, J., Taylor, L. P., and Walbot, V. (1987). Electroporation of DNA and RNA into plant protoplasts. *Methods Enzymol.* **153,** 351–366.

Jefferson, R. A. (1987). Assaying chimeric genes in plants: The GUS gene fusion system. *Plant Mol. Biol. Reporter* **5,** 387–405.

Klöti, A., Iglesias, V. A., Wünn, J., Burkhardt, P. K., Swapan, K. D., and Potrykus, I. (1993). Gene transfer by electroporation into intact scutellum cells of wheat embryos. *Plant Cell Rep.* **12,** 671–675.

Labarca, C., and Paigen, K. (1980). A simple, rapid, and sensitive DNA assay procedure. *Anal. Biochem.* **102,** 344–352.

Nagy, J. I., and Maliga, P. (1976). Callus induction and plant regeneration from mesophyll protoplasts of *Nicotiana sylvestris. Z. Pflanzenphysiol. Bd.* **78,** 453–455.

Rogers, S. T., Horsch R. B., and Fraley, R. T. (1986). Gene transfer in plants: production of transformed plants using Ti plasmid vectors. *Methods in Enzymol.* **118,** 627–640.

Shillito, R. D., Paszkowski, J., and Potrykus, I. (1983). Agarose plating and a bead type culture technique enable and stimulate development of protoplast-derived colonies in a number of plant species. *Plant Cell Rep.* **2,** 244–247.

Tsong, T. Y. (1989). Electroporation of cell membranes: Mechanisms and applications. *In* "Electroporation and Electrofusion in Cell Biology" (E. Neumann, A. E. Sowers, and C. A. Jordan, eds.), pp. 149–163. New York: Plenum Press.

CHAPTER 27

Particle Bombardment

Paul Christou

Laboratory for Transgenic Technology and Metabolic Pathway Engineering
John Innes Center
Norwich, United Kingdom

I. Introduction

The importance of particle bombardment technology in agricultural biotechnology and its impact on basic studies in plant molecular biology is vividly illustrated by the fact that prior to its development only a few plants, primarily *Solanaceous* species, could be engineered using conventional gene transfer methods. Following the first report on particle bombardment (Klein *et al.*, 1987) most major crops, including the previously recalcitrant legumes and cereals, can now be engineered, efficiently, often in a variety-independent fashion.

Microprojectile bombardment employs high-velocity metal particles to deliver biologically active DNA into plant cells. The concept has been described in detail by Sanford (Sanford *et al.*, 1987). Christou *et al.* (1988) demonstrated that the process could be used to deliver biologially active DNA into living cells and result in the recovery of stable transformants. The ability to deliver foreign DNA into regenerable cells, tissues, or organs appears to provide the best method for achieving truly genotype-independent transformation in many agronomic crops, bypassing *Agrobacterium* host specificity- and tissue culture-related regeneration difficulties. Due to the physical nature of the technique, there is no biological limitation to the actual DNA delivery process; thus genotype is not a limiting factor. Combining the relative ease of DNA introduction into plant cells with an efficient regeneration protocol avoiding protoplast or suspension culture, particle bombardment appears to be the optimum system for transformation. The ability to engineer organized and potentially regenerable tissue permits introduction of foreign genes into elite germplasm. Consequently, backcrossing is not required to restore the original line compared to other transformation methods limited by genotype and host specificity. Transient gene expression has been demonstrated in numerous tissues representing many different species. A number of reviews describing recent advances in particle bombardment have been published (Sanford *et al.,* 1993; Christou, 1993a,b).

II. Materials

A number of different instruments based on various accelerating mechanisms are currently in use. These include the original gunpowder device (Klein *et al.,* 1987), an apparatus based on electric discharge (Christou *et al.,* 1988), a microtargeting apparatus (Sautter *et al.,* 1991), a pneumatic instrument (Iida *et al.,* 1990; Oard, 1993), an instrument based on flowing helium (Vain *et al.,* 1993), and an improved version of the original gunpowder device utilizing compressed helium (Russell Kikkert, 1993). Hand-held devices for both the original Biolistics device and the Accell device are also in use. The most widely used instrument is the one currently marketed by Bio-Rad, Inc. (Biolistics), but Accell-based methodology has been particularly useful in developing variety-independent gene transfer methods for the more recalcitrant cereals and legumes. Detailed descriptions of the various acceleration devices, principles of operation, and other particulars may be found in the primary references.

III. Methods

Methods for preparing DNA/metal mixtures have now been standardized. The only exception is transformation utilizing the microtargeting device in which DNA is not bound onto the metal particles prior to bombardment (Sautter *et al.,* 1991).

In a standard procedure in which gold is used as the accelerating particle, DNA is typically loaded onto 1.5- to 3-μm gold beads (Alpha Chemicals, Inc.) at a rate of up to 40 μg DNA/mg of gold using PEG [100 μl of a 25% solution (MW 1300–1600)] and spermidine, 100 μl of a 0.1 M solution (Klein *et al.,* 1987), to precipitate the DNA onto the gold. The coated beads are centrifuged gently and resuspended in 100% ethanol, then pipetted in 162-μl aliquots onto the carrier sheets (18 × 18-mm squares of one-half mil metalized mylar; Dupont 50MMC). After a brief period of settling, the ethanol is drained away and the sheet dried.

In typical procedures in which tungsten is used (Sanford *et al.,* 1993) 60 mg of particles are washed extensively in 1 ml of 70–100% ethanol. The particles are soaked in ethanol for 15 min, pelleted by a 15-min centrifugation (15,000 rpm), decanted, washed three times with sterile distilled water, and brought to a final volume of 1 ml in a 50% (v/v) glycerol solution. The particles can be stored at room temperature for up to 2 weeks. DNA used for biolistic experiments should be free of protein. Aliquots (25 μl) of the tungsten suspension are transferred into microcentrifuge tubes (vortexed continuously while removing aliquots of the suspension to avoid nonuniform sampling) and 2.5 μl DNA (1 μg/μl), 25 μl $CaCl_2$ (2.5 M), and 10 μl spermidine (0.1 M) are added in that order, while the microcentrifuge tube is continuously being vortexed. The mixture is allowed to react for several minutes with continuous vortexing. The coated particles are then gently pelleted by pulse centrifugation. It is recommended that the tungsten/DNA complex be used as soon as it is made due to the fact that such mixtures have been shown to be unstable. For the helium-driven system, all of the supernatant is removed and the pellet is washed in 70% ethanol. The particles are then gently pelleted and brought up in 24 μl of 100% ethanol. Six microliters of the mixed suspension is loaded onto the carrier.

IV. Critical Aspects of the Procedures

A number of critical variables have been identified and need to be considered very carefully in experiments involving transformation using particle bombardment. These include the chemical and physical properties of the metal particles used to carry the foreign DNA, the method of preparation and binding of DNA onto the particles, and the target tissue. Particles should be of high enough mass to possess adequate momentum to penetrate into the appropriate tissue. Suitable metal particles include gold, tungsten, palladium, rhodium, platinum, iridium, and possibly other second- and third-row transition metals. Metals should be chemically inert, to prevent adverse reactions with the DNA or cell components, and they should also be able to form organometallic complexes with the DNA possessing the correct stereochemistry that will allow optimal dissociation of the metal–DNA entity once the coated particle enters the target cell. Additional desirable properties for the metal include size and shape as well as agglomeration and dispersion properties. The nature, form, and concentration of the DNA need also be carefully considered. In the process of coating the metal particles with exogenous DNA, certain additives such as spermidine and calcium chloride appear to be useful. It is very important to target the appropriate cells that are competent for both transformation and regeneration. It is apparent that different tissues have different requirements; extensive histology needs to be performed in order to ascertain the origin of regenerating tissue in a particular transformation study. Depth of penetration thus becomes one of the most important variables and the ability to tune a system to achieve particle delivery to specific cell layers may be the difference between success and failure in recovering transgenic plants from a number of different tissues. Environmental variables include such parameters as the conditions of temperature, photoperiod, and relative humidity under which the plants are raised and the explants and bombarded tissues are manipulated and cultured. These variables can have a direct effect on the physiology of tissues and therefore on the success of transformation. We have found, in general, that when donor plants are stressed, the frequency of transformation, expressed in terms of independently derived transgenic plants, is significantly reduced compared to explants isolated from nonstressed plants. Interestingly enough, differences in transient expression were minimal. We have also observed seasonal effects, with much higher levels of transformation being obtained in summer than in winter months; this can be corrected by supplemental lighting (high-pressure sodium or halogen lighting).

Many investigators have overstressed the significance of transient expression data. Transient expression studies should only be used as a guide to develop systems for the stable transformation of a given species. In some cases exhaustive experiments were performed using transient expression data in an attempt to achieve complete protocol optimization for the recovery of stable transformants. This, however, may be unwise, as optimization or maximization of transient activity does not necessarily result in optimal or any stable transformation. Therefore, studies involving numbers of transiently expressing cells and foci per unit mass or volume of recipient cells may be meaningless and in a lot of cases irrelevant to the final outcome, particularly when the object is recovery of transgenic plants. It is important to utilize data from stable transformation experiments to draw conclusions pertaining to stable transformation. Of course, if no transient activity is observed following a bombardment experiment, the likelihood of obtaining stable transformants is practically zero.

V. Results and Discussion

Experiments involving transformation through particle bombardment can only be successful if optimization of cellular and gene delivery parameters progress in a concerted fashion. The two examples selected here to illustrate this point involve variety-independent gene transfer into immature rice embryos, a system that has been very well characterized developmentally and histologically, and recently published experiments involving gene transfer into immature wheat embryos from a variety that exhibits a prolific regeneration response.

A. Transformation of Rice Immature Embryos Using the Accell Instrument (Christou *et al.*, 1991, 1992)

Twelve- to fifteen-day-old rice immature embryos are excised from greenhouse-grown plants. These serve as target tissues for transformation experiments. The Accell "gun" is loaded by placing a 10-μl drop of water between the points and covering the spark chamber with the reflecting cap. The carrier sheet is subsequently placed over the top of the reflection chamber and the retaining screen put in place. The target is prepared and positioned in a way that will allow the desired area to be exposed as it is inverted above the retaining screen. The distance to the target tissue is always fixed with the electric discharge gun (about 10 cm). The assembly is evacuated to 600 mb before the discharge is activated. The scutellar region of the embryo is bombarded by charging the capacitor to 10–12 kV. Bombarded tissue is then plated on regeneration media (basal medium supplemented with 2,4D and appropriate selective agents; in the references given, hygromycin at 50 mg/liter provides efficient selection in culture). Continous selection of the proliferating tissue results in transformed embryogenic callus. The callus is transferrerd to shooting media, in the presence of the selective agent for recovery for transgenic plantlets. This procedure is applicable to all rice cultivars, including elite indica and japonica varieties. Transgenic plants are recovered from selected, clonally derived embryogenic callus, forming on the scutellum of bombarded explants. Consequently, no chimerism is encountered in this system. In addition, as a result of the very short dedifferentiation phase, somaclonal variation and other adverse effects associated with prolonged periods in culture are not prevalent.

B. Transformation of Wheat Immature Embryos Using the Biolistic PDS/He Device (Troy Weeks *et al.*, 1993)

Success with this method depends on the identification of a wheat variety that can be regenerated efficiently from immature embryos. Wheat plants from the cultivar "Bobwhite" were used to establish embryogenic callus cultures that were bombarded using the Biolistics PDS-1000/He device. Approximately 25 embryos were placed in the center of a petri dish (15 × 100 mm) containing MS medium supplemented with 2,4D (1.5 mg/liter) and sucrose (2%), solidified with 0.35% Phytagel. After 5 days in culture, the embryo-derived calli were bombarded under vacuum, with DNA-coated gold particles. Distance from the stopping plate to the target was 13 cm, and the rupture disc strength was 1100 psi. Immediately after bombardment, calli were transferred to MS selection medium containing bialaphos (1 mg/liter). Frequent visual selection and subculturing of proliferating callus resulted in the establishment of transgenic callus lines that could be regenerated to plants upon transfer to basal media supplemented with dicamba (0.5 mg/liter). Regenerated shoots were then rooted on half-strength hormone-free medium.

VI. Conclusions and Perspectives

Until recently the key barrier in achieving effective transformation of agronomically important species was the DNA delivery method. Microprojectile bombardment has had a tremendous impact on this limitation. The challenge now is shifting back to the biology of the explant used in bombardment experiments. It is apparent that the conversion frequency of transient to stable transformation events is very low. This makes recovery of large numbers of independently derived transformants labor-intensive and rather expensive. More attention needs to be paid to the biology of explants prior to, and following bombardment. We need to identify how more cells can be induced to become competent for stable DNA uptake and regeneration. Optimization of biological interactions between physical parameters and target tissue needs to be better studied and understood. Not much is known about the fate of DNA from the time particles are introduced into plant cells. Recipient tissue variation and variability due to bombardment conditions complicate the picture even further. Additional issues such as irregular particle size and uniformity as well as improvements in hardware design need also be addressed.

Particle bombardment has been shown to be the most versatile and effective way for the creation of transgenic organisms, including microorganisms, mammalian cells, and a large number of plants species. Table I provides a comprehensive listing of major crop plants that have been successfully engineered, to date, using this technology.

Table I
Species for Which Transformation Has Been Achieved

Species	Explant	Genes introduced	Reference
Glycine max	Meristems	*gus, npt II, bar, epsps*	Christou *et al.*, 1990
Arachis hypogaea	Meristems	*bar, gus*, tomato spotted wilt virus coat protein	Brar *et al.*, 1992
Phaseolus vulgaris	Meristems	*bar, gus*, bean golden mosaic virus coat protein	Russell *et al.*, 1993
Zea mays	Suspension cultures	*bar, gus*	Gordon-Kamm *et al.*, 1990
	Suspension cultures	*bar, gus, lux*, chlorsulfuron resistance	Fromm *et al.*, 1990
	Immature embryos	*bar, bt*	Koziel *et al.*, 1993
Oryza sativa	Immature embryos	*gus, bar, hmr*	Christou *et al.*, 1991, 1992
Triticum aestivum	Embryogenic callus		
	Immature embryos	*bar, gus*	Vasil *et al.*, 1992
	Immature embryos	*bar, gus*	Troy-Weeks *et al.*, 1993
Hordeum vulgare	Immature embryos		
	Embryogenic callus	*bar, gus*, barley yellow dwarf virus coat protein	Wan and Lemaux, 1993
Saccharum	Embryogenic callus	*npt II*	Bower and Birch, 1992
Gossypium hirsutum	Meristems	*gus, bar*	McCabe and Martinelli, 1993

References

Bower, R., and Birch, R. G. (1992). Transgenic sugarcane plants via microprojectile bombardment. *The Plant J.* **2,** 409–416.

Brar, G. S., Cohen, B. A., and Vick, C. L. (1992). Germline transformation of peanut (*Arachis hypogaea* L.) utilizing electric discharge particle acceleration (ACCELL®) technology. Proceedings of the American Peanut Research and Education Soc., Inc. Norfolk, Virginia, Vol. 24, p. 21.

Christou, P., McCabe, D. E., and Swain, W. F. (1988). Stable transformation of soybean callus by DNA-coated gold particles. *Plant Physiol.* **87,** 671–674.

Christou, P., McCabe, D. E., Martinell, B. J., and Swain, W. F. (1990). Soybean genetic engineering—Commercial production of transgenic plants. *Trends Biotechnal.* **8,** 145–151.

Christou, P., Ford, T., and Kofron, M. (1991). Production of transgenic rice. (*Oryza sativa* L.) plants from agronomically important indica and japonica varieties via electric discharge particle acceleration of exogenous DNA into immature zygotic embryos. *Bio/Technology* **9,** 957–962.

Christou, P., Ford, T., and Kofron, M. (1992). The development of a variety-independent gene-transfer method for rice. *Trends Biotechnol.* **10,** 239–246.

Christou, P. (1993a). Philosophy and practice of variety-independent gene transfer into recalcitrant crops. *In vitro Cellular Dev. Biol* **29,** 119–124.

Christou, P. (1993b). Particle gun mediated transformation. *Current Opinion Biotechnol.* **4,** 135–141.

Fromm, M. E., Morrish, F., Armstrong, C., Williams, R., Thomas, J., and Klein, T. M. (1990). Inheritance and expression of chimeric genes in the progeny of transgenic maize plants. *Bio/Technology* **8,** 833–839.

Gordon-Kamm, W. J., Spencer, T. M., Mangano, M. L., Adams, T. R., Daines, R. J., Start, W. G., O'Brien, J. V., Chambers, S. A., Adams, W. R., Willetts, N. G., Rice, T. B., Mackey, C. J., Krueger, R. W., Kausch, A. P., and Lemaux, P. G. (1990). Transformation of maize cells and regeneration of fertile transgenic plants. *The Plant Cell* **2,** 603–618.

Iida, A., Seki, M., Kamada, M., Yamada, Y., and Morikawa, H. (1990). Gene delivery into cultured plant cells by DNA-coated gold particles accelerated by a pneumatic particle gun. *Theor. Appl. Genet.* **80,** 813–816.

Koziel, M. G., Beland, G. L., Bowman, C., Carozzi, N. B., Crenshaw, R., Crossland, L., Dawson, J., Desai, N., Hill, M., Kadwell, S., Launis, K., Lewis, K., Maddox, D., McPherson, K., Meghji, M. R., Merlin, E., Rhodes, R., Warren, G. W., Wright, M., and Evola, S. V. (1993). Field performance of elite transgenic maize plants expressing an insecticidal protein derived from *Bacillus thuringiensis*. *Bio/Technology* **11,** 194–200.

Klein, T. M., Wolf, E. D., Wu, R., and Sanford, J. C. (1987). High-velocity microprojectiles for delivering nucleic acids into living cells. *Nature* **327,** 70–73.

McCabe, D. E., and Martinell, B. J. (1993). Transformation of elite cotton cultivars via particle bombardment of meristems. *Bio/Technology* **11,** 596–598.

McCabe, D. E., and Christou, P. (1993). Direct DNA transfer using electric discharge particle acceleration (ACCELL® technology). *Plant Cell Tissue Organ Culture* **33,** 227–236.

Oard, J. (1993). Development of an airgun device for particle bombardment. *Plant Cell Tissue Organ Culture* **33,** 247–250.

Russell, D. R., Wallace, K. M., Bathe, J. H., Martinell, B. J., and McCabe, D. E. (1993). Stable transformation of *Phaseolus vulgaris* via electric-discharge mediated particle acceleration. *Plant Cell Rep.* **12,** 165–169.

Russell Kikkert, J. (1993). The Biolistic PDS-100/He device. *Plant Cell Tissue Organ Culture* **33,** 221–226

Sanford, J. C., Klein, T. M., Wolf, E. D., and Allen N. J. (1987). Delivery of substances into cells and tissues using a particle bombardment process. *J. Particulate Sci. Technol.* **6,** 559–563.

Sanford, J., Smith, F. D., and Russell, J. A. (1993). Optimizing the biolistic process for different biological applications. *Methods Enzymol.* **217,** 483–509.

Sautter, C., Waldner, H., Neuhaus-Url, G., Galli, A., Neuhaus, G., and Potrykus, I. (1991). Micro-targeting: High efficiency gene transfer using a novel approach for the acceleration of micro-projectiles. *Bio/Technology* **9,** 1080–1085.

Troy Weeks, J., Anderson, O. D., and Blechl, A. E. (1993). Rapid production of multiple independent lines of fertile transgenic wheat (*Triticum aestivum*). *Plant Physiol.* **102,** 1077–1084.

Vain, P., Keen, N., Murillo, J., Rathus, C., Nemes, C., and Finer, J. J. (1993). Development of the particle inflow gun. *Plant Cell Tissue Organ Culture* **33,** 327–336.

Vasil, V., Castillo, A. M., Fromm, M. E., and Vasil, I. K. (1992). Herbicide resistant fertile transgenic wheat plants obtained by microprojectile bombardment of regenerable embryogenic callus. *Bio/Technology* **10,** 667–674.

Wan, Y., and Lemaux, P. (1994). Generation of large numbers of independently-transformed fertile barley plants. *Plant Physiol.* **104,** 37–48.

CHAPTER 28

Preparation and Transformation of Monocot Protoplasts

C. Maas, C. Reichel, J. Schell, and H.-H. Steinbiß

Abteilung Genetische Grundlagen der Pflanzenzüchtung
Max-Planck-Institut für Züchtungsforschung
D-50829 Köln, Germany

I. Introduction

Efficient introduction of chimeric genes by direct DNA-transfer techniques is crucial in the manipulation of monocot cells. Several conceptually and technically different methods for introduction of DNA into plant cells have been developed. Besides injection of DNA into plant cells (Crossway *et al.,* 1986), or soaking dried embryos in a DNA solution (Töpfer *et al.,* 1990), only PEG-mediated gene transfer, electroporation, and introduction of DNA by microprojectile bombardment have been established as efficient and routine methods, and have been successfully applied to many mono- and dicot species. Microprojectile bombardment has successfully been used for stable transformation and transient expression of chimeric genes but turned out to have limited efficiency. This is mainly because only a small fraction of the competent cells of a culture are targeted. The development of efficient systems for protoplast isolation and plant regeneration tends to favor the use of PEG-mediated gene transfer or electroporation, with DNA-uptake frequencies of about 50% (Bower and Birch, 1990). Of those two systems, PEG-mediated gene transfer appears to be the most convenient. It is highly reproducible for routine gene transfer into plant protoplasts (review: Steinbiß and Davidson, 1991; Potrykus, 1991).

A detailed description of a universal PEG-mediated gene transfer protocol, applicable to all mono- and dicot protoplasts tested so far, will be given in this chapter. This is followed by a typical comparative analysis of transient gene expression using various different promoters and marker genes in both mono- and dicot protoplasts.

II. Materials and Methods

A list of chemicals and equipment needed, as well as sources of these items, will be given in each part. The method and protocols, as far as possible, are presented as detailed flow charts.

A. Propagation of Plant Cell Suspension Cultures

Establishment of a monocot plant cell suspension culture normally starts from isolated immature embryos, which rapidly begin to form undifferentiated callus. This callus material is then transferred to liquid medium and is incubated on a gyrotory shaker. This forms the primary culture. Cell aggregates of the primary culture are heterogeneous in size. Smaller clumps of cells removed from this primary cultures by sieving are used as inoculum for fresh secondary cultures to form a fine and homogenous plant cell suspension culture. These latter cultures are then used to prepare protoplasts.

The barley cell suspension culture used here (*Hordeum vulgare* L. cv. Golden Promise, kindly provided by P. Lazzeri and H. Lörz, see Fig. 1A) and the media composition have already been described by Lührs and Lörz (1988). The suspension culture is grown at 25°C with constant agitation (120 rpm) in 100-ml Erlenmeyer flasks. Subculturing is done twice a week by transfer of 3 ml of culture into 10 ml of culture medium. For protoplasting, cells are collected after a total of 7 to 10 days of exponential growth and 2 days after subculture.

B. Preparation of Monocot Suspension Culture Protoplasts

Monocot plant protoplasts for analysis of gene expression can be obtained from different sources such as mesophyll, endosperm, or aleurone tissue (Lee

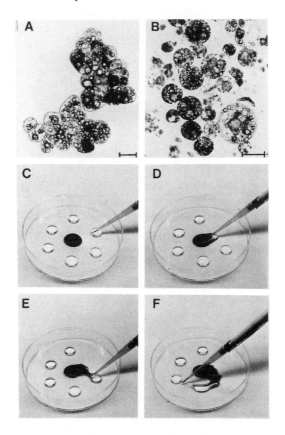

Fig. 1 (A) Barley suspension culture cells suitable for protoplast preparation (bar, 30 μm). (B) Barley protoplasts in enzyme solution prior to purification. Digestion was for 15 h (bar, 30 μm). (C) Petri dish configuration prior to DNA transfer. The droplet in the middle, containing 1×10^6 barley protoplasts (300 μl), is surrounded by 6 droplets of PEG solution (100 μl each). (D,E,F) Starting protoplast transfection: DNA (50 μl) was added to the protoplast droplet and each of the surrounding PEG droplets is sequentially mixed with the protoplasts using an Eppendorf pipette.

et al., 1989; Diaz *et al.,* 1993). However, they are difficult to obtain and the yield of protoplasts often is low. This limitation can be overcome by using a finely divided and rapidly growing cell suspension culture. The protocol described here for preparing large amounts of barley cell suspension protoplasts (see Fig. 1B) was originally used for preparing protoplasts from a maize cell suspension culture (Maas and Werr, 1989). It should be applicable to other cell suspension cultures.

1. Flow Chart for Preparation and Purification of Monocot Protoplasts

1. Cell material is used after 7 to 10 days of exponential growth, subcultured 2 days before protoplasting.

2. Pipette 2–3 g fresh weight of suspension culture into a petri dish (9 cm). Remove suspension culture medium with a pipette and add 10 ml of enzyme solution.

3. Incubate overnight (15–20 h) at 28°C.

4. Before starting the process of protoplast purification, agitate the petri dishes carefully two to three times.

5. For purification, the protoplast suspension must be transferred into 12.5-ml glass tubes by using pipettes with a large outflow diameter.

6. Centrifuge the protoplasts for 5 min at $35 \times g$; remove supernatant.

7. Add 10 ml of artificial seawater (SW) and resuspend the pellet with gentle agitation.

8. Centrifuge again for 5 min at $35 \times g$; remove supernatant.

9. Add 1 ml of SW to each tube, resuspend by gentle agitation, and pool the protoplasts into one glass tube.

10. Fill the glass tube with SW, mix by inverting two to three times, and estimate the protoplast concentration using hemocytometry.

11. Add SW to give a final protoplast concentration of $1–2 \times 10^6$ in a final volume of 300 μl.

2. Calculating Protoplast Concentrations

Calculation of protoplast yield is done using a Fuchs–Rosenthal hemocytometer that has a chamber depth of 0.2 mm. The ruling consists of 16 large squares of 1 mm each. Each large square is subdivided into 16 minisquares each with a side of 0.25 mm and an area of 0.0625 mm. For counting protoplasts, take a drop of the protoplast suspension and gently transfer it into the chamber. Cover with the coverslip and count the numbers of protoplasts using a light microscope with a 10×10 objective. Several minisquares have to be counted and the average number per minisquare is then calculated. Protoplast (pp) concentration corresponds to the following:

$$\text{number of pp} \times 16 \times 10,000 = \text{number of pp/ml.}$$

For 10 pp per minisquare (a typical value), the corresponding concentration is 1.6×10^6 pp/ml. Two glass tubes each containing 12 ml of protoplast suspension would give a total yield of 3.8×10^7 protoplasts.

3. Protocols

All solutions are passed through a sterile filter. We used Millipore Sterivex GS filtration units (0.22 μm) with a prefilter. The enzyme solution can be prepared and stored for 1–2 weeks at 4°C. SW medium is stored at room temperature and is stable for several months.

4. Enzyme Solution

> 0.66% Cellulase TC (0.5 U/mg)
> 0.33% Macerozyme R 10
> 0.033% Pectolyase Y 23
> 0.033% MES buffer

Dissolve the enzymes in two-thirds volume of Millipore-filtered water; stir for 2–3 h. Add 1 ml/per 100 ml enzyme solution of a $CaCl_2$ solution (7.3 g dissolved in 100 ml H_2O). Add 1 ml/per 100 ml enzyme solution of a NaH_2PO_4 solution (0.8 g dissolved in 100 ml H_2O). Add mannitol to adjust osmolarity (e.g., 11.5 g/100 ml should give 730 mosM). Adjust pH to 5.6 using $1N$ HCl.

5. SW Medium

> 25 mM $CaCl_2$ (in H_2O)
> Adjust osmolarity with mannitol (e.g., 11.8 g/100 ml gives 720 mosM).

6. Material and Chemicals

Cellulase TC (Serva), macerozyme R10 (Serva), pectolyase Y23 (Seishin/Japan), salts (Merck/Serva), MES buffer [2(N-morpholino)-ethane-sulfonic acid; Sigma], petri dish (non-tissue-culture grade; Greiner), glass pipettes, sterile plastic pipettes (2 ml/10 ml; Greiner), Sterivex GS sterile filters (0.22 μm; Millipore), 12.5-ml glass tubes with aluminum caps, sterile hood (BSE 4S; Gelaire), pipettes (Hirschmann/Germany), tissue culture centrifuge, tweezers, protoplast chamber (Fuchs–Rosenthal/Germany).

C. PEG-Mediated Gene Transfer

Introduction of chimeric genes into plant protoplasts by PEG-mediated gene transfer is a reproducible and widely used method. Deeper insight into the mechanism of this gene transfer method suggests a purely physico-chemical process, which appears to be independent of the protoplasts used. Data by Maas and Werr (1989) clearly showed that extracellular DNA is precipitated efficiently by the combined action of PEG together with divalent cations. The DNA seems to be taken up by the plant protoplast in the precipitated form. The particle size of the precipitate is strongly affected by the pH of the PEG solution. At the optimal pH 6.0–6.5 a very fine and homogeneous DNA precipitate forms in the presence of Ca^{2+} and Mg^{2+} ions, which is efficiently incorporated and leads to a high level of gene expression in maize protoplasts. This mechanism of gene transfer is very similar to the well-established $Ca(PO_4)$/DNA coprecipitate gene transfer into animal cells (Graham and van der Eb, 1973). Here the efficiency of DNA uptake is dependent on the quality of the DNA precipitate, which is greatly affected by the pH and bivalent cation concentration of the DNA solution.

The protocol described here for gene transfer into barley protoplasts has already been applied to protoplasts from other monocot species, such as maize, rice, and wheat (Maas and Werr, 1989; Maas *et al.*, 1990; Maas *et al.*, 1992b). Gene expression in dicot protoplasts from tobacco mesophyll (see Fig. 3) and petunia cell suspension cultures has also been analyzed using this PEG gene transfer method, suggesting that the mechanism is independent of protoplast source. Comparison of the protocol presented here with the widely used PEG protocol of Negrutiu *et al.* (1987) revealed similar DNA uptake efficiencies. The only modification necessary is that the osmolarity of all media must be individually adjusted for each batch of protoplasts.

1. Preparation of DNA

The plasmid DNA used for DNA-transfer protocols is sterilized by precipitation with ethanol overnight. This avoids phenol extraction for sterilization and simplifies pipetting since plasmid DNA solutions of different constructs (and different concentrations) can be easily resuspended to the desired concentration ratio. For DNA transfer, CsCl gradient DNA twice purified is used after extensive

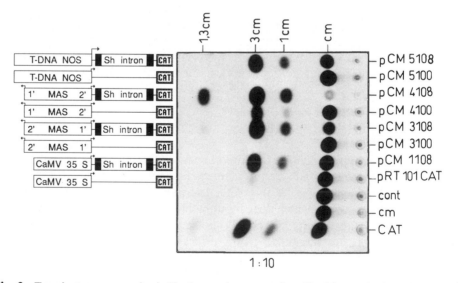

Fig. 2 Transient gene expression in *Hordeum vulgare* protoplasts. Double transfections with 1×10^6 protoplasts and 25 μg DNA were done for each plasmid. Protoplasts were cultured for 19 h prior to reporter gene analysis. CAT activity analysis of the pooled protoplast samples was done with equivalent protein amounts. Due to the extremely high CAT-activities, protein extracts were diluted successively 1 : 10-fold prior to determination of reporter gene expression. CAT activity of the 1 : 10 dilution representing 1×10^5 protoplasts is shown here. Abbreviations are as in Fig. 3.

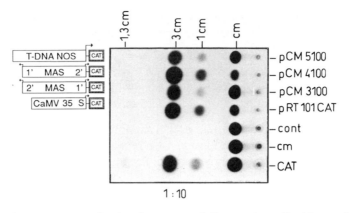

Fig. 3 Transient gene expression in tobacco mesophyll protoplasts. Double transfections with 2×10^5 protoplasts and 10 μg plasmid DNA were done for each construction. Protoplasts were cultivated 19 h prior to reporter gene analysis. CAT activity analysis of the pooled protoplast samples was done with equivalent amounts of protein. CAT activity of the 1 : 10 dilution representing 2×10^5 protoplasts is shown here. Cont, proteins extracted from untransfected control protoplasts; CAT, chloramphenicol acetyltransferase enzyme; cm, chloramphenicol alone; 1, 3, 1.3 cm, chloramphenicol actylated at the positions indicated.

dialysis. Plasmid DNA (e.g., 25 μg per monocot transfection) together with 100 μg of calf thymus (CT) carrier DNA is precipitated overnight from 0.3 M ammonium acetate by addition of three volumes of ethanol. CT DNA is prepared by intensive sonication of a CT DNA stock solution (10 μg/μl). The average size of CT DNA after sonication should be approximately 0.3 to 3.0 kbp. The overnight ethanol precipitate is then centrifuged and the supernatant is removed. The DNA is dried under sterile conditions; using a nonsterile Speedvac does not pose significant sterility problems. The DNA (25 μg plasmid/100 μg CT) is dissolved in 50 μl of sterile water.

2. Flow Chart for PEG–mediated Gene Transfer

1. Pipette 300 μl of protoplasts ($1–2 \times 10^6$) into the middle of a sterile plastic petri dish (non-tissue-culture grade).

2. Add 6 drops of Mg_2Cl–PEG solution (100 μl each) around the 300-μl protoplast drop. For the configuration, see Fig. 1C.

Several transfections can be done in parallel by following the sequence given below, separating the individual transfections by intervals of 1 or 2 min. The DNA solutions are previously prepared and are stored on ice until starting the transfections.

3. Add 50 μl of DNA solution to the protoplast drop; mix it by gentle stirring using the Eppendorf tip.

4. Sequentially move each PEG drop into the protoplast–DNA solution, and mix by moving the Eppendorf tip in circles from outside to inside (see Figs. 1D–1F).

5. Incubate for 30 min in the laminar flow hood.

6. Add 5 ml of $Ca(NO_3)_2$ wash solution.

7. Incubate for 90 s; add 5 ml of medium SW.

DNA transfer is now accomplished. If several transfections are to be performed, the scheme can be repeated. Those already completed can wait until the whole series is finished. Prepare glass tubes and petri dishes needed later on.

8. Pipette the transfection mixture (~11.5 ml) into presterilized glass tubes. Centrifuge for 5 min at $35 \times g$.

9. Remove the supernatant. This can be done by decanting, but this type of manipulation requires practice. Alternatively, use a pasteur pipette.

10. Resuspend the protoplast pellet by gentle agitation in the small volume (100–200 μl) remaining in the tube after the supernatant is decanted.

11. Add 4 ml of the appropriate medium; the protoplast concentration should be around 2–4×10^5/ml.

12. Transfer the protoplast suspension into a small petri dish (6 cm), seal with parafilm, and incubate overnight (19 h) at 28°C.

3. Protocols

All solutions have to be sterilized by filtration. We use Millipore Sterivex GS units (0.22 μm), or one-way sterile filtration units (Millipore). The $MgCl_2$–PEG solution is stored at room temperature and is stable over several months. On occasion, part of the PEG crystallizes from solution. The precipitate can be redissolved by warming the solution to 50°C. The Ca $(NO_3)_2$ wash solution is also stored at room temperature; it is stable for several months.

$MgCl_2$–PEG solution	$Ca(NO_3)_2$–wash solution
100 mM $MgCl_2$·$6H_2O$	275 mM $Ca(NO_3)_2$·$4H_2O$
450 mM mannitol	44 mM mannitol
20 mM Hepes	20 mM Hepes
25% PEG 1500	
pH 6 with 1 N KOH	pH 6 with KOH

4. Materials and Chemicals

Sterile and tissue culture equipment are as mentioned before, PEG 1500 (Merck), salts (Merck), Hepes buffer (Sigma), petri dishes (non-tissue-culture grade, Greiner), CT DNA (Sigma).

D. Analysis of Transient Gene Expression

Reporter genes have been extensively used to study gene expression by monitoring enzymatic activity of the gene product. Widely used reporter genes include those encoding β-glucuronidase (uidA), firefly luciferase (LUC), chloramphenicol acetyltransferase (CAT), and neomycin phosphotransferase II (NPTII). In this chapter we describe promoter studies that employ CAT and NPTII as reporters. The CAT and NPTII assays will be described only very briefly; for a detailed description, the reader is referred to Chapter 31 (this volume).

Typically, protoplasts are harvested 15 h after PEG-mediated gene transfer. Transfection experiments with Black Mexican Sweet (BMS) maize and barley protoplasts indicated that, although reporter gene expression can be detected in as little as 30 min after PEG-mediated gene transfer, maximal expression occurs 12–15 h after DNA uptake (Maas *et al.*, 1990; Maas *et al.*, 1992a).

1. Flow Chart for Harvesting Protoplasts

1. The protoplasts are cultivated for the appropriate time at 28°C.

2. Pipette the protoplast suspension into a glass tube; add an equal volume of SW medium.

3. Centrifuge for 5 min at 35 × *g;* remove the supernatant, leaving 200–300 μl of solution. Gently resuspend the protoplast pellet.

4. Transfer the protoplast suspension using a 2-ml plastic pipette (large outflow diameter) to an Eppendorf tube.

5. Centrifuge the protoplasts for 1 min at 4000 rpm using an Eppendorf Biofuge A.

6. Remove the supernatant, freeze the protoplast pellet in liquid nitrogen, and store at −70°C.

7. Resuspend the samples in the extraction buffer appropriate for the assay.

The estimation of CAT activity and protein concentration follows the protocol described previously (Maas and Werr, 1989). NPTII activity was assayed according to Reiss *et al.* (1984). Blocking protease activity of the protoplast extracts is essential for the assays. This can be achieved by addition to the extraction buffer of phenylmethylsulfonyl fluoride (PMSF; Serva) to a final concentration of 1 mM, and of bovine serum albumin (100 μg/assay sample, from a 10 mg/ml stock solution in H_2O).

It is important to standardize carefully the transfections. This is normally done by cotransfection with a second marker gene. Firefly luciferase is useful in this respect, since the LUC assay is highly sensitive and can be performed quickly in a nonradioactive manner. A constant amount of luciferase standardization plasmid is added to each ethanol precipitation. After transformation and incubation, the protoplasts are split and assayed in parallel for the two marker genes. Differences in luciferase activity are used to calculate a correction factor, which later is used to correct the activity of the primary marker gene. Several years of experience with transfections has led to the observation that the variation between individual transformations in one series is factor of ± 1.4. Therefore, we routinely employ two separate transfections for each construct, which later are pooled and analyzed as a combination. Standardization using the second marker gene is only done for randomly-selected samples from each transfection series.

III. Critical Aspects of Preparation, Manipulation, and Transfection of Monocot Protoplasts

A. Propagation of Suspension Cultures

It is critical to obtain finely divided, homogeneous (see Fig. 1A), and rapidly growing cell material for use in protoplast preparation. Maintenance of high cell densities during cell suspension culture is important. Furthermore, sieving (300 μm mesh) the cultures from time to time to remove larger cell aggregates is helpful.

B. Protoplast Preparation

Under optimal growth conditions, essentially all of the cells within the suspension cultures are converted to protoplasts after an overnight cellulase digestion (see Fig. 1B), and sieving to remove undigested cells is not necessary. Crucial to step B is that mechanical stresses to the protoplasts be minimized or eliminated; i.e., sieving, high-speed centrifugation, violent agitation in glass tubes, and shearing during pipetting (use of normal pipettes having small outflow diameters). This is particularly important if the protoplasts are observed to contain granules of any kind. We have observed examples of accumulation of starch granules in both maize and barley suspension culture cells. For both types of cultures, we have also experienced long-term (1–2 year) alterations that led to a reduced osmotic stability of the protoplasts. This problem was solved by increasing the initial osmolarity of protoplast solution by 200 mosM using mannitol; this suggests changes to the cell cultures that result in the intracellular accumulation of osmotically active as well as osmotically inactive compounds (cf. starch).

C. DNA Uptake

Physical parameters are crucial to optimizing DNA uptake by the protoplasts. A pH value of 6 for the PEG solution is essential. In addition, the PEG concentration experienced by the protoplast/DNA suspension should be increased slowly. This process gradually decreases hydration of the DNA and leads to the production of a fine DNA precipitate. Therefore the directions regarding moving each of the 6 PEG droplets successively into the central protoplast/DNA droplet and mixing (see Figs. 1C–1F) should be followed literally. Checking the variation between each transfection using the second marker gene should be done. When comparing several promoters, a constant and reliable DNA uptake efficiency is crucial, and should be carefully checked as part of the experiment.

For critical evaluation of the parameters influencing NPTII and CAT assay in step D, see Chapter 31 (this volume). We stress the importance of blocking protease activity by addition of PMSF and/or BSA.

IV. Promoter Analysis in Protoplasts

While establishing methods for transient gene expression in barley protoplasts, we discovered that the CaMV 35S RNA promoter is poorly expressed in monocot plant cells. Insertion of intron 1 of the maize *Shrunken 1* (*Sh1*) gene into the 5'-untranslated region of the reporter genes increases expression from CaMV 35 promoter constructs in monocot cells by a factor of about 100-fold (Maas *et al.*, 1990). Further data indicated that this stimulation is independent of the promoter that is used (Maas *et al.*, 1992a). The following experiments illustrate, for the barley protoplast system, the independence of the stimulatory effect of the *Sh1* intron 1 with respect to the promoter used and provide a comparison to the situation observed for dicot cells.

A. Materials and Methods

1. Chimeric Constructs

All chimeric genes are based on pRT cassettes (Töpfer *et al.*, 1993) containing the polyadenylation signal from the CaMV 35S RNA gene. Construction of the chimeric gene pCM1108 has already been described (Maas *et al.*, 1992b). Plasmids pRT101 CAT and pRT 103 neo were kindly provided by Reinhard Töpfer (Max-Planck Institut für Züchtungsforschung, Cologne). Plasmid pCM3100 and pCM 4100 were obtained by inserting a 500-bp *Eco*RI/*Hind*III restriction fragment, representing the 1'-2' MAS promoter from pOP443 (Velten *et al.*, 1984), in both orientations into *Hinc*II/*Xho*I cut pRT101 CAT vector (Klenow fill-in). The chimeric construct pCM5100 was obtained by inserting a 620-bp *Hind*III/*Bgl*II NOS promoter restriction fragment from plasmid pBIG HPT (Becker, 1990) into the pRT101 CAT vector also used for pCM3100 and pCM4100 cloning. To obtain the intron containing constructs pCM108, pCM4108, and pCM5108, a unique *Sma*I restriction site in the 5'-untranslated region of the CAT gene was used to insert the 1-kbp *Sh1* intron 1 isolated as a *Hinc*II restriction fragment from the chimeric plasmid pSP1076+1084 (Maas *et al.*, 1991). The NPTII construct pCM4200 was obtained by inserting the 500-bp *Eco*RI/*Hind*III 1'-2' MAS promoter fragment as isolated for pCM3100 and pCM4100 into *Hinc*II/*Xho*I-cut pRT 103neo (Töpfer *et al.*, 1993) after Klenow fill-in. The maize ubiquitin promoter plasmid pUBI CAT (Christensen *et al.*, 1992), kindly provided by Peter Eckes (Hoechst AG/Calgene), was used to isolate a 2-kbp *Hin*dIII ubiquitin promoter restriction fragment. It was inserted into *Hin*cII/*Xho*I-digested pRT101 CAT or *Hin*cII/*Xho*I-digested pRT103neo as described before, to give plasmids pCM6100 and pCM6200, respectively.

2. Transient Expression Analysis

Barley protoplasts were prepared and transfected as detailed in previous sections. For cultivation of *Nicotiana tabacum* cv. Petit Havana SR1 (Maliga *et al.*, 1973) and preparation of leaf mesophyll protoplasts the reader is referred to Pröls *et al.* (1988) and Chapter 29 (this volume). For transfection, tobacco mesophyll protoplasts were resuspended in SW medium, adjusted to 620 mosM; further steps were identical to that in Pröls *et al.* (1988). The tobacco protoplasts were finally resuspended in K3 (0.4 *M* sucrose; Nagy and Maliga, 1976; Chapter 29, this volume) and cultivated for the time indicated. Analysis of CAT and NPTII expression has already been described (Maas *et al.*, 1990) and partially in this chapter.

B. Results

As few data for promoter activity in monocot cells are available for the promoter from the T-DNA nopaline synthase (NOS) as well as for the bidirectional 1′–2′ promoter from the T-DNA mannopine synthase gene (MAS), we tested those promoters in monocot cells. Both directions of the 1′–2′ MAS promoter and the NOS promoter were fused to an expression cassette containing the CAT-coding region as well as the *Sh1* intron 1 inserted into the 5′-untranslated region of the CAT gene. Transient gene expression of these constructs in barley cells, in comparison with the CaMV 35S constructions, is shown in Fig. 2. Low-level activity of the CaMV 35S promoter (see pRT101CAT) is obvious. Transcriptional activities in the 1′ direction from the MAS promoter, and from the NOS promoter are barely detectable (see pCM100, pCM5100; Fig. 2), whereas activity from the 2′ direction of the MAs promoter appears to be high (see pCM4100; Fig. 2). The same trend is obvious from Fig. 2, in comparing the analogous constructs containing the *Sh1* intron. In all cases, insertion of the *Sh1* intron 1 strongly stimulates gene expression. As expected highest activity can be seen with the 2′ direction of the MAS promoter combined with the *Sh1* intron 1 (see pCM4108; Fig. 2).

As a control, we monitored activity of our CaMV 35S-, 1′-2′ MAS- and NOS-promoter constructs in tobacco mesophyll protoplasts using the PEG-gene transfer protocol described above. As is obvious from Fig. 3, there is no significant difference in promoter activity between these constructs in tobacco cells. These data are different from published data with transgenic tobacco, where the 2′

direction was found to be lower in activity compared to the CaMV 35S, followed by much lower activities of the 1' direction of the MAS and the NOS promoter (Harpster *et al.,* 1988).

As the 2' direction of the MAS promoter shows high level activity in barley cells, when compared to the CaMV 35S promoter, we wondered whether this promoter might allow high level expression of a selectable marker gene. This would permit more stringent selection of monocot cells, thus improving cereal transformation protocols. In initial experiments, we checked NPTII activity with the CaMV 35S-, 2' MAS as well as the promoter of the maize ubiquitin 1 gene (Christensen *et al.,* 1992). CAT expression data in Fig. 4A again reveal much higher activity of the 2' MAS promoter (pCM4100) in barley protoplasts when compared to the CaMV 35S promoter (pRT101 CAT). However, activity of the maize ubiquitin promoter seems to be extremely stimulated (see pCM 6100; Fig. 4A). As can be seen in Fig. 4B, the same differences in promoter activity are obvious when they are combined with the selectable marker gene NPTII. As

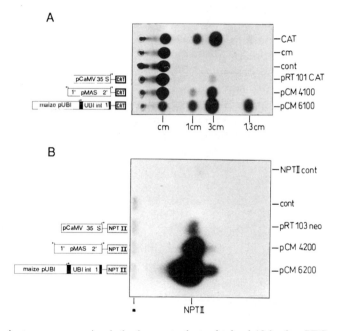

Fig. 4 Transient gene expression in barley protoplasts obtained 19 h after PEG-mediated gene transfer. (A) For determination of activity of CAT-expressing constructs double transfections with 1×10 protoplasts were done with 25 μg of plasmid DNA. CAT activity of the 1:10 dilution representing 1×10^5 protoplasts is shown here. Abbreviations are as in Fig. 2. (B) NPTII activity of double transfections (each 1×10^6 protoplasts, 5 μg plasmid DNA) 19 h after gene transfer is shown. Comparable protein amounts representing 2×10^5 protoplasts were assayed for NPTII activity. Exposure time of the autoradiograph was 90 min. Cont, proteins extracted from untransfected control protoplasts; NPTIIcont, proteins isolated from NPTII transgenic tobacco plants; *, unspecific phosphorylation activity.

expected, NPTII gene expression from the 2′ MAS promoter in barley protoplasts is much higher compared to the CaMV 35S promoter (pCM4200, pRT 103neo; Fig. 4B). Even higher NPTII activity is observed with the maize ubiquitin promoter (pCM6200; Fig. 4B).

C. Discussion

Considerable success in tissue culture and transformation of recalcitrant cereals such as wheat (Vasil *et al.*, 1992; Weeks *et al.*, 1993) and barley (Wan and Lemaux, 1994) has been achieved. Still, high-level expression of transgenes must be achieved. In this study, different promoters and promoter–intron combinations were tested for their activity. Evaluation of promoter activity clearly reveals low-level expression of the CaMV 35S RNA promoter in barley protoplasts (pRT101CAT; Figs. 2 and 4A). However, a dramatic increase in expression is observed for the 2′ direction of the MAS promoter in barley cells (pCM 4100, Figs. 3 and 4A). Insertion of *Sh1* intron 1 into the 5′-untranslated region of the CAT gene in all cases strongly stimulated gene expression. However, combination with the 2′ MAS promoter superseded expression levels of the other intron combinations (pCM4108, Fig. 2). These data suggest that stimulation of gene expression by inclusion of *Sh1* intron 1 is independent of the promoter used. This extends previous observations in barley protoplasts transfected with promoterless marker genes containing *Sh 1* intron 1, in which expression was driven by an endogenous barley promoter (Maas *et al.*, 1992a). Furthermore, it is obvious from Figure 2 that the enhanced levels of gene expression by intron 1 qualitatively reflects the differences in promoter activity in constructs lacking the intron. This indicates a constant level of stimulation independent of the promoter used.

To check activity of the maize ubiquitin 1-promoter (Christensen *et al.*, 1992) in the barley system, we also included this promoter in the analysis. As can be seen in Fig. 4A, activity of the ubiquitin 1-promoter (pCM6100) exceeds that from the 2′ direction of the MAS promoter. Whether this activity is comparable to that of the CaMV 35S–intron combination (pCM1108) has to be checked. Interestingly, the maize ubiquitin 1-gene is one of the few cases in monocots where the AUG start codon is located in exon 2, and is separated by an intron from the transcription start and exon 1. For other monocot genes with the same configuration, including the maize sucrose synthase gene (Vasil *et al.*, 1989; Maas *et al.*, 1990) and the rice actin 1-gene (Mc Elroy *et al.*, 1990), the first intron has been found to have stimulatory capacity. Whether this is also true for the intron 1 of the maize ubiquitin 1-gene remains to be established.

Enhanced expression of the CAT marker gene suggests increased expression of selectable marker genes, thus improving the selection procedure in cereal transformation. A combination of the maize ubiquitin 1 promoter with the bar gene, conferring resistance against the herbicide phosphinotricin, has already been constructed by Christensen and Quail (pAHC25, unpublished) and successfully used for rice (Cornejo *et al.*, 1993), wheat (Weeks *et al.*, 1993), and barley transformation (Wan and Lemaux, 1994). As can be seen in the case of the NPTII-gene (Fig. 4B), use of the maize ubiquitin1 promoter leads to enhanced levels of gene expression in barley cells compared to the CaMV 35S promoter (pRT103neo, pCM6200). An intermediate level of NPT II gene expression can be observed for the 2′ direction of the MAS promoter (pCM4200, Fig. 4B), which fits well with the expression data for the CAT-gene in Fig. 4A. It may be that increased levels of NPTII gene expression in barley cells could lead to an improved selection procedure.

High-level expression of the 2′ direction of the MAS promoter in barley cells as well as comparably high levels of activity of the 1′–2′ MAS, NOS and CaMV 35S promoter in tobacco cells (pRT101 CAT, pCM3100, pCM4100, pCM5100; Fig. 2) initially appears surprising. In dicots (e.g., tobacco, sugarbeet, and oilseed rape), activity normally is highest with the CaMV 35S promoter in all species, followed by the 2′-MAS and then by the NOS and 1′ MAS promoter (Harpster et al., 1988). However, Velten et al. (1984) found that in tobacco calli the activity of CaMV 35S, NOS and both directions of 1′–2′ MAS promoter is comparably high. Dekeyser et al. (1989) demonstrated an even higher activity of the 2′ MAS promoter, over that of the CaMV35S promoter, both for tobacco mesophyll and rice protoplasts. These apparent contradictions may be explained by the observation that the NOS and the 1′–2′ MAS promoters are induced by exogenously supplied auxin and by wounding (Langridge et al., 1989; An et al., 1990; Leung et al., 1991). High-level activity of these promoters in callus or protoplasts may be a consequence of cultivation at high levels of auxins and/or wounding during protoplast preparation.

Acknowledgments

The authors thank Sabine Schulze for excellent technical assistance. We are grateful to Elisabeth Schell for critical reading of the manuscript. We also appreciate collaboration with the Hoechst AG, who provided us with the maize Ubiquitin 1 promoter C. Reichel was supported by a fellowship from the Volkswagen Stiftung (I/66041).

References

An, G., Costa, M. A., and Ha, S.-B. (1990). Nopaline synthase promoter is wound inducible and auxin inducible. *The Plant Cell* **2**, 225–233.

Becker, F. (1990). Binary vectors which allow the exchange of plant selectable markers and reporter genes. *Nucl. Acids Res.* **18**, 203.

Bower, D., and Birch, R. G. (1990). Competence for gene transfer by electroporation in a subpopulation of protoplasts from uniform carrot cell suspension cultures. *Plant Cell Rep.* **9**, 386–389.

Christensen, A. H., Sharrock, R. A., and Quail, P. H. (1992). Maize polyubiquitin genes: Structure, thermal perturbation of expression and transcript splicing, and promoter activity following transfer to protoplasts by electroporation. *Plant Mol. Biol.* **18**, 675–689.

Cornejo, M.-S., Luth, D., Blankenship, K. M., Anderson, O. D., and Blechel, A. E. (1993). Activity of a maize ubiquitin promoter in transgenic rice. *Plant Mol. Biol.* **23**, 567–581.

Crossway, A., Oakes, V. J., Irvine, M. J., Ward, B., Knauf, C. V., and Shewmaker, K. C. (1986). Integration of foreign DNA following microinjection of tobacco mesophyll protoplasts. *Mol. Gen. Genet.* **202,** 179–185.

Dekeyser, R., Claes, B., Marichal, M., Van Montague, M., and Caplan, A. (1989). Evaluation of selectable markers for rice transformation. *Plant Physiol.* **90,** 217–223.

Diaz, I., Royo, J., and Carbonero, P. (1993). The promoter of barley trypsin-inhibitor BTI-CMe, discriminates between wheat and barley endosperm protoplasts in transient expression assays. *Plant Cell Rep.* **12,** 698–701.

Graham, F. L., and Van der Eb, A. J. (1973). A new technique for the assay of infectivity of human adenovirus 5 DNA. *Virology* **52,** 456–467.

Harpster, M. H., Townsend, J. A., Jones, J. D. G., Bedrook, J., and Dunsmuir, P. (1988). Relative strengths of the 35S cauliflower mosaic virus, 1′, 2′, and nopaline synthase promoters in transformed tobacco, sugarbeet and oilseed rape callus tissue. *Mol. Gen. Genet.* **212,** 182–190.

Langridge, W. H. R., Fitzgerald, K. J., Koncz, C., Schell, J., and Szalay, A. A. (1988). Dual promoter of Agrobacterium tumefaciens mannopine synthase genes is regulated by plant growth hormones. *Proc. Natl. Acad. Sci. U.S.A.* **86,** 3219–3233.

Lee, B., Murdoch, K., Topping, J., Kreis, M., and Jones, G. K. (1989). Transient gene expression in aleurone protoplasts isolated from developing caryopses of barley and wheat. *Plant Mol. Biol.* **13,** 21–29.

Leung, J., Fukuda, H., Wing, D., Schell, J., and Masterson, R. (1991). Functional analysis of cis-elements, auxin response and early developmental profiles of the mannopine synthase bidirectional promoter. *Mol. Gen. Genet.* **230,** 463–474.

Lührs, R., and Lörz, H. (1988). Initiation of morphogenic cell-suspension and protoplast cultures of barley (*Hordeum vulgare* L.). *Planta* **175,** 71–81.

Maas, C., and Werr, W. (1989). Mechanism and optimized conditions for PEG mediated DNA transfection into plant protoplasts. *Plant Cell Rep.* **8,** 148–151.

Maas, C., Schaal, S., and Werr, W. (1990). A feedback control element near the transcription start site of the maize Shrunken gene determines promoter activity. *EMBO J.* **9,** 3447–3452.

Maas, C., Laufs, J., Grant, S., Korfhage, C., and Werr, W. (1991). The combination of a novel stimulatory element in the first exon of the maize Shrunken-1 gene with the following intron 1 enhances reporter gene expression up to 1000-fold. *Plant Mol. Biol.* **16,** 199–207.

Maas, C., Schell, J., and Steinbiβ H.-H. (1992a). Applications of an optimized monocot expression vector in studying transient gene expression and stable transformation of barley. *Physiol. Plant.* **85,** 367–373.

Maas, C., Reichel, C., Schulze, S. C., Matzeit, V., Schell, J., and Steinbiβ, H.-H. (1992b). Modular structure of monocot expression vectors: A novel stimulatory element combined with the Sh 1 intron 1. *In* "Technological Impacts and Limiting Factors of Plant Genetic Transformation." Le conseil regional de picardie et ministere de la recherche. pp. 78–88, ISBN 2-908589-03-6.

Maliga, P., Sz.-Breznovitis, A., and Marton, L. (1973). Streptomycin resistant plants from callus culture of haloid tobacco. *Nature* **244,** 29–30.

Mc Elroy, D., Zhang, W., Cao, J., and Wu, R. (1990). Isolation of an efficient actin promoter for use in rice transformation. *The Plant Cell* **2,** 163–171.

Nagy, J. I., and Maliga, P. (1976). Callus induction and plant regeneration from mesophyll protoplasts of *Nicotiana sylvestris*. *Z. Pflanzenphys.* **78,** 453–455.

Negrutiu, I., Shilito, R., Potrykus, I., Biasini, G., and Sala, F. (1987). Hybrid genes in the analysis of transformation conditions. I. Setting up a simple method for direct gene transfer in plant protoplasts. *Plant. Mol. Biol.* **8,** 363–373.

Potrykus, I. (1991). Gene transfer to plants—Assessment of published approaches and result. *Ann. Rev. Plant. Phys.* **42,** 205–226.

Pröls, M., Töpfer, R., Schell, J., and Steinbiβ H.-H. (1988). Transient gene expression in tobacco protoplasts: I. Time course of CAT appearance. *Plant Cell Rep.* **7,** 221–224.

Reiss, B., Sprengel, R., Will, H., and Schaller, H. (1984). A new sensitive method for qualitative and quantitative assay of neomycinphosphotransferase in crude cell extracts. *Gene* **3,** 217–223.

Steinbiß, H.-H., and Davidson, A. (1991). Transient gene expression of chimeric genes in cells and tissues of crops. *Subcell. Biochem.* **17,** 143–166.

Töpfer, R., Gronenborn, B., Schell, J., and Steinbiß, H.-H. (1990). Uptake and transient expression of chimeric genes in seed-derived embryos. *The Plant Cell* **1,** 133–139.

Töpfer, R., Maas, C., Höricke-Grandpierre, C., Schell, J., and Steinbiß, H.-H. (1993). Improved sets of expression vectors for high level gene expression in dicotyledonous and monocotyledonous plants. *In* "Methods in Enzymology: Rec. DNA" (R. Wu, ed.), Vol. 217, pp. 66–78. New York: Academic Press.

Vasil, V., Clancy, M., Ferl, R. J., and Vasil, I. K. (1989). Increased gene expression by the first intron of maize Shrunken-1 locus in grass species. *Plant Physiol.* **91,** 1575–1579.

Vasil, V., Brown, S. M., Re, D., Fromm, M. E., and Vasil, I. K. (1992). Herbicide resistant fertile transgenic wheat plants obtained by micro-projectile bombardment of regenerable embryogenic callus. *Bio/Technology* **10,** 667–674.

Velten, J., Velten, L., Hain, R., and Schell, J. (1984). Isolation of a dual plant promoter fragment from the Ti plasmid of Agrobacterium tumefaciens. *EMBO J.* **3,** 2723–2730.

Wan, Y., and Lemaux, P. G. (1994). Generation of large numbers of independently transformed fertile barley plants. *Plant Physiol.* **104,** 37–48.

Weeks, J. T., Anderson, O. D., and Blechel, A. E. (1993). Rapid production of multiple independent lines of fertile transgenic wheat (*Triticum aestivum*). *Plant Physiol.* **102,** 1077–1084.

CHAPTER 29

Tobacco Protoplast Transformation and Use for Functional Analysis of Newly Isolated Genes and Gene Constructs

Regina Fischer and Rüdiger Hain

PF-F/Biotechnologie
Bayer AG
51368 Leverkusen, Germany
Institute of Plant Physiology
University of Hohenheim
D-70593 Stuttgart, Germany

I. Introduction

For about two decades, it has been known that under certain conditions plant protoplasts can take up viruses, microorganisms, and nucleic acids. After the development of chimeric genes (e.g., Herrera-Estrella *et al.,* 1983a,b), direct gene transfer (DGT) and the generation of transgenic plants were reported (Paszkowski *et al.,* 1984; Hain *et al.,* 1985). Since then, numerous reports of DGT into plant protoplasts have been published (for a review see Potrykus, 1990). DGT is not restricted to certain plant species, but the breadth of its application depends on the ability of protoplasts to divide subsequently and regenerate. DNA uptake methods are a valuable tool for the rapid analysis of promoter/reporter gene constructs, and can be used for studying promoter or other regulatory elements that influence transcription. If protoplast regeneration is possible, expression of the transferred DNA can also be analyzed at the level of the regenerated callus or plant (stable expression). DGT also offers the possibility of cotransformation of nonlinked DNA molecules, and this can facilitate the analysis of newly isolated genes. For instance, isolated and partially characterized λ phage DNA can be cotransferred into tobacco protoplasts along with plasmid DNA containing a marker gene selectable in the plant, for example, kanamycin resistance (nptII). About 30% of the kanaycin-resistant transgenic calli are found to express genes located on the cotransferred phage DNA. The method presented here leads reproducibly to transgenic calli and plants in sufficient amounts for biochemical, molecular, and genetic analysis and has been easily reproduced by students and beginners in the field.

II. Materials

A. Plant Material

Leaf protoplasts of *Nicotiana tabacum* cv. Petit Havana SR1 (Maliga *et al.,* 1973) are isolated from sterile shoot cultures grown on 1/2 LS medium containing 1% sucrose, 50 mg/liter myoinositol, 0.2 mg/liter thiamine-HCl, solidified with 0.8% agar–agar (Linsmaier and Skoog, 1965). Sterile shoots are maintained at 24–26°C at 3000 lux with a 16-h photoperiod.

B. Chemicals

Macerozyme R10 and Cellulase R10 are from Serva (Heidelberg, FRG), agar–agar is from Merck (Darmstadt, FRG), Seaplaque LMT agarose is from FMC Biozym (Rockland, USA), kanamycin acid sulfate is from Sigma (Munich, FRG), LS medium basal salt mixture is from Serva (Heidelberg, FRG), and B5 medium is from Flow (Costa Mesa, USA). K3 and B5 media are prepared as described in Section II,C.

C. Media

K3 medium (Nagy and Maliga, 1976) consists of (per liter):

a. Macro elements
NH_4NO_3 0.25 g

KNO_3	2.5 g
$CaCl_2 \cdot 2\ H_2O$	0.9 g
$MgSO_4 \cdot 7\ H_2O$	0.25 g
$(NH_4)_2SO_4$	0.134 g
$NaH_2PO_4 \cdot H_2O$	0.15 g
$CaHPO_4 \cdot 2\ H_2O$	0.05 g

b. Micro elements

H_3BO_3	3 mg
$MnSO_4 \cdot H_2O$	10 mg
$ZnSO_4 \cdot 7\ H_2O$	2 mg
$Na_2MoO_4 \cdot 2\ H_2O$	0.25 mg
KI	0.75 mg
$CuSO_4 \cdot 5\ H_2O$	0.025 mg
$CoCl_2 \cdot 6\ H_2O$	0.025 mg

c. Vitamins

Nicotinic acid	1 mg
Thiamine-HCl	10 mg
Pyridoxine-HCl	1 mg

d. Xylose   0.25 g

Inositol   0.1 g

e. Hormones

3-Naphthalene acetic acid	1 mg
Kinetin	0.2 mg

f. $FeSO_4 \cdot 7\ H_2O$ 27.8 mg

Na_2-EDTA   37.2 mg

g. Sucrose

0.4 M	137 g (580 mOsm)
0.3 M	103 g (430 mOsm)
0.2 M	68.5 g (320 mOsm)
0.1 M	34.25 g (200 mOsm)

pH 5.6, filter-sterilized

LS medium (Linsmaier and Skoog, 1965) basal salt mixture can be bought from Serva (Heidelberg, FRG) or other suppliers and consists of (per liter):

a. Macro elements

KNO_3	1.9 g
NH_4NO_3	1.65 g
$CaCl_2 \cdot 2\ H_2O$	0.44 g

$MgSO_4 \cdot 7\ H_2O$	0.37 g
KH_2PO_4	0.17 g

b. Micro elements

H_3BO_4	6.2 mg
$MnSO_4 \cdot H_2O$	16.9 mg
$ZnSO_4 \cdot 7\ H_2O$	10.6 mg
$Na_2MoO_4 \cdot 2\ H_2O$	0.25 mg
$CoCl_2 \cdot 6\ H_2O$	0.025 mg
KI	0.83 mg
$CuSO_4 \cdot 5\ H_2O$	0.025 mg

c. Thiamine-HCl   0.4 mg

d. Inositol   0.1 g

   Sucrose   10 g

e. $FeSO_4 \cdot 7\ H_2O$ 27.8 mg

   Na_2EDTA 37.2 mg

   pH 6, autoclaved

Gamborg's B5 medium (Gamborg *et al.,* 1968; bought from Flow, Costa Mesa, USA) basal salt mixture consists of:

a. Macro elements

KNO_3	2.5 g
$CaCl_2 \cdot 2\ H_2O$	0.15 g
$MgSO_4 \cdot 7\ H_2O$	0.25 g
$NaH_2PO_4 \cdot H_2O$	0.17 g
$(NH_4)_2SO_4$	0.134 g

b. Micro elements

H_3BO_4	3 mg
$MnSO_4 \cdot H_2O$	10 mg
$ZnSO_4 \cdot 7\ H_2O$	2 mg
$Na_2MoO_4 \cdot 2\ H_2O$	0.25 mg
$CoCl_2 \cdot 6\ H_2O$	0.025 mg
KI	0.75 mg
$CuSO_4 \cdot 5\ H_2O$	0.025 mg

c. Sucrose   20 g

   Inositol   0.1 g

d. FeNaEDTA   40 mg

e. Vitamins

   Thiamine-HCl   10 mg

Nicotinic acid   1 mg

Pyridoxine-HCl   1 mg

pH 5.5, autoclaved

III. Procedures

A. Start Up: Establishing Tobacco Shoot Cultures

1. Surface-Sterilization of Tobacco Seeds

Tobacco seeds are washed briefly in sterile water in a petri dish. After incubation in 70% ethanol for 2 min, the seeds are sterilized in 3% sodium hypochlorite (13% active chlorine; from Merck, Darmstadt, FRG) for 10 min. The seeds are washed three times in sterile H_2O, are dried under a laminar flow hood, and are either stored at 4°C or germinated on LS medium containing 2% sucrose, 100 mg/liter myoinositol, 0.4 mg/liter thiamine-HCl, solidified with 0.8% agar–agar at 24–26°C.

2. Shoot Culture Initiation

After 3 weeks, seedlings are transferred to sterile culture pots containing LS medium with 2% sucrose. When plants have reached a size of 5 cm (approximately 3 weeks later), the shoot tips are excised and transferred to new LS Medium in sterile culture boxes. Stem segments are cut into pieces containing at least one internode and placed on solid LS medium without hormones. Axillary shoots will grow from the nodes.

B. Protoplast Preparation from Sterile Shoot Cultures

Leaf protoplasts are prepared from 6-week-old sterile shoots growing on half-concentrated basal LS medium at pH 6 (Linsmaier and Skoog, 1965) containing 1% (w/v) sucrose, 50 mg/liter myoinositol, 0.2 mg/liter thiamine-HCl, solidified with 0.8% agar–agar. Leaves from a 6-week-old shoot culture (about 2 to 4 g = all leaves of one to two shoots) are sliced into segments of about 1 cm^2 under sterile conditions using a new razor-blade or a scalpel and the segments are incubated in 20 ml of the filter-sterilized K3 medium (Section II,C; Nagy and Maliga, 1976; Wullems *et al.*, 1981), 0.4 *M* sucrose, pH 5.5, containing 0.25% (w/v) macerozyme and 1% (w/v) cellulase for cell wall removal. The remaining shoot tip and stem segments of the shoot cultures can be used to generate new shoot cultures (Section III,A). Incubation is done for 14–16 h at room temperature in darkness. To release the protoplasts from the leaves, the mixture is slowly rotated 5–10 times and further incubated for 30 min. The protoplast digest is subsequently filtered through a 100-μm and a 50-μm mesh steel filter to remove cellular debris. The suspension is then centrifuged at $100 \times g$ for 10 min in glass or plastic tubes.

During this centrifugation, intact protoplasts float and are concentrated in a ring at the surface of the solution, from which they are collected. The protoplasts can be collected using a pasteur pipette, or the underlying solution can be removed using a fine glass capillary connected to a peristaltic pump. The protoplasts are gently resuspended in 10 ml of fresh K3 medium, 0.4 M sucrose, and are washed by centrifugation at $100 \times g$ for 10 min. The medium is again discarded and the yield of purified protoplasts is determined using a Fuchs–Rosenthal hemocytometer ($0.25 \times 0.25 \times 0.2$ mm; Brand, Germany). A sample ($10 \mu l$) of the purified protoplast preparation is diluted $1:10$ in K3 medium containing0.4 M sucrose and placed under the lid of the chamber. Protoplasts are counted under the microscope and the number of protoplasts (of round shape and containing chloroplasts) on four double framed fields consisting of 16 small squares is determined. The average number of protoplasts under the double-framed fields is calculated and multiplied by 5000 (chamber factor) and subsequently by 10 (dilution factor). The number represents protoplasts/per milliliter prior to transformation.

C. Transformation of Protoplasts

The concentration of the protoplasts is adjusted to 2.5×10^6/ml in K3 medium containing 0.4 M sucrose, and a single 200-μl droplet of this solution (5×10^5 protoplasts) is transferred into a sterile plastic Petri dish (85-mm-diameter). The same volume of a filter-sterilized PEG solution, pH 9, containing 25% polyethylenglycol-6000, 0.45 M mannitol, 0.1 M Ca(NO$_3$)$_2$, is placed next to the protoplast droplet but without contacting it. Ten to fifty micrograms of the DNA to be transformed (e.g., DNA containing a chimeric kanamycin resistance gene $\pm$ DNA to be analyzed for expression; do not exceed a volume of 20 μl) is added drop-wise to the protoplast solution, which is mixed gently afterward with the pipette tip (Fig. 1). The PEG droplet is then mixed with the protoplast/DNA mixture, and incubated for 20 min at room temperature. After this treatment, the PEG is diluted by addition of 5 ml of 0.275 M Ca(NO$_3$)$_2$, pH 6, dropwise to the border of the protoplast mixture. Incubation is continued for 10 min at room temperature. Protoplasts are transferred into a centrifuge tube and 5 ml of W5 medium is added. A further 5-ml aliquot of W5 medium is used to rinse gently the

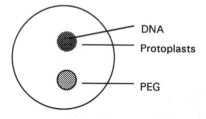

Fig. 1 Diagrammatic representation of the configuration employed for DGT into protoplasts.

remaining protoplasts from the petri dish and these are added to the protoplasts in the centrifuge tube. W5 medium, pH 5.6, consists of 0.125 M CaCl$_2$, 0.155 M NaCl, 5 mM KCl, and 5 mM glucose and is filter-sterilized. Protoplasts are pelleted by centrifugation at 65 $\times$ g for 5 min and are resuspended in 3 ml of K3 medium containing 0.4 M sucrose. Protoplasts (approximately 10^5 surviving protoplasts/ml) are incubated within a 40-mm-diameter petri dish for 3 days at 24–26°C in the dark and after that period they are cultured at the same temperature at 3000 lux.

D. Culture of Protoplasts and Selection of Kanamycin-Resistant Calli

One week after DGT, 3 ml K3 medium (containing 0.4 M sucrose and 1.6% Seaplaque agarose cooled to about 37°C) is mixed with the protoplast suspension within the 40-mm-diameter petri dish (bead type culture, Shillito et al., 1983). The agarose disc (bead), containing the protoplasts (approximately 5 $\times$ 10^4/ml), is then transferred to an 85-mm-diameter petri dish, thereby inverting the bead, and is sliced into quarters. Ten milliliters of liquid K3 medium containing 0.3 M sucrose and 75 mg/liter kanamycin acid sulfate is added, and incubation is continued at 24–26°C at 3000 lux. Each week the liquid medium is replaced with freshly prepared medium containing antibiotic. The osmotic pressure of the K3 medium is reduced step-wise every second week of culture: K3 medium containing 0.2 M sucrose is used 2 weeks after embedding of the protoplasts in agarose and K3 medium containing 0.1 M sucrose is used 4 weeks after embedding. Kanamycin-resistant calli are then transferred and maintained on LS medium [2% sucrose (w/v), 100 mg/liter myoinositol, 0.4 mg/liter thiamine-HCl] supplemented with 1 mg/liter naphthylacetic acid (NAA) and 0.2 mg/liter kinetin and 75 mg/liter kanamycin acid sulfate.

E. Callus Culture and Regeneration of Transgenic Plants

Calli are grown on solid LS [2% sucrose (w/v), 100 mg/liter myoinsitol, 0.4 mg/liter thiamine-HCl, 1 mg/liter NAA and 0.2 mg/liter kinetin and in case of kanamycin-resistant transgenic calli 75 mg/liter kanamycin acid sulfate is added]. Regeneration of transgenic plants is carried out on selective (75 mg/liter kanamycin acid sulfate) LS medium [2% sucrose (w/v), 100 mg/liter myoinsitol, 0.4 mg/liter thiamine-HCl) supplemented with 0.5 mg/liter benzylaminopurine (BAP). Shoots are rooted on hormone-free LS medium [2% sucrose (w/v), 100 mg/liter myoinsitol, 0.4 mg/liter thiamine-HCl].

F. Establishing Tobacco Suspension Cultures

Suspension cultures are initiated from approximately 4 g of soft tobacco callus material grown on LS according to Section III,E. Callus material is cut into small pieces and incubated in liquid Gamborg's B5-medium containing 0.1 mg/liter kinetin and 2 mg/liter 2,4-dichlorophenoxyacetic acid on a rotary shaker for 2 weeks at 120 rpm agitation under low light conditions (500 lux). Using a serological pipette, 10 ml of the fine cells suspended are then transferred to 40 ml of fresh medium. This is repeated each week and after 4–6 weeks a fine tobacco suspension culture is established.

G. Transient Assay

Leaf protoplasts are prepared as described in Section III,A, but 1×10^6 instead of 5×10^5 protoplasts are used for DNA uptake. Transformation is performed as described in Section III,B, with the exception that the protoplasts are left in the centrifuge tube after resuspension in 3 ml of K3 containing 0.4 M sucrose, and are incubated at 25°C in the dark for 24 h, with the tube slightly tilted. After the incubation period, 7 ml of W5-medium (Section II,B) is added and protoplasts are pelleted by centrifugation at $65 \times g$ for 5 min. The protoplasts can be applied directly to assays for gene expression (see Chapter 31, this volume).

IV. Critical Aspects of the Procedures

The DGT procedure described is simple and is not critical, provided that all media, solutions, and materials are sterilized and all operations are carried out under sterile conditions in a laminar flow hood.

V. Results

The DGT method routinely yields 100–500 transformants per petri dish and has been used over 10 years by us. About 50% of the initial protoplasts survive the DGT procedure and, on average, result in 0.1% of stable transformants. The minimum DNA amount yielding about five transformants is 100 ng of plasmid DNA (e.g., pLGVneo2103; Hain *et al.,* 1985). This method can be used for transient assays as well as for the generation of stable transgenic tobacco plants. Furthermore, bacteriophage λ DNA clones containing foreign genes (with sizes up to 42 kb) have been transferred routinely together with a plasmid containing a chimeric kanamycin resistance gene (e.g., pLGVneo2103; Hain *et al.,* 1985). About 30% of the kanamycin-resistant calli express the cotransferred genes located on the λ clone. In the case of recently isolated disease resistance genes from grapevine (Hain *et al.,* 1993), DGT experiments proved that the genes located on the clone were stilbene synthase (STS) genes and not the homologous chalcone synthase genes (CHS). Six to eight weeks after DGT, the elicitor-inducible synthesis of stilbene synthase mRNA in tobacco callus was shown (Fig. 2). Stilbene synthase activity was demonstrated in crude extracts of these tobacco

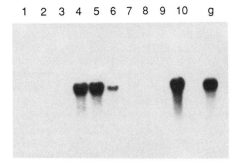

Fig. 2 Screening of 10 randomly chosen kanamycin-resistant calli for the inducible synthesis of stilbene synthase mRNA by Northern blot hybridization. Of these cotransformants, 4 accumulate stilbene synthase mRNA upon elicitor treatment. Signals are compared to the accumulation of stilbene synthase mRNA in grapevine (g, right lane).

calli (Hain *et al.,* 1993), indicating that the cloned DNA within the bacteriophage contained at least one functional stilbene synthase gene. (Hain *et al.,* 1990, 1993)

If DGT is employed for the generation of stable transgenic plants, it should be noted that the number of copies integrated can vary, and that concatemerization and changes in the structure of the input DNA can occur. Furthermore, complex DNA arrays are integrated at a presumably low, but yet undetermined, number of integration sites (Czernilofsky *et al.,* 1986a,b). Irrespectively, DGT is of particular importance for plants that are outside the normal host range of *A. tumefaciens*. *A. tumefaciens*-mediated transfer, however, remains the method of choice if foreign genes are to be integrated into the host genome as defined structures and with low copy number or if higher transformation frequencies are required.

References

Czernilofsky, A. P., Hain, R., Herrera-Estrella, L., Lörz, H., Goyvaerts, E., Baker, B. J., and Schell, J. (1986a). Fate of selectable marker DNA integrated into the genome of *Nicotiana tabacum*. *DNA* **5,** 101–113.

Czernilofsky, A. P., Hain, R., Baker, B. J., and Wirtz, U. (1986b). Studies on the structure and functional organization of foreign DNA integrated into the genome of *Nicotiana tabacum*. *DNA* **5,** 473–482.

Gamborg, O. L., Miller, R. A., and Ojima, K. (1968). Nutrient requirements of suspension cultures of soybean root cells. *Exp. Cell. Res.* **50,** 151–158.

Hain, R., Stabel, P., Czernilofsky, A. P., Steinbiß, H. H., Herrera-Estrella, L., and Schell, J. (1985). Uptake, integration, expression and genetic transmission of a selectable chimaeric gene by plant protoplasts. *Mol. Gen. Genet.* **199,** 161–168.

Hain, R., Bieseler, B., Kindl, H., Schröder, G., and Stöcker, R. (1990). Expression of a stilbene synthase gene in *Nicotiana tabacum* results in synthesis of the phytoalexin resveratrol. *Plant Mol. Biol.* **15,** 325–335.

Hain, R., Reif, H.-J., Krause, E., Langebartels, R., Kindl, H., Vornam, B., Wiese, W., Schmelzer, E., Schreier, P. H., Stöcker, R. H., and Stenzel, K. (1993). Disease resistance results from foreign phytoalexin expression in a novel plant. *Nature* **361**, 153–156.

Herrera-Estrella, L., De Block, M., Messens, E., Hernalsteens, J.-P., van Montagu, M., and Schell, J. (1983a). Chimeric genes as dominant selectable markers in plant cells. *EMBO J.* **2**, 987–995.

Herrera-Estrella, L., Depicker, A., Van Montagu, M., and Schell, J. (1983b). Expression of chimeric genes transferred into plant cells using a Ti-plasmid derived vector. *Nature* **303**, 209–213.

Linsmaier, E. M., and Skoog, F. (1965). Organic growth factor requirements of tobacco tissue cultures. *Physiol. Plant.* **18**, 100–127.

Maliga, P. S., Breznovitis, A., and Marton, L. (1973). Streptomycin-resistant plants from callus culture of haploid tobacco. *Nature New Biol.* **244**, 29–30.

Nagy, J. I., and Maliga, P. (1976). Callus induction and plant regeneration from mesophyll protoplasts of *Nicotiana sylvestris*. *Z. Pflanzenphysiol.* **78**, 453–455.

Paszkowski, J., Shillito, R. D., Saul, M., Mandak, V., Hohn, T., and Potrykus, I. (1984). Direct gene transfer to plants. *EMBO J.* **3**, 2717–2722.

Potrykus, I. (1990). Gene transfer to plants: Assessment and perspectives. *Physiol. Plant.* **79**, 125–134.

Shillito, R. D., Paszkowski, J., and Potrykus, I. (1983). Agarose plating and bead type culture technique enable and stimulate development of protoplast-derived colonies in a number of plant species. *Plant Cell Rep.* **2**, 244–247.

Wullems, G. J., Molendijk, L., Ooms, G., and Schilperoort, R. A. (1981). Differential expression of crown gall tumor markers in transformants obtained after in vitro *Agrobacterium tumefaciens*-induced transformation of cell wall regenerating protoplasts derived from *Nicotiana tabacum*. *Proc. Natl. Acad. Sci. U.S.A.* **78**, 4344–4348.

CHAPTER 30

Novel Inducible/Repressible Gene Expression Systems

Christiane Gatz

Institut für Genetik
Universität Bielefeld
33501 Bielefeld
Germany

I. Introduction

We describe a method that allows the regulated expression of a transgene upon a specific chemical stimulus. The control mechanism works at the level of initiation of mRNA synthesis. Transcriptional regulation is generally mediated through the interaction of proteins with certain DNA sequences within a promoter. Plant promoters—like other eukaryotic promoters—contain two functionally distinct classes of elements that affect transcription. The TATA motif, which is typically located 30 nucleotides upstream of the transcription start site, makes up the first class. Within this region the basal transcription factors and RNA polymerase II assemble to form a transcription initiation complex. The second class of promoter elements is composed of small DNA sequences recognized specifically by DNA-binding proteins that synergistically regulate the activity of the transcription initiation complex. Proteins that activate transcription via these elements are modular, containing a DNA binding domain and an activation domain (Drapkin et al., 1993).

Endogenous plant promoters that respond to external stimuli such as heat (Schöffl et al., 1989), wounding (Keil et al., 1989), nitrate (Back et al., 1991), or light (Gilmartin et al., 1990) can be used to control expression of transgenes. A potential disadvantage is that these stimuli affect transcription of a whole range of endogenous genes. An alternative approach is to place the general plant transcription machinery under the control of regulatory elements taken from other organisms, which respond to signals usually not encountered by plants. In this way, highly specific inductive conditions can be obtained.

Based on this idea, two different concepts of gene control can be realized, namely promoter-repressing systems and promoter-activating systems. The glucocorticoid receptor from mammalian cells activates transcription in transiently transformed plant cells. The target promoter contains the recognition sites for the receptor upstream of the TATA box. Transcription strictly depends on addition of dexamethasone (Schena et al., 1991). Similarly, the copper-dependent transcription factor from yeast activates in transgenic plants a suitable target promoter, containing the corresponding recognition sites. Transcription is severely reduced in the absence of copper (Mett et al., 1993). Other laboratories have used bacterial repressors to block transcription of the normally constitutive cauliflower mosaic virus (CaMV) 35S promoter by engineering operators into various locations of the promoter. The repressor–operator complex sterically interferes with the access of factors required for transcription to the promoter region. Using the Lac repressor–operator system, IPTG-inducible transcription was obtained in protoplasts from transgenic plants (Wilde et al., 1992). We have succeeded in constructing a tightly repressible expression system by using the Tn10-encoded Tet repressor (TetR) in combination with a suitably engineered CaMV 35S promoter (Gatz et al., 1992). As the DNA-binding affinity of TetR can be abolished by low amounts of tetracycline (Tc) addition of this inducer leads to a 200- to 500-fold induction of promoter activity throughout intact tobacco plants. At this point, the Tc-inducible promoter is the best characterized system.

In this chapter we will describe how Tc-dependent expression can be achieved in transgenic plants. In a number of enteric bacteria, the Tn10-encoded Tet repressor regulates the expression of the Tc resistance operon by binding to two nearly identical 19-bp operator sequences that overlap with three divergent promoters (Hillen *et al.*, 1984). The genes of the *tet* operon are only transcribed in the presence of the inducer Tc, which prevents the repressor from binding to its operator sequences. We have chosen the Tet repressor for regulation of a plant promoter for three reasons: (1) with a native molecular mass of 47 kDa (Hillen *et al.*, 1983) the protein can enter the nucleus as the nuclear pore allows passive diffusion of macromolecules below 45–60 kDa (Lang *et al.*, 1986); (2) the high equilibrium association constant of about 10^{-9} M for the repressor–inducer complex ensures efficient induction at very low Tc concentrations (Takahashi *et al.*, 1986); (3) Tc, like other antibiotics, diffuses readily through membranes and can therefore be taken up by plant cells.

The Tc-dependent expression system functions in transiently transformed protoplasts as well as in stably transformed cells. To efficiently work in transgenic plants, the system requires two consecutive transformation steps (Fig. 1). First, explants are transformed with the repressor gene under the control of the CaMV 35S promoter. Transformants showing high *tetR* expression are chosen for the second transformation step with the gene of interest under the control of a modified CaMV 35S promoter. This modified promoter contains three *tet* operator sequences in the vicinity of the TATA box. Tc can be applied through the roots to induce expression of the transgene.

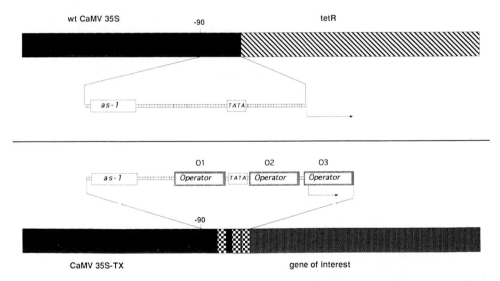

Fig. 1 Chimeric genes needed for Tc-inducible expression of transgenes. The *tetR* coding region was put under the control of the wt CaMV 35S promoter. The location of the TATA box and the most proximal activation sequence 1 (*as-1*) is indicated (Benfey *et al.*, 1989). The location of the Tet operators (ACTCTATCAGTGATAGAGT) relative to the promoter sequences is depicted in the lower panel. The arrow indicates the location of the start site of transcription. One square represents one base pair.

II. Materials

A. Plasmids

Five plasmids are used in setting up the Tc-inducible expression system in plants (Fig. 2).

1. pUC–Tet1 and pBin–Tet1

These constructs contain the *tetR* coding region under the control of the CaMV 35S promoter (Gatz *et al.,* 1991). The octopine synthase (*ocs*) polyadenylation signal provides the signals for processing the mRNA at its 3′ end. pUC-Tet1 is used for transient expression experiments, because it can be obtained in milligram amounts. The vector confers ampicillin resistance to *Escherichia coli.* pBin-Tet1 is used for stable transformation experiments. The plasmid contains the chimeric *tetR* gene as well as a chimeric neomycin phosphotransferase (*npt*) gene between the borders of the T-DNA. The remainder of the plasmid contains a broad host range replicon, which allows replication in *E. coli* and *Agrobacterium tumefaciens* as well as a *npt II* gene for selection in bacteria (Bevan, 1984). The *tetR* coding

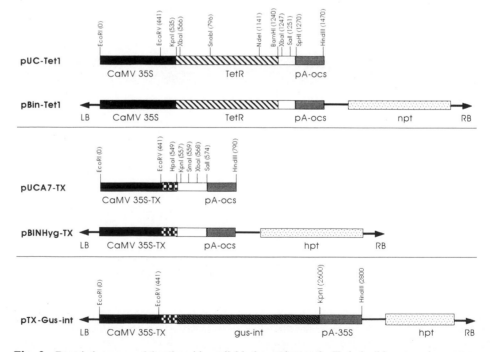

Fig. 2 Restriction maps of the plasmids available for setting up the Tc-inducible expression system. In BINHyg-TX, sites from *Hpa*I to *Sal*I can be used for inserting genes. pA, polyadenylation signal; LB, left border; RB, right border.

region (700 bp) can be cut out with *Xba*I. The fragment may be used for Northern blot experiments to identify transgenic plants with high *tetR* expression levels.

2. pBIN-HygTX and pUC-A7TX

These plasmids contain the CaMV 35S promoter with three *tet* operator sites. The *ocs* polyadenylation signal provides the signals for processing the mRNA at its 3′ end. A multiple cloning site between the promoter and the polyadenylation signal allows convenient insertion of DNA fragments. pUC-A7TX is used for transient expression experiments, because it can be obtained in milligram amounts. The vector confers ampicillin resistance to *E. coli*. pBIN-HygTX is used for stable transformation experiments. The plasmid contains the chimeric expression cassette as well as a chimeric hygromycin phosphotransferase (*hpt*) gene between the borders of the T-DNA (Becker, 1990). The remainder of the plasmid contains a broad host range replicon that allows replication in *E. coli* and *A. tumefaciens* as well as a *npt II* gene for selection in bacteria.

3. pTX-Gus-int

This plasmid is provided as a control. The ß-glucuronidase (*gus*) gene (Jefferson *et al.*, 1987) with a portable intron (Vancanneyt *et al.*, 1990) was cloned into pBIN-HygTX. It allows early staining of explants, because the intron interferes with expression of the *gus* gene in *A. tumefaciens*.

4. Control Plasmids

Plasmids (pUC derivatives for transient analysis and/or binary vectors for stable transformation) encoding the *gus* gene with the portable intron under the control of the wt CaMV 35S promoter (Benfey *et al.*, 1989) should be used for control experiments.

B. Plants

The system has been shown to function in *N. tabacum* (W38 and SNN) and *Solanum tuberosum* cv. *Desiree*. It is currently being tested in tomato and *Arabidopsis thaliana* (see also Critical Aspects). Seeds from stably transformed *N. tabacum* (SNN and W38) and *S. tuberosum* cv. *Désirée* that express sufficient amounts of TetR are available.

C. Equipment

The only special equipment needed for this protocol is the design for keeping plants in hydroponic culture. Figure 3 outlines the setup. An aquarium pump is used to aerate the nutrient solution.

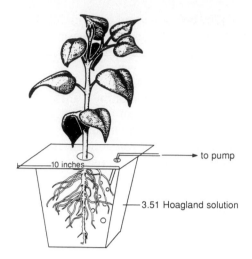

to pump

10 inches

3.51 Hoagland solution

Fig. 3 Setup for hydroponic manipulation of the plants.

D. Buffers

1. Hoagland Solution

Hoagland solution is prepared freshly from the following stock solutions:

1 M NH$_4$H$_2$PO$_4$	1 ml
1 M KNO$_3$	6 ml
1 M Ca(NO$_3$)$_2$	4 ml
1 M MgSO$_4$	2 ml
Sol. III	1 ml
Fe-EDTA	0.2 ml
H$_2$O	to one liter, pH 5.5

Sol. III:

H$_3$BO$_3$	2.86 g/liter
MnCl$_2$.2H$_2$O	1.81 g/liter
ZnCl$_2$	0.11 g/liter
CuCl$_2$	0.05 g/liter
Na$_2$MoO$_4$.2H$_2$O	0.025 g/liter

Fe-EDTA

Na$_2$EDTA·2H$_2$O	37.3 mg/liter
FeSO$_4$·7H$_2$O	27.5 mg/liter

Store at 4°C in the dark. For efficient induction chlor-tetracycline should be used.

III. Methods

The methods described here cannot cover all the experimental steps involved. Standard transformation and tissue culture procedures, Northern blot analysis, and the use of reporter genes are found in other chapters in this volume and elsewhere (Gelvin *et al.*, 1991).

A. Protocol 1: Detection of Functional Tet Repressor in Transgenic Plants

For efficient repression of the CaMV 35S-TX promoter, high levels of TetR (ca. 1×10^6 molecules per cell) are essential. pBin-Tet1 can be used to stably transform plants via *A. tumefaciens* or through direct gene transfer (for *A. thaliana*, see Critical Aspects). Regenerated plants can be analyzed by Northern blot analysis, using the *Xba*I fragment of pUC-Tet1 as a probe. The mRNA is clearly detectable at the level of total RNA.

There are two ways to analyze whether the transgenic tetR$^+$ plants produce sufficient amounts of TetR for efficient repression of the CaMV 35S-TX promoter.

1. Protocol 1a: Transient Expression

Briefly, a reporter gene is placed under the control of the CaMV 35S-TX promoter and transiently introduced into cells from tetR$^+$ plants. We have done this using PEG-mediated DNA uptake into protoplasts (Gatz *et al.*, 1991).
Steps of the procedure:

1. Prepare protoplasts from tetR$^+$ plants.

2. Transform protoplasts (0.5×10^6 protoplasts per batch) according to the scheme shown in Table I.

3. After the washing steps combine batches $1 + 2, 3 + 4, 5 + 6, 7 + 8$

Table I
Transient Expression Analysis to Test tetR$^+$ Plants

Experiment:	1	2	3	4	5	6	7	8
pUCA7TX-Rep	10	10	10	10	—	—	—	—
pCaMV-Rep	—	—	—	—	10	10	—	—
pUC-Tet1	—	—	90	90				—
HSP	90	90	—	—	90	90	100	100

Note: Numbers indicate the amount of transfected DNA in milligrams. Concentrations of DNA stocks should not be below 3 mg/ml. Rep, any reporter gene; HSP, herring sperm DNA.

4. Divide each sample of protoplasts again into two batches. Add Tc to a concentration of 1 mg/liter into the medium of batches, 2, 4 and 6, and 8. Tc is first dissolved in 50% ethanol (10 mg/liter, prepare freshly each time), then diluted in protoplast incubation medium (1 mg/ml). Five microliters of this solution is then added to a 5-ml protoplast batch.

5. Incubate for 1 to 2 days in the dark.

6. Prepare plant extracts for enzyme assays of the reporter gene.

2. Protocol 1b: Detection of Gus Activity Early after *Agrobacterium tumefaciens*-Mediated Gene Transfer

The *gus* is often used as a reporter gene because its activity can be easily detected *in situ. A. tumefaciens,* which does not contain endogenous Gus activity, expresses the gene from the binary vector. This complicates the monitoring of early events, when the Agrobacteria are not yet fully removed. Therefore, Vancanneyt *et al.* (1990) have engineered an intron into the bacterial *gus* gene, which is efficiently spliced in plants, but prevents expression in *A. tumefaciens.* pTX-gus-int contains this modified *gus* gene under the control of the CaMV 35S TX promoter on a binary vector conferring hygromycin resistance.
Steps of the procedure:

1. Take axenic leaves from tetR$^+$ plants.

2. Cocultivate leaves with *A. tumefaciens* pGV2260/pTX-gus-int and for control purposes with a corresponding strain containing the *gus-int* gene under the control of a wt CaMV 35S promoter (For pGV2260 see Deblaere *et al.,* 1985).

3. Incubate half of the explants on selective regeneration medium containing 1 mg/liter Tc.

4. Stain for Gus activity after 7 days.

B. Protocol 2: Tc Induction Procedures

Chlor-tetracycline (Sigma) is a more efficient inducer than tetracycline.

1. Protocol 2a: Infiltration of Single Leaves

After regeneration of tetR$^+$ plants with a gene of interest under the control of the CaMV 35S-TX promoter, plants expressing the transgene must be identified. The most efficient way to do this is by infiltration of single leaves.
Steps of the procedure:

1. Take a leaf from a transgenic plantlet and place it into a sterile beaker containing 1 mg/l Tc in 50 mM Na-citrate (pH 5.5). Construct a design to prevent

the leaf from floating to the surface. We usually use metal sieves of the type commonly employed for filtration of protoplasts.

2. Put the beaker into a desiccator. Apply vacuum for 2 min. Bubbles should appear; these derive from the air within the intercellular spaces. Upon removal of the vacuum the buffer is sucked into the leaf, which turns dark green.

3. Incubate leaves on liquid 2MS containing 1 mg/l Tc for up to 14 days.

4. Analyze for gene expression either at the level of RNA or protein.

2. Protocol 2b: Induction in Hydroponic Culture

Steps of the procedure:

1. Take two cuttings of one plantlet and put it into the hydroponic device (see Fig. 3)

2. Add 1 mg/l Tc to the Hoagland buffer.

3. Every other day expression of the transgene can be monitored. Disks from various leaves should be punched out to get an induction profile that gives the mean value for the whole plant.

4. Every other day, fresh Tc should be applied. Every fourth day, the buffer solution should be changed. A bacterial slime sometimes develops on the root system, presumably due to a Tc-invoked change in the bacterial population. Roots should be rinsed whenever the medium is changed.

IV. Critical Aspects of the Procedures

A. Expression of TetR

High expression of TetR (1×10^6 molecules per cell) is critical for stringent repression. Preliminary data from other labs suggest that no regulation can be observed, when the chimeric *tetR* gene is positioned on the same T-DNA as the target promoter (T. Schmülling and T. Altmann, personal communication). Other labs have reported that *A. thaliana*, which has been transformed with pBin-Tet1, does not accumulate as much *tetR* mRNA as tobacco.

B. Side Effects of Tc on Plant Growth

The Tc concentration should not be higher than 1 mg/liter. Tobacco explants regenerate on Tc containing medium as fast as on Tc-free medium. Rooting is not affected at 1 mg/liter Tc, but is strongly inhibited at 5 mg/liter Tc. In hydroponic culture, plants sometimes grow slightly slower in the presence of Tc if slime adhering to the roots is not carefully rinsed off.

C. Induction in Whole Plants and Seedlings

In tissue culture, Tc induction of whole plantlets is not efficient due to insufficient transpiration. This can be helped by opening the lid of the tissue culture container once a day, but plants start to suffer under this treatment. Moreover, Tc is not very stable. Therefore, plants must be put onto fresh medium every 3 or 4 days. This interferes with rooting. Seedlings germinated on Tc do not show full induction of the transgene. Here, again, growth in hydroponic solution or germination in a beaker on a rotary shaker containing 2MS and Tc is the most efficient way for Tc uptake. Spraying of whole plants does not lead to efficient induction.

D. Local Induction

Tc treatment of one leaf can be achieved by infiltrating Tc into the intercellular spaces using a syringe. However, due to the efficient transport of Tc, the antibiotic will eventually lead to a systemic induction of the transgene.

E. Half-Life of Tc in Plants

With respect to RNA levels, we have observed a decrease already at Day 2 after application of Tc. We do not know how Tc gets inactivated in the plant, but we recommend the addition of fresh Tc at least every other day.

V. Results and Discussion

A. Transient Expression

Repression of the CaMV 35S promoter is very stringent in transiently transformed tetR$^+$ protoplasts. Figure 4 shows the results of three independent assays

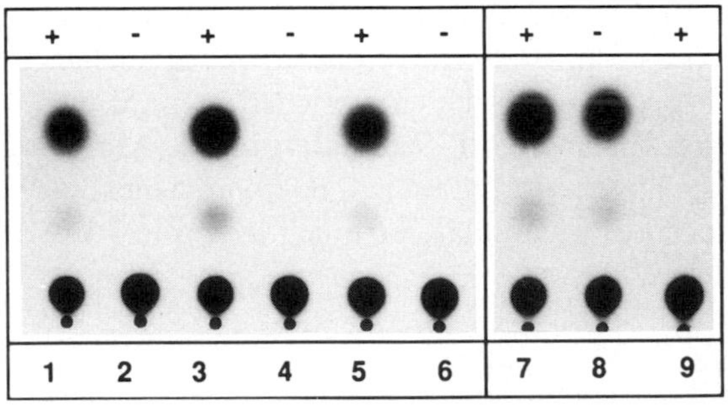

Fig. 4 Chloramphenicol acetyltransferase activity in transiently transformed tobacco protoplasts that constitutively express the *tetR* gene (Gatz *et al.,* 1991). Lanes 1 to 6 correspond to batches 1 and 2 (see Table I), lanes 7 and 8 correspond to batches 5 and 6 (see Table I), and lane 9 corresponds to batch 8 (Table I). Protoplasts were either incubated with (+) or without (−) Tc. Copyright Blackwell Science Ltd.

using chloramphenicol acetyltransferase as a reporter system. In transient expression systems, the influence of a certain gene product on the expression of a reporter gene is sometimes difficult to evaluate due to variability between samples. This can be due to differences in the PEG treatment, or due to different qualities of the input DNA. The Tc-inducible expression system allows the generation of a homogeneous population of protoplasts, which can be subdivided and incubated in the presence or absence of Tc. The effect on a reporter system can thus be clearly attributed to the induction of a gene product.

B. Infiltration of Single Leaves

Using the *gus* gene as a reporter gene, we observed full induction of mRNA synthesis after 0.5 h. Steady-state levels of the protein may be reached later. Figure 5 shows RNA analyses of six independent plants using this procedure.

C. Induction in Hydroponic Culture

Tc uptake through the roots seems to be very efficient. A clear induction of mRNA synthesis can be observed after 2 days in plants that contain five leaves. Using the *Agrobacterium rhizogenes*-encoded *rolB* gene as a marker gene, induction of the phenotype by Tc can be observed (Fig. 6) throughout the plant (Roeder *et al.*, 1994). The *rolB* gene product causes necrosis, delayed growth, and wrinkling of the leaves. By varying Tc concentrations in the nutrient solution, the transgene can be induced to different degrees.

VI. Conclusions and Perspectives

Expression of foreign genes in transgenic plants is a widely used tool to confer new characters to different species (Schell, 1987; Willmitzer, 1988). In addition, enhancing or reducing the expression of endogenous genes helps in the under-

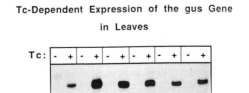

Fig. 5 Northern blot analysis of plants containing the *gus* gene under the control of an operator-containing promoter. RNA was prepared from leaves directly taken from plants [lanes marked (−)], or from leaves which were infiltrated with Tc (+). From Gatz *et al.* (1991).

Fig. 6 Induction of the *rolB* phenotype (reduced growth, necrosis) using different amounts of Tc in hydroponic culture. Tc concentrations from left to right: 0 mg/liter Tc, 0.1 mg/liter, 0.5 mg/liter, 1 mg/liter. From Roeder *et al.* (1994).

standing of the contribution of a defined gene product to the phenotype. The ability to control expression of a gene via a highly specific mechanism offers unique opportunities to study the physiological functions of certain gene products at different stages of development. The correlation of the phenotype with the kinetics of induction allows differentiation between primary and secondary consequences, which generates another advantage of a regulated expression system. Moreover, a stringently regulated promoter is absolutely required if the expression of a gene product of interest interferes with the regeneration process.

The Tc-inducible system described here can serve these purposes. Repression is stringent enough to completely suppress the *rolB* phenotype. The regulation is based on the competition of the repressor with endogenous transcription factors for binding in the vicinity of the TATA box. As the transcription initiation complex is stabilized by multiple protein–protein interactions, high levels of TetR are needed for efficient displacement of these factors. Therefore, a certain background level of transcription in the absence of Tc is observed, although three operators are located in the promoter. Each operator is located in a position where it contributes to the repression. A careful analysis of different promoter derivatives, containing single operator sites in different positions (Frohberg *et al.,* 1991; Heins *et al.,* 1992), has shown that the repressor interferes only with transcription when located close to the TATA box. There is no further location where a fourth operator could be positioned to increase the efficiency of repression. The regulation amounts to a factor of 500, which might be sufficient for most experiments.

Recently, we have developed another novel regulated expression system. We have expressed in transgenic plants a fusion protein consisting of TetR and the activation domain of the viral transcription factor VP16 (Gossen and Bujard, 1992). This chimeric activator is able to activate transcription from a target promoter containing several *tet* operators upstream of a TATA box. Upon addition of Tc the TetR-VP16 fusion protein dissociates from the DNA and transcription is completely turned off. The stringency of the regulation of this system is higher than that of the inducible system (Weinmann *et al.*, 1994).

Although Tc does not seem to affect plant growth at the concentrations used, a derivative that functions as an inducer, but does not act as an antibiotic, would be desirable. Studies on structural requirements on the Tc–TetR interaction have already indicated that TetR recognizes the drug in a manner different than that of the ribosome (Degenkolb *et al.*, 1991). In addition the crystal structure of the Tc–TetR complex has been solved (Hinrichs *et al.*, 1994) so that detailed information on the synthesis of a non-antibiotic inducer might eventually become available (W. Hillen, personal communication).

References

Back, E., Dunne, W., Schneiderbauer, A., de-Framond, A., Rostagi, R., and Rothstein, S. J. (1991). Isolation of the spinach nitrite reductase gene promoter which confers nitrate inducibility on GUS gene expression in transgenic tobacco. *Plant Mol. Biol.* **17**, 9–18.

Becker, D. (1990). Binary vectors which allow the exchange of plant selectable markers and reporter genes. *Nucleic Acids Res.* **18**, 203.

Benfey, P. N., Ren, L., and Chua, N.-H. (1989). The CaMV 35S enhancer contains at least two domains which can confer different developmental and tissue-specific expression patterns. *EMBO J.* **8**, 2195–2202.

Bevan, M. (1984). Binary *Agrobacterium* vectors for plant transformation. *Nucleic Acids Res.* **12**, 8711–8721.

Deblaere, R., Bytebier, B., De Greve, H., Debroeck, F., Schell, J., van Montagu, M., and Leemans, J. (1985). Efficient octopine Ti plasmid derived vectors for *Agrobacterium*-mediated gene transfer to plants. *Nucleic Acids Res.* **13**, 4777–4788.

Degenkolb, J., Takahashi, M., Ellestad, G., and Hillen, W. (1991). Structural requirements of tetracycline-Tet repressor interaction: Determination of equilibrium binding constants for tetracycline analogs with Tet repressor. *Antimicrob. Agents Chemother.* **35**, 1591–1592.

Drapkin, R., Merino, A., and Reinberg, D. (1993). Regulation of RNA polymerase II transcription. *Curr. Opinion Cell Biol.* **5**, 469–476.

Frohberg, C., Heins, L., and Gatz, C. (1991). Characterization of the interaction of plant transcription factors using a bacterial repressor protein. *Proc. Natl. Acad. Sci. U.S.A.* **88**, 10470–10474.

Gatz, C., Kaiser, A., and Wendenburg, R. (1991). Regulation of a modified CaMV 35S promoter by the Tn*10*-encoded Tet repressor in transgenic tobacco. *Mol. Gen. Genet.* **227**, 229–237.

Gatz, C., Frohberg, C., and Wendenberg, R. (1992). Stringent repression and homogeneous derepression by tetracycline of a modified CaMV 35S promoter in intact transgenic tobacco plants. *Plant J.* **2**, 397–404.

Gelvin, S. B., Schilperoort, R. A., and Verma, D. P. S. (1991). "Plant Molecular Biology Manual." Dordrecht, The Netherlands: Kluwer.

Gilmartin, P. M., Sarokin, L., Memelink, J., and Chua, N.-H. (1990). Molecular light switches for plant genes. *Plant Cell* **2**, 369–378.

Gossen, M., and Bujard, H. (1992). Tight control of gene expression in mammalian cells by tetracycline responsive promoters. *Proc. Natl. Acad. Sci. U.S.A.* **89,** 5547–5551.

Heins, L., Frohberg, C., and Gatz, C. (1992). The Tn*10* encoded Tet repressor blocks early but not late steps of assembly of the RNA polymerase II initiation complex *in vivo. Mol. Gen. Genet.* **232,** 328–331.

Hillen, W., Gatz, C., Altschmied, L., Schollmeier, K., and Meier, I. (1983). Control of expression of the Tn*10*-encoded tetracycline resistance genes. Equilibrium and kinetic investigation of the regulatory reactions. *J. Mol. Biol.* **169,** 707–721.

Hillen, W., Schollmeier, K., and Gatz, C. (1984). Control of expression of the Tn*10*-encoded tetracycline resistance operon. II. Interaction of RNA polymerase and Tet repressor with the *tet* operon regulatory region. *J. Mol. Biol.* **172,** 185–201.

Hinrichs, W., Kisker, C., Düvel, M., Tovar, K., Hillen, W., and Sanger, W. (1994). Structure of the Tet repressor–tetracycline complex and regulation of antibiotic resistance. *Science* **264,** 418–420.

Jefferson, R. A., Kavanagh, R. H., and Bevan, M. W. (1987). GUS fusions: ß-Glucuronidase as a sensitive and versatile gene fusion marker in higher plants. *EMBO J.* **6,** 3901–3907.

Keil, M., Sanchez-Serrano, J. J., and Willmitzer, L. (1989). Both wound-inducible and tuber specific expression are mediated by the promoter of a single member of the potato proteinase inhibitor II gene family. *EMBO J.* **8,** 1323–1330.

Lang, I., Scholz, M., and Peter, R. (1986). Molecular mobility and nucleocytoplasmic flux in hepatoma cells. *J. Cell. Biol.* **102,** 1183–1190.

Mett, V. L., Lochhead, L. P., and Reynolds, P. H. S. (1993). Copper controllable gene expression system for whole plants. *Proc. Natl. Acad. Sci. U.S.A.* **90,** 4567–4571.

Roeder, F. T., Schmülling, T., and Gatz, C. (1994). Efficiency of the tetracycline dependent expression system: Complete suppression and efficient induction of the *rolB* phenotype in transgenic plants. *Mol. Gen. Genet.* **243,** 32–38.

Schell, J. (1987). Transgenic plants as tools to study the molecular organization of plant genes. *Science* **237,** 1176–1183.

Schena, M., Lloyd, A. M., and Davis, R. (1991). A steroid-inducible gene expression system for plant cells. *Proc. Natl. Acad. Sci. U.S.A.* **88,** 10421–10425.

Schöffl, F., Rieping, M., Baumann, G., Bevan, M. W., and Angermüller, S. (1989). The function of plant heat shock promoter elements in the regulation of chimaeric genes in transgenic tobacco. *Mol. Gen. Genet.* **217,** 246–253.

Takahashi, M., Altschmied, L., and Hillen, W. (1986). Kinetic and equilibrium characterization of the Tet repressor-tetracycline complex by fluorescence measurements. Evidence for divalent matal ion requirement and energy transfer. *J. Mol. Biol.* **187,** 341–348.

Vancanneyt, G., Schmidt, R., O'Connor-Sanchez, A., Willmitzer, L., and Rocha-Sosa, M. (1990). Construction of an intron-containing marker gene: Splicing of the intron in transgenic plants and its use in monitoring early events in *Agrobacterium*-mediated plant transformation. *Mol. Gen. Genet.* **220,** 245–250.

Weinmann, P., Gossen, M., Hillen, W., Bujard, H., and Gatz, C. (1994). A chimeric transactivator allows tetracycline-responsive gene expression in whole plants. *Plant J.* **5,** 559–569.

Wilde, R. J., Shufflebottom, E., Cooke, S., Jasinska, I., Merryweather, A., Beri, R., Brammar, W. J., Bevan, M. W., and Schuch, W. (1992). Control of gene expression in tobacco cells using a bacterial operator-repressor system. *EMBO J.* **11,** 1251–1259.

Willmitzer, L. (1988). The use of transgenic plants to study plant gene expression. *Trends Genet.* **4,** 13–18.

CHAPTER 31

Reporter Genes

Clemens Suter-Crazzolara, Manfred Klemm, and Bernd Reiss

Max-Planck-Institut für Züchtungsforschung
Carl-von-Linné-Weg 10
50829 Köln, Germany

I. Introduction

Differential gene expression is essential for the development of higher organisms. Transcriptional regulation appears to play a key role in this process. However, the relevant gene products may be difficult to detect, due to low abundance or chemical properties. Therefore, the analysis of developmental processes is greatly simplified if promoters or other regulatory sequences are linked to reporter genes. Likewise, reporter enzymes that remain active when fused to other peptides allow the study of protein trafficking in the cell.

To be optimal, reporter genes should have the following characteristics:

1. the genetic organization should be well described,

2. the gene product should not be present in the organism or tissue under study,

3. the gene product should be well characterized with regard to biochemical activity, substrate dependence, and stability, and

4. the product of the reaction catalyzed by the reporter gene product should be stable, easily detectable, and quantifiable.

Several reporter genes are currently in use in plants. The four most widely used systems, GUS, NPTII, CAT and luciferase, are described here. Methods to assay other reporters are described elsewhere. These include phosphinothricin acetyltransferase (De Block *et al.*, 1989), catechol oxygenase (Buell and Anderson, 1993), gentamycin acetyltransferase (Hayford *et al.*, 1988), and cytosine deaminase (Stougaard, 1993).

GUS (β-glucuronidase, EC 3.2.1.31) is encoded by the *Escherichia coli uidA* gene (Jefferson *et al.*, 1986, 1987) and has become the most widely used reporter gene in plants. GUS hydrolyzes β-glucuronides, of which a wide variety of synthetic substrates are commercially available. The enzyme is very stable under a range of physiological conditions. The frequent use of GUS is mainly justified because it allows histochemical localization of enzyme activity in tissues or complete plantlets. Also, after the initial investment in a fluorimeter, a fluorogenic assay provides a relatively inexpensive and highly sensitive method to quantify GUS activity. Both assays will be described here.

NPTII (neomycin phosphotransferase II, EC 2.7.1.95), derived from transposon Tn5, is mainly known as a selectable marker, but the gene is also used as a reporter (Beck *et al.*, 1982; Reiss *et al.*, 1984a). NPTII belongs to a group of enzymes which, among other amino-glycosides, phosphorylate kanamycin. Here we describe a method to assay NPTII in crude extracts that uses gel electrophoresis to separate interfering and nonspecific activities from NPTII. Specific activity is detected by *in situ* phosphorylation of kanamycin. (Reiss *et al.*, 1984b).

CAT (chloramphenicol acetyltransferase, EC 2.3.1.28), derived from transposon Tn9, is widely used as a reporter. CAT transfers an acetyl group from acetyl-coenzyme A (Ac-CoA) to chloramphenicol to yield acetylated derivatives. The classical assay, developed by Gorman *et al.* (1982), is the most widely performed variant and uses radioactively labeled chloramphenicol as the substrate. Acetylated products and substrate are separated by silica gel thin-layer chromatography and visualized by autoradiography.

Luciferase (luciferin 4-monooxygenase, EC 1.13.12.7) is derived from the firefly *Photinus pyralis* (De Wet *et al.*, 1987; Luehrsen *et al.*, 1992). Yellow-green light (540 nm) is emitted upon ATP-dependent oxidation of D-luciferin and can be measured directly in a luminometer. Luciferase combines easy and rapid detection with great sensitivity. If activities are high enough, light emission can be monitored by scintillation counting (Nguyen *et al.,* 1988). Luciferase can also be localized *in situ* in a nondestructive manner. However, quantification of luciferase activity is cumbersome, since luciferase becomes bound to oxyluciferin during the reaction and becomes inactive (Brasier and Ron, 1992).

II. Materials

A. General Materials

1. Instrumentation

Homogenization pestle adapted to microcentrifuge tubes (Ncolab or custom-made glass) connected to a rotary electric drive.

2. Chemicals and Solutions

Unless otherwise indicated, all chemicals were purchased from Merck, Sigma or Serva. Seasand (Merck, Cat. No. 7712). For all reagents and buffers, double distilled H_2O was used. X-ray films were purchased from Kodak. Protein assay according to Bradford (BioRad, Cat. No. 500-002)

B. GUS Assays

1. Instrumentation

Fluorimeter: Perkin–Elmer LS-2B (for fluorogenic assay only)

2. Chemicals and Solutions

Gus extraction buffer:
50 mM NaPO$_4$, pH 7
10 mM 2-mercaptoethanol
10 mM Na$_2$EDTA
0.1% sodium N-lauroylsarcosine (Serva, Cat. No. 27570)
0.1% Triton X-100 (Serva, Cat. No. A 46252)

Histochemical staining buffer (X-Gluc):
To prepare a stock solution, 100 mg X-Gluc (5-bromo-4-chloro-3-indolyl glucuronide, Sigma, Cat. No. B9401) is dissolved in 1 ml dimethylformamide.
Standard staining solution:
Add 50 μl X-Gluc stock solution to 10 ml GUS extraction buffer (final concentration is 0.5 mg/ml).

Enhanced staining solution (optional):

Mix 50 μl X-Gluc stock solution with 9.8 ml GUS extraction buffer, 100 μl 50 mM K$_3$Fe(CN)$_6$, and 100 μl 50 mM K$_4$Fe(CN)$_6$.

Fluorogenic assay buffer (MUG):

To prepare the assay buffer, dissolve 22 mg MUG (4-methyl umbelliferyl glucuronide, Sigma, Cat. No. M9130) in 50 ml GUS extraction buffer (final concentration is 0.44 mg/ml). The solution is stable at 4°C in the dark for up to 1 month.

Stop buffer:

Dissolve 21.2 g Na$_2$CO$_3$ in 1 liter of water (final concentration is 0.2 M)

Polyclar solution:

5 mg/ml polyclar (Serva, Cat. No. 33162) in GUS extraction buffer.

C. NPTII Assay

1. Instrumentation

Vertical electrophoresis system BRL Model V16, including 1.5-mm spacers and combs, glass plates (14 × 17 cm), power supply (Pharmacia).

2. Chemicals and Solutions

Isopropanol
Agarose (low melting, FMC, Cat. No. 50082), 2% in water
Ammonium persulfate (APS)
TEMED (*N,N,N',N'*-tetramethylethylenediamine)
[γ-^{32}P]ATP (>5000 Ci/μmol, Amersham, Cat. No. PB10218)
Phosphocellulose paper P81 (Whatman)
Blotting paper (Schleicher & Schuell, Cat. No. GB002)
Optional: RepelSilane (Pharmacia)

Lower Tris:
1.5 M Tris/HCl, pH 8.8

Upper Tris:
0.5 M Tris/HCl, pH 6.8

Sample buffer:
10 ml glycerol
5 ml 2-mercaptoethanol
12.5 ml upper Tris

20 mg xylene xyanole FF
10 mg bromophenol blue
1 ml 10% SDS
ad 100 ml water

Tris/glycine:
30 g Tris
144 g glycine
ad 1 liter water.
Running buffer is a 1:50 dilution in water.

Acrylamide (29 + 1):
290 g acrylamide (Serva, 2× crystallized)
10 g bisacrylamide (Serva)
ad 1 liter water.

NPTII assay buffer:
8 g Tris
8.4 g $MgCl_2$
21.2 g NH_4Cl
add 800 ml water, adjust to pH 7.1 with 1 M maleic acid.
Adjust volume to 1 liter.

Kanamycin solution: Kanamycin sulfate (Boehringer, Cat. No. 106801), 20 mg/ml in water.

D. CAT Assay

1. Instrumentation

Chamber (TLC glass developing tank, Sigma).
TLC plastic sheets silica gel 60 F254, precoated (0.2-mm thickness, Merck, Cat. No. 5735).

2. Chemicals and Solutions

D-*threo*-[dichlorocetyl-1-^{14}C]Chloramphenicol, 56 mCi/mmol (Amersham, Cat. No. CFA 515), store at 4°C.
Acetyl-coenzyme A (Sigma A-2056), 10 ml/ml in water, store at −70°C.
Ethyl acetate (Merck Cat. No. 9623)
Extraction buffer: 0.25 M Tris/Hcl, pH 7.5
Chloroform/methanol, 95:5

E. Luciferase Assay

1. Instrumentation

Luminometer (Berthold "Biolumat LB9500C[2] or LKB" 1251 luminometer[2]).

2. Chemicals and Solutions

Both of the following buffers should be stored in aliquots at $-20°C$. Add 2-mercaptoethanol just prior to use.

Luciferase extraction buffer:
100 mM HK$_2$PO$_4$/H$_2$KPO$_4$, pH 7.8
1 mM EDTA (ethylenediaminetetraacetic acid)
10% (v/v) glycerol
7 mM 2-mercaptoethanol

Luciferase assay buffer:
25 mM Tricine, pH 7.8
15 mM MgCl$_2$
5 mM ATP
0.5 mg/ml bovine serum albumin
7 mM 2-mercaptoethanol

Luciferin solution: Dissolve 28 mg of D-luciferin (Sigma, Cat. No. L9504, Boehringer, Cat. No. 411400) in 10 ml of water and store this stock solution (10 mM) at 4°C in the dark. For actual use, this stock is diluted to 250 μM with water. Only freshly prepared solution should be used.

III. Methods

A. General Remarks

The homogenization method depends on the type of tissue or cultured cells to be analyzed. Also, many alternatives for homogenization exist. We use a microcentrifuge tube-adapted pestle exclusively. Basically, protoplasts or tissues are mixed with seasand and extraction buffer and ground with the sample pestle and electric drive. The exact quantities are given with each individual method. Cell debris is removed by centrifugation at room temperature for 5 min. The supernatant is transferred to a new tube and kept on ice water until processed. In order to compare enzyme activities, standardization of extracts is necessary. The most convenient method is determination of the protein content of extracts, for example, using the assay described by Bradford (1976). Enzyme kinetics are characterized by a lag, linear, and stationary phase. During the linear range, substrate accumulation is proportional to enzyme concentration. Enzyme assays are quantitative only within the linear phase, which may have to be determined initially. As a control, standard extracts should be included, which may be prepared from bacterial cells or plant tissues with known activity.

Detailed protocols to assay GUS (histochemically and fluorogenically), NPTII, CAT, and luciferase are described below.

B. Histochemical Localization of GUS

Fresh tissue is directly stained in X-Gluc solution. This colorless substrate is hydrolyzed by GUS to yield finally a blue precipitate at the site of enzyme activity. To reduce the amount of X-Gluc to be used, plant material or plantlets are placed in a vial of minimal dimensions. We routinely collect plant material in Eppendorf tubes and use 100 μl of histochemical (X-Gluc) staining buffer. After vacuum infiltration for 1 min (an aspirator is sufficient for this purpose), we stain 16 h at 37°C. To facilitate detection of the blue color, chlorophyll can then be removed by incubation in 70% ethanol for several hours. Tissues may be kept in this solution at room temperature for some time, but signals are lost eventually.

C. Fluorogenic GUS Assay

This is used for the quantification of GUS activity in extracts. GUS hydrolyzes MUG to form the fluorogenic compound MU (4-methyl umbelliferone). The protocol described below is designed for Eppendorf tubes, but can easily be scaled down for microtiter plates. Extracts are prepared, as described under general remarks, by grinding 10^6 cells or 100 mg tissue with 100 mg seasand and 100 μl GUS extraction buffer. It is important to remove carefully all cell debris, since this may block the cuvette of the fluorimeter. To start the actual enzyme assay, mix 1 to 50 μl of extract with 500 μl MUG assay buffer (equilibrated at 37°C). A time course of enzymatic activity can be obtained by removing 100-μl aliquots after 1, 30, 60, and 120 min. To terminate the reaction, aliquots are immediately added to 900 μl of stop buffer. This also adjusts the pH to optimal conditions for measuring MU fluorescence.

Set the fluorimeter to an excitation wavelength of 365 nm and emission wavelength of 465 nm, and calibrate the instrument with known MU standards. Measure fluorescence and calculate the MU concentration of the samples. GUS activity is expressed as nanomoles MU per minute. Linearity of the assay is excellent for concentrations between 1 nM MU and 5–10 μM MU. Samples exceeding the upper limit of the fluorimeter can simply be diluted in stop buffer, as long as activities are within the linear range. Tissues and extracts can be stored at −70°C for several months without loss of activity, or at 4°C, with very little loss (dependent on intrinsic protease levels). Storage at −20°C in extraction buffer appears to reduce enzymatic activity.

D. NPTII Assay

For this assay, the proteins are separated first by native gel electrophoresis and NPTII activity is determined later *in situ* after overlay with an agarose gel containing the substrates. The assay is carried out at room temperature, but rapid preparation and loading of extracts on the gel is important. To start, a discontinuous, native 10% acrylamide gel is prepared. This gel system is essentially according to Laemmli (1970), but SDS is omitted from all solutions. Since detailed protocols to prepare protein gels are described (Piccioni *et al.,* 1982), we only give a short description for the BRL apparatus used here. For the separation gel, 40 ml of solution (5 ml lower Tris, 13.3 ml acrylamide solution, 21.7 ml water) is prepared. After addition of 12 mg APS, polymerization is initiated by 40 μl TEMED and the gel is poured and overlaid with isopropanol. After polymerization and removal of isopropanol, the stacking gel (2.5 ml upper Tris, 2 ml acrylamide solution, 5.5 ml water, 4 mg APS, 10 μl TEMED) is cast on top of the separation gel.

Extracts are prepared as described in Section IIIA. Usually, 100 mg leaf tissue are ground with 100 mg seasand and 80 μl NPTII sample buffer. After removal of cell debris, 40 μl of supernatant are applied to the gel. For electrophoresis, a current of 30 mA is applied until the dye has entered the separating gel. Continue electrophoresis until the bromophenol blue dye reaches the end of the separating gel. To run the gel overnight, we adjust the current to 9 mA; however, currents up to 20 mA are tolerated by this system. NPTII migrates approximately at the position of the xylene xyanole FF dye.

The next step in the NPTII assay is *in situ* phosphorylation of kanamycin. After marking its orientation, the separation gel is washed twice in water for 15 min and equilibrated in NPTII assay buffer for 30 min. Meanwhile, the radioactive agarose gel should be prepared. For this purpose, 20 ml NPTII assay buffer, 50 μl kanamycin solution, 10 μl [γ-^{32}P]-ATP (10 μCi/μl), and 20 ml 2% agarose (approximately 65°C) are mixed and cast in a gel tray, which can be constructed with adhesive tape and the small glass plate used for the protein gel. Decontamination is facilitated if the glass is treated with RepelSilane. After gelling, the agarose gel is placed on top of the protein gel. After 30 min, place a piece of P81 paper, a piece of blotting paper, a stack of paper towels, and a weight (1 kg) on top of the agarose gel. Capillary transfer of kanamycin to the P81 paper is allowed for 3 h. Wash the P81 paper with hot water (80°C, 2 × 5 min) and briefly with cold water. Expose the air-dried filter to X-ray film with an intensifying screen. Overnight exposures are usually sufficient. For exact quantification, radioactive spots can be excised from the filter and radioactivity measured by liquid scintillation counting, using common scintillation cocktails.

E. CAT Assay

For this assay, 100 mg of leaf tissue is ground with 100 mg of seasand and 105 μl of buffer as described in Section IIIA. After removal of cell debris, the supernatant is incubated for 10 min at 65°C in order to inactivate background activities and acetylases competing for acetyl-CoA. A precipitate will form during this step, which is removed by centrifugation. To perform the enzymatic reaction, 100 μl of supernatant is mixed with 10 μl of 10 mM acetyl-CoA and 0.5 μCi [^{14}C]chloramphenicol and incubated at 37°C for 1 h. The incubation time may vary depending on the activity of the extract. To stop the reaction, an equal volume of ethyl acetate is added. Chloramphenicol is extracted from the aqueous phase by thorough mixing and centrifugation. Extraction is repeated twice and the organic phases are combined. Ethyl acetate is evaporated at 96°C and the residue taken up in 20 μl of ethyl acetate. The starting positions for chromatography are marked on the TLC plate by pencil (distance 1.5 cm, 2 cm from the bottom). The extracts are applied in small aliquots and the spots are allowed to dry in air. The TLC plate is placed in the chamber and the bottom immersed 1 cm deep in chloroform/methanol. Chromatography is continued until the solvent reaches the top of the plate. The chromatogram is removed from the chamber, air-dried, wrapped in Saran wrap, and exposed to X-ray film with an intensifying screen. Overnight exposures are usually sufficient. Quantification is possible by liquid scintillation counting of radioactive spots scraped off the plate.

F. Luciferase Assay

For this assay, 100 mg of leaf tissue is ground with 100 mg of seasand and 100 μl extraction of buffer as described in Section IIIA. These extracts are stable at -70°C for up to 3 months. To perform the assay, add 5–50 μl of extract to 200 μl of assay buffer in a luminometer cuvette. After the luciferin solution and the samples are adjusted to room temperature, the cuvette is placed in the luminometer and the enzymatic reaction is initiated by injection of 100 μl of luciferin solution. The number of photons emitted within the first 10 s of the reaction is determined and integrated over this period. Luciferase activity is expressed as light units/10 s.

IV. Critical Aspects of the Procedures

General remarks: Many tissues (for instance, tobacco leaves) contain factors that inactivate reporter gene products. If this is the case, it is advisable to modify the appropriate extraction buffer by addition of protease inhibitors (for instance, 0.13 mg/ml leupeptin or 0.1% PMSF; phenylmethylsulfonyl flouride), antioxidants (for instance, 5 mM cysteine), or 5 mM EDTA. Bovine serum albumin may also be added after determination of the protein concentration.

The quality of histochemical localization of GUS is affected by numerous variables. These include all aspects of sample preparation, fixation, and the reaction itself (reviewed by Jefferson, 1987). A frequently observed problem is uneven staining of tissue that is expected to express GUS homogeneously. One reason for this problem is that the product of X-Gluc hydrolysis is an indoxyl derivative that itself is not colored. After oxidative dimerization, an insoluble, highly colored indigo dye is formed. This process is stimulated by oxygen. Therefore, local variations in oxygen availability may be the cause of heterogeneous staining. This problem is minimized by the addition of an oxidation catalyst such as the ferricyanide/ferrocyanide mixture in the enhanced staining solution. Moreover, without such a catalyst localized peroxidase may enhance the apparent localization of GUS activity. Another factor that may lead to uneven staining of tissues is local underrepresentation of staining solultion. In this case, the vacuum infiltration conditions described here may have to be optimized.

The fluorogenic GUS assay is quantitative and more sensitive than histochemical GUS localization. Background problems may be caused by fluorogenic compounds within the samples. Tissues or cells can be ground in GUS extraction buffer with Polyclar to reduce high levels of fluorogenic polyphenolic substances (Jefferson, 1987). Prehydrolyzed substrate in commercial preparations may also cause background fluorescence. To prevent condensation, which may cause MUG hydrolysis, keep the solid stock frozen and desiccated and only open prewarmed bottles.

The NPTII assay as described here is reliable and sensitive. Failures are usually related to the quality of the extract. For instance, extracts prepared from tobacco leaves rapidly lose activity, even when kept on ice. Therefore, extracts should be loaded onto the gel as quickly as possible. Alternatively, protecting agents like those discussed above may be added to the extraction buffer. Another critical point is the detergent in the NPTII sample buffer. NPTII is sensitive to SDS, which should be omitted from the NPTII sample buffer if the protein concentration of the extract is expected to be low (for instance with protoplasts). Alternatively, Triton X-100 (1%) or sodium-deoxycholate (2%) may be used.

The CAT assay as described here has relatively few associated problems. Factors in extracts interfering with the determination of CAT activity may cause problems, which may only partially be overcome by the addition of protease inhibitors and anti-oxidizing agents, or by removal of interfering activities by heat treatment. However, heat treatment leads to precipitation of proteins from the extract, which may also cause variability in the assay.

The Luciferase assay is rapid and easy to perform. However, the reaction product oxyluciferin remains associated with the active site of the luciferase enzyme. As a result, luciferase activity drops rapidly after addition of luciferin and a peak flash of photons (0.3 s) that decays within the next few seconds is observed (Gould and Subramani, 1988). In addition, some luminometers may require 1–2 s to stabilize after the sample is introduced. Therefore, exact quantification of luciferase activity is difficult.

V. Results and Discussion

The advantages and pitfalls of histochemical GUS staining are illustrated in Fig. 1. In the example shown, callus and shoot organogenesis at the basal end of an *Antirrhinum majus* hypocotyl explant was induced after *Agrobacterium tumefaciens*-mediated transformation. The X-Gluc-stained explant shows regions with high level of GUS activity that can be clearly distinguished from nonstained tissues. In our example, callus tissue (Fig. 1a) is chimeric and stains heterogeneously (which is not clearly visible in this photographic reproduction). The developing shoot (Fig. 1b) is homogeneously stained, and originated presumably from one transformed progenitor cell. In both of these examples, GUS activity is due to the introduced reporter gene. In contrast, GUS staining in trichomes (Fig. 1c) is also visible in untransformed explants; it is caused by endogenous GUS activity.

A typical result of a NPTII assay is shown in Fig. 2. The advantage of using electrophoresis to separate nonspecific activities from the actual NPTII signal is demonstrated. Cells contain activities, for instance, autophosphorylating kinases, which also bind phosphocellulose paper and produce a signal. These do not interfere with detection and quantification of NPTII since they can be identified using proper controls. Nonspecific activities can also be used to advantage as an indication of the amount and quality of the extract loaded on the assay gel.

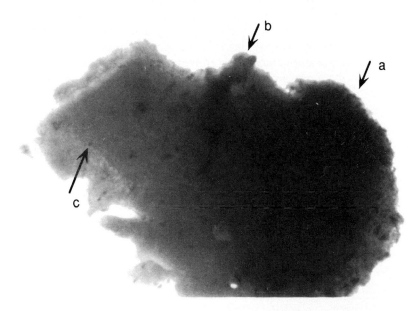

Fig. 1 Histochemical staining with X-Gluc. *A. majus* hypocotyl explant, transfected by *A. tumefaciens* with a GUS-encoding construct. a, callus tissue; b, young shoot; c, trichomes, visible as small dots at the explant surface.

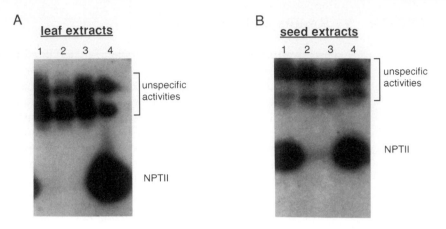

Fig. 2 NPTII assay. NPTII activity in leaves (A) and seed (B) of tobacco plants was determined. Lane 1: Plant transformed with a construct leading to seed-specific expression of NPTII. Lane 2: untransformed control. Lane 3: Plant transformed by the vector only. Lane 4: Plant expressing NPTII constitutively. The autoradiogram was exposed overnight.

Results of a CAT assay are shown in Fig. 3. Extracts of tobacco tissues expressing different amounts of CAT were analyzed for enzymatic activity. Radioactively labeled substrate and products were separated by thin-layer chromatography. Endogenous CAT activity in tobacco is very low, as demonstrated by the untransformed control (lane 2). In samples separated in lanes 3 and 4, most of the chloramphenicol was converted to its acetylated forms and therefore extracts were presumably measured out of the linear range. To allow exact quantification, these extracts should have been diluted or shorter reaction times chosen.

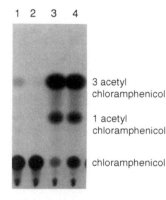

Fig. 3 CAT assay. Results shown are from extracts of tobacco tissues expressing CAT at different levels. Lane 1: Extract with low CAT activity. Lane 2: Extract prepared from a plant not transgenic for CAT. Lanes 3 and 4: Extracts with high CAT activities. The autoradiogram was exposed overnight.

VI. Conclusions and Perspectives

All reporter genes described here have specific advantages. GUS is a stable, easily detectable, and quantifiable reporter, which can also be localized *in situ*. However, endogenous GUS activities have been reported in several plant species and tissues (Plegt and Bino, 1989; Tör *et al.,* 1992). These appear to have a distinctly different pH optimum and a reduced heat stability and can selectively be repressed using appropriate conditions (Hodal *et al.,* 1992). In addition, the GUS gene seems to influence transcription patterns of promoters (Uknes *et al.,* 1993). NPTII is more frequently used as a selectable marker than as a reporter gene. No activities in plants are described that might phosphorylate kanamycin. However, a simplified nonradioactive assay is not available. In contrast to the NPTII assay, several variations of the classical CAT assay have been described. Fluorescence-labeled chloramphenicol (Stratagene) makes a nonradioactive variant of this method possible. Thin-layer chromatography may be replaced by HPLC. Furthermore, a CAT diffusion assay has been developed for which an adaptation for plant extracts is available (Peach and Velten, 1992).

A reporter that is becoming more popular is luciferase, especially since light is easy to detect, and no radioactivity is required. However, this reporter protein is difficult to quantify. An assay that helps to circumvent this problem was developed by Promega. The availability of nondestructive *in situ* detection systems for luciferase activity makes this reporter particularly attractive.

As an alternative to assays based on enzyme activity, ELISAs will gain importance in the future. These assays are nonradioactive, simpler to perform, faster, and cheaper than many enzymatic assays. In addition, with ELISA virtually any gene can be chosen as a reporter gene.

Acknowledgments

We thank Petra Becker, Gisela Simons, and Edelgard Wendeler for excellent technical assistance.

References

Beck, E., Ludwig, G., Auerswald, E. A., Reiss, B., and Schaller, H. (1982). Nucleotide sequence and exact localization of the neomycin phosphotransferase gene from transposon Tn5. *Gene* **19,** 327–336.

Bradford, M. M. (1976). A rapid and sensitive method for the quantitation of microgram quantities of protein utilizing the principle of protein-dye binding. *Anal. Biochem.* **72,** 248–254.

Brasier, A. R., and Ron, D. (1992). Luciferase reporter gene assay in mammalian cells. *Methods Enzymol.* **216,** 386–397.

Buell, C. R., and Anderson, A. J. (1993). Expression of the aggA locus of *Pseudomonas putida in vitro* and *in planta* as detected by the reporter gene, xylE. *M.P.M.I.* **6,** 331–340.

De Block, M., De Brouwer, D., and Tenning, P. (1989). Transformation of *Brassica napus* and *Brassica oleracea* using *Agrobacterium tumefaciens* and the expression of the bar and neo genes in the transgenic plants. *Plant Physiol.* **91,** 695–701.

De Wet, J. R., Wood, K. V., DeLuca, M., Helinski, D. R., and Subramani, S. (1987). Firefly luciferase gene: Structure and expression in mammalian cells. *Mol. Cell. Biol.* **7,** 725–737.

Gould, S. J., and Subramani, S. (1988). Firefly luciferase as a tool in molecular and cellular biology. *Anal. Biochem.* **175,** 5–13.

Gorman, C. M., Moffat, L. F., and Howard, B. H. (1982). Recombinant genomes which express chloramphenicol acetyl transferase in mammalian cells. *Mol. Cell. Biol.* **2,** 1044–1051.

Hayford, M. B., Medford, J. I., Hoffmann, N. L., Rogers, S. G., and Klee, H. J. (1988). Development of a plant transformation selection system based on expression of genes encoding gentamicin acetyltransferases. *Plant Physiol.* **86,** 1216–1222.

Hodal, L., Bochardt, A., Nielsen, J., Mattsson, O., and Okkels, F. (1992). Detection, expression and specific elimination of endogenous β-glucuronidas activity in transgenic and nontransgenic plants. *Plant Sci.* **87,** 115–122.

Jefferson, R. (1987). Assaying chimeric genes in plants: The GUS gene fusion system. *Plant Mol. Biol. Rep.* **5,** 387–405.

Jefferson, R., Kavanagh, T., and Bevan, M. (1987). GUS fusion: β-glucuronidase as a sensitive and versatile gene fusion marker in higher plants. *EMBO J.* **6,** 3901–3907.

Jefferson, R., Burgess, S., and Hirsch, D. (1986). β-Glucuronidase from *Escherichia coli* as a gene-fusion marker. *Proc. Natl. Acad. Sci. U.S.A.* **83,** 8447–8451.

Laemmli, U. K. (1970). Cleavage of the structural proteins during the assembly of the head of the bacteriophage T4. *Nature* **227,** 680–685.

Luehrsen, K. R., De Wet, J. R., and Walbot, V. (1992). Transient expression analysis in plants using the firefly luciferase reporter gene. *Methods Enzymol.* **216,** 386–397.

Nguyen, V. T., Morange, M., and Bensaude, O. (1988). Firefly luciferase luminescence assays using scintillation counters for quantitation in transfected mammalian cells. *Anal. Biochem.* **171,** 404–408.

Peach, C., and Velten, J. (1992). Application of the chloramphenicol acetyltransferase (CAT) diffusion assay to transgenic plant tissues. *Biotechniques* **12,** 181–184.

Piccioni, R., Bellemare, G., and Chua, N-H. (1982). Methods of polyacrylamide gel electrophoresis in the analysis and preparation of plant polypeptides. *In* "Methods in Chloroplast Molecular Biology" (M. Edelmann, R. B. Hallick, and N. H. Chua, eds.), Elsevier Biomedical Press. pp. 985–1014.

Plegt, L., and Bino, R. (1989). β-Glucuronidase activity during development of the male gametophyte from transgenic and non-transgenic plants. *Mol. Gen. Genet.* **216,** 321–327.

Reiss, B., Sprengel, R., Will, H., and Schaller, H. (1984a). A new sensitive method for qualitative and quantitative assay of neomycin phosphotransferase in crude cell extracts. *Gene* **30,** 211–218.

Reiss, B., Sprengel, R., and Schaller, H. (1984b). Protein fusions with the kanamycin resistance gene from transposon Tn5. *EMBO J.* **55,** 3317–3322.

Stougaard, J. (1993). Substrate-dependent negative selection in plants using a bacterial cytosine deaminase gene. *The Plant J.* **3,** 755–761.

Tör, M., Mantell, S., and Ainsworth, C. (1992). Endophytic bacteria expressing β-glucuronidase cause false positives in transformation of *Dioscorea* species. *Plant Cell Rep.* **11,** 452–456.

Uknes, S., Dincher, S., Friedrich, L., Negrotto, D., Williams, S., Thompson-Taylor, H., Potter, S., Ward, E., and Ryals, J. (1993). Regulation of pathogenesis-related protein-1a gene expression in tobacco. *Plant Cell* **5,** 159–169.

CHAPTER 32

Cell-Specific Ablation in Plants

Mary K. Thorsness* and June B. Nasrallah†

*Department of Molecular Biology
University of Wyoming
Laramie, Wyoming 82071-3944
†Section of Plant Biology
Division of Biological Sciences
Cornell University
Ithaca, New York 14853

I. Introduction

Ablation of specific cell types provides a powerful analytical tool in eukaryotic developmental biology. This technique has been used in a variety of organisms to analyze the origin and fate of specific cell types, to study the interaction of and communication between different cell types, and to probe cellular function. Traditionally, ablation has been achieved physically using methods such as microdissection or laser ablation, or genetically through the characterization of specific mutations that affect cell lineage. These approaches have been particularly successful in *Drosophila* and *Caenorhabditis elegans,* but are difficult to apply to more complex organisms that are less amenable to physical and genetic manipulation. More recently, genetic ablation has been achieved by directing expression of a toxic gene product from a tissue-specific promoter. By this method any tissue can, in theory, be targeted for killing using a promoter specifically expressed in that tissue. This toxigenic approach has been used to ablate a diverse group of tissues in mammals (reviewed in Breitman and Bernstein, 1992), plants (Kandasamy *et al.,* 1993; Koltunow *et al.,* 1990; Mariani *et al.,* 1990; Thorsness *et al.,* 1991; Thorsness *et al.,* 1993), and *Drosophila* (Bellen *et al.,* 1992; Kalb *et al.,* 1993; Kunes and Steller, 1991; Moffat *et al.,* 1992). In contrast to physical ablation methods, this genetic approach allows disruption of tissues inaccessible to physical disruption, as well as of tissues that are functionally related but physically dispersed within an organism. It thus provides a powerful tool for analyzing cell fate and function in all organisms amenable to transformation, including plants.

II. Methods of Genetic Ablation

Several toxic gene products have been used to direct cell-specific ablation: the diphtheria toxin A chain polypeptide (DT-A; see Section II,A), the ricin toxin A chain polypeptide (Landel *et al.,* 1988; Moffat *et al.,* 1992), RNase T1 and the Barnase RNase (Mariani *et al.,* 1990), and thymidine kinase (tk) of the herpes simplex virus (Heyman *et al.,* 1989). DT-A and the RNases have been used successfully to direct ablation in plants, and strategies for using these toxins are described in detail. Ricin, a plant toxin isolated from castor bean seeds, is variably toxic to plants (Olsnes and Pihl, 1982). Thus, it is likely to be a poor choice for plant ablation experiments. Expression of herpes tk is conditionally toxic in dividing mammalian cells in the presence of specific drugs (Heyman *et al.,* 1989). Because ablation using herpes tk exploits catalytic differences between the mammalian and viral tk enzymes, and requires delivery of the drug to the tissue expressing herpes tk, it may not be applicable to plants. Strategies using DT-A and RNases should, however, be sufficient for ablation studies in plants, as outlined below.

A. Ablation with Diphtheria Toxin A Chain

The diphtheria toxin A chain polypeptide was the first toxin used to direct cell-specific ablation in mice (Behringer *et al.,* 1988; Breitman *et al.,* 1987; Palmiter *et al.,* 1987) and has subsequently been used in plants (Kandasamy *et al.,* 1993; Koltunow *et al.,* 1990; Thorsness *et al.,* 1991; Thorsness *et al.,* 1993) and *Drosophila* (Bellen *et al.,* 1992; Kalb *et al.,* 1993; Kunes and Steller, 1991). Diphtheria toxin is the naturally occurring toxin of *Corynebacterium diphtheriae* (reviewed in Collier, 1977). It is synthesized as a precursor polypeptide that is subsequently cleaved into two fragments, the A and B chains, that are joined by a disulfide bond. DT-A catalyzes the ADP-ribosylation of elongation factor 2, causing an inhibition of protein synthesis and subsequent cell death. The DT-B chain directs internalization to most eukaryotic cells via a specific membrane receptor. Additionally, the DT-A chain does not include a signal peptide and is not secreted. Thus, DT-A is cell autonomous and directs killing only in those cells in which it is expressed. This allows a strict analysis of a specific cell type in ablation experiments via expression of the DT-A chain.

Initial ablation experiments using DT-A utilized tightly regulated tissue-specific promoters, as spurious expression of this toxic product was expected to inhibit transformation. One molecule of diphtheria toxin is sufficient to direct cell killing (Yamaizumi et al., 1978). Thus, a low level of toxin expression in an unexpected tissue, due to either natural promoter activity or position effects at the site of integration, could sabotage a transformation experiment. Inhibition of transformation through spurious expression has been a significant problem in *Drosophila* ablation studies (Kunes and Steller, 1991). To circumvent this problem, altered versions of DT-A have been developed for use in ablation experiments. Attenuated, less active forms of DT-A exist that require significantly higher levels of expression to kill a cell. These alleles may allow transformation with toxic constructs that are weakly expressed in unintended tissues, and can also be used to generate transformants with useful variations in ablation phenotypes (Breitman et al., 1990; Kalb et al., 1993). Additionally, conditional alleles of DT-A have been isolated that allow activation of DT-A activity after transformation (Bellen et al., 1992; Kunes and Steller, 1991). The most useful of these to plant scientists is a temperature sensitive allele of DT-A (Bellen et al., 1992). With this allele, DT-A-directed killing is activated after transformation by shifting the growth temperature of transformants. These altered versions of DT-A should allow ablation studies to be carried out using promoters that are not absolutely tissue-specific, as well as in tissues that are vital to the survival of the organism.

B. Ablation with RNases

RNase T1 of *Aspergillus oryzae* and the Barnase RNase of *Bacillus amyloliquefaciens* have been used to direct cell-specific ablation in several plant species (Mariani et al., 1990). The toxic effect of these genes is apparently due to the intracellular degradation of RNAs required for cell survival. Indeed, in tissues where these RNases are expressed, specific mRNAs and their products are not detected. In tobacco, Barnase was a significantly better toxin than RNase T1, although this difference was not observed in *Brassica napus*. This suggests that Barnase may be a better choice for plant ablation experiments. This choice is reinforced by the availability of Barstar, a specific inhibitor of Barnase (Hartley, 1989). When Barnase and Barstar are expressed in the same cell types, the toxic effects of Barnase are reversed (Mariani et al., 1992). Attenuated and conditional alleles of these RNases have not been reported, however, and so their use in ablation experiments requires expression from tightly regulated tissue-specific promoters.

III. Genetic Ablation in Plants

The specific experimental techniques used in tissue-specific ablation experiments are largely described elsewhere (Gelvin and Schilperoort, 1988; Sambrook et al., 1989), or in other chapters of this volume. Thus, we do not provide a detailed description of how to carry out an ablation experiment, but rather discuss the important factors that should be considered in planning an ablation experiment. To provide an additional framework for designing an ablation study in plants, we include a brief description of our experiments using the *Brassica* S-locus glycoprotein (SLG) gene promoter to direct tissue-specific ablation.

A. Promoter Choice

Promoter choice is of utmost importance in planning a tissue-specific ablation experiment. A minimal promoter fragment that is expressed at a high level in a single tissue is an ideal choice for directing toxic gene expression to that tissue. Such an ideal promoter may, however, be difficult to identify. The following steps can be taken to characterize a promoter and assess its utility for cell-specific ablation.

Initial identification of a highly expressed tissue-specific promoter can be achieved through standard techniques. A strong promoter is required to ensure effective killing, although expression from a weak promoter can also disrupt cellular function. Promoters that are expressed in nonessential tissues are easiest to use in ablation experiments, as they are unlikely to direct expression that would inhibit transformation or regeneration. Alternatively, mutant toxin alleles may allow transformation with promoters that are expressed in essential tissues. Thorough characterization of the sites of gene expression directed by the promoter of choice using *in situ* hybridization, RNA blot analysis, and reporter gene expression may identify sites of expression that could inhibit transformation by a toxic gene. Although toxic gene expression can technically be used as a reporter of gene expression, a more traditional reporter such as the *Escherichia coli* ß-glucuronidase gene (GUS; Jefferson *et al.,* 1987) is easier to analyze. Additionally, a nontoxic reporter construct will serve as a positive control during toxic gene transformation.

If a promoter is expressed in more than one tissue, deletion analysis may identify a fragment that is expressed in only a single tissue for use in ablation experiments. In some cases, convenient restriction sites within a promoter can be used to construct toxic gene fusions. In others, introduction of a cloning site via site-directed mutagenesis will facilitate toxic fusion construction. Finally, a plant polyadenylation signal is typically included at the 3′ end of chimeric gene constructs to ensure proper RNA maturation in plants. The polyadenylation signal of the nopaline synthase gene of the *Agrobacterium* Ti plasmid (Bevan *et al.,* 1983a; Bevan *et al.,* 1983b) can be isolated from pBI121 (Jefferson *et al.,* 1987) or other plant vectors.

B. Toxin Choice

Cell-specific killing in plants has been achieved via expression of either RNases or wild-type DT-A. If the chosen promoter is tightly regulated and expressed only in a nonessential tissue, either of these toxins can be used to direct ablation.

The T1 and Barnase RNases both direct plant cell killing; however, Barnase may be more generally toxic, and its killing effects can be reversed by coexpression of Barstar, a specific inhibitor of Barnase (Hartley, 1989; Mariani *et al.*, 1992). Thus, Barnase is a better choice for ablation experiments. If promoter activity is not limited to a single cell type, expression of Barnase and Barstar from different promoters with overlapping cell specificities can be used to direct ablation to a single cell type. Using this strategy, Barstar expression would negate the Barnase-directed killing in unintended cell types; however, Barnase activity would be uninhibited in the desired cell type (Goldberg *et al.*, 1993).

DT-A has been engineered for expression in mammalian cells (Maxwell *et al.*, 1986; Palmiter *et al.*, 1987), and this cassette also directs ablation in plants (Kandasamy *et al.*, 1993; Koltunow *et al.*, 1990; Thorsness *et al.*, 1991; Thorsness *et al.*, 1993). Mutant alleles of DT-A are a valuable addition to the repertoire of toxins available for use in ablation experiments. First, a temperature-sensitive DT-A allele that is active at 18°C but not at 24, 28, or 30°C has been isolated (Bellen *et al.*, 1992). Transformation can be carried out at the higher temperatures, followed by activation of cell killing by shifting transformants to 18°C. Use of this allele in *Drosophila* allowed transformation with a promoter that was spuriously expressed during transformation, thus blocking the ability to obtain transformants. This conditional allele should also allow transformation with fusions to promoters that are highly expressed in essential tissues, although this application has not yet been tested. Second, attenuated DT-A alleles that are less active than wild-type can be used in conjunction with wild-type DT-A to generate a collection of transformants with different DT-A-induced phenotypes (Breitman *et al.*, 1990; Kalb *et al.*, 1993). These differences can then be exploited in the characterization of cellular function. Transformation with wild-type DT-A can generate different phenotypes as well, as the site and number of integrations affect the severity of killing; however, attenuated alleles may broaden the spectrum of phenotypes observed. Attenuated alleles may also allow transformation with promoters that are weakly expressed in essential tissues, if the threshold level of expression required for killing with the less active DT-A is not achieved.

C. Introduction into Plants

Toxic gene fusion constructs should be carried on an appropriate vector, such as a binary vector, that includes a selectable marker for plant transformation to allow introduction into plants (see, for example, Bevan, 1984). Toxic gene constructs have been successfully introduced to plants via *Agrobacterium*-mediated transformation, but could presumably be introduced by other transformation methods as well. Transformation experiments should include a positive control for transformation, such as a reporter gene fusion that links the promoter of interest to GUS. Comparison of transformation frequencies obtained with GUS and DT-A or RNase constructs will provide a basis for determining whether a toxic gene construct is detrimental to transformation. Ultimately, transformants can be analyzed for the presence of the toxic gene construct by DNA blot hybridization analysis, and the phenotypic consequences of toxin gene expression can be assessed by microscopic analysis and functional assays.

D. SLG : DT-A Fusion Gene Construction

The *Brassica* SLG genes encode the S-locus-specific glycoproteins that are involved in the recognition of self-pollen during the self-incompatibility response (reviewed in Nasrallah and Nasrallah, 1989). The SLG genes are naturally expressed in specific floral tissues in *Brassica,* as well as in transgenic *Nicotiana tabacum* and *Arabidopsis.* Expression of DT-A from the S_{13} allele SLG promoter was used to direct floral cell ablation in these species in order to analyze the role of SLG-expressing tissues in floral development, fertilization, and self-incompatibility.

At the time we made the SLG : DT-A fusion, a minimal promoter fragment required for expression had not been identified. In order to ensure that our construct contained all the information required for tissue-specific expression, we used a large (3.65 kb) fragment in our toxic gene constructs. The 5' end of the SLG promoter fragment was defined by a naturally occurring *Bam*HI site; the 3' end was defined by a second *Bam*HI site introduced by site-directed mutagenesis at a position 8 nucleotides upstream of the initiating ATG. This site was inside the start of transcription that had been mapped to nucleotide -18. A DT-A gene cassette, modified for expression in mammalian cells, is carried on an 0.8-kb *Bgl*II fragment (Maxwell *et al.,* 1986; Palmiter *et al.,* 1987). This fragment was fused to the compatible *Bam*HI site of the SLG promoter. Both sense and antisense constructs were retained for analysis. The polyadenylation sequence of the nopaline synthase gene (from pBI121, Jefferson *et al.,* 1987) was inserted downstream of the DT-A sequences to ensure correct processing of the chimeric RNA in plants. The constructs were carried on the binary vector BIN19 (Bevan, 1984). Toxic gene constructs were introduced into *N. tabacum* (Thorsness *et al.,* 1991), *Arabidopsis* (Thorsness *et al.,* 1993), and *Brassica* (Kandasamy *et al.,* 1993) by *Agrobacterium*-mediated transformation, and the consequences of DT-A expression were analyzed microscopically and functionally.

IV. Results and Discussion

In tobacco, the SLG promoter is expressed in the stigma and transmitting tissue of the style throughout floral development, and in pollen grains (Kandasamy *et al.,* 1990; Moore and Nasrallah, 1990; Thorsness *et al.,* 1991). Expression of the SLG:DT-A fusion caused necrosis of the pistil tissues that express the promoter, abnormal pistil development, female sterility, and pollen inviability (Figs. 1A–1C and Thorsness *et al.,* 1991). In *Arabidopsis* (Toriyama *et al.,* 1991) and *Brassica*

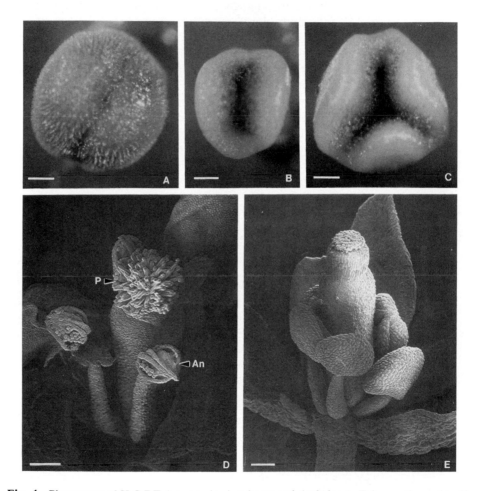

Fig. 1 Phenotypes of SLG:DT-A expression in tobacco and *Arabidopsis.* Tobacco stigmas (A–C) are analyzed by light microscopy. Bar, 400 μm. *Arabidopsis* flowers (D and E) are analyzed by scanning electron microscopy. Bar, 100 μm. (A) Stigma from young bud of untransformed control tobacco has two rounded lobes covered with papillar cells. (B) Stigma from young bud of SLG:DT-A tobacco transformant. The central transmitting tissue is dark and necrotic, and the stigma surface lacks papillar cells and the normal green coloration. (C) Stigma of mature bud of SLG:DT-A tobacco transformant. The stigma is trilobed, due to abnormal carpel fusion, and has dark necrotic tissue lining the central furrows. (D) Flower from untransformed control *Arabidopsis* has elongated papillae at the stigma surface, and dehiscent anthers. P, papillar cell; An, anther. (E) Flower from SLG:DT-A *Arabidopsis* has stunted papillae and non-dehiscent anthers.

(Sato *et al.,* 1991), the SLG promoter is strongly expressed in the stigma papillae of mature flowers and is weakly expressed in the anther tapetum and microspores. SLG : DT-A expression in these species generated flowers with stunted, biochemically inactive papillae, and reduced pollen viability (Figs. 1D and 1E; Kandasamy *et al.,* 1993; Thorsness *et al.,* 1993). In these experiments, the timing of toxic gene expression led to significant differences in the ablation phenotype. Expression of DT-A early in floral development in tobacco led to obliteration of specific cell types. In contrast, DT-A expression late in development in the papillar cells of *Arabidopsis* and *Brassica* inactivated these cells, but did not eliminate them. The presence of the stunted papillae, complete with a cell wall, allowed us to demonstrate that papillae play a passive role in fertilization in *Arabidopsis,* and an active role in *Brassica.* In general, ablation phenotypes obtained using wild-type DT-A or RNases will depend on the timing of promoter expression. With some promoters it may be possible to control the timing of DT-A expression using the temperature-sensitive allele of DT-A. Activation of DT-A at different developmental stages may alter the severity of DT-A-directed killing and thus generate different phenotypes.

V. Conclusions and Perspectives

Ablation of a specific cell type by directed expression of a toxic gene product can provide valuable information on cellular origin, fate, and function. Additionally, this technique has been used to engineer new traits in plants. Ablation of the tapetal cell layer of the anther by expression of Barnase generates male-sterile mutants; fertility is restored via coexpression of Barstar in the tapetum (Mariani *et al.,* 1990; Mariani *et al.,* 1992). These traits are agriculturally desirable for generation of hybrid seed. Ablation technology may thus be applied to many areas of plant biology, including studies of plant development and cellular function, as well as to crop development.

Acknowledgments

We thank Dr. M. K. Kandasamy for performing scanning electron microscopy of *Arabidopsis* flowers.

References

Behringer, R. R., Mathews, L. S., Palmiter, R. D., and Brinster, R. L. (1988). Dwarf mice produced by genetic ablation of growth hormone-expressing cells. *Genes Dev.* **2,** 453–461.

Bellen, H. J., D'Evelyn, D., Harvey, M., and Elledge, S. J. (1992). Isolation of temperature-sensitive diphtheria toxins in yeast and their effects on *Drosophila* cells. *Development* **114,** 787–796.

Bevan, M. (1984). Binary *Agrobacterium* vectors for plant transformation. *Nucleic Acids Res.* **12,** 8711–8721.

Bevan, M. W., Barnes, W. M., and Chilton, M.-D. (1983a). Structure and transcription of the nopaline synthase gene region of T-DNA. *Nucleic Acids Res.* **11,** 369–385.

Bevan, M. W., Flavell, R. B., and Chilton, M.-D. (1983b). A chimaeric antibiotic resistance gene as a selectable marker for plant cell transformation. *Nature (London)* **304,** 184–187.

Breitman, M. L., and Bernstein, A. (1992). Engineering cellular deficits in transgenic mice by genetic ablation. *In* "Transgenic Animals" (F. Grosveld and G. Kollias, eds.), pp. 127–146. London: Academic Press.

Breitman, M. L., Clapoff, S., Rossant, J., Tsui, L.-C., Glode, M., Maxwell, I. H., and Bernstein, A. (1987). Genetic ablation: Targeted expression of a toxin gene causes microphthalmia in transgenic mice. *Science* **238,** 1563–1565.

Breitman, M. L., Rombola, H., Maxwell, I. H., Klintworth, G. K., and Bernstein, A. (1990). Genetic ablation in transgenic mice with an attenuated diphtheria toxin A gene. *Mol. Cell. Biol.* **10,** 474–479.

Collier, R. J. (1977). Inhibition of protein synthesis by exotoxins from *Corynebacterium diphtheriae* and *Pseudomonas aeruginosa. In* "The Specificity and Action of Animal, Bacterial and Plant Toxins" (P. Cuatrecasas, ed.), pp. 67–98. London: Chapman & Hall.

Gelvin, S. B., and Schilperoort, R. A. (eds.) "Plant Moleculer Biology Manual." Dordrecht: Kluwer Academic Publishers.

Goldberg, R. B., Beals, T. P., and Sanders, P. M. (1993). Anther development: Basic principles and practical applications. *Plant Cell* **5,** 1217–1229.

Hartley, R. W. (1989). Barnase and Barstar: Two small proteins to fold and fit together. *Trends Biochem. Sci.* **14,** 450–454.

Heyman, R. A., Borrelli, E., Lesley, J., Anderson, D., Richman, D. D., Baird, S. M., Hyman, R., and Evans, R. M. (1989). Thymidine kinase obliteration: Creation of transgenic mice with controlled immune deficiency. *Proc. Natl. Acad. Sci. U.S.A.* **86,** 2698–2702.

Jefferson, R. A., Kavanagh, T. A., and Bevan, M. W. (1987). GUS fusions: β-Glucuronidase as a sensitive and versatile gene fusion marker in higher plants. *EMBO J.* **6,** 3901–3907.

Kalb, J. M., DiBenedetto, A. J., and Wolfner, M. F. (1993). Probing the function of *Drosophila melanogaster* accessory glands by directed cell ablation. *Proc. Natl. Acad. Sci. U.S.A.* **90,** 8093–8097.

Kandasamy, M. K., Dwyer, K. G., Paolillo, D. J., Doney, R. C., Nasrallah, J. B., and Nasrallah, M. E. (1990). *Brassica* S-proteins accumulate in the intercellular matrix along the path of pollen tubes in transgenic tobacco. *Plant Cell* **2,** 39–49.

Kandasamy, M. K., Thorsness, M. K., Rundle, S. J., Goldberg, M. L., Nasrallah, J. B., and Nasrallah, M. E. (1993). Ablation of papillar cell function in *Brassica* flowers results in the loss of stigma receptivity to pollination. *Plant Cell* **5,** 263–275.

Koltunow, A. M., Truettner, J., Cox, K. H., Wallroth, M., and Goldberg, R. B. (1990). Different temporal and spatial gene expression patterns occur during anther development. *Plant Cell* **2,** 1201–1224.

Kunes, S., and Steller, H. (1991). Ablation of *Drosophila* photoreceptor cells by conditional expression of a toxin gene. *Genes Dev.* **5,** 970–983.

Landel, C. P., Zhao, J., Bok, D., and Evans, G. A. (1988). Lens-specific expression of recombinant ricin induces developmental defects in the eyes of transgenic mice. *Genes Dev.* **2,** 1168–1178.

Mariani, C., DeBeuckeleer, M., Truettner, J., Leemans, J., and Goldberg, R. B. (1990). Induction of male sterility in plants by a chimaeric ribonuclease gene. *Nature* **347,** 737–741.

Mariani, C., Gossele, V., DeBeuckeleer, M., DeBlock, M., Goldberg, R. B., DeGreef, W., and Leemans, J. (1992). A chimaeric ribonuclease-inhibitor gene restores fertility to male sterile plants. *Nature* **357,** 384–387.

Maxwell, I. H., Maxwell, F., and Glode, L. M. (1986). Regulated expression of a diphtheria toxin A-chain gene transfected into human cells: Possible strategy for inducing cancer cell suicide. *Cancer Res.* **46,** 4660–4664.

Moffat, K. G., Gould, J. H., Smith, H. K., and O'Kane, C. J. (1992). Inducible cell ablation in *Drosophila* by cold-sensitive ricin A chain. *Development* **114,** 681–687.

Moore, H. M., and Nasrallah, J. B. (1990). A *Brassica* self-incompatibility gene is expressed in the stylar transmitting tissue of transgenic tobacco. *Plant Cell* **2,** 29–38.

Nasrallah, J. B., and Nasrallah, M. E. (1989). The molecular genetics of self-incompatibility in *Brassica. Annu. Rev. Genet.* **23,** 121–139.

Olsnes, S., and Pihl, A. (1982). Toxic lectins and related proteins. *In* "Molecular Action of Toxins and Viruses" (P. Cohen and S. Van Heyningen, eds.), pp. 51–105. Amsterdam: Elsevier Biomedical Press.

Palmiter, R. D., Behringer, R. R., Quaife, C. J., Maxwell, F., Maxwell, I. H., and Brinster, R. L. (1987). Cell lineage ablation in transgenic mice by cell-specific expression of a toxin gene. *Cell* **50,** 435–443.

Sambrook, J., Fritsch, E. F., and Maniatis, T. (1989). "Molecular Cloning: A Laboratory Manual." Cold Spring Harbor, NY: Cold Spring Harbor Laboratory.

Sato, T., Thorsness, M. K., Kandasamy, M. K., Nishio, T., Hirai, M., Nasrallah, J. B., and Nasrallah, M. E. (1991). Activity of an S-locus gene promoter in pistils and anthers of transgenic *Brassica. Plant Cell* **3,** 867–876.

Thorsness, M. K., Kandasamy, M. K., Nasrallah, M. E., and Nasrallah, J. B. (1991). A *Brassica* S-locus gene promoter targets toxic gene expression and cell death to the pistil and pollen of transgenic *Nicotiana. Dev. Biol.* **143,** 173–184.

Thorsness, M. K., Kandasamy, M. K., Nasrallah, M. E., and Nasrallah, J. B. (1993). Genetic ablation of floral cells in *Arabidopsis. Plant Cell* **5,** 253–261.

Toriyama, K., Thorsness, M. K., Nasrallah, J. B., and Nasrallah, M. E. (1991). A *Brassica* S-Locus gene promoter directs sporophytic expression in the anther tapetum of transgenic *Arabidopsis. Dev. Biol.* **143,** 427–431.

Yamaizumi, M., Mekada, E., Uchida, T., and Okada, Y. (1978). One molecule of diphtheria toxin fragment A introduced into a cell can kill the cell. *Cell* **15,** 245–250.

CHAPTER 33

Ribozymes

Peter Steinecke[*] and Peter H. Schreier[*,†]

[*]Department of Genetic Principles of Plant Breeding
Max-Planck-Institut für Züchtungsforschung
D-50829 Köln, Germany
[†]Bayer AG
PF-F/Biotechnologie
D-51368 Leverkusen, Germany

I. Introduction

RNA molecules that are capable of catalyzing endonucleolytic cleavage reactions are called ribozymes. They have the potential to create phenotypic mutants or to inhibit viral gene expression, as has been shown for antisense genes. In nature, some viroids, virusoids, and satellite RNAs of plant viruses perform self-cleavage reactions (Symons, 1992). The cleavage occurs within a consensus structure called the "hammerhead" motif 3' to a GUX triplet, where X can be C, U, or A (Haseloff and Gerlach, 1988). The nucleotide region directing the catalysis of the cleavage reaction can by physically separated from the region where cleavage occurs. In principle, recognition of any target RNA could be envisaged by linking the complementary nucleotide sequences of the regions flanking the cleavage site to the catalytic domain. As a consequence ribozymes may catalyze cleavage reactions *in trans* and may be directed to many different target RNAs. Until recently, this promising concept was very difficult to prove *in vivo.*

In vitro tests, commonly used to predict ribozyme activity *in vivo,* have been unreliable in predicting ribozyme activity in a living cell (Mazzolini *et al.,* 1992; Steinecke *et al.,* 1994). Selection of the ribozymes and their target sites by computer-aided analysis and transient expression of the ribozymes with their target in cotransfected cells (as described below) appear to be, at present, the most suitable approach to study ribozyme activity in transgenic systems. In a previous study (Steinecke *et al.,* 1992), the *in vivo* cleavage of a target mRNA by a ribozyme *in trans* was shown in plant protoplasts for the first time. Expression of this ribozyme in transgenic plants (Wegener *et al.,* 1994) indicates that the results from this transient expression system are valid. In addition a ribozyme directed against *fushi tarazu* causes the anticipated mutation and shows the expected phenotypic mutation (Zhao and Pick, 1993).

Below, we briefly outline the approach used in our laboratory to design ribozymes. This involves computer simulation of the RNA secondary structure and the predicted bimolecular ribozyme–target RNA interaction. This is followed by protocols for the isolation and transfection of protoplasts prepared from leaves of *Nicotiana tabacum* SR1 and for RNA mapping with the RNase protection assay.

II. Materials

A. Computers and Software

A Vax/VMS digital station and the sequence analysis software package of the University of Wisconsin Genetics Computer Group (UWGCG) (Devereux *et al.,* 1984) are used. Minimal free energy calculations and predicted secondary structures of ssRNA and of RNA complexes between target RNA and ribozymes are performed using the program LINALL (Steger *et al.,* 1984; Schmitz and Steger, 1992; Steinecke *et al.,* 1994). This program is able to handle secondary structure calculations of linear RNA sequences up to 840 nucleotides. Graphic representations were obtained with SQUIGGLES, a program within the above-mentioned software package.

B. Chemicals

Abbreviations: BPB, bromphenol blue; Pipes, 1,4-piperazinediethanesulfonic acid; Hepes, *N*-2-hydroxyethylpiperazine-*N*′-2-ethanesulfonic acid; MES, 4-morpholinoethanesulfonic acid, Tris, tris[hydroxymethyl]aminomethane; EDTA, ethylenediamine tetraacetic acid; SDS, sodium dodecyl sulfate; NTPs, ribonucleoside triphosphates.

Deionized formamide, SDS (BDH Biochemicals Ltd., Poole/UK), urea (BRL Life Technologies, Gaithersburg), NTPs (Boehringer Mannheim, FRG), Gamborg's B5 medium, MS Plant Basal Mixture and Vitamins (Duchefa, Haarlem, The Netherlands), phenol polyacrylamide, (Roth, Karlsruhe, FRG), PEG 4000, macerozyme, cellulase, kinetin (Serva, Heidelberg, FRG), Hepes, Tris (Sigma Chemie, Deisenhofen, FRG). All chemicals are of the highest purity and obtained from Merck (Darmstadt, FRG), unless otherwise noted.

C. Plant Material

N. tabacum c.v. Petit Havana SR1 (Maliga *et al.,* 1973) seeds were surface-sterilized for 2 min in 70% ethanol, for 10 min in 7% sodium hypochlorite, and washed four times for 2 min with sterile H_2O. The seeds were germinated on 1/2 MS medium (MS medium at one-half concentration) (Murashige and Skoog, 1962) supplemented with 1.5% (w/v) sucrose and 0.8% (w/v) agar–agar; the final pH is adjusted to 5.8. Plants were grown in sterile glass jars (WECK) at 22°C with a 16-h photoperiod.

III. Methods

A. Selection of the Target Site and the Ribozymes

When ribozymes are expressed as part of a functional gene, additional sequences at the 5′- and the 3′-terminus of the ribozyme domain are required. In the case described here, the expression cassettes employed contain a multiple cloning site separating the 35S promoter from the transcription termination and polyadenylation signals of cauliflower mosaic virus (*CaMV*), such as encoded in the pRT series (Töpfer *et al.,* 1987). Thus, the *in vivo* ribozyme transcripts are processed to RNA molecules of about 250–300 nucleotides (depending on the expression vector and cloning strategy). These supplementary sequences not only affect the length of the ribozyme molecule, but may also alter the secondary and tertiary structure, which is essential to the enzymatic activity of the ribozyme domain. Furthermore, the performance of an ideal ribozyme molecule may be reduced by a rigid secondary structure in the target region, which could preclude complex formation with the ribozyme. Therefore, both the target RNA and the ribozymes with the surrounding sequences are analyzed, using the program

LINALL, to predict potential intramolecular structure and bimolecular interactions.

1. The Target Sequence

It is worthwhile to choose GUC sequences within the target RNA that are likely to be free from rigid secondary structures, i.e., the region around the AUG codon. In addition, the nascent folding of RNA prevents secondary structure formation at the 5'-end from being influenced by the rest of the target RNA. Therefore, GUC sites within the first 500 nucleotides of the target RNA are preferred for analysis. For the GUC sites predicted to be accessible for a ribozyme–target interaction, the ribozymes are then designed.

2. The Ribozyme

The catalytic domain within the hammerhead is used as described (Steinecke *et al.,* 1992), since its endonucleolytic activity *in vivo* has been demonstrated.

Thus, the ribozymes consist of the catalytic domain flanked by 10 and 11 nucleotides, respectively, complementary to the 5'- and 3'-neighboring sequences of the selected GUC in the target RNA (see Fig. 1). The next step is to embed the ribozymes into the sequences contributed by the *CaMV* 35S expression cassette as encoded in the pRT series and calculate the secondary structure of the expected *in vivo* ribozyme transcripts.

3. Prediction of Ribozyme–Target Interaction

After selection of the target site and the ribozyme sequences, the predicted interaction of the first 500 nucleotides from the target RNA with the potential

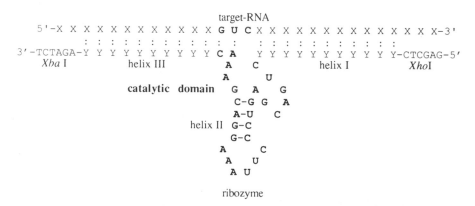

Fig. 1   Scheme of the ribozyme used in this study associated with the target RNA.

Table I
Free Energy $\Delta G^0_{27°C}$ (kJ/mol) of the Secondary Structures of the Target and Ribozyme Transcripts, and of the Target + Ribozyme Complexes, As Predicted by LINALL

$\Delta G^0_{27°C}$ [kJ/mol] of structure formation	Target	Rz1	Rz5
Intramolecular structure	−326.9	−164.3	−167.9
Target + ribozyme complex		−646.9	−575.2
Energy gain due to complex formation		−155.7	−80.4

ribozyme RNAs is determined. The data on energy gains by complex formation of the ribozymes with their target RNA are collected and compared (two examples are shown in Table I). The temperature value is given with respect to the cell culture conditions (here 27°C).

The protocol used is as follows:

1. Determine the secondary structure and $\Delta G^0_{27°C}$ of the first 500 nucleotides of the target sequence and examine free GUC target sites within that region.

2. Plan to appropriate ribozymes according to Fig. 1 and assemble the ribozyme with the vector sequences expected to be transcribed as the functional RNA in the cell.

3. Determine the secondary structure and $\Delta G^0_{27°C}$ of the ribozyme sequences and analyze if helix II is built and the flanking sequences are available for interaction with the target RNA.

4. $\Delta G^0_{27°C}$ and the structure of the target + ribozyme RNA complexes are calculated by combination of the sequences used for their intramolecular structure prediction.

5. Compare the energy gains by complex formation of the different ribozymes with their target RNA:

$$\Delta \Delta G^0 + \Delta G^0_{27°C} \text{ (target + ribozyme complex)} - \Delta G^0_{27°C} \text{ (target)} - \Delta G^0_{27°C} \text{ (ribozyme)}.$$

6. The ribozymes giving the highest energy gain in complex formation with the target RNA are used in further studies.

The target sequence and the ribozymes are then cloned into expression vectors; it is advisable to first choose the same vector to ensure expression of the genes simultaneously and in the same compartment when used in the transient expression assay. Furthermore, the target and ribozyme sequences are cloned into transcription vectors to allow *in vitro* transcription of the antisense probes.

B. Isolation and Transfection of Protoplasts

N. tabacum SR1 protoplasts were chosen because of their high and reproducible uptake of DNA and their viability during and after transfection.

1. All manipulations are carried out in laminar sterile airflow hood.

2. Leaf tissue (4–5 g) from leaves from 4- to 6-week-old vegetative plants is excised, sliced into 1×1-cm segments, and incubated in sterile 500-ml Erlenmeyer flasks containing 40 ml enzyme solution composed of 1.5% cellulase Onozuka R10 and 0.5% macerozyme R10 in 0.4 MB5 medium.

B5 medium "0.4 M" (before sterilization) consists of:

> Gamborg's BT salt (Gamborg, 1968)
> 136.92 g/liter sucrose
> 1 mg/liter naphthalene acetic acid
> 0.2 mg/liter kinetin
> the pH is adjusted with KOH to 5.8.

The enzyme solution is made as follows:

1.5 g cellulase and 0.5 g macerozyme are gently stirred (1–3 h at room temperature or overnight at 4°C) until dissolved in 100 ml B5 medium and sterile-filtered through a 0.22-μm Nalgene filter unit (Nalge Co., Rochester, NY). Storage is safe for 1 month at 4°C.

3. After incubation for 16 h at 27°C in the dark, the suspension of protoplasts is subjected to gentle orbital agitation for 30 min.

4. The cell suspension is filtered through a 250-μm and a 100-μm sieve (Wilson Sieves, Nottingham, UK) and 10 ml B5 medium is added to the filtrate.

5. The protoplast suspension is transferred to 12-ml centrifuge tubes (Nunc) (10 ml/tube) and centrifuged at 60 g for 5 min (Hettich Universal 2S) (protoplasts will be floating on top in 0.4 B5 medium, and a green pellet of broken cells can be noticed).

6. The lower phase (including pellet) underneath the layer of protoplasts is removed using a glass capillary (50–100 $\varnothing$ μm) and a Micro Tube Pump (EYELA, Verder, Düsseldorf, FRG), leaving the protoplasts in a volume of about 1 ml B5 medium. The protoplasts are washed with 10 ml B5 medium 0.4 M, to remove the cellulytic enzymes and centrifuged as indicated above.

7. The protoplasts are collected in 10 ml W5 medium (154 mM NaCl, 125 mM CaCl$_2$, 5 mM glucose, 5 mM KCl, adjusted to pH 5.6).

8. Protoplasts are counted (i.e., Fuchs–Rosenthal chamber).

9. After centrifugation and removal of W5 medium the protoplasts are resuspended in MaMg solution (0.45 M mannitol, 15 mM MgCl$_2$, 0.1% MES, pH 5.6) to a final concentration of 3×10^6 protoplasts/ml. The protoplasts should be used immediately. The typical yield is 6×10^6 cells. To obtain a sufficient number of protoplasts, it is advisable to start the procedure with two or more samples. Transfection by the polyethylene glycol method is carried out according to Negrutiu et al. (1987).

9. After a heat shock for 5 min at 45°C followed by 45 s on ice, 0.35 ml protoplasts are distributed over 10-ml centrifuge tubes (Nunc).

10. Each sample is mixed with 100 μg of plasmid DNA (in 100 μl H_2O).

11. After 10 min, 0.35 ml PEG solution (0.4 M mannitol, 0.1 M Ca(NO$_3$)$_2$, 40% PEG 4000, pH 6–9) is slowly added.

The PEG solution is made as follows:

40 g PEG, MW 4000, is dissolved in a total volume of 70 ml of a solution containing 0.1 M Ca(NO$_3$)$_2$ and 0.4 M mannitol, and 2 M KOH is added drop by drop with a micropipette under continuous stirring until a pH of 9 is reached. Sterilize the solution with a Nalgene filter and store in aliquots at $-20°C$. Before use, thaw for 20 min at 45°C in a water bath, and reassure that the pH is between 6 and 9.

12. After 20 min (shake the tubes from time to time), the mixture is transferred into 40-mm-diameter sterile plastic Petri dishes and diluted with 4 ml B5 medium.

13. The protoplasts are incubated at 27°C for 7–24 h in the dark.

14. The protoplasts are transferred into 12-ml centrifuge tubes (Nunc) filled with 6.5 ml W5 medium and centrifuged at 100 × g for 3 min.

15. The upper phase is removed (the protoplasts form a pellet at the bottom of the tube in W5 medium) until 1 ml W5 medium is left. The protoplasts are then resuspended, transferred into Eppendorf tubes, and centrifuged for 1 min at 13,000 × g.

16. RNA is isolated from the protoplasts by the acid guanidinium–phenol chloroform extraction method and treated with DNase I as detailed by Goodall and collaborators (Goodall *et al.*, 1990).

C. RNase Protection Assay (RPA)

The RNase protection assay is used to analyze quantitatively the steady-state levels of the target (L RNA) RNA. In this assay the RNA isolated from transfected protoplasts is hybridized with complementary, uniformly ^{32}P-labeled RNA probes, and the mixture digested with single strand-specific RNases A, T1, and T2. The double-stranded regions of the RNA hybrid are protected from RNase digestion, and the labeled fragments can be separated by polyacrylamide gel electrophoresis. The advantage of the RPA is the small amount of RNA required to analyze gene expression, and the potential to detect the cleavage product. It is recommended that the RPA II kit from Ambion (Ambion, Inc., Austin, TX) be used, since a detailed description of the RNase protection assay, a troubleshooting guide, and some reagents mentioned below are provided.

1. 5 μg total protoplast RNA is coprecipitated with a ^{32}P-labeled, complementary RNA probe as follows:

    x μl H_2O
    x μl RNA (5μg)
    <u>x μl probe (100,000 cpm)</u>
    50 μl total, add

    5.5 μl 3 M NaAc, pH 4.8

    140 μl 100% EtOH

The antisense probe for the assay is made as follows:

  Plasmid DNAs are digested by the appropriate restriction enzymes, phenol–chloroform-extracted and ethanol-precipitated.

    2 μg plasmid DNA

    x μl H$_2$O

    2.5 μl 10 mM NTPs (−CTP)

    2 μl 10x transcription buffer (Boehringer)

    x μl = 10 units RNase inhibitor (Boehringer)

    x μl = 50 μCi[α-P^{32}]CTP (>800 Ci/mmol, Amersham)

    <u>x μl = 20 units T7, T3, or SP6 RNA polymerase (Boehringer)</u>

    20 μl total, incubate for 60 min at 37°C

The reaction is stopped by adding 10 μl 80% formamide, 0.1% xylene cyanol, 0.1% bromphenol blue, 2 mM EDTA. The *in vitro* transcripts are separated on a 6% polyacrylamide gel (PAG) containing 7 M urea and autoradiographed for 2 min. Bands containing the full-length RNAs are cut out from the gel, crushed-up in an Eppendorf tube, and shaken for 1 h at room temperature in a solution of 300 μl 0.5 M NH$_4$OAc, 1 mM EDTA, 0.1% SDS. After centrifugation for 10 min, the supernatant is collected and 2 μl is subjected to Cerenkov counting.

  2. After 15 min at −70°C and centrifugation for 15 min at 4°C, the supernatant is removed and the pellet is air-dried.

  3. The pellet is redissolved in 20 μl of 40 mM Pipes, pH 6.4, 80% formamide, 0.4 M NaOAc, and 1 mM EDTA.

  4. Samples are heated for 5 min at 95°C, and then incubated overnight at 45–48°C.

  5. Add 200 μl of RNase digestion mixture containing 50 U/ml RNase A, 400 U/ml RNase T$_1$(Ambion), and 150 U/ml RNase T$_2$ (GIBCO BRL) in 5 mM EDTA, 0.3 M NaCl, 10 mM Tris–HCl, pH 7.5, to each sample, and incubate for 40 min at 37°C.

  6. After RNase digestion, 10 μl proteinase K (10 mg/ml in 100 mM NaCl, 10 mM Tris–HCl, pH 7.6, and 0.1 mg/ml yeast RNA) and 10 μl 20% SDS are added, followed by further incubation for 15 min at 37°C.

  7. The RNA is transferred into an Eppendorf tube containing 250 μl phenol–chloroform–isoamyl alcohol (50:48:2, v/v/v), extensively mixed, and then centrifuged for 5 min.

  8. The aqueous phase is transferred into a new Eppendorf tube containing 625 μl 100% EtOH to precipitate the RNA. After 15 min at −70°C and centrifugation for 15 min at 4°C, the supernatant is removed and the pellet is air-dried.

  9. The pellet is resuspended in 8 μl formamide plus dye (see step 1) and RNA fragments are separated by electrophoresis on a 6% PAG with 7 M urea (0.5 mm thick).

10. The gel is transferred onto 3MM paper (Whatman, Clifton, NJ), covered in Saran Wrap, and dried by application of a vacuum (vacuum drier, Zabona, AG, Basel/Switzerland).

11. The gel is exposed to X-ray films at −70°C with an intensifying screen (Super Rapid, Eastman, Rochester, NY).

To compare the effect of different ribozymes, reduction of target gene expression is calculated by scanning the autoradiographs using a laser densitometer. A more convenient method of quantitation is to scan the dried gel directly with a β-scanner such as the FUJI Radioanalytic Imaging System.

IV. Critical Aspects of the Procedure

1. Plasmid DNA used for transfection and *in vitro* transcription is isolated from CsCl gradients to avoid contamination with RNA or chromosomal DNA.

2. Concentration and purity of plasmid DNA and isolated protoplast RNA should be inspected on an agarose gel.

3. For control transfection (target without ribozyme) the relevant expression vector or a mutant ribozyme should be used as carrier plasmid to ensure that the observed reduction of gene expression is not the result of a competition for cellular transcription factors.

4. Use antisense probes of about 100–300 nucleotides.

5. A proteinase K digestion and a phenol extraction step are added to the experimental setup described in the RPA II kit (Ambion). Omitting these steps may cause problems when dissolving the final pellet and therefore loss of product is frequently observed.

6. The RPA is not always sensitive enough, and one must adjust the conditions for detecting the cleavage products with *in vitro*-cleaved target RNA. Even then the detection of the cleavage products may not be possible.

V. Results and Discussion

A. Prediction of Secondary Structure of the Target RNA, of the Ribozymes, and of the Target RNA + Ribozyme Complexes

Ribozymes have been tested as antiviral agents in plants to inhibit gene expression and replication of the tomato spotted wilt virus (TSWV), a single-stranded RNA-virus. On the basis of its genome organization and morphology, TSWV has been classified a member of the Bunyaviridae, a large family of arthropod-borne animal viruses, being unique in its property to infect plants (Matthews, 1991). The examples discussed here relates to ribozymes against potential target sites within the viral complementary LRNA (de Haan *et al.*, 1992) of TSWV.

The sequences are analyzed by computer simulation to predict their bimolecular interactions with the target RNA. Two different ribozymes, Rz5 and Rz1, and their GUC sites (nucleotide positions 89 and 297) were examined in detail. The ribozyme sequences have been analyzed with respect to the anticipated *in vivo* RNA when cloned in pRT100 (Töpfer *et al.,* 1987).

The ribozymes Rz5 and Rz1 form helix II and leave the flanking sequences open and therefore should allow interaction with the target RNA to produce helix 1 and helix III.

Calculation of secondary structures of the ribozyme and the target in their bimolecular interaction favors complex formation (see Table I); i.e., helix I and helix III are formed in both combinations. The energy gain by complex formation with Rz1 is, however, twofold higher than in the case of the complex with Rz5, predicting an elevated increase of the inhibitory effect of Rz1 on the target gene expression.

B. Cotransfection of Ribozyme and Target–Producing Plasmids Results in Reduced Expression of the Target Gene

The ribozyme-encoding oligonucleotides were synthesized and cloned into plasmid pRT100. The target sequence comprising the first 500 nucleotides of the viral complementary LRNA was amplified by PCR and cloned into pRT100. A 10-fold molar excess of the plasmids containing the ribozyme genes was cotransfected with the plasmid producing the viral target RNA into plant protoplasts.

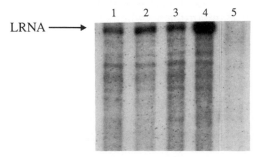

Fig. 2 Ribozyme-mediated reduction of LRNA expressed in cotransfected tobacco protoplasts. Protoplasts were isolated and transfected with a 10-fold molar excess of the ribozyme gene encoding plasmids (pRz1 and pRz5) over target LRNA-expressing vector (pLRNA). Total RNA was isolated after 7 h and total cellular RNA (5 μg) was hybridized with ^{32}P-labeled antisense probe (100,000 cpm) of 250 nucleotides complementary to the target LRNA. After RNase digestion, the protected fragments were separated on a 7 M urea–6% polyacrylamid gel and autoradiographed. The migration of the full-length LRNA antisense probe protected from RNase digestion is indicated on the left. The results indicate reduction of the amount of LRNA by ribozyme Rz1 (lane 1) and Rz5 (lane 3) compared to control transfection (pRT100, lane 4). Analyses were also carried out with nontransfected tobacco protoplasts (lane 5). The cotransfection of a mixture of pRz5 and pRz1; i.e., a 5-fold molar excess of Rz1 and Rz5 each resulting in a 10-fold molar excess of ribozyme-coding plasmids, leads to intermediate reduction of the amount of the target LRNA (lane 2).

Production of the target RNA was analyzed 7 h after transfection via RNase protection with an antisense probe to the target RNA.

As demonstrated in Fig. 2, the amount of target RNA is reduced by the ribozymes Rz1 (Fig. 2, lane 1) and Rz5 (Fig. 2, lane 2) compared to a control transfection (no ribozyme, lane 4). The results shown in Table I suggest that Rz1 may be a better candidate to suppress the target gene expression, Indeed the observed reduction of gene expression by Rz1 is more pronounced than by Rz5.

The LRNA amount is reduced by approximately 85% by Rz1 but only by 70% with Rz5 (Fig. 3). When both Rz1 and Rz5 were each cotransfected in a fivefold molar excess to LRNA, the target vector gene expression is reduced by 80%. Thus, the experimental data are in agreement with the calculated target–ribozyme interaction in the computer simulation.

VI. Conclusion and Perspectives

We anticipate developing this concept as an alternative and improved strategy to the antisense approach for engineering gene regulation. Ribozymes not only

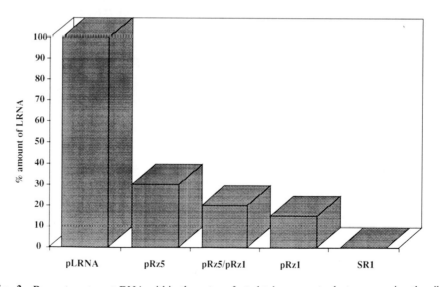

Fig. 3 Percentage target RNA within the cotransfected tobacco protoplasts, expressing the ribozyme and the target, both under the control of the 35S promotor. Protoplasts were isolated and transfected as described in Fig. 2. The amount of protected target LRNA was quantified by scanning the autoradiographies with a laser densitometer. The amount of LRNA obtained by the transfection with pLRNA and carrier plasmid was set to 100%. The amount of LRNA in cells cotransfected with the ribozyme plasmids was measured relative to that obtained by pLRNA. The cotransfected plasmids containing the ribozymes are indicated; SR1 designates untransfected protoplasts. In the case of pRz1/pRz5 equal amounts of plasmids coding for ribozymes Rz1 and Rz5 are mixed and cotransfected with pLRNA (see Fig. 2).

allow modulation of gene expression but also the generation of conditional, preconceived mutants (Zhao and Pick, 1993). Another aspect of using ribozymes as an antiviral strategy, which remains to be addressed, is to produce target RNA and ribozyme in different cellular compartments; this is not simulated in the transient expression assay. Expression of engineered ribozymes in transgenic plants, and their challenge with TSWV, will show whether results obtained in the transient expression system are transferable into practice.

Acknowledgments

We thank Jeff Schell for constant support and encouragement of our ribozyme work and H. van Eijl for helpful comments for the manuscript. The award of a fellowship from Boehringer Ingelheim Fonds (to P. Steinecke) is gratefully acknowledged.

References

De Haan, P., Kormelink, R., Resende, R. de O., Van Poelwijk, F., Peters, D. (1991). Tomato spotted wilt virus L RNA encodes a putative RNA polymerase. *J. Gen. Virol.* **71**, 2207–2216.

Devereux, J., Haeberli, P., and Smithies, O. (1984). A comprehensive set of sequence analysis programs for the VAX. *Nucleic Acids Res.* **12**, 387–395.

Gamborg, O. L. (1968). Nutrient requirements of suspension culture of soybean root cells. *Exp. Cell Res.* **50**, 151–158.

Goodall, G. J., Wiebauer, K., and Filipowicz, W. (1990). Analysis of Pre-mRNA processing in transfected plant protoplasts. *Methods Enzymol.* **181**, 148–161.

Haseloff, J., and Gerlach, W. (1988). Simple RNA enzymes with new and highly specific endoribonuclease activities. *Nature* **334**, 585–591.

Maliga, P., Sz.-Breznovits, A., and Marton, L. (1973). Streptomycin-resistent plants from callus culture of haploid tobacco. *Nature* **244**, 29–30.

Matthews, R. E. F. (1991). "Plant Virology," 3rd ed. San Diego: Academic Press.

Mazzolini, L., Axelos, M., Lescure, N., and Yot, P. (1992). Assaying synthetic ribozymes in plants: High level expression of a functional hammerhead structure fails to inhibit target gene activity in transiently transformed protoplasts. *Plant Mol. Biol.* **20**, 715–731.

Murashige, T., and Skoog, F. (1962). A revised medium for rapid growth and bioassays with tobacco tissue cultures. *Physiol. Plant.* **15**, 473–497.

Negrutiu, I., Shillito, R., Potrykus, I., Biasini, G., and Sala, F. (1987). Hybrid genes in the analysis of transformation conditions. *Plant Mol. Biol.* **8**, 363–373.

Schmitz, M., and Steger, G. (1992). Base-pair probability profiles of RNA secondary structure. *Comp. Appl. Biosci.* **8**, 389–399.

Steger, G., Hofman, H., Förtsch, F., Gross, H. J., Randles, J. W., Sänger, H. L., and Riesner, D. (1984). Conformatial transitions in viroids and virusoids: Comparison of results from energy minimization algorithm and from experimental data. *J. Biomol. Struct. Dynam.* **2**, 543–572.

Steinecke, P., Herget, T., and Schreier, P. H. (1992). Expression of a chimeric ribozyme gene results in endonucleolytic cleavage of target mRNA and a concomitant reduction of gene expression *in vivo*. *EMBO J.* **11**, 1525–1530.

Steinecke, P., Steger, G., and Schreier, P. H. (1994). A stable hammerhead structure is not required for endonucleolytic activity of a ribozyme *in vivo*. *Gene* **149**, 47–54.

Symons, R. H. (1992). Small catalytic RNAs. *Annu. Rev. Biochem.* **61**, 641–671.

Töpfer, R., Matzeit, V., Gronenborn, B. Schell, J., and Steinbiss, H.-H. (1987). A set of plant expression vectors for transcriptional and translational fusions. *Nucleic Acids Res.* **15**, 5890–.

Wegener, D., Steinecke, P., Herget, T., Petereit, I., Philipp, C., and Schreier, P. H. (1994). Expression of a reporter gene is reduced by a ribozyme in transgenic plants. *Mol. Gen. Genet.* **245**, 465–470.

Zhao, J. J. G., and Pick, L. (1993). Generation of loss-of-function phenotypes of the *fushi tarazu* gene with a targeted ribozyme in *Drosophila*. *Nature* **365**, 448–451.

CHAPTER 34

Expression of Plant Proteins in Baculoviral and Bacterial Systems

Reinhard Kunze, Heidi Fußwinkel, and Siegfried Feldmar

Institute of Genetics
University of Cologne
Weyertal 121
50931 Cologne, Germany

PART I EXPRESSION OF PLANT PROTEINS IN *Escherichia coli*

  I. Introduction
 II. Methods
      A. Subcloning into the Expression Plasmid
      B. Protein Expression and Harvest
      C. Purification of Inclusion Bodies
      D. Denaturation (Solubilization) of Inclusion Bodies
      E. Purification of Recombinant Protein by Denaturing Gel Filtration
      F. Purification of Recombinant Protein by Ni Affinity Chromatography
      G. Renaturation
III. Discussion

PART II EXPRESSION OF PLANT PROTEINS IN BACULOVIRUS SYSTEMS

  I. Introduction
 II. Materials and Methods
      A. Maintenance of Sf9 Cell Cultures
      B. Preparation of Large Stocks of Wild-Type AcNPV
      C. Preparation of Genomic AcNPV-DNA
      D. Cotransfection into Sf9 Cells and Selection (Screening)
         for Recombinants
      F. Plaque Assay to Identify and Isolate Recombinant Virus
      F. Rescreening of Isolated Virus by PCR and Sequence Analysis
      G. Amplification of Recombinant Virus
      H. Expression of Recombinant Protein and Cell Harvest
      I. Preparation of Nuclear and Cytoplasmic Protein Extracts

III. Discussion
    References

PART I EXPRESSION OF PLANT PROTEINS IN *Escherichia coli*

I. Introduction

   Escherichia coli is widely used for the expression of heterologous proteins. The advantages of *E. coli* are ease of handling, speed, low cost, high yields, and the choice of a variety of versatile vector/strain systems. Potential disadvantages of *E. coli* are the lack of eukaryote-specific protein modifications and the phenomenon that heterologous proteins frequently do not fold correctly, particularly at high expression levels (Verburg *et al.*, 1993), and accumulate in insoluble form. Unfavorable codon usage (i.e., several consecutive AGG and AGA codons; Rosenberg *et al.*, 1993), premature termination of translation, and/or instability of the overexpressed protein can also create problems.

   The T7 RNA polymerase system has become one of the most widely used expression systems in *E. coli* (Studier and Moffatt, 1986). It relies on the very selective recognition of a bacteriophage T7 promoter by the T7 RNA polymerase, but not the *E. coli* RNA polymerases. The T7 RNA polymerase is usually provided by induction of a chromosomally integrated copy of the T7 RNA polymerase gene, but infection with bacteriophage is also possible. The T7 system combines the benefits of low background expression in noninduced conditions, which allows the cloning and expression of proteins toxic for *E. coli*, and very high expression levels after induction. A number of T7 plasmid vectors are available which allow the synthesis of unfused protein or protein fused to a variety of different (poly)peptides, allowing immunological recognition and/or affinity purification (Studier *et al.*, 1990; Studier, 1991; Sullivan and Vierstra, 1991).

   Many functionally different plant proteins have been expressed in the T7 system. A number of them were soluble and similar, if not identical, to the authentic protein (Galili, 1989; Mori *et al.*, 1991; Anuntalabhochai *et al.*, 1991; Angenent *et al.*, 1992; Koberstaedt *et al.*, 1992; Mori *et al.*, 1993; Hatfield and Vierstra, 1992; Rolland *et al.*, 1993; Grima-Pettenati *et al.*, 1993; Ng *et al.*, 1993). However, there are only a few reports concerning the successful renaturation of aggregated, inactive plant proteins.

   We have used the T7 system to produce large amounts of wild-type and mutant derivatives of the transposase (TPase) protein of maize transposable element *Activator* (*Ac*). After PCR-aided *in vitro* mutagenesis of the TPase-ORF to enclose the ATG start codon within a *Nco*I restriction site, the coding region was inserted into the pET-3d plasmid (Feldmar and Kunze, 1991). The recombinant proteins were used to generate polyclonal antisera and, after renaturation, to perform DNA binding studies. The *Ac* TPase synthesized in *E. coli* is completely insoluble. However, after applying a de- and renaturation step (Jaenicke and Rudolph, 1989), a fraction of the protein remains soluble and has DNA-binding properties indistinguishable from those of TPase expressed in a baculovirus system (see Section II) (Kunze and Starlinger, 1989).

The following protocol, which was developed for the expression and renaturation of the maize *Ac* TPase, is applicable for the expression and, if required, renaturation of many kinds of heterologous proteins in *E. coli*. Potential technical problems or modifications of individual protocol steps, marked by numbers in braces {1}, are illustrated in Section III.

II. Methods

A. Subcloning into the Expression Plasmid

The subcloning steps for integrating the coding region of the protein of interest into a T7 expression vector of the pET or pT7 family should usually be performed in a recA *E. coli* strain lacking the T7 RNA polymerase gene, like HMS174 (Studier *et al.*, 1990) or DH5. Only the final construct is transformed into the *E. coli* lysogen BL21(DE3), which carries a chromosomal copy of the T7 RNA polymerase gene (Studier and Moffatt, 1986). Standard procedures for most kinds of DNA manipulations and transformation of bacteria are described in Sambrook *et al.* (1989).

B. Expression of the *Ac* TPase and Harvest

A variety of growth media are suitable for growth of the described strains and expression of protein (Studier *et al.*, 1990). We routinely grow the bacteria in ZB (10 g N-Z-amine A and 5 g NaCl per liter) or M9ZB medium. Preparation of M9ZB: dissolve 10 g N-Z-amine A, 5 g NaCl, 6 g Na_2HPO_4, 3 g KH_2PO_4 and 1 g NH_4Cl in 1 liter of water. After autoclaving add 20 ml sterile 20% glucose and 1 ml sterile 1 M $MgSO_4$.

1. Inoculate 100 ml medium (ZB + 200 μg/ml ampicillin) with a single colony and incubate overnight at 37°C.

2. Inoculate 2× 0.5 liter prewarmed fresh M9ZB medium (200 μg/ml ampicillin) in 2-liter flasks with 1 ml of the overnight culture and shake at 37°C until the culture reaches an OD_{600} of 0.6–1 (ca. 5 h). {1}

3. Induce expression by making the cultures 0.4 mM in IPTG, and continue shaking for 2–3 h at 37°C. {2}

4. Harvest the bacteria by centrifugation at 4000× g for 20 min at room temperature. Discard the supernatant and keep the pellet on ice.

5. Resuspend bacteria in 20 ml of wash buffer comprising 10% glycerol, 150 mM NaCl, 0.5 mM EDTA, buffered with 50 mM Tris–HCl, pH 8.5, at 4°C. Sediment bacteria by centrifugation at 4000× g for 10 min. The pellet may be stored frozen at −70°C for a few days.

C. Purification of Inclusion Bodies

1. Resuspend bacteria (1- to 3-g pellet wet weight) at 0–4°C in 15 ml IB1 buffer (20 mM Tris–HCl, pH 8.5, containing 1 mM EDTA, and 0.2% Triton X-100) and adjust to 6 mM MgCl$_2$ and 15 U/ml BenzonaseTM-nuclease (E. Merck, Darmstadt, Germany).

2. Disrupt the bacteria with two passes at 4°C through a French pressure cell at 15,000 psi.

3. Add 150 μl PMSF (10 mg/ml in ethanol) and 15 μl proteinase inhibitor cocktail (1 mg/ml each of aprotinin, leupeptin, pepstatin A, and antipain) to the homogenate and incubate for 30 min at 4°C.

4. Centrifuge the lysate for 25 min at 20,000× g and 4 °C. Store the supernatant for controls.

5. Resuspend the pellet in 20 ml ice-cold IB2 buffer (20 mM Tris–HCl, pH 8.5, containing 0.5 M NaCl, 5 mM EDTA, 0.5% Triton X-100, and 10% glycerol) and add 150 μl PMSF and 15 μl proteinase inhibitor cocktail. {3}

6. Centrifuge the suspension for another 25 min at 15,000× g and 4°C, discard the supernatant, resuspend the inclusion body pellet in 5 ml IB2 buffer, and add 40 μl PMSF and 4 μl proteinase inhibitor cocktail.

7. Determine the protein concentration of the suspension and adjust with IB2 buffer to ca. 5 mg/ml. Store aliquots (100 μl, 1 ml) at −70°C. {4}

D. Denaturation (Solubilization) of Inclusion Bodies (Or Proceed to E or F)

1. Sediment 0.5 mg inclusion bodies by centrifugation for 5 min at 10,000× g and 4°C.

2. Solubilize the pellet in 200 μl denaturation buffer [6 M guanidinium hydrochloride (GdmCl), , 100 mM DTT, 2 mM EDTA, 100 mM Tris–HCl, pH 8.5] by shaking for 2 h at room temperature, followed by 15 min at 50°C. Clear the protein solution by centrifuging for 20 min at 20,000× g and 10°C and transfer the supernatant to a fresh tube. Continue with renaturation.

E. Purification of Recombinant Protein by Denaturing Gel Filtration

The inclusion body preparations contain some contaminating bacterial protein and/or recombinant protein fragments. If these contaminants differ significantly in molecular mass, they can easily and efficiently be removed by gel filtration under denaturing conditions {5}. The following protocol is carried out on a FPLC system (Pharmacia), but can be adapted to most other liquid chromatography devices. Amounts and volumes are given for the "Superose 6 HR10/30" column and, in parentheses, for the "HiLoad" column.

1. Sediment 1.5 (10) mg inclusion bodies by centrifugation for 5 min at 10,000× g and 4°C.

2. Solubilize the pellet in 0.25 (1.2) ml denaturation buffer by shaking for 2 h at room temperature, followed by 15 min at 50°C. Clear the protein solution by centrifuging for 20 min at 20,000× g and 10°C, and transfer the supernatant to a fresh tube.

3. Equilibrate a Superose 6 HR10/30 (HiLoad 16/60 Superdex 200 prep grade) column (Pharmacia) with running buffer (6 M GdmCl, 50 mM Tris–Cl, 10 mM DTT, pH 8.5).

4. Load the column with 0.2 (1) ml solubilized inclusion bodies and run it with 0.3 (0.75) ml/min. Collect 0.25 (1)-ml fractions.

5. Withdraw 10 (50)-μl aliquots from each fraction and store the rest at −70°C. Dilute the aliquots with 300 μl water. Mix with 350 μl 10% TCA and incubate 15 min on ice.

6. Collect the protein precipitates by centrifuging 20 min at 14,000 rpm in a tabletop centrifuge. Wash the pellets with 500 μl ice-cold ethanol and dry them under vacuum.

7. Dissolve the protein pellets in 50 μl Laemmli sample buffer (0.14 M Tris–Cl, 4% SDS, 1% DTT, 20% glycerol, 0.01% bromophenol blue) by boiling for 10 min. After boiling, add immediately 1 μl 1 M DTT to each sample and cool on ice.

8. Analyze the samples by denaturing SDS–polyacrylamide gel electrophoresis. Pool the eluates containing full-length recombinant protein and proceed to step G or F.

F. Purification of Recombinant Protein by Ni Affinity Chromatography

Another very efficient procedure for the purification of proteins under denaturing conditions is Ni affinity chromatography (Hochuli et al., 1987). For this technique no expensive chromatography equipment is required, but a $(His)_6$ peptide must be fused to the protein reading frame (preferably at the carboxy terminus) as the affinity tag. The required affinity resin (Ni-NTA resin) can be purhased from QIAGEN, Inc. Denaturing Ni affinity and gel filtration chromatography can be performed successively. An example with the maize Ac TPase protein is shown in Fig. 1.

1. Sediment 5–10 mg IBs by centrifuging 5 min at 10,000× g at 4°C.

2. Solubilize the pellet in 1 ml NTA buffer A (6 M GdmCl, 0.1 M sodium phosphate, 10 mM Tris–Cl, 5 mM 2-mercaptoethanol, pH 8) by shaking 1 h at

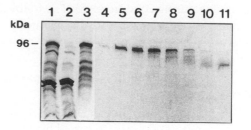

Fig. 1 Purification of a bacterially expressed 96-kDa derivative of the maize *Ac* TPase carrying an amino-terminal (His)$_6$-tag, TPase(H6-103-807), by successive Ni affinity and gel filtration chromatography steps under denaturing conditions. The proteins were fractionated by SDS–PAGE and Coomassie-stained. Lane 1: solubilized inclusion bodies; lane 2: NTA resin flow-through; lane 3: NTA–resin eluate with elution buffer (pH 4.5). This eluate was applied to a "HiLoad 16/60 Superdex 200 pg" column. Lanes 4–11: fractions from the HiLoad-column. The TPase(H6-103-807) protein in lane 5 is nearly homogeneous.

room temperature, followed by 15 min at 50°C. Clear the protein solution by centrifuging 20 min at 20,000× *g* and 15°C; transfer the supernatant to a fresh tube. {6}

3. Sediment 1 ml Ni-NTA resin by centrifuging for 5 min at 1000× *g*. Discard supernatant and resuspend resin in 1 ml NTA buffer A.

4. Repeat step 3.

5. Add the solubilized IBs (step 2) to the NTA resin. Swirl gently for 45 min at room temperature to keep the resin in the suspension.

6. Transfer the slurry into a 1-ml disposable column closed at the bottom and let the resin settle. Place a frit on top of the resin and drain the column.

7. Wash in 1-ml steps with 5 column volumes NTA buffer A.

8. Wash in 1-ml steps with 5–10 column volumes NTA buffer B (6 *M* GdmCl, 0.1 *M* sodium phosphate, 10 m*M* Tris–Cl, 5 m*M* 2-mercaptoethanol, pH 6.3). If possible, monitor the flow-through by reading the OD280 nm, which should drop to below 0.02 before proceeding to the next step. Collect the protein-containing eluates.

9. Elute in 1-ml steps with 5–10 column volumes NTA buffer C (6 *M* GdmCl, 0.1 *M* sodium phosphate, 10 m*M* Tris–Cl, 5 m*M* 2-mercaptoethanol, pH 5.9). {7}

10. Prepare 10-μl samples of each fraction for analysis by SDS–PAGE: Follow steps 5–8.

G. Renaturation

1. Renature the protein by diluting 1:50–1:100 with ice-cold renaturation buffer (50 m*M* Tris–Cl, 50 m*M* NaCl, 3 m*M* MgCl$_2$, 5 μ*M* ZnCl$_2$, 0.2% Triton X-100, 5 m*M* glutathione-reduced, 0.5 m*M* glutathione-oxidized, 10% glycerol, pH 8.5) and incubating overnight to 3 days at 4°C. {8}

2. Clear the protein solution by centrifuging 30 min at 25,000× *g* and 4°C. Transfer supernatant to a fresh tube. The renatured protein may be stored in aliquots at −70°C.

3. If required, the renatured protein may be concentrated and/or the buffer may be exchanged by ultrafiltration [i.e., in Centricon concentrators (Amicon, Inc.)].

III. Discussion

{1} In cases when the expressed protein is toxic for *E. coli,* problems of plasmid instability can arise. Since the selective antibiotic ampicillin is degraded quickly in the culture medium by the secreted β-lactamase, cells that have lost the plasmid may overgrow the plasmid-containing cells. If unusually low yield of protein is obtained, the fraction of cells containing the plasmid should be determined before induction and the precautions described by Studier *et al.* (1990) should be taken to prevent overgrowth of the culture by plasmid-free cells.

If the expressed protein is toxic to *E. coli,* the basal activity of the T7 RNA polymerase in the BL21(DE3) cells can be reduced by the coexpression of T7 lysozyme which is an inhibitor of T7 RNA polymerase. T7 lysozyme is provided by plasmids pLysS, pLysL, or pLysE, which confer resistance to chloramphenicol and are compatible with the pET vectors (Studier *et al.,* 1990; Studier, 1991). A different approach is to block the target T7 promoter with the *lac* repressor (Dubendorff and Studier, 1991). If the expression plasmid cannot be maintained in the BL21(DE3) cells in the presence of pLysS or pLysE, it must be propagated in T7 RNA polymerase-free *E. coli* strain like HMS174. Expression is induced by infecting the culture with the lambda derivative CE6 carrying the gene for T7 RNA polymerase (Studier *et al.,* 1990). A similar strategy, but using a readthrough transcription-suppressing expression vector and an M13 derivative for supplying the T7 RNA polymerase was used by Brown and Campbell (1993).

{2} Many proteins form insoluble protein aggregates (inclusion bodies) during heterologous expression in *E. coli.* Since the foreign protein is the main component of these aggregates, their isolation provides an excellent first step in the purification of the protein. However, in cases when the protein renaturation efficiency is very low, it may be desirable to increase the fraction of soluble protein in the bacteria. Schein and Noteborn (1988) reported for three mammalian proteins that at a growth temperature of 23–30°C, 30–90% of the recombinant protein was soluble, whereas at 37°C these proteins are exclusively found in aggregates.

{3} The contamination of the IB preparation with *E. coli* proteins can be reduced by raising the stringency of the washing steps, i.e., by raising the NaCl concentration and/or by including moderate concentrations of urea (~ 1 *M*). For each individual protein the washing conditions must be optimized.

{4} All protein determination systems are very sensitive against certain substances other than proteins. The buffer components we use are compatible with the MicroBCA reagent (Pierce). If you use other systems, test their compatibility first.

{5} The purification of the solubilized protein by denaturing gel filtration we have described here is applicable for most proteins, but it requires chromatography equipment. If only small amounts (a few micrograms) of the protein are required for subsequent experiments, elution of the protein from SDS-containing gels and subsequent renaturation may be successfull (Hager and Burgess, 1980).

{6} Avoid using the strong reducing agents DTT or DTE, as they will leach the Ni ions from the resin. However, 1–10 mM 2-mercaptoethanol is usually tolerated, but its effect should be tested.

{7} The binding of proteins to the Ni-NTA resin is destabilized by lowering the pH. For a good purification of the modified protein the pH of the buffers must be optimized, or the column should be eluted with a pH gradient. We have had the experience that some unmodified proteins, lacking the (His)$_6$ tag, are quite stably bound by the resin, and some proteins carrying the (His)$_6$ tag do not elute at pH 5.9. The Ac (His)$_6$ TPase, for instance, elutes at pH 4.5.

{8} Appropriate renaturation conditions are crucially important for a good yield of active protein! These conditions must be optimized for each protein (Jaenicke and Rudolph, 1989; Buchner and Rudolph, 1991). A higher glycerol content in the renaturation buffer sometimes stabilizes the protein. In case of the Ac TPase, Triton X-100 is essential to prevent precipitation of the protein during renaturation.

PART II EXPRESSION OF PLANT PROTEINS IN BACULOVIRUS SYSTEMS

I. Introduction

During the last decade the insect baculovirus *Autographa californica* nuclear polyhedrosis virus (AcNPV) has been successfully developed as an expression vector system. Foreign genes are usually expressed from the strong polyhedrin promoter, which is highly transcribed in *Spodoptera frugiperda* (Sf) cells during the late stages of infection. The baculovirus system has properties that make it particularly suitable for the expression of foreign eukaryotic genes: (a) the strength of the polyhedrin promoter that is used to transcribe the gene of interest; (b) the capability of the infected insect cells to accomplish most eukaryotic post-translational protein modifications (processing of transit peptides, nuclear import, phosphorylation, glycosylation, acylation, assembly of oligomers, disulfide cross-linking); (c) the biological safety of the vector system, since the virus cannot replicate or express its DNA in mammalian cells (Tjia *et al.,* 1983; Luckow and

Summers, 1988). Recently, improved expression systems, i.e., p10 promoter-based transfer vectors and suitable viruses carrying *lacZ* as an indicator gene have been developed, that allow easier screening for recombinant viruses and/or enable the simultaneous expression of several genes.

Comprehensive reviews of the biology of baculoviruses, baculovirus expression vectors, materials, procedures, and equipment are found in the two recently published baculovirus laboratory manuals of O'Reilly *et al.* (1992) and King and Possee (1992).

However, compared with *E. coli* the baculovirus system also has considerable disadvantages: the work is much slower and more laborious, equipment for cell culture work is required, and media and other materials are expensive. Therefore, the baculovirus system usually is the expression system of choice if the protein of interest expressed in *E. coli* is not functional.

Several plant proteins have been successfully expressed in the baculovirus system, some of which were inactive in *E. coli* (Andrews *et al.*, 1988; Bustos *et al.*, 1988; Hauser *et al.*, 1988; Vernet *et al.*, 1990; Tessier *et al.*, 1991; Nagai *et al.*, 1992; Doan *et al.*, 1993; Korth and Levings III, 1993; Verburg *et al.*, 1993). Most of the baculovirus-expressed proteins were functional, but post-translational processing (i.e., signal sequence removal) was in some cases incomplete or not authentic (i.e., incorrect structure of the carbohydrate chains added during glycosylation).

We have expressed the *Ac* transposase (TPase) as an unfused protein in the baculovirus system. The *Ac* cDNA was inserted into the transfer plasmids pAc373 (Smith *et al.*, 1985) and pAc610 (Luckow and Summers, 1988), respectively, using standard procedures (Sambrook *et al.*, 1989). The resulting plasmids were cotransfected with wild-type genomic AcNPV isolate E-DNA into cultured Sf9 cells and recombinant, polyhedrin-negative (occ-) plaques were selected. The *Ac* TPase accumulated in the nuclei of infected *S. frugiperda* cells (Hauser *et al.*, 1988). Although the main fraction of the TPase was insoluble, total nuclear extracts from infected Sf9 cells contained sufficient TPase protein to perform gel mobility shift assays without further fractionation (Kunze and Starlinger, 1989).

The following protocol represents a brief overview of the stages involved in constructing a recombinant baculovirus. For details and alternative procedures the reader must consult a laboratory manual, i.e., O'Reilly *et al.* (1992). A flowchart of the individual steps is shown in Fig. 2. Potential technical problems or modifications of individual protocol steps marked by numbers in braces {1} are illustrated in Section IV.

II. Materials and Methods

In addition to the standard laboratory equipment the following items are essential: Laminar flow hood, temperature-controlled incubator, inverted tissue

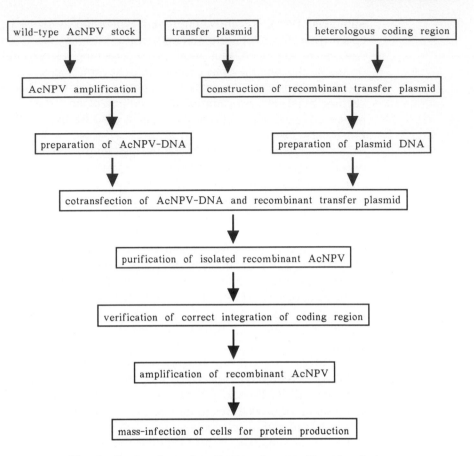

Fig. 2 Construction and purification of a recombinant baculovirus.

culture microscope, 75-cm^2 tissue-culture flasks, 60-mm tissue-culture dishes, and 96-well tissue-culture plates (Greiner, Falcon, Corning).

Cell culture medium (in the following protocol termed "medium") is TC100 medium (commercially available from GIBCO) supplemented with 10% fetal calf serum and 0.26% TB (tryptose broth; GIBCO). Transfection medium is IPL-41-medium (GIBCO) with 10% FCS and 0.26% TPB (tryptose phosphate broth; GIBCO). For composition of the media see O'Reilly *et al.* (1992).

Sf9 cells, AcNPV strains, and transfer vectors may be requested from several laboratories, whose addresses are given in O'Reilly *et al.* (1992). Baculovirus expression kits are also commercially available from Invitrogen, Clontech, and PharMingen.

A. Maintenance of Sf9 Cell Cultures

Sf9 cells are propagated in monolayers in 75-cm^2 tissue-culture flasks at 27°C. They have a generation time of 18 to 24 h and are subcultured two to three times weekly. When the cells are confluent they must be subcultured. When the cells begin to float freely their density is too high. {1}

1. Detach the cells from the surface of a 75-cm² flask by knocking against the flask.

2. Pipet 10 ml fresh, prewarmed medium into a new flask. Add 1.5 ml of the cell suspension (ca. 6×10^6 cells), mix carefully, and incubate at 27°C. Subculture after 3 to 4 days.

B. Preparation of Large Stocks of Wild-Type AcNPV

All virus stocks should be initiated with thoroughly purified inoculum!

1. Determination of the Titer of the Starter AcNPV Stock

1. Pipet 4 ml medium into 60-mm tissue-culture plates. Add 0.5 ml Sf9 cell suspension (ca. 2×10^6 cells), mix gently, incubate 1 h at 27°C to allow the cells to attach to the plate.

2. Dilute the virus stock 10^4-, 10^5-, 10^6-, and 10^7-fold in medium.

3. Carefully discard the medium from the plates and overlay the cells with 0.5 ml of each virus dilution. Incubate 1 h at room temperature with very gentle rocking.

4. Prepare a 5% standard DNA-grade agarose solution in water, cool it to 60°C. Dilute the agarose 1:10 with tissue culture medium prewarmed to 50°C in a waterbath and cool the diluted agarose to 40°C.

5. Remove the virus dilution from the cells by aspiration and immediately overlay the cells with 4 ml agarose dilution per plate. Let the agarose solidify at room temperature for 5–10 min. Incubate the plates at 27°C until plaques just become visible (3–6 days).

6. Prepare a 0.5% agarose solution as in step 4. Add 1/20 vol of a neutral red-solution (1 mg/ml) and overlay each plate with 3 ml of the stain/agarose solution.

7. After overnight incubation count the plaques and calculate the titer of the stock (number of plaques × dilution factor × 2 = plaque forming units (pfu)/ml.

2. Amplification of AcNPV Stock

1. Pipet 10 ml fresh, prewarmed medium in a 75-cm² flask. Add 2 ml of Sf9 cell suspension (ca. 8×10^6 cells), mix carefully by swirling, and allow the cells to attach by incubating 1 h at 27°C.

2. Remove the medium and overlay the cells with 1.6×10^8 pfu (20 pfu/cell). Rock very gently for 1 h at room temperature. If less than 1.5 ml virus stock was used, simply add 10 ml prewarmed medium. If more than 1.5 ml virus stock was used, remove the fluid by aspiration and add 12 ml prewarmed medium. Incubate at 27°C for 3 days.

3. Remove the tissue-culture medium and centrifuge it for 5 min at $1000 \times g$. Sterilize the supernatant by filtration through a 0.45-μm filter and determine the virus titer. This virus stock may be stored at 4°C for more than 6 months.

C. Preparation of Genomic AcNPV-DNA

1. Transfer 33 ml of virus stock ($>5 \times 107$ pfu/ml) to a Beckman SW28 polyallomer tube. Underlay the solution with 3 ml 25% sucrose, 5 mM NaCl, 10 mM EDTA. Centrifuge for 1 h at $90,000 \times g$.

2. Remove the supernatant by aspiration. Resuspend the virus pellet in 4 ml 20 mM Tris–Cl, 5 mM EDTA, pH 7.5 by pipetting up and down. Adjust the suspension to 200 μg/ml proteinase K and 0.5% SDS. Incubate for 5 h at 50°C.

3. Extract the solution sequentially with 1 vol phenol, phenol/chloroform, and chloroform by inverting the tube. {2}

4. Add exactly 1 g/ml CsCl and dissolve the CsCl by gentle shaking at 37°C. Adjust the solution to 100 μg/ml ethidium bromide (10 mg/ml stock). Transfer the solution to a Beckman VTi65 tube and centrifuge in the VTi65 vertical rotor at 20°C for 12 h at 45,000 rpm.

5. Pierce an 18-gauge needle into the centrifuge tube about 2 mm below the (lower) DNA band. Collect the DNA by allowing the CsCl solution to drip out slowly. {2}

6. Remove the ethidium bromide by extracting with 2-propanol saturated with 5 M NaCl. Repeat the extractions until no pink color is detectable in the 2-propanol phase. {2}

7. Dialyze the sample once for 1 h at 4°C against 1 liter TE, change the TE, and continue the dialysis over night at 4°C. Determine the DNA concentration photometrically.

D. Cotransfection into Sf9 Cells and Selection (Screening) for Recombinants

The following protocol is an example of the calcium phosphate coprecipitation procedure. This is still the most commonly used method for delivering virus and transfer plasmid DNA to the Sf9 cells, although lipofection has become an efficient alternative (O'Reilly *et al.*, 1992). To ensure that at least two (better three) independent recombinants will be obtained, two (or three) parallel cotransfections should be done and the resulting virus stock be kept separately.

1. Pipet 4 ml medium into a 60-mm dish. Seed with 2×10^6 cells (0.5 ml) and allow cells to attach for 1 h. Replace medium with 0.8 ml transfection medium (IPL-41 (GIBCO) with 10% FCS and 0.26% TPB).

2. Mix 1 μg of circular virus DNA and 2 μg of transfer plasmid with 0.8 ml filter-sterilized transfection buffer (25 mM Hepes, 140 mM NaCl, 125 mM CaCl$_2$, pH 7.1). {3}

3. Add this solution dropwise to the cells while gently rocking. Incubate for 4h at 27°C.

4. Remove all liquid from the cells by aspiration. Rinse the cells once with complete medium. Add 4 ml fresh medium and incubate at 27°C until many cells contain polyhedral inclusion bodies (PIB), which should be after 4–5 days.

5. Collect the medium, clear it by centrifuging for 5 min at 1000 × g. Store the virus stock at 4°C.

6. Determine the titer of the stock as in Section B,1 (it should be >107 pfu/ml).

E. Plaque Assay to Identify and Isolate Recombinant Virus

The virus stock from the previous steps contains a large excess of wild-type and abberant virus over the correct recombinant virus (0.2–1%), except when linearized virus DNA was used for cotransfection. Individual clones are isolated by plaque purification and subsequently characterized by DNA hybridization, PCR, immunoblotting or other methods. {4}

1. Seed ca. 20 60-mm plates with 2×10^6 cells each as in step 1, Section II,B,1. {5}

2. Dilute the virus stock from step 5, Section II,D to ca. 200 pfu/ml.

3. Remove medium from plates. Distribute 0.5 ml diluted virus stock to each plate. Adsorb the virus for 1 h at room temperature while gently rocking.

4. Prepare 0.5% agarose as in step 4, Section II,B,1.

5. Thoroughly aspirate virus inoculum from the cells and overlay them with 4 ml 0.5% agarose solution. Incubate for 5–7 days at 27°C.

6. Screen the inverted plates under the stereo dissecting microscope. Mark putative recombinant plaques and, if possible, reinspect them with a phase-contrast tissue culture microscope. {6}

7. Isolate putative recombinant plaques by punching out small plugs of the agarose.

F. Rescreening of Isolated Virus by PCR and Sequence Analysis

Rescreening of the primary virus isolates by PCR is a quick and easy way to confirm the correct insertion of the heterologous gene by double crossover (Malitschek and Schartl, 1991; Webb et al., 1991). Four primers should be chosen, two of which are flanking the cloning site in the transfer vector, and the other two located within the inserted gene close to its ends. By performing the PCR reaction with different combinations of these primers and direct sequencing of the PCR products the correct insertion can be unambiguously verified.

Viral DNA may be eluted from the agarose plugs of the primary screen into 100 μl FCS- and TD-free medium and PCR-amplified directly from ca. 10 μl of the eluates. However, it is recommended to amplify the virus in 96-well tissue-culture plates before PCR according to the following protocol. {7}

1. Dilute Sf9 cells 1:10 with fresh medium to 4×10^5 cells/ml. Pipet 50 μl diluted cells into each well of a 96-well plate. If available, use a multichannel pipet.

2. Inoculate the cells with plaque-containing agarose plugs. Inoculate two wells with control virus. Incubate for 2 days at 27°C.

3. Remove the supernatants and store them at 4°C.

4. Pipet 100 μl 1 N NaOH into each well. Rock the plate for 2 min at room temperature. Add 40 μl 5 M ammonium acetate to each well. Mix the contents of each well by pipetting up and down.

5. Take 2–5 μl of the DNA solution for each PCR reaction.

6. At least two independent isolates of the desired recombinant virus should be purified to homogeneity by one or two additional plaque assays from the supernatants. Elute the final isolated plaque into 1 ml FCS- and TB-free medium.

G. Amplification of Recombinant Virus

1. Pipet 4 ml medium into a 50-mm tissue-culture plate. Add 0.5 ml of Sf9 cell suspension (ca. 2×10^6 cells, mix gently, and incubate 1 h at 27°C to allow the cells to attach to the plate.

2. Discard the medium from the plates and overlay the cells with 0.5 ml of the virus eluate. Incubate 1 h at room temperature with gentle rocking.

3. Add 4 ml medium and incubate at 27°C for 4 days.

4. Remove the medium (primary inoculum) and clear it by centrifuging for 5 min at $1000 \times g$ (4°C). Store the primary inoculum for up to a few months at 4°C.

5. Pipet 10 ml fresh, prewarmed medium in a 75-cm^2 flask. Add 2 ml of Sf9 cells (ca. 8×10^6 cells), mix carefully by swirling, and allow the cells to attach by incubating 1 h at 27°C.

6. Remove the medium and overlay the cells with 0.5 ml primary inoculum. Rock gently for 1 h at room temperature. Add 10 ml prewarmed medium and incubate at 27°C for 4 days.

7. Harvest the medium (secondary inoculum) and centrifuge it for 5 min at $1000 \times g$. Sterilize the supernatant by filtration through a 0.45-μm filter and determine the virus titer. This virus stock may be stored at 4°C for more than 6 months. For long-term storage freeze 1-ml aliquots at −80°C. The secondary inoculum may be used for mass amplification of recombinant virus.

H. Expression of Recombinant Protein and Cell Harvest

Polyhedrin expression in infected cells is detectable 24 h post infection (pi), and the protein accumulates to very high levels at 40–50 h pi. Heterologous protein expression from the polyhedrin promoter usually follows a similar time course. However, the optimal time point after infection for harvesting the cells (maximal protein accumulation before proteolysis and cell lysis begins) should be determined separately for each independent recombinant virus isolate.

The volumes in the following protocol are calculated for a 75-cm^2 flask. If required, they can easily be scaled upward.

1. Pipet 10 ml medium in a 75-cm^2 flask. Add 2 ml of Sf9 cells (ca. 8×10^6 cells), mix carefully by swirling, and allow the cells to attach by incubating for 1 h at 27°C.

2. Remove the medium and overlay the cells with secondary inoculum at 20 pfu/cell. Distribute the inoculum and incubate for 1 h at room temperature. Replace the inoculum with 12 ml prewarmed medium and incubate about 48 h at 27°C.

3. Detach the cells from the bottom of the plate by knocking and swirling and collect the cell suspension in a centrifuge tube. Sediment the cells by centrifugation (5 min 1000 $\times$ g, 4°C).

4. Discard the supernatant. Wash the cells by resuspending them in 5 ml cold Tris saline (25 mM Tris–Cl, 140 mM NaCl, 5 mM KCl, 0.7 mM Na$_2$HPO$_4$, 0.1% glucose, pH 7.4) and collect them by centrifugation (5 min 1000 $\times$ g, 4°C). The cell pellet may be stored at -70°C for several weeks.

I. Preparation of Nuclear and Cytoplasmic Protein Extracts

The following protocol is based on the procedure developed by Dignam *et al.* (1983). It is particularly useful for the preparation of native nuclear protein extracts, and the isolated proteins can readily be used for activity tests, i.e., enzymatic or DNA-binding activity. All steps of the protocol should be performed at 4°C or on ice. {8}

1. Resuspend cells in 3 ml ice-cold buffer A (10 mM Hepes, pH 7.9, 50 mM KCl, 3 mM MgCl$_2$, 0.3 mM PMSF, 1 μg/ml each of aprotinin, leupeptin, pepstatin A, and antipain). Sediment the cells by centrifugation (5 min 1000 $\times$ g, 4°C).

2. Resuspend the cells in 2 ml ice-cold buffer A. Transfer the suspension to a cold Dounce homogenizer and break the cells by ca. 10 strokes with a D-type pestle. Avoid generating foam! Check for complete cell breakage by inspection of an aliquot of the homogenate under the microscope.

3. Sediment the nuclei by centrifuging for 10 min at 1000 $\times$ g, 4°C. Thoroughly remove the supernatant (consisting of the cytoplasmic protein fraction and the organelles).

4. Wash the nuclei by resuspending them in 1 ml buffer A. Collect them by centrifuging for 15 min at 20,000 $\times$ g and 4°C. Discard the supernatant.

5. Resuspend nuclei in 1.5 ml ice-cold buffer B (10 mM Hepes, pH 7.9, 0.42 M NaCl, 50 mM KCl, 0.2 mM EDTA, 0.2% NP40, 20% glycerol, 0.3 mM

PMSF, 1 μg/ml each of aprotinin, leupeptin, pepstatin A, and antipain). Transfer the suspension to a cold Dounce homogenizer.

6. Lyse the nuclei by 10 strokes with a D-type pestle. Avoid generating foam! Incubate the homogenate for 30 min on ice. Mix every 5 min with two strokes.

7. Remove insoluble material by centrifuging for 30 min at $25,000 \times g$ (4°C). [9]

8. Dialyze the supernatant over night at 4°C against two changes or buffer C (10 mM Hepes, pH 7.9, 100 mM KCl, 0.2 mM EDTA, 20% glycerol, 0.5 mM DTT, 0.1 mM PMSF).

9. Remove precipitated material by centrifuging for 30 min at $25,000 \times g$ (4°C). Adjust the protein solution to 1 μg/ml each of aprotinin, leupeptin, pepstatin A, and antipain. {9}

III. Discussion

{1} Before passaging cells, carefully inspect the culture by microscopy for bacterial or fungal contamination! If a contamination has occurred in one of the cultures, the noncontaminated cells should be subcultured once in fresh medium with 50 μg/ml gentamycin.

{2} As the genome size of AcNPV is 128 kb, its DNA is very sensitive to mechanical shear. Never vortex AcNPV-DNA during phenol and 2-propanol extractions, and avoid excessive suction when removing the DNA band from the CsCl gradient.

{3} Recently, transfer vectors and AcNPV derivatives have been developed that allow the generation of recombinant baculovirus at frequencies approaching 100%. For this system, linearized virus DNA deleted for the polyhedrin gene and part of an essential downstream gene in combination with a transfer vector providing the deleted gene are used for transfection (Kitts and Possee, 1993).

Another strategy was developed by Luckow *et al.* (1993). A baculovirus shuttle vector (bacmid) was constructed that can replicate in *E. coli* and allows the insertion of the heterologous gene to be done in *E. coli*. By this means the time to construct and purify a recombinant baculovirus can be reduced from 4–6 weeks to several days (Luckow *et al.,* 1993).

{4} Visual screening for the occurrene of polyhedral inclusion body (PIB)-free plaques among many PIB-containing plaques (or vice versa, depending on the used transfer vector and parent virus) is frequently used as the first purification step. PIB-free and PIB-containing plaques are sometimes difficult to distinguish. It is therefore helpful to prepare as controls two plates infected with known occ⁻ and occ⁺ virus, respectively.

{5} When linearized virus DNA was used for cotransfection fewer plates (about five) are sufficient.

{6} Due to the high refraction of the (PIBs), PIB-containing plaques are easily recognized as yellowish and very bright cell groups against the background of uninfected cells. Cells infected with occ⁻ recombinant virus appear also more refractive (brighter) than the surrounding uninfected cells, which have a more grayish appearance.

{7} Plaque hybridization is also frequently used to identify recombinant virus.

{8} A simpler but slightly less efficient procedure for nuclei preparation from insect cells is the following: resuspend the cells in 5 ml 10 mM Tris–Cl, pH 8.1, 100 mM NaCl, 5 mM MgCl$_2$, 1 mM EDTA, adjust the suspension to 0.4% NP40, vortex for 10–20 s, incubate 5–10 min on ice, and centrifuge for 2 min with 8000 rpm in a (cooled) tabletop centrifuge. The pellet consists of the nuclei.

{9} Most nuclear proteins are soluble. However, there are exceptions, i.e., the *Ac* TPase. If the yield of your recombinant protein after extraction of nuclei is low, check whether it sediments during centrifugation of the nuclear homogenate. If can also occur that the protein precipitates during dialysis, depending on the buffer composition.

References

Andrews, D. L., Beames, B., Summers, M. D., and Park, W. D. (1988). Characterization of the lipid acyl hydrolase activity of the major potato (*Solanum tuberosum*) tuber protein, patatin, by cloning and abundant expression in a baculovirus vector. *Biochem. J.* **252**, 199–206.

Angenent, G. C., Busscher, M., Franken, J., Mol, J. N. M., and van Tunen, A. J. (1992). Differential expression of two MADS box genes in wild-type and mutant petunia flowers. *Plant Cell* **4**, 983–993.

Anuntalabhochai, S., Terryn, N., Van Montagu, M., and Inze, D. (1991). Molecular characterization of an *Arabidopsis thaliana* cDNA encoding a small GTP-binding protein, Rhal. *Plant J.* **1**, 167–174.

Brown, W. C., and Campbell, J. L. (1993). A new cloning vector and expression strategy for genes encoding proteins toxic to *Escherichia coli*. *Gene* **127**, 99–103.

Buchner, J., and Rudolph, R. (1991). Renaturation, purification and characterization of recombinant Fab-fragments produced in *Escherichia coli*. *Bio/Technology* **9**, 157–162.

Bustos, M. M., Luckow, V. A., Griffing, L. R., Summers, M. D., and Hall, T. C. (1988). Expression, glycosylation and secretion of phaseolin in a baculovirus system. *Plant Mol. Biol.* **10**, 475–488.

Dignam, J. D., Lebowitz, R. M., and Roeder, R. G. (1983). Accurate transcription initiation by RNA polymerase II in a soluble extract from isolated mammalian nuclei. *Nucleic Acids Res.* **11**, 1475–1489.

Doan, D. N. P., Hoj, P. B., Collins, A., Din, N., Hoogenraad, N. J., and Fincher, G. B. (1993). Post-translational processing of barley a-glucan endohydrolases in the baculovirus-insect cell expression system. *DNA Cell Biol.* **12**, 97–105.

Dubendorff, J. W., and Studier, F. W. (1991). Controlling basal expression in an inducible T7 expression system by blocking the target T7 promoter with lac repressor. *J. Mol. Biol.* **219**, 45–59.

Feldmar, S., and Kunze, R. (1991). The ORFa protein, the putative transposase of maize transposable element Ac, has a basic DNA binding domain. *EMBO J.* **10**, 4003–4010.

Galili, G. (1989). Heterologous expression of a wheat high molecular weight glutenin gene in Escherichia coli. *Proc. Natl. Acad. Sci. U.S.A.* **86**, 7756–7760.

Grima-Pettenati, J., Feuillet, C., Goffner, D., Borderies, G., and Boudet, A. M. (1993). Molecular cloning and expression of a *Eucalyptus gunnii* cDNA clone encoding cinnamyl alcohol dehydrogenase. *Plant Mol. Biol.* **21**, 1085–1095.

Hager, D. A., and Burgess, R. R. (1980). Elution of proteins from sodium dodecyl sulfate-polyacrylamide gels, removal of sodium dodecyl sulfate, and renaturation of enzymatic activity: Results with sigma subunit of *E. coli* RNA polymerase, wheat germ DNA topoisomerase, and other enzymes. *Anal. Biochem.* **109**, 76–86.

Hatfield, P. M., and Vierstra, R. D. (1992). Multiple forms of ubiquitin-activating enzyme E1 from wheat: Identification of an essential cysteine by in vitro mutagenesis. *J. Biol. Chem.* **267,** 14799–14803.

Hauser, C., Fußwinkel, H., Li, J., Oellig, C., Kunze, R., Müller-Neumann, M., Heinlein, M., Starlinger, P., and Doerfler, W. (1988). Overproduction of the protein encoded by the maize transposable element Ac in insect cells by a baculovirus vector. *Mol. Gen. Genet.* **214,** 373–378.

Hochuli, E., Döbeli, H., and Schacher, A. (1987). New metal chelate adsorbent selective for proteins and peptides containing neighbouring histidine residues. *J. Chromatogr.* **411,** 177–184.

Jaenicke, R., and Rudolph, R. (1989). *In* "Protein Structure, a Practical Approach" (T. E. Creighton, ed.), pp. 191–223. Oxford: IRL Press.

King, L. A., and Possee, R. D. (1992). "The Baculovirus Expression System: A Laboratory Guide." New York: Chapman & Hall.

Kitts, P. A., and Possee, R. D. (1993). A method for producing recombinant baculovirus expression vectors at high frequency. *BioTechniques* **14,** 810–817.

Koberstaedt, A., Lenz, M., and Retey, J. (1992). Isolation, sequencing and expression in *E. coli* of the urocanase gene from white clover (*Trifolium repens.*). *FEBS Lett.* **311,** 206–208.

Korth, K. L., and Levings III, C. S. (1993). Baculovirus expression of the maize mitochondrial protein URF13 confers insecticidal activity in cell cultures and larvae. *Proc. Natl. Acad. Sci. U.S.A.* **90,** 3388–3392.

Kunze, R., and Starlinger, P. (1989). The putative transposase of transposable element Ac from *Zea mays* L. interacts with subterminal sequences of Ac. *EMBO J.* **8,** 3177–3185.

Luckow, V. A., and Summers, M. D. (1988). Trends in the development of baculovirus expression vectors. *Bio/Technology* **6,** 47–55.

Luckow, V. A., Lee, S. C., Barry, G. F., and Olins, P. O. (1993). Efficient generation of infectious recombinant baculoviruses by site-specific transposon-mediated insertion of foreign genes into a baculovirus genome propagated in *Escherichia coli. J. Virol.* **67,** 4566–4579.

Malitschek, B., and Schartl, M. (1991). Rapid identification of recombinant baculoviruses using PCR. *Biotechniques* **11,** 177–178.

Mori, H., Tanizawa, K., and Fukui, T. (1991). Potato tuber Type H phosphorylase isozyme: molecular cloning, nucleotide sequence, and expression of a full-length cDNA in Escherichia coli. *J. Biol. Chem.* **266,** 18446–18453.

Mori, H., Tanizawa, K., and Fukui, T. (1993). A chimeric alpha-glucan phosphorylase of plant type L and H isozymes: Functional role of a 78-residue insertion in type L isozyme. *J. Biol. Chem.* **268,** 5574–5581.

Nagai, A., Suzuki, K., Ward, E., Moyer, M., Hashimoto, M., Mano, J., Ohta, D., and Scheidegger, A. (1992). Overexpression of plant histidinol dehydrogenase using a baculovirus expression vector system. *Arch. Biochem. Biophys.* **295,** 235–239.

Ng, J. D., Ko, T.-P., and McPherson, A. (1993). Cloning, expression, and crystallization of jack bean (*Canavalia ensiformis*) canavalin. *Plant Physiol.* **101,** 713–728.

O'Reilly, D. R., Miller, L. K., and Luckow, V. A. (1992). "Baculovirus Expression Vectors: A Laboratory Manual." New York: Freeman.

Rolland, N., Droux, M., Lebrun, M., and Douce, R. (1993). O-Acetylserine(thiol)lyase from spinach (*Spinacia oleracea* L.) leaf: cDNA cloning, characterization, and overexpression in *Escherichia coli* of the chloroplast isoform. *Arch. Biochem. Biophys.* **300,** 213–222.

Rosenberg, A. H., Goldman, E., Dunn, J. J., Studier, F. W., and Zubay, G. (1993). Effects of consecutive AGG codons on translation in *Escherichia coli,* demonstrated with a versatile codon test system. *J. Bacteriol.* **175,** 716–722.

Sambrook, J., Fristsch, E. F., and Maniatis, T. (1989). "Molecular Cloning; A Laboratory Manual," 2nd ed. Cold Spring Harbor, NY: Cold Spring Harbor Laboratory Press.

Schein, C. H., and Noteborn, M. H. M. (1988). Formation of soluble recombinant proteins in *Escherichia coli* is favored by lower growth temperature. *Bio/Technology* **6,** 291–294.

Smith, G. E., Ju, G., Ericson, B. L., Moschera, J., Lahm, H.-W., Chizzonite, R., and Summers, M. D. (1985). Modification and secretion of human interleukin 2 produced in insect cells by a baculovirus expression vector. *Proc. Natl. Acad. Sci. U.S.A.* **82,** 8404–8408.

Studier, F. W. (1991). Use of bacteriophage T7 lysozyme to improve an inducible T7 expression system. *J. Mol. Biol.* **219,** 37–44.

Studier, F. W., and Moffatt, B. A. (1986). Use of T7 RNA polymerase to direct selective high-level expression of cloned genes. *J. Mol. Biol.* **189,** 113–130.

Studier, F. W., Rosenberg, A. H., Dunn, J. J., and Dubendorff, J. W. (1990). Use of T7 RNA polymerase to direct expression of cloned genes. *Methods Enzymol.* **185,** 60–89.

Sullivan, M. L., and Vierstra, R. D. (1991). Cloning of a 16-kDa ubiquitin carrier protein from wheat and *Arabidopsis thaliana:* Identification of functional domains by in vitro mutagenesis. *J. Biol. Chem.* **266,** 23878–23885.

Tessier, D. C., Thomas, D. Y., Khouri, H. E., Laliberte, F., and Vernet, T. (1991). Enhanced secretion from insect cells of a foreign protein fused to the honeybee melittin signal peptide. *Gene* **98,** 177–183.

Tjia, S. T., Meyer zu Altenschildesche, G., and Doerfler, W. (1983). *Autographa californica* nuclear polyhedrosis virus (AcNPV) does not persist in mass cultures of mammalian cells. *Virology* **125,** 107–117.

Verburg, J. G., Rangwala, S. H., Samac, D. A., Luckow, V. A., and Huynh, Q. K. (1993). Examination of the role of tyrosine-174 in the catalytic mechanism of the *Arabidopsis thaliana* chitinase: Comparison of variant chitinases generated by site-directed mutagenesis and expressed in insect cells using baculovirus vectors. *Arch. Biochem. Biophys.* **300,** 223–230.

Vernet, T., Tessier, D. C., Richardson, C., Laliberte, F., Khouri, H. E., Bell, A. W., Storer, A. C., and Thomas, D. Y. (1990). Secretion of functional papain precursor from insect cells: Requirement for N-glycosylation of the pro-region. *J. Biol. Chem.* **265,** 16661–16666.

Webb, A., Bradley, M., Phelan, S., Wu, J., and Gehrke, L. (1991). Use of the polymerase chain reaction for screening and evaluation of recombinant baculovirus clones. *Biotechniques* **11,** 512–519.

CHAPTER 35

Expression and Localization of Plant Membrane Proteins in *Saccharomyces*

Ramón Serrano* and José-Manuel Villalba†

*Departamento de Biotecnología
Esc. Tec. Sup. Ing. Agrónomos
Universidad Politécnica
46022 Valencia, Spain
†Departamento de Biología Celular
Facultad de Ciencias
Universidad de Córdoba
14004 Córdoba, Spain

I. Introduction

The expression of cloned genes in heterologous systems has two major goals: (a) production of antigen for antibody generation; (b) expression of functional protein for study of its activity. Antigen production is usually accomplished by massive expression in *Escherichia coli* as insoluble, denatured fusion proteins with either β-galactosidase (*lac*Z gene; Stanley, 1988) or anthranilate synthase (*trpE* gene; Yansura, 1990).

E. coli is also for most purposes the best host for the expression of soluble proteins in active form (Gold, 1990). In the case of membrane proteins, however, problems are usually found with the bacterial expression system (Bitter, 1987). Hydrophobic membrane proteins are usually quite toxic to *E. coli* and are expressed at very low levels (Bitter, 1987). This toxicity is probably due to depolarization of the bacterial membrane by improperly assembled membrane proteins (Eraso and Serrano, 1990). The baculovirus expression system could provide a useful alternative, but it seems that in some cases the membrane proteins are inactive (Klaasen *et al.*, 1993).

The yeast expression system described in this chapter can produce large amounts of plant membrane proteins in active form (Villalba *et al.*, 1992). Plant organelle and secreted proteins may also benefit from this system because the targeting machinery is in many cases conserved between yeast and plants (Bowler *et al.*, 1989; Edens *et al.*, 1984; Learned and Fink, 1989; Denecke *et al.*, 1991; Bozak *et al.*, 1992). Table I provides a collection of examples of the functional expression of plant proteins in different yeast compartments.

In this chapter we will describe the basic principles and methods for the expression of plant proteins in yeast cells. Some specific procedures such as

Table I
Examples of Functional Expression of Plant Proteins in Different Yeast Compartments

Protein	Compartment	Reference
H^+-ATPase	ER	Villalba *et al.*, 1992
	PM and ER	Palmgren and Christensen 1993
K^+ channel	PM	Anderson *et al.*, 1992
		Sentenac *et al.*, 1992
Sugar permeases	PM	Sauer and Stadler, 1993
		Reismeier *et al.*, 1992
Amino acid permease	PM	Frommer *et al.*, 1993
Mn-superoxide dismutase	MT	Bowler *et al.*, 1989
Rieske iron–sulfur protein	MT	Huang *et al.*, 1991
T-URF13 protein	MT	Glab *et al.*, 1990
Thaumatin	SC	Edens *et al.*, 1984
Cytochrome *P*-450	ER	Bozak *et al.*, 1992
BiP	ER	Denecke *et al.*, 1991
3-Hydroxy-3-methylglutaryl-coenzyme A reductase	ER	Learned and Fink, 1989

Note. ER, endoplasmic reticulum; PM, plasma membrane; MT, mitochondria; SC, secretory compartment.

transformation of yeast cells with shuttle plasmids are described in Chapter 29, Vol. 49 of this series, and will not be repeated here. Expression of a plant organelle or membrane protein within yeast cells does not result necessarily in its targeting to the same membrane as seen within the plant cell (Villalba *et al.*, 1992). Therefore two methodologies to localize recombinant proteins in yeast, subcellular fractionation and indirect immunfluorescence, are also described.

II. Requirements for Efficient Transgene Expression in Yeast

We divide the requirements for expression in yeast into five important aspects:

1. yeast plasmids and strains
2. promoters and transcriptional termination sequences
3. codon usage and introns
4. mRNA leader sequences
5. secretion of extracellular proteins

A brief discussion will provide the basic criteria to construct efficient yeast expression plasmids.

A. Yeast Plasmids and Strains

Yeast autonomous plasmids contain either a chromosomal centromere and origin of replication (YCp plasmids, *y*east *c*entromeric *p*lasmids) or a region of the natural yeast episome "2-μm circle," which includes the origin of replication and the *REP3* locus controlling partition (YEp plasmids, *y*east *e*pisomal *p*lasmids). YCp plasmids exist in one to two copies per cell whereas YEp plasmids have a copy number of 10–40 and are the vectors of choice for efficient expression (Rose and Broach, 1990). Yeast plasmids are shuttle plasmids, and contain an *E. coli* origin of replication and partition and antibiotic resistance genes. Selection in yeast is affected by nutritional requirements, with the plasmid carrying the *URA3, LEU2, HIS3, LYS2,* or *TRP1* genes and the yeast strain containing stable, nonreverting mutations in these genes (*ura3-52* or *ura3-251,328,372,leu2-3,112, his-Δ200, lys-201,* or *trpl-Δ901*) (Stearns *et al.*, 1990). Continuous selection for the plasmid in minimal medium without the corresponding nutritional requirement is important.

B. Promoters and Transcriptional Termination Sequences

Typical yeast promoters contain a TATA box (consensus: TATAAA) located 60–120 bp upstream of the transcription initiation site and a variety of regulatory sequences located 20–500 bp upstream of the TATA box (Schneider and Guarente, 1991). Promoters of highly expressed glycolytic enzymes are the most utilized for heterologous expression. For relatively constitutive expression, the promoter of the phosphoglycerate kinase gene (*PGK1*) is the most widely utilized (Kingsman *et al.,* 1990; see Chapter 29, Vol. 49 of this series). Recently, the promoter of the plasma membrane H⁺-ATPase gene (*PMA1*) has also been successfully utilized (Villalba *et al.,* 1992). Both promoters have only a small degree of regulation (up to 10-fold change) by carbon source, with maximum expression in glucose media and reduced expression in galactose or ethanol media. For tightly regulated expression, the promoter of the galactokinase gene (*GAL1*) is the best choice (Schneider and Guarente, 1991), with expression in glucose media being 1000-fold less than that in galactose media. This regulation is essential if the expressed protein is toxic to yeast cells.

The presence of DNA sequences required for efficient transcription termination is also important to maximize expression (Zaret and Sherman, 1982). Expression plasmids usually contain yeast 3'-flanking sequences to provide for this function.

Table II provides a listing of some of the most useful plasmids for expression of heterologous proteins in yeast.

Table II
Some Useful Plasmids for Expression of Plant Proteins in Yeast

Vector	Promoter	Cloning sites	Marker	Reference
pMA91	*PGK1*	*Bgl*II	*LEU2*	Kingsman *et al.,* 1990
pFL61	*PGK1*	*Not*I	*URA3*	See Chapter 29[a]
		*Bst*XI		
pRS-1024	*PMA1*	*Xho*I	*LEU2*	Villalba *et al.,* 1992
pRS-699	*PMA1*	*Xho*I	*URA3*	This chapter
pEMBLyex4	*GAL1*	SacI	*URA3*	Cesareni and Murray, 1987
		*Sma*I	*LEU2*	
		*Bam*HI		
		*Xba*I		
		*Sal*I		
		*Pst*I		
		*Hind*III		
pYES 2	*GAL1*	*Hind*III	*URA3*	Invitrogen
		*Kpn*I		
		SacI		
		*Bam*HI		
		*Eco*RI		
		*Not*I		
		*Xho*I		
		*Sph*I		
		*Xba*I		

Note. All have the 2 μm origin of replication and partition sequences.
[a] In Vol. 49 of this series.

C. Codon Usage and Introns

Highly expressed yeast genes have a strong codon bias for 26 of the 61 possible triplets (Hinnebusch and Liebman, 1991; see also Table III). It has been reported that the expression level of a mouse immunoglobulin kappa chain in yeast increases 50-fold if a synthetic gene with only yeast-preferred codons is employed (Kotula and Curtis, 1991). This approach should be taken into consideration for maximizing expression of plant proteins in yeast.

The splicing of mRNAs from higher organisms produced in yeast is inefficient and aberrant (Bitter, 1987). Therefore, heterologous gene expression in yeast, as in *E. coli*, is limited to cDNAs, chemically synthesized genes, and genomic clones lacking introns.

D. mRNA Leader Sequences

Translation in yeast is particularly sensitive to secondary structures in the 5'-noncoding region of the mRNAs (Donahue and Cigan, 1990; Hinnebusch and

Table III
The 26 Preferred Triplet Codons in Yeast Protein Biosynthesis

Amino acid	Preferred codons
Ala	GCT, GCC
Arg	AGA
Asn	AAC
Asp	GAT, GAC
Cys	TGT
Gln	CAA
Glu	GAA
Gly	GGT
His	CAC
Ile	ATT, ATC
Leu	TTG
Lys	AAG
Met	ATG
Phe	TTC
Pro	CCA
Ser	TCT, TCC
Thr	ACT, ACC
Trp	TGG
Tyr	TAC
Val	GTT, GTC

Note. Adapted From Hinnebusch and Liebman (1991).

Liebman, 1991). Stem–loop structures of 4 bp or more result in drastic inhibition of gene expression. This contrasts with the situation in animal cells, where inhibition is only observed with stem–loop structures of more than 20 bp. Therefore, for efficient expression in yeast the leaders of plant genes should be inspected to avoid (G + C)-rich regions and sequences of dyad symmetry that could be introduced during cDNA constructions. For example, a construction where a *Xho*I linker (CCTCGAGG) was ligated into an *Alu*I site (AG'CT) in the leader region reduced the expression of the yeast *PMA1* gene more than 10-fold (P. Eraso, unpublished). This construction generated a palindrome of 12 bases, which results in a stem–loop of 4–5 bp (allowing 2–4 bases for the loop).

Another important feature to be avoided in the leader is the presence of upstream ATG's. They strongly reduce expression at the correct, downstream ATG. The degree of inhibition may range from 80% to more than 99.9% (Hinnebusch and Liebman, 1991). Lower inhibition is obtained when there are in-frame stop codons that terminate translation of the upstream open reading frame.

Less important for the efficiency of expression are the length of the leader and the context of the start AUG codon (Hinnebusch and Liebman, 1991). Although highly expressed yeast genes have an A at position −3 from the AUG (AXXAUG), changing this base has only minor effects on expression levels.

E. Secretion of Extracellular Proteins

Extracellular proteins usually contain essential disulfide bonds. As the conditions inside cells are too reducing to permit their formation, cytoplasmic expression may produce inactive proteins (Smith *et al.*, 1985). Yeast or heterologous secretion signals (signal peptides) may be utilized for secretion of heterologous proteins in yeast (Edens *et al.*, 1984; Smith *et al.*, 1985). However, there are two major problems with secretion in yeast. Heterologous proteins are often poorly secreted in yeast and most of them remain trapped within the endoplasmic reticulum (Smith *et al.*, 1985). In addition, glycosylation in yeast is very different from that in higher organisms: there is a less extensive trimming of the "core" carbohydrates and long outer chains of more than 50 mannose residues are appended (Brake, 1990). As a result heterologous proteins secreted from yeast may be inactive or antigenically different from the natural proteins. Both problems can be ameliorated by certain yeast mutations (Smith *et al.*, 1985; Hitzeman *et al.*, 1990).

III. Materials and Methods

A. Growth of Yeast Transformants

1. Preparation of Growth Media

Synthetic media should be employed for continuous selection of yeast plasmids. The most convenient preparation is the yeast nitrogen base w/o amino acids from Difco (Catalog No. 0919-15-3). This dehydrated formulation contains all the vitamins and minerals required for growth of wild-type yeast (including ammonium sulfate as nitrogen source). It needs to be supplemented with a carbon source (usually glucose or galactose) and with the auxotrophic requirements of the particular yeast strain. To obtain optimal yeast growth it should also be supplemented with a buffer at pH 4–6 and with adenine, uracil, and all the amino acids (with the exception of those nutrients provided by the plasmid marker; see above).

Convenient stock solutions for yeast media (dilution factor in the final media between brackets) are:

Yeast Nitrogen Base ($\times$10): 6.7% in water, warmed to dissolve, sterilized by filtration, and stored at 4°C.

Glucose ($\times$10): 20% in water, autoclaved, and stored at 4°C.

Galactose ($\times$10): 20% in water, autoclaved, and stored at 4°C.

Succinate buffer ($\times$10): 0.5 M succinic acid adjusted to pH 5 with Tris base, autoclaved, and stored at 4°C.

Adenine sulfate and uracil ($\times$100): 5 mg/ml, autoclaved, and stored at 4°C.

Leucine, serine, and threonine ($\times$100): 20 mg/ml, autoclaved, and stored at 4°C.

Methionine and tryptophan ($\times$100): 10 mg/ml, sterilized by filtration and stored frozen.

Other amino acids ($\times$100): 10 mg/ml, autoclaved, and stored at 4°C.

Media are prepared by mixing the required amounts of stock solutions with sterilized water. In the case of solid media the sterilized water should contain the required amount of bacteriological agar to provide a 2% final concentration.

2. Growth of Yeast Cultures

Growth is started from single colonies of yeast transformants. First, small-scale glucose cultures (5 ml) are grown to saturation for 2 days (OD_{660} about 3). Then the 5 ml of saturated culture is utilized to inoculate 500 ml of glucose medium in one flask of 2-liter capacity. After 14–18 h of growth at 28°C with shaking, the culture should be at late-exponential phase (OD_{660} about 0.4–0.6). Stationary phase is to be avoided because of increased protein degradation. Duration of growth may need to be adjusted for particular strains in order to harvest the culture at the convenient OD_{660} of about 0.5. This corresponds to about 1.3 mg fresh weight of cells/ml, 0.4 mg dry weight of cells/ml, 13×10^6 cells/ml, and 0.16 mg total protein/ml. Therefore, about 0.65 g of cells and 80 mg of total yeast protein can be obtained from 500 ml of yeast culture. Recombinant proteins produced using strong promoters (Table II) may represent on the order of 1% of total protein. Therefore, a yield of about 1 mg recombinant protein per 500 ml of yeast culture may be expected.

In the case of galactose-dependent expression, a 25-ml culture (in 100-ml flask) should be grown in glucose medium for 2 days to reach saturation and deplete the glucose. Then the 25-ml cultures should be added to 500 ml of galactose medium (in a 2-liter flask) and grown for about 24 h to an OD_{660} of about 0.5. Duration of growth may need to be adjusted for particular strains in order to harvest the culture at the desired absorbance.

B. Subcellular Fractionation

We will describe methods for the homogenization of whole yeast cells by vigorous shaking with glass beads, and for separation of organelles by differential and sucrose gradient centrifugation (Serrano, 1978; 1988). Alternatives not discussed here are the osmotic lysis of protoplasts and the utilization of other gradient materials (Percoll, metrizamide, etc.) for equilibrium centrifugation. It must be emphasized that a scheme for the complete resolution of yeast organelles by subcellular fractionation has not yet been established (Franzusoff et al., 1991). Therefore, only partial correlations between distributions of markers and of heterologous expressed proteins can be made (Villalba et al., 1992). In this respect, immunofluorescence is an important alternative to biochemical approaches (see next section).

Concerning markers, the traditional biochemical markers for organelles (Serrano, 1985) have been superseded by immunological markers (Franzusoff et al., 1991). A panel of some of the most useful antigenic markers for yeast organelles is given at Table IV. This immunological approach has the advantage of allowing identification of organelles both by subcellular fractionation and by immunofluorescence. A systematic comparison of biochemical and immunological markers, however, has not yet been made.

For small amounts of yeast cells (up to 1 g) a convenient homogenization procedure is to shake by hand in a vortex mixer a tube containing the yeast

Table IV
Immunological Markers for Yeast Organelles

Organelle	Antigenic marker	Reference
Plasma membrane	*PMA1* ATPase	Monk et al., 1991
Endoplasmic reticulum	*KAR2* protein	Pringle et al., 1991
Mitochondria	F1 ATPase	Franzusoff et al., 1991
Vacuoles	*VMA1* protein	Hirata et al., 1990
	Dipeptidyl aminopeptidase B	Franzusoff et al., 1991
Golgi	*KEX2* protein	Pringle et al., 1991
Secretory vesicles	*SEC4* protein	Pringle et al., 1991
Peroxisomes	Thiolase	Pringle et al., 1991
Nuclei	p110/p95 envelope protein	Franzusoff et al., 1991

suspension and glass beads. For example, the cells from 500 ml of culture harvested at an OD_{660} of about 0.5 (0.65 g) may be homogenized as follows. After centrifugation (10 min at $700\times g$) and washing with cold water, the cells are resuspended with cold water to a final volume of 3 ml in 40-ml centrifuge tubes placed on ice. The following additions are made:

1.5 ml extraction buffer $\times 3$ (30% sucrose, 0.3 M Tris, 0.15 M K_2SO_4, 15 mM EDTA, pH 8, with HCl, stored frozen)

18 μl dithiothreitol 0.5 M (in water, stored frozen)

18 μl phenylmethylsulfonyl fluoride 0.1 M (dissolved in isopropanol, stored at $-20°C$; warm to redissolve)

18 μl chymostatin 5 mg/ml (dissolved in dimethlyl sulfoxide, stored at $-20°C$; warm to redissolve)

9 ml cold glass beads 0.5-mm diameter (acid washed and stored frozen)

The tube is vigorously shaken by hand in a vortex mixer at top speed for three periods of 1 min with 2–3 min of cooling in between.

Glass beads can be removed by filtration under vacuum through a glass-sintered filter. The beads should be washed with 4.5 ml of extraction buffer ($\times 1$, mixture of 3 ml water and 1.5 ml extraction buffer $\times 3$). Alternatively, the washing mixture is added before the beads are removed, the tube is shaken for a few seconds, and the beads are removed in the first centrifugation step (see below). This avoids filtration but results in loss of about one-third of the homogenate trapped between the beads.

Up to four batches (3 g cells, 0.3 g total protein) of homogenization as described above can be simultaneously handled with two vortex mixers. For large-scale work (more than 10 g of cells or 1 g total protein) a Vibrogen Cell Mill (E. Bühler, Tübingen, Germany) or any similar device for shaking refrigerated containers should be used.

Differential centrifugation proceeds at 4°C in three steps (Serrano, 1978). First a debris fraction is removed by centrifugation for 10 min at $700\times g$. This pellet includes nonruptured cells, cell wall remnants, and some nuclei and attached endoplasmic reticulum. It is discarded. The supernatant is centrifuged for 20 min at $20,000\times g$ to obtain a crude mitochondrial fraction. This includes, in addition to mitochondria and peroxisomes, some nuclei and attached endoplasmic reticulum not removed with the debris, and vesicles comprising most of the plasma membrane and part of the tonoplast. The supernatant is centrifuged for 30 min at $120,000\times g$ to obtain a microsomal fraction. This contains some mitochondrial and endoplasmic reticulum fragments, some plasma membrane and tonoplast vesicles, and most of the Golgi and secretory vesicles. The pellets are resuspended with 0.5 ml extraction buffer supplemented with 2 μl phenylmethylsulfonyl fluoride (0.1 M, see above), 2 μl chymostatin (5 mg/ml, see above), and 2 μl dithiothreitol (0.5 M, see above), homogenized with a glass–glass homogenizer, and stored at $-70°C$.

Continuous sucrose gradients are prepared by horizontal diffusion of sucrose layers. In a 12-ml centrifuge tube (for example, those for the Beckman SW40 rotor), three 4-ml portions of 50, 35, and 20% sucrose (w/w) are carefully layered. These solutions contain 10 mM Tris–HCl (pH 7.6), 1 mM EDTA, and 1 mM dithiothreitol. The tubes are closed with parafilm and slowly tilted to a horizontal position in the cold room. After 3 h of diffusion they are slowly returned to a vertical position, 1 ml is removed from the top, and the sample of mitochondrial or microsomal membranes is applied. After overnight centrifugation at 30,000 rpm and 4°C to reach equilibrium density, fractions of 0.7 ml are collected.

Sucrose concentration is determined with a refractometer and protein concentration with the Bio-Rad protein assay system. Marker antigens (Table IV) are analyzed by Western blotting.

Cell wall fragments and nuclei are the most dense organelles, equilibrating at about 60–65% sucrose. They will pellet in the above gradient. They are followed in density by peroxisomes and plasma membrane vesicles, which equilibrate at 45–50% sucrose. Then follow mitochondria, tonoplast vesicles, and rough endoplasmic reticulum (35–40% sucrose), Golgi vesicles (25–30% sucrose), and smooth endoplasmic reticulum (20–25% sucrose). These sucrose values (% w/w) should be viewed as orientative because, as indicated above, a scheme for the complete resolution of yeast organelles by subcellular fractionation has not been established.

C. Indirect Immunofluorescence of Yeast Cells to Detect Expressed Proteins

Detection of recombinant proteins expressed in yeast by indirect immunofluorescence is a simple and sensitive method for identification of the subcellular compartment within which the expressed protein is located (Pringle *et al.*, 1991). The method relies on the fixation and permeabilization of cells with formaldehyde, followed by incubation in the presence of a primary antibody directed against the protein of interest. Detection is achieved through incubation of the cells in the presence of a fluorescently labeled secondary antibody directed against the primary antibody. Methods for production and purification of mono- and polyclonal antibodies are beyond the scope of this chapter. Since general methods for immunofluorescence are described elsewhere in this volume (Chapter 7, Vol. 49 of this series), only those specific for yeast cells are described here.

1. Materials

Phosphate-buffered formaldehyde: 50 mM potassium phosphate buffer (pH 6.5) with 0.5 mM MgCl$_2$ and 0.1 vol of concentrated formaldehyde (35–40%, Merck).

Washing solution: 0.1 M potassium phosphate buffer (pH 7.4) with 1.2 M sorbitol.

Zymolyase-100T (Miles, No. 32-093-1): 1 mg/ml in water, store frozen.

Multiwell slides (Flow laboratories, No. 60-408-05)

Coverslips (24 × 60 mm)

Polylysine (Sigma, No. P-1524): 1 mg/ml in water, stored frozen.

Phosphate-buffered saline (PBS): 0.14 M NaCl and 10 mM potassium phosphate (pH 7.4).

Blocking buffer: PBS with 1 mg/ml bovine serum albumin.

DAPI stock solution: DAPI (4', 6'-diamidino-2-phenylindole, Sigma, No. D-1388) 1 mg/ml in water and stored frozen protected from light).

Primary antibody: The polyclonal antisera or monoclonal antibodies are stored either precipitated with ammonium sulfate at 4°C or in 50% glycerol at −20°C. For precipitation, mix one part of antibody solution with one part of saturated ammonium sulfate. Before use, mix gently to resuspend the precipitate and dilute in blocking buffer as determined empirically. The working dilutions for immunofluorescence are about 10 times more concentrated than those for Western blots.

Secondary antibody: Anti-rabbit IgG-fluorescein (No. 1238833), anti-rabbit IgG-rhodamine (No. 1238841), anti-mouse Ig-fluorescein (No. 821462), and anti-mouse Ig-rhodamine (No. 1214594) can be obtained from Boehringer Mannheim. They can be stored at 4°C for several weeks. For long-term storage aliquots should be frozen at −80°C. Always manipulate the fluorophore conjugates in low light. The working dilution is usually 1/1000.

Mounting Medium: Either *p*-phenylenediamine or *N*-propyl gallate may be added to retard photobleaching. We prefer the latter because *p*-phenylenediamine is reported to be carcinogenic. Our mounting medium contains 20 mg/ml *N*-propyl gallate (Sigma No. P-3130) in a mixture of 3 vol of 0.1 M Tris–HCl buffer (pH 9) and 7 vol glycerol. For nuclear staining add 2.5 μl of DAPI stock solution per 100 ml mounting medium. Store frozen and protected from light.

Photographic film: Fujichrome 1600 for color slides and Kodak T-MAX p3200 (400 ASA) for black and white prints.

2. Procedure

Yeast cells (5 ml culture; OD$_{600}$ approximately 0.5) are fixed in growth medium by rapid addition of 0.1 vol of concentrated formaldehyde solution (35–40%, Merck). After 30 min of incubation at room temperature, the cells are recovered in 0.5 ml of phosphate-buffered formaldehyde and fixed for an additional period of up to 2 h at room temperature. When fixation is completed, yeast cells are washed three times and resuspended in 0.25 ml of washing solution. Then 0.5 μl β-mercaptoethanol and 5 μl of zymolyase solution are added and the incubation is continued for 30 min at 37°C. Cells are recovered by low-speed centrifugation and gently resuspended in 0.25 ml of washing solution.

Multiwell slides are treated with polylysine as follows. Add 10–15 μl of polylysine solution per well and leave 30 s. Remove the polylysine solution, wash three times with water, and air-dry. Aliquots of 10 μl of fixed cell suspensions are added to each well and after 30 s the liquid is removed. Check under the microscope for an adequate density of cells attached to the wells. Slides are air-dried and then immersed sequentially in cold methanol (6 min at −20°C) and cold acetone (30 s at −20°C). Once dried, nonspecific binding sites are saturated by incubation in blocking buffer for 30 min. Incubations are carried out at room temperature in a moist chamber. Remove the solution, add 10 μl of an appropriate dilution of primary antibody, and incubate for 1 h. Wash slides four times with blocking buffer and then incubate for 1 h in the dark with 10 μl of an appropriate dilution of fluorophore-conjugated secondary antibody. Wash again four times with PBS and drain the slides. Place a drop of mounting medium per four wells and cover with coverslips. Wipe excess medium with filter paper and seal the edges with clear nail polish. Once the polish has hardened preparations can be visualized or stored at −20°C in the dark.

IV. Critical Aspects of the Procedures

The most crucial aspect in Western blot and immunofluorescence analysis is the specificity of the primary antibody. As animal sera are able to recognize yeast proteins, only monoclonal and affinity-purified polyclonal antibodies are adequate. Occasionally, crude antisera of high titer may be specific enough at high dilutions (more than 1/10,000). Control experiments with yeast cells that do not express the recombinant protein should be run in parallel to confirm the specificity of the primary antibody. Commercial secondary antibodies, which are usually affinity-purified, usually pose no problems.

For Western blot of the yeast and plant plasma membrane H$^+$-ATPases, samples cannot be prepared by the standard solubilization procedure of boiling with the SDS buffer because this induces aggregation of the ATPases (Serrano, 1988). Solubilization with SDS at 30–40°C avoids aggregation but, under these conditions, endogenous proteases remain activate and may degrade proteins in the sample. The ATPase is especially susceptible and high concentrations of protease inhibitors should be included in the SDS solubilization buffer.

V. Results and Discussion

The plasma membrane H$^+$-ATPase of *Arabidopsis thaliana* (isoform 1, *AHA1* gene) has been expressed in yeast utilizing plasmid pRS-1024 of Table II (Villalba *et al.*, 1992). The level of expression was relatively high (1% of total yeast protein) and the enzyme was fully functional. Both subcellular fractionation and immunofluorescence indicated that the plant protein accumulated in the endoplasmic reticulum, which in yeast is mostly located surrounding the nucleus (Fig. 1). This was confirmed by immunogold electron microscopy, which at high magnification showed that perinuclear staining was due to the accumulation of stacked endoplasmic reticulum membranes surrounding the nucleus (Villalba *et al.,* 1992). This phenomenon has also been observed upon overproduction of 3-hydroxy-3-methylglutaryl coenzyme A reductase in yeast (Wright *et al.,* 1988), and is apparently due to the stimulation of membrane biogenesis by the overproduction of membrane proteins (Browse and Somerville, 1991).

It is not clear why a plasma membrane protein from plants is retained within the yeast endoplasmic reticulum and does not reach the plasma membrane. Recent experiments indicate that a deletion of the carboxyl terminus of plant ATPase results in partial localization of the enzyme at the yeast plasma membrane (Palmgren and Christensen, 1993). This suggests that the terminal domain of the plant enzyme is responsible in part for retention within the yeast endoplasmic reticulum.

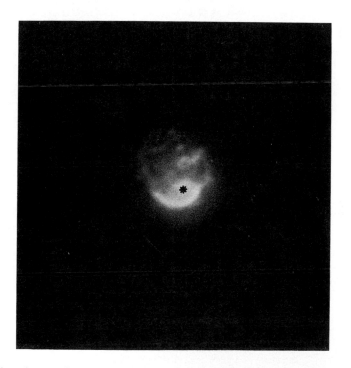

Fig. 1 Indirect immunofluorescence of yeast cells expressing *Arabidopsis* plasma membrane H^+-ATPase. Cells of strain RS-934 grown in glucose medium were labeled with monoclonal antibody 46E5/B11C10 against plant ATPase and fluorescein-conjugated antimouse antiserum as secondary antibody (Villalba *et al.,* 1992). Recombinant plant ATPase is localized intracellularly in the endoplasmic reticulum, mostly surrounding the yeast nucleus (asterisk).

===== **VI. Conclusions and Perspectives**

Plant sugar and amino acid permeases transport substrates into yeast cells (Sauer and Stadler, 1993; Riesmeier *et al.,* 1992; Frommer *et al.,* 1993), and therefore some of these proteins are located at the plasma membrane. However, the quantitative distribution of these plant proteins in different yeast membranes was not investigated. Future studies should clarify the requirements for plant membrane proteins to move within the yeast endomembrane secretory pathway. However, the studies performed to date suggest that yeast cells may be a very convenient system for the functional expression of plant membrane proteins.

References

Anderson J. A., Huprikar, S. S., Kochian, L. V., Lucas, W. J., and Gaber, R. F. (1992). Functional expression of a probable *Arabidopsis thaliana* potassium channel in *Saccharomyces cerevisiae. Proc. Natl. Acad. Sci. U.S.A.* **89,** 3736–3740.

Bitter, G. A. (1987). Heterologous gene expression in yeast. *Methods Enzymol.* **152,** 673–684.

Bowler, C., Alliote, T., Van den Bulcke, M., Bauw, G., Vandekerckhove, J., Van Montagu, M., and Inzé, D. (1989). A plant manganese superoxide dismutase is efficiently imported and correctly processed by yeast mitochondria. *Proc. Natl. Acad. Sci. U.S.A.* **86,** 3237–3241.

Bozak, K. R., O'Keefe, D. P., and Christoffersen, R. E. (1992). Expression of a ripening-related avocado (*Persea americana*) cytochrome P450 in yeast. *Plant Physiol.* **100,** 1976–1981.

Brake, A. J. (1990). α-Factor leader-directed secretion of heterologous proteins from yeast. *Methods Enzymol.* **185,** 408–421.

Browse, J., and Somerville, C. (1991). Glycerolipid synthesis: Biochemistry and regulation. *Annu. Rev. Plant Physiol. Plant Mol. Biol.* **42,** 467–506.

Cesareni, G., and Murray, A. H. (1987). Plasmid vectors carrying the replication origin of filamentous single-stranded phages. *In* "Genetic Engineering. Principles and Methods" (J. K. Setlow and A. Hollaender, eds.), Vol. 9, pp. 135–154. New York: Plenum.

Denecke, J., Goldman, M. H. S., Demolder, J., Seurinck, J., and Botterman, J. (1991). The tobacco luminal binding protein is encoded by a multigene family. *Plant Cell* **3,** 1025–1035.

Donahue, T. F., and Cigan, A. M. (1990). Sequence and structural requirements for efficient translation in yeast. *Methods Enzymol.* **185,** 366–372.

Edens, L., Bom, I., Ledeboer, A. M., Maat, J., Toonen, M. Y., Visser, C., and Verrips, C. T. (1984). Synthesis and processing of the plant protein thaumatin in yeast. *Cell* **37,** 629–633.

Eraso, P., and Serrano, R. (1990). Expression of yeast plasma membrane H⁺-ATPase in *Escherichia coli* deenergizes the bacterial membrane. *In* "Trends in Biomembranes and Bioenergetics" (A. Jacob, ed.), Vol. 1, pp. 1–10. Sreekanteswaram, India: Council of Scientific Research Integration.

Franzusoff, A., Rothblatt, J., and Schekman, R. (1991). Analysis of polypeptide transit through yeast secretory pathway. *Methods Enzymol.* **194,** 662–674.

Frommer, W. B., Hummel, S., and Riesmeier, J. W. (1993). Expression cloning in yeast of a cDNA encoding a broad specificity amino acid permease from *Arabidopsis thaliana. Proc. Natl. Acad. Sci. U.S.A.* **90,** 5944–5948.

Glab, N., Wise, R. P., Pring, D. R., Jacq, C., and Slonimski, P. (1990). Expression in *Saccharomyces cerevisiae* of a gene associated with cytoplasmic male sterility from maize: respiratory disfunction and uncoupling of yeast mitochondria. *Mol. Gen. Genet.* **223,** 24–32.

Gold, L. (1990). Expression of heterologous proteins in *Escherichia coli. Methods Enzymol.* **185,** 11–14.

Hinnebusch, A. G., and Liebman, S. W. (1991). Protein synthesis and translational control in *Saccharomyces cerevisiae. In* "The Molecular and Cellular Biology of the Yeast *Saccharomyces.* Volume

I: Genome Dynamics, Protein Synthesis and Energetics" (J. R. Broach, J. R., Pringle, and E. W. Jones, eds.), pp. 627–735, Cold Spring Harbor, NY: Cold Spring Harbor Laboratory Press.

Hirata, R., Ohsumi, Y., Nakano, A., Kawasaki, H., Suzuki, K., and Anraku, Y. (1990). Molecular structure of a gene, *VMA1,* encoding the catalytic subunit of H⁺-translocating adenosine triphosphatase from vacuolar membranes of *Saccharomyces cerevisiae. J. Biol. Chem.* **265,** 6726–6733.

Hitzeman, R. A., Chen, C. Y., Dowbenko, D. J., Renz, M. E., Liu, C., Pai, R., Simpson, N., Kohr, W. J., Singh, A., Chisholm, V., Hamilton, R., and Chang, C. N. (1990). Use of heterologous and homologous signal sequences for secretion of heterologous proteins from yeast. *Methods Enzymol.* **185,** 421–440.

Huang, J., Struck, F., Matzinger, D. F., and Levings III, C. S. (1991). Functional analysis in yeast of cDNA coding for the mitochondrial Rieske iron-sulfur protein of higher plants. *Proc. Natl. Acad. Sci. U.S.A.* **88,** 10716–10720.

Kingsman, S. M., Cousens, D., Stanway, C. A., Chambers, A., Wilson, M., and Kingsman, A. J. (1990). High-efficiency yeast expression vectors based on the promoter of the phosphoglycerate kinase gene. *Methods Enzymol.* **185,** 329–341.

Klaasen, C. H. W., Van Uem, T. J. F., De Moel, M. P., De Caluwe, G. L. J., Swarts, H. G. P., and De Pont, J. J. H. H. M. (1993). Functional expression of gastric H⁺, K⁺-ATPase using the baculovirus expression system. *FEBS Lett.* **329,** 277–282.

Kotula, L., and Curtis, P. J. (1991). Evaluation of foreign gene codon optimization in yeast: Expression of a mouse Ig kappa chain. *Bio/Technol.* **9,** 1386–1389.

Learned, R. M., and Fink, G. R. (1989). 3-Hydroxy-3-methylglutaryl-coenzyme A reductase from *Arabidopsis thaliana* is structurally distinct from the yeast and animal enzymes. *Proc Natl. Acad. Sci. U.S.A.* **86,** 2779–2783.

Monk, B. C., Montesinos, C., Ferguson, C., Leonard, K., and Serrano, R. (1991). Immunological approaches to the transmembrane topology and conformational changes of the carboxy-terminal regulatory domain of yeast plasma membrane H⁺-ATPase. *J. Biol. Chem.* **266,** 18097–18103.

Palmgren, M. G., and Christensen, G. (1993). Complementation in situ of the yeast plasma membrane H⁺-ATPase gene *pma1* by an H⁺-ATPase gene from a heterologous species. *FEBS Lett.* **317,** 216–222.

Pringle, J. R., Adams, A. E. M., Drubin, D. G., and Haarer, B. K. (1991). Immunofluorescence methods for yeast. *Methods Enzymol.* **194,** 565–602.

Riesmeier, J. W., Willmitzer, L., and Frommer, W. B. (1992). Isolation and characterization of a sucrose carrier cDNA from spinach by functional expression in yeast. *EMBO J.* **11,** 4705–4713.

Rose, A. B., and Broach, J. R. (1990). Propagation and expression of cloned genes in yeast: 2-μm circle based vectors. *Methods Enzymol.* **185,** 234–279.

Sauer, N., and Stadler, R. (1993). A sink-specific H⁺-monosaccharide co-transporter from *Nicotiana tabacum:* Cloning and heterologous expression in baker's yeast. *Plant J.* **4,** 601–610.

Schneider, J. C., and Guarente, L. (1991). Vectors for expression of cloned genes in yeast: Regulation, overproduction and underproduction. *Methods Enzymol.* **194,** 373–388.

Sentenac, H., Bonneaud, N., Minet, M., Lacroute, F., Salmon, J.-M., Gaymard, R., and Grignon, C. (1992). Cloning and expression in yeast of a plant potassium ion transport system. *Science* **256,** 663–665.

Serrano, R. (1978). Characterization of the plasma membrane ATPase of *Saccharomyces cerevisiae. Mol. Cell. Biochem.* **22,** 51–63.

Serrano, R. (1985). "Plasma Membrane ATPase of Plants and Fungi." pp. 83–89. Boca Raton, FL: CRC Press.

Serrano, R. (1988). H⁺-ATPase from plasma membranes of *Saccharomyces cerevisiae* and *Avena sativa* roots: Purification and reconstitution. *Methods Enzymol.* **157,** 533–544.

Smith, R. A., Duncan, M. J., and Moir, D. T. (1985). Heterologous protein secretion from yeast. *Science* **229,** 1219–1224.

Stanley, K. K. (1988). Production of antibodies of predetermined specificity from *Escherichia coli* hybrid proteins. *In* "Methods in Molecular Biology" (J. M. Walker, ed.), Vol. 4, pp. 343–349. Clifton, New Jersey: Humana Press.

Stearns, T., Ma, H., and Botstein, D. (1990). Manipulating yeast genome using plasmid vectors. *Methods Enzymol.* **185,** 280–297.

Villalba, J. M., Palmgren, M. G., Berberian, G. E., Ferguson, C., and Serrano, R. (1992). Functional expression of plant plasma membrane H⁺-ATPase in yeast endoplasmic reticulum. *J. Biol. Chem.* **267,** 12341–12349.

Wright, R., Basson, M., D'Ari, L., and Rine, J. (1988). Increased amounts of HMG-CoA reductase induce "karmellae": A proliferation of stacked membrane pairs surrounding the yeast nucleus. *J. Cell. Biol.* **107,** 101–114.

Yansura, D. G. (1990). Expression as *trpE* fusion. *Methods Enzymol.* **185,** 161–166.

Zaret, K. S., and Sherman, F. (1982). DNA sequences required for efficient transcription termination in yeast. *Cell* **28,** 563–573.

CHAPTER 36

Synthesis of Plant Proteins in Heterologous Systems: *Xenopus laevis* Oocytes

Gad Galili*, **Yoram Altschuler,*** **and Aldo Ceriotti**[†]

*Department of Plant Genetics
The Weizmann Institute of Science
Rehovot 76100, Israel
[†]Istituto Biosintesi Vegetali
Consiglio Nazionale delle Ricerche
Via Bassini 15, 20133, Milano Italy

I. Introduction

The recent advent in recombinant DNA and gene transfer technologies, which enables the construction of mutant genes and their subsequent expression in plant and animal cells, has provided an impetus for a wide array of studies on the synthesis, assembly, and transport of plant proteins. An important outcome of these studies is that many processes, although not all, are independent of the eukaryotic cell type in which the protein of interest is synthesized. It appears, for instance, that plant secretory proteins can fold, assemble, and acquire transport competence when expressed in animal cells. This suggests that animal systems can aid in elucidating the signals and mechanisms that regulate these processes in plant proteins. The major advantage of such heterologous systems is the ease of analysis and the fast results, compared to the use of transgenic plants. Moreover, in specific cases, heterologous systems can be used for specific experiments that are either difficult or impossible to perform in plant cells. In the present report, we describe the use of *Xenopus laevis* oocytes as a heterologous system for the expression of plant proteins, using seed storage proteins as an example.

Over the last several decades, *Xenopus* oocytes have been proven to be an excellent heterologous system for efficient translation of foreign mRNAs and analysis of various post-translational modifications of the encoded proteins (Colman, 1984; Gurdon and Wickens, 1983; Kawata *et al.,* 1988; Smith *et al.,* 1991; Soreq and Seidman, 1992). Examples of foreign proteins that have been translated in oocytes include receptors and ion-transport proteins (see Chapter 37, this volume), ER proteins (Ceriotti and Colman, 1988), and immunoglobulins (Colman *et al.,* 1982), as well as secreted enzymes (Simon and Jones, 1988) and storage proteins (Altschuler and Galili, 1993; Altschuler *et al.,* 1993; Bassuner *et al.,* 1983; Hurkman *et al.,* 1981; Simon *et al.,* 1990; Vitale *et al.,* 1986; Wallace *et al.,* 1988). In these studies, oocytes were successfully used to study various aspects of the assembly and co- and post-translational modifications as well as transport of the foreign proteins, suggesting that the factors regulating these processes have been highly conserved in evolution (Colman, 1984; Colman *et al.,* 1984; Kawata *et al.,* 1988). Due to their large size compared to other cells used as expression systems, *Xenopus* oocytes also offer several additional advantages: (i) these oocytes can accommodate exogenous macromolecules and even organelles, after administration by injection (Richter *et al.,* 1982; Richter and Smith, 1981); (ii) analysis can be performed at a level of a single cell; (iii) the expression level of the protein of interest may be easily controlled by the amount of injected mRNA; (iv) since oocytes can be dissected at the equator, it is possible to study diffusion of macromolecules from one hemisphere to the other (Ceriotti and Colman, 1988; Drummond *et al.,* 1985); and (v) cotranslation of different proteins can be easily achieved by injection of mixtures of mRNAs. In the present chapter, seed storage proteins from wheat and beans are used as examples in describing methods for analyzing the assembly and transport of plant proteins synthesized in *Xenopus* ooctyes. We discuss the advantages and disadvantages of the oocyte system and provide several critical comments for successful utilizaton of this system.

================ **II. Materials and Methods**

A. Preparation of Oocytes

For general details regarding the experimental manipulation of *Xenopus* oocytes, readers are encouraged to refer to Volume 36 of this series (1991), as well as to Colman (1984), Kawata *et al.* (1988), and Soreq and Seidman (1992). As procedures for using oocytes for translating animal or plant mRNAs are essentially the same, we will describe them only briefly in the present chapter.

After arrival, we generally let the frogs recover for a period of 3 to 4 months before use. All further procedures are done at ambient temperature, unless otherwise noted. The ovarian lobes are transferred into dissection media and, using watch-maker forceps, are dissected into clumps containing not more than 50 oocytes; the use of small clumps improves gas exchange. Two dissection media are commonly used, either Modified Barth's saline (MBS) or OR-2 medium (Kawata *et al.*, 1988). Modified Barth's saline comprises 88 mM NaCl, 1 mM KCl, 2.4 mM NaHCO$_3$, 15 mM Hepes–NaOH (or Tris–HCl), pH 7.6, 0.30 mM Ca(NO$_3$)$_2$·4H$_2$O, 0.41 mM CaCl$_2$·6H$_2$O, and 0.82 mM MgSO$_4$·7H$_2$O. OR-2 medium comprises 82.5 mM NaCl, 2.5 mM KCl, 1 mM CaCl$_2$, 1 mM MgCl$_2$, 1 mM Na$_2$HPO$_4$, 5 mM Hepes, and 3.8 mM NaOH, pH 7.8.

The clumps of oocytes can next be defolliculated. Defolliculation reduces considerably the resistance of the oocytes to the penetration of the micropipette, and also reduces the incorporation of radiolabeling into the follicle cells (see later). Defolliculation can be performed manually (Kawata *et al.*, 1988; Smith *et al.*, 1991), a somewhat difficult procedure for an untrained person or, alternatively, by the use of collagenase (Sigma, Type 1A, C-9891). For enzymatic defolliculation, the oocyte clumps are immersed in a small petri dish in calcium-free modified Barth's saline, or OR-2 medium, containing 2 mg/ml collagenase. The ooctyes are incubated at room temperature with gentle shaking for 1 h and then free oocytes are removed. The rest of the oocytes are again incubated with a fresh collagenase solution for one further hour. This generally induces release of most oocytes. The oocytes are then washed several times in excess medium in order to remove completely the collagenase and are allowed to recover from the enzymatic treatment in Modified Barth's saline or OR-2 medium containing calcium for at least 8 h before injection (Smith *et al.*, 1991). It should be noted that not every batch of collagenase is active; hence new batches should be tested before used (personal experience).

In order to obtain efficient protein synthesis, it is important to select oocytes at stage VI because of their higher translational capacity (Smith *et al.*, 1991). Although oocytes of stages V and VI are similar in morphology, stage VI oocytes can be distinguished by their larger size (generally over 1.2 mm in diameter) and by the presence of a relatively unpigmented equatorial band (Dumont, 1972; Smith *et al.*, 1991). Following selection of stage VI oocytes, they are transferred to MBS or OR-2 media containing horse or calf serum (5% v/v). This increases the viability of the oocytes and reduces seasonal variability (Quick *et al.*, 1992). The oocytes can optimally be used for injection after one day. All solutions are made sterile, but the oocytes are not maintained under sterile conditions.

B. Preparation of mRNA

The translation of heterologous proteins in the oocytes is most commonly achieved by injecting total RNAs or mRNAs. Until a decade ago, only natural mRNAs were used; recently, following the development of various simple procedures for *in vitro* transcription of large amounts of synthetic mRNAs from cloned cDNAs (or intronless genes), such mRNAs have been extensively used with very successful results (Kawata *et al.*, 1988; Melton, 1987; Wormington, 1991). With the advent of synthetic mRNA technology, it is also possible to express antisense constructs as well as modified mRNAs derived from cDNAs that have been subjected to mutagenesis *in vitro*.

Methods and plasmids for *in vitro* transcription of synthetic mRNAs using phage promoters (for example, SP6, T7, or T3) have been reviewed in detail (Kawata *et al.*, 1988; Melton, 1987; Wormington, 1991). Supplementary information about plasmids for *in vitro* transcription as well as a detailed transcription protocol are given elsewhere in this volume (Chapters 21 and 37). Reagent kits and protocols for *in vitro* transcription are now available from a large number of companies, and therefore our comments here are restricted to some important notes regarding the quality of the synthetic mRNAs that are suitable for expression in the oocytes.

C. Critical Aspects Involving mRNA Production

As developing oocytes are actively engaged in the synthesis of endogenous proteins, the injected mRNA competes with the endogenous mRNAs. Thus, in order to obtain efficient translation of the protein of interest, the injected mRNA should be of high competence for protein synthesis. The following aspects are critical for the *in vitro* transcription of high-quality mRNAs.

1. In order to maintain the integrity of the mRNA, all solutions should be RNase-free, and preferably treated with diethyl pyrocarbonate (DEPC).

2. The plasmid used for *in vitro* transcription should not be linearized with restriction enzymes creating a 3′ overhang. This is because transcription of templates prepared by digestion with such restriction enzymes may result in the production of long transcripts containing extraneous sequences derived from both strands of the template (Schenborn and Mierendorf, 1985).

3. In order to eliminate traces of RNase in the sample, the linearized DNA (1 mg/ml) should be treated with 2 μl of proteinase K (50 μg/ml) for 30 min at 37°C. Then, the DNA should be treated with phenol/chloroform, EtOH-

precipitated, and rinsed well with 70% EtOH to remove any remains of salt. Residual salt can inhibit the phage RNA polymerases.

4. The mRNA should be capped at the 5′ end in order to obtain efficient translation within the oocytes (Drummond *et al.*, 1985; Kawata *et al.*, 1988; Wormington, 1991). Although capping may be performed enzymatically using guanylyltransferase, a simpler procedure is to use a cap analog recognized by the phage polymerases and therefore added at the first position of the RNA (Krieg and Melton, 1984). As SP6, T7, and T3 polymerases initiate transcription at a G residue, the cap structure should be 7m(5′)Gppp(5′)G. Such an analog can be added intact, although it is also possible to use the less expensive analog (5′)Gppp(5′)G since the oocytes can methylate this compound *in vivo* (Furuichi *et al.*, 1977). Cap analogs are commercially available (Boehringer; Pharmacia, etc.). In order to get efficient incorporation of the cap analog, the concentration of GTP is generally reduced by about 10-fold from 500 to 50 μM. Reducing the level of GTP reduces the total amount of RNA synthesized. Thus, some groups recommend transcription reactions that include both 500 μM of the cap analog and 500 μM GTP (Matthews and Colman, 1991). We have empirically found that the injection of mRNA synthesized under either set of conditions results in similar levels of expression.

5. Messenger RNAs containing 3′ poly(A) tail are translated much more efficiently within the oocytes, compared to nonadenylated mRNAs (Drummond *et al.*, 1985; Galili *et al.*, 1988). Although the poly(A) tail can be added enzymatically (Wormington, 1991), a more convenient way is to introduce an oligo(dA) into the coding strand of the DNA downstream to the insert, followed by a unique restriction enzyme site (Galili *et al.*, 1988; Hoffman and Donaldson, 1988). In this case, linearization of the plasmid 3′ to the oligo(dA) will result in runoff transcription of a poly(A^+) mRNA. The minimal size of the poly(A) tail that is active in translation is yet unknown. However, a tail of 20 A residues was found to be sufficient (Krieg and Melton, 1984).

A PCR-based method for transcription of poly(A^+) mRNA has been also reported (Cestari *et al.*, 1993). Although this method is rapid and simple, care should be taken to avoid possible nucleotide errors introduced by the PCR reaction.

6. The 5′ noncoding region of the mRNA should be planned carefully, as this region may have a large influence on translation efficiency. Hairpin structures and G-rich sequences in this region can reduce translation efficiency (Galili *et al.*, 1986; Pelletier and Sonenberg, 1985), whereas specific 5′ noncoding sequences, like the Ω sequences of viral RNAs, can be used to enhance translation (Gallie *et al.*, 1987; Mager *et al.*, 1993).

7. At the termination of transcription, the DNA template should be removed using RNase-free DNase. This avoids injection problems associated with viscosity.

8. Following transcription, it is recommended to test the quality and approximate quantity of the mRNA by fractionating a small sample (1 to 2 μl) of the

reaction on a regular RNA gel. The RNA can be easily detected by staining with ethidium bromide and radioactive RNA can also be detected by autoradiography. Using this method it is possible to recognize the intact RNA by its size and test whether truncated or longer RNAs have been produced due to nonoptimal transcription conditions (Wormington, 1991).

D. DNA Constructs for *in vivo* Transcription and Translation within Oocytes

Synthesis of heterologous proteins within oocytes can also be achieved through injection of DNA constructs into the oocyte nucleus. In these constructs, the coding DNA sequence of the gene of interest is fused to promoters and polyadenylation signals that can function with the oocytes. A variety of promoters and polyadenylation signals from various orrnganisms are found to function within the oocytes, including ones of plant origin (Ballas *et al.*, 1989; Pfaff *et al.*, 1990; Soreq and Seidman, 1992; Swick *et al.*, 1992). These regulatory elements can be used for expression of cDNAs in the oocytes by transcription and translation. The major advantage of DNA injection is that it overcomes costs and problems associated with *in vitro* RNA synthesis. The major disadvantage is that microinjection of DNA into the nucleus is more difficult than microinjection of RNA into the cytoplasm even though the nucleus, which is located in the germinal hemisphere, is relatively large. This can increase the problem of variability between occytes, as not all of the injections actually hit the nucleus. In the following section, we outline a means for dealing with this problem (Swick *et al.*, 1992).

E. Accounting for Variation between Individual Oocytes

One of the major problems associated with *Xenopus* oocytes is the relatively high variation in translation capacity between individual oocytes (Gurdon and Wickens, 1983). Two major factors account for this variability: (i) biological and environmental factors and (ii) variation in the actual amount of nucleic acids that is administered into the cells due to differences in the injected volume or due to partial leakage of the injected material. One way to overcome this problem at least partially is to use a large number of oocytes (generally five oocytes or more per treatment, thereby performing the study in a statistically controlled manner). However, this may not be sufficient when studying protein assembly within the oocytes inasmuch as protein assembly may be critically dependent on the concentration of the protein within the cell or even within a given organelle inside the cell (see examples later in this chapter). Thus, even if the average amount of protein synthesis by a large group of oocytes is fair, it is possible that this may reflect the sum of several high-expressing and several low-expressing oocytes. In such a case, it is possible that no protein assembly will take place in the low-expressing oocytes, leading to an overall observation of incomplete assembly. This problem may be overcome by a recently published procedure, which is based on the translation of a secreted form of the enzyme alkaline phosphatase (Swick *et al.*, 1992; Tate *et al.*, 1990). When oocytes are injected with mRNA or DNA encoding this enzyme, the amount of the expressed enzyme secreted from a single oocyte into the medium can yield detectable enzymatic activity. Thus, oocytes are coinjected with the DNA or mRNA of interest as well as with the alkaline phosphatase DNA or mRNA as a control. The efficiency of translation of each oocyte is then monitored by the level of alkaline phosphatase activity in the medium.

1. Protocol for Measuring Alkaline Phosphatase Activity

1. Oocytes are coinjected with the mRNA or DNA encoding the protein of interest plus mRNA or DNA encoding the alkaline phosphatase (the amounts of injected DNAs or RNAs depend on the transcription and translation competence of the constructs, but both generally range between 1 and 100 ng).

2. The injected oocytes are individually placed in separate wells of a 96-well tissue culture plate (precoated with 1 mg/ml BSA) containing 200 μl of medium, and the plate is incubated at 18°C for 16–20 h. The following alkaline phosphatase assay (Tate *et al.*, 1990) is then performed on a sample of the extracellular medium of each well.

3. A sample (100 μl) from each well is transferred into a corresponding well of a second tissue culture plate containing 100 μl of an assay buffer (1 M diethanolamine, pH 9.8, 0.5 M MgCl$_2$, 0.02 mM ZnSO$_4$, and 20 mM L-homoarginine (Sigma H1007).

4. The plate is incubated at 37°C for 10–20 min and then the assay is initiated by addition of 40 μl of 60 mM nitrophenylphosphate (Sigma 104-40) dissolved in assay buffer and prewarmed to 37°C. The plate is shaken well and further incubated at 37°C.

5. Every 30 min, the A_{405} absorbance is measured using an ELISA reader. The reaction is continued until a linear increase in absorbance with time is obtained (this generally requires 1–2 h).

6. Individual oocytes are then divided into groups possessing relatively low, medium, and high translation levels, based on the A_{405} absorbance per minute in the linear range. It is important to note that some oocytes may have low A_{405} readings following 30 min of incubation and later on possess high readings in the linear range. These oocytes appear to be of high translation competence.

F. Radiolabeling of Oocytes

Individual oocytes can synthesize proteins rapidly (ng amounts per h), and this is a sufficient quantity to be detected by Western blots,. However, the most common way to detect the expressed proteins is through radioactive labeling. For this purpose, it is advisable to select amino acids with relatively small endogenous pool sizes, such as methionine or leucine (Colman, 1984). As oocytes are permeable to amino acids, labeling can be done by addition of the radioisotopes to the medium. In this case, generally, groups of about five oocytes are submerged in a single well of a 96-well plate containing 25–50 μl of medium (Modified Barth's or OR-2) containing about 1–5 μCi/μl of radioactive amino acid at the highest available specific activity. An alternative way of labeling is to microinject the radioactive amino acid into the oocytes either together with the nucleic acids or by a separate injection following that of the nucleic acids (Colman, 1984; Galili *et al.*, 1988; Kawata *et al.*, 1988; Smith *et al.*, 1991; Soreq and Seidman, 1992; Wormington, 1991). The radioisotope is dried down using a Speedvac, and is dissolved in the injection solution. In this case, only about 5–10 μCi is generally injected into each oocyte. The labeling period may vary from as short as 30 min to several days. However, for radioactive amino acids with small pool size and high specific activity, the labeling may be essentially complete after several hours (Colman *et al.*, 1984). Longer periods of pulse labeling can be obtained by reducing the specific activity of the radioactive amino acid (Colman, 1984). Chasing can be done by transfer of the oocytes into media containing 5 m*M* of the appropriate nonradioactive amino acid.

The advantages of labeling via the culture medium are (i) this procedure is simpler to perform and (ii) the label can be easily added at any time after microinjection of the mRNA with no need for subsequent injections. Allowing an interval of several hours between mRNA microinjection and the microinjection of radiolabeled amino acid is preferable in order to obtain good mRNA recruitment onto polysomes. The advantages of microinjection-derived labeling include (i) protein labeling is more efficient and hence shorter pulse labeling periods can be performed; (ii) injection of labeling is generally more economical; and (iii) most of the label is incorporated into the oocyte proteins (if the oocytes are not defolliculated the follicular cells can incorporate a significant amount of any radioactive amino acids presented in the medium).

Labeling of proteins synthesized in the oocytes can also be achieved using radioactive precursors other than amino acids. Oligosaccharide side chains can be labeled with various sugar residues, such as D-[2-³H]mannose (Vitale *et al.*, 1989), or L-[6-³H]fucose (Vitale *et al.*, 1986). Glycosaminoglycans can be labeled with [³⁵S]-labeled sulfate (Leaf *et al.*, 1990). Protein phosphorylation or acetylation can also be monitored by labeling with [³²P]phosphate or [³H]acetate, respectively (Woodland, 1979).

G. Detection and Characterization of Heterologous Radiolabeled Proteins over the Background of Endogenous Oocyte Proteins

As both the injected oocytes and the surrounding follicle cells are active in the synthesis of endogenous proteins, the expressed polypeptide must be detected above an endogenous background. The background of endogenous proteins within oocytes is far more intense than that found in the incubation medium.

Various methods can be used to detect heterologous proteins that are translated within the oocytes and the choice mainly depends on the characteristics of the protein of interest and on the final purpose of the study. The methods include (i) separation of the expressed protein from the endogenous oocyte proteins by SDS–PAGE; (ii) Western blots; (iii) immunoprecipitation; (iv) enzyme-linked immunosorbent assay; (v) activity or electrophysiological assays; (vi) characterization of the oligomeric state of the expressed proteins; and (vii) complete protein purification. Immunoselection of the radiolabeled translation product by immunoprecipitation followed by SDS–PAGE analysis and characterization of the oligomeric state of these expressed products will be described in detail below.

1. Immunoprecipitation

Several different immunoprecipitation protocols can be used to detect proteins expressed in the oocytes. In general, the various methods differ in the composition of the immunoprecipitation and the wash buffer. Some buffers, which allow preservation of the native state of the antigen, may be of particular importance if one aims to detect noncovalent associations between subunits of oligomeric proteins, or if partial denaturation is accompanied by enhanced susceptibility to proteases. We have found that the method described below, which preserves protein conformation, consistently yields clean results when applied to both oocyte homogenate and incubation medium.

The method by which the oocytes are homogenized is not important. This can be done by various regular ways such as mortar and pestle or by pippeting up and down in an Eppendorf tube together with the homogenization buffer (20 mM Tris–HCl, pH 7.6, 0.1 M NaCl, 1% Triton X-100, 1 mM PMSF). The homogenate is clarified (2 min at 10,000× g) and an aliquot is diluted to a final volume of 1 ml with a buffer comprising 50 mM Tris–HCl, pH 7.5, 150 mM NaCl, 0.1% NP-40, 1 mM EDTA and 0.02% sodium azide (NET buffer) to which has been added 0.25% gelatin. The gelatin is prepared as a 2.5% solution, sterilized by autoclaving, made 0.02% sodium azide and stored at 4°C—this solution is melted in a microwave oven just before use. Sufficient antibody is then added to the resuspended homogenate to achieve complete precipitation of the target protein. The tube is then incubated for 1–2 h on ice. The amount needed must be determined empirically, but generally 1 to 5 μl of a polyclonal antiserum is required. If a monoclonal antibody is used which does not bind efficiently to protein A, an appropriate anti-immunoglobulin preparation should be added and the incubation continued for an additional 1 h period on ice. Prior to use, protein A–Sepharose (e.g., Pharmacia) is prepared by swelling for 3 h in NET buffer. It is then incubated for 1 h in 1 M Tris–HCl, pH 7.5, and finally equilibrated in NET buffer as a 10% suspension, prior to storage at 4°C. The prepared protein A–Sepharose is then added (100 μl of a 10% suspension), and the tubes are gently swirled for 1 h at 4°C. This quantity should be sufficient unless very large amounts of antiserum are used. The beads are then collected by centrifugation at 10,000× *g* for 30 s at 4 °C, the supernatant is removed and the protein A–Sepharose is resuspended with gentle vortexing in 1 ml of NET buffer containing gelatin. The beads are collected as above and the washing step is repeated twice. If background is found to be a problem, the stringency of the washing procedure can be raised by performing the first wash in NET buffer containing both gelatin and 0.5 *M* NaCl. Removal of the washing buffer can be conveniently performed with a drawn-out Pasteur pipette attached to a vacuum line. After the last wash, any remnant of the washing buffer is carefully removed from the sides of the tube and 30 μl of standard 1.5 × SDS–PAGE loading buffer (Kawata *et al.,* 1988) is added. The sample is vortexed, briefly spun, and placed for 3 min a boiling water bath. The beads are finally pelleted by centrifugation (30 s at 10,000× *g*) and the supernatant is loaded on the gel. In some cases, it might be helpful to preclear the lysate by incubating the extract with preimmune serum and protein–A Sepharose.

H. Detection of Protein Oligomerization

Assembly of oligomeric protein complexes following biosynthesis within oocytes is followed using sedimentation velocity centrifugation on sucrose gradients. The exact conditions employed depend on the sedimentation characteristics of the proteins under study. For example, for oocytes synthesizing bean phaseolins, oocyte homogenate (10 oocyte equivalents), or incubation medium is layered onto 12-ml linear 5–25% linear sucrose gradients in 20 m*M* Tris–HCl, pH 7.5, 100 m*M* NaCl, 0.1% Triton X-100. The gradients are centrifuged in a Beckman SW40 rotor at 40,000 rpm for 16 h at 20°C. The gradients are then fractionated into 0.6-ml fractions.

I. EndoH Susceptibility

ß-Endoglycosidase H removes the oligosaccharide chains from denatured glycoproteins, leaving one N-acetylglucosamine residue linked to the glycosylated Asn, unless these chains have been modified by Golgi-resident enzymes. For endoH treatment, the radiolabeled protein is first immunoprecipitated using the protocol described above. The protein A–Sepharose beads are washed twice in distilled water and the bound proteins are eluted at 95°C for 4 min in 50 ml of 100 m*M* Tris–HCl, pH 8, 1% 2-mercaptoethanol, 0.5% SDS. The beads are removed by centrifugation for 5 min at 10,000× *g* and the supernatant is transferred to a fresh tube, diluted with 950 μl of 0.1 *M* Na-citrate, pH 5.5, and further supplemented with 5 μl of a 10% (w/v) BSA solution, 5 μl of 100 m*M* PMSF, and 1 μl of anti-protease mix (antipain, chymostatin, leupeptin, pepstatin, bestatin, 2 mg/ml each, aprotinin 10 TI U/ml in 50% DMSO). The sample is split in two aliquots, one of which is supplemented with 20 mU of endo H. After incubation for 18 h at 37°C, the protein is recovered by precipitation in 15% trichloroacetic acid (30 min on ice) and centrifugation (20 min at 10,000× *g*). The protein pellets are washed twice with cold acetone, dried, resuspended in SDS–PAGE loading buffer, and analyzed by SDS–PAGE and fluorography.

J. Inhibitors of Protein Processing and Secretion

The regulation of protein processing and secretion can be studied either by expression of modified proteins or by using drugs that inhibit these processes. To study whether proteins are transported via the oocyte Golgi, the oocytes are incubated in a medium containing micromolar concentrations of the ionophor monensin, which inhibits protein transport via this organelle (Simon *et al.,* 1990). Monensin is generally prepared as a 2 *M* stock solution dissolved in ethanol and can be stored at −20°C. Another commonly used inhibitor, tunicamycin, inhibits protein glycosylation and is also known to induce the expression of ER-resident chaperones. Injection of 40 μg/ml tunicamycin into oocytes not only inhibits protein glycosylation (Vitale *et al.,* 1986) but also retards the secretion of proteins that apparently interact with the molecular chaperones within the ER (Simon *et al.,* 1990). The drug can be coinjected with the mRNA several hours prior to labeling. The lag between injection and labeling allows the tunicamycin-induced depletion of the endogenous pool of oligosaccharide precursors. Tunicamycin is prepared as a 2 mg/ml stock solution dissolved either in DMSO or in distilled water, and adjusted to pH 10 with NaOH. It is also possible to block the Golgi-mediated processing of N-linked oligosaccharide side chains in the oocytes. For this purpose, oocytes are preincubated for 3 h in MBS or OR-2 medium containing 5 mM 1-deoxymannojirimycin (prepared as a stock solution in water) before labeling in the same medium (Fabbrini *et al.,* 1988).

K. Protein Diffusion

The strategy to study the comparative mobility of different proteins within the lumen of the ER involves coinjection or injection of their mRNAs along with a radioactive amino acid tag into one hemisphere of the oocyte. Protein labeling is continued for 1.5 h, then is followed by various periods of chasing using excess nonradioactive amino acid. The oocytes are then dissected at the equator and proteins from each hemisphere are analyzed separately. Under such conditions, movement of the mRNA during the pulse labeling is very restricted (Drummond *et al.,* 1985), and the radioactive secretory protein can appear in the other hemisphere only if it is able to diffuse within the lumen of the ER.

1. Protocol for Protein Diffusion

[^{35}S]Methionine (75 μCi) is dried in a microcentrifuge tube using a Speedvac and is resolubilized in 3 μl of water containing the synthetic mRNA. The mixture is then coinjected into either the animal or the vegetal hemisphere of the oocytes.

The needle is positioned parallel to the animal/vegetal axis at the far end of the hemisphere, so that the injected material is released as far as possible from the opposite hemisphere (Fig. 1, stage 1). At that point, the oocytes are placed in medium for 1.5 h (pulse labeling). To initiate the chase period, the medium is replaced by identical medium which contains 5 mM L-methionine and the oocytes are incubated for increasing time periods. For subsequent dissection at the equator, the oocytes are placed on a transparency (or a used X-ray film), with the axis between hemispheres perpendicular to the transparency. Excess medium is eliminated by soaking with 3MM filter paper and the transparency is then placed on a flattened piece of dry ice, under a stereoscope. Although the frozen oocytes can be stored at $-80°C$ before use, we generally dissect them right away, as

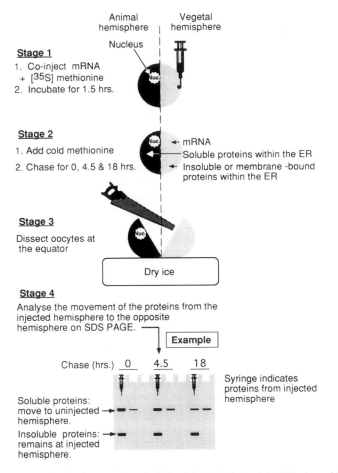

Fig. 1 Schematic representation of the procedure for studying protein diffusion within the ER of the oocytes. The injection of mRNA, diffusion during the pulse chase period, dissection of the oocytes at the equator, and SDS–PAGE analysis are shown in stages 1 to 4, respectively.

condensation built on top of them strongly interferes with the dissection. Each dissection experiment should preferably be done with a new razor blade and the blade should be replaced every 30–50 oocytes. During the dissection, individual hemispheres may occasionally escape from the work area. To avoid this, we cut out a large hole on a lid of a 35-mm petri dish and use it to cover the oocyte. The large opening enables to reach the oocyte with the razor blade through the lid, but prevents escape of the dissected hemispheres. Vegetal and animal hemispheres are pooled separately and stored at −80°C.

III. Results and Discussion

Upon sequestration into the ER, the newly synthesized proteins follow several steps of maturation including folding, assembly, and various co- and post-translational modification processes that render them competent for transport to their final destination. The ability of *Xenopus* oocytes faithfully to carry out these maturation events can be exploited to study the relationships among the structure, maturation, and function of the protein of interest. However, the suitability of the oocyte expression system to study protein maturation should be evaluated separately for each particular case. The two following examples clarify this issue.

1. Assembly and transport of proteins via the ER: It appears that the quality control system within the ER operates faithfully in the oocytes by facilitating the correct folding, assembly, and transport of secretory proteins and by preventing the transport of malfolded and malassembled proteins. However, the efficiency of assembly and hence subsequent transport of wild-type oligomeric proteins may depend also on the concentration of the individual subunits within the ER (Braakman *et al.*, 1991; Ceriotti and Colman, 1990; Ceriotti *et al.*, 1991). The concentration of a given subunit can (intentionally or nonintentionally) be significantly different between microinjected oocytes and the cell type where the protein is normally expressed.

2. Protein Glycosylation: In glycoproteins, the commitment of oligosaccharide chains to processing mediated by enzymes of the Golgi seems to be cell type-independent. However, the kind of processing and therefore the final structure of the chains may be different between the oocytes and the original cells expressing the protein of interest (Hsich *et al.*, 1983; Sheares and Robbins, 1986; Vitale *et al.*, 1986; Yet *et al.*, 1988).

Taken together the advantages and limitations, *Xenopus* oocytes appear to be a very convenient system to study many aspects of the co- and post-translational maturation events of secretory proteins. As examples, we will next describe the ues of this expression system to study the assembly and the intracellular transport of wheat and bean storage proteins.

A. Secretion of Wheat Storage Proteins from the Oocytes

The major storage proteins in wheat are the alcohol-soluble gliadins, and the glutenins. These storage proteins accumulate in dense protein bodies within the vacuole of the endosperm cells. However, they appear to use two distinct routes to the final site of deposition (Galili *et al.*, 1993). Following sequestration into the ER, part of these storage proteins are retained and assemble into dense protein bodies within this organelle and some escape this retention process and are transported via the Golgi complex to vacuoles (Galili *et al.*, 1993). Protein bodies that are initially formed within the ER are subsequently internalized into vacuoles by an autophagy-like process (Galili *et al.*, 1993). The initial retention of the wheat storage proteins and their assembly into protein bodies within the ER apparently occur by virtue of specific rentention and assembly signals that prevent the export of these proteins from the ER to the Golgi complex.

The point of branching of wheat storage proteins for the two separate routes occurs inside the ER. Yet, the question whether a given storage protein is transported from the ER to vacuoles via the Golgi or by the autophagy-like route cannot be studied directly in wheat endosperm due to the following problems: (i) wheat storage proteins are not glycosylated and therefore it is not possible to trace their transport via the Golgi by testing for Golgi-specific modifications of glycan side chains; (ii) as storage proteins transported by the two distinct routes end up accumulating in the same vacuole, it is very difficult to study the dynamics of the transport process taken by individual storage proteins by using conventional electron microscopy and cell fractionation techniques. This problem can be overcome using *Xenopus* oocytes. As these cells lack vacuoles, vacuolar proteins that are exported from the ER to the Golgi complex are eventually secreted from the oocytes into the medium by the default pathway. Thus, it is easily possible to discriminate between wheat storage proteins that are retained within the ER (inside the oocytes) and those that are exported from the ER to the Golgi complex and are eventually secreted into the medium.

To study the secretion efficiency of gliadins from the oocytes, the cells were injected either separately or together with synthetic mRNAs encoding γ and aggregated types of gliadins. Following injection, the oocytes were incubated with medium supplemented with [^{35}S]methionine for increasing time periods, and the oocytes and medium were then harvested separately. In order to detect the gliadins present in the oocytes or incubation medium, we took advantage of the fact the wheat gliadins are freely soluble in 70% EtOH, a solution that hardly solubilizes any endogenous oocyte protein. Thus, the oocytes or medium were homogenized in 70% EtOH and incubated for 30 min at 60°C, and nonsoluble proteins were removed by centrifugation for 15 min in an Eppendorf centrifuge. The supernatants were dried by Speedvac centrifugation and then subjected to SDS–PAGE and fluorography. As shown in Fig. 2, the γ-gliadin was secreted efficiently, whereas the aggregated gliadin did not. Moreover, the extent of secretion of the two gliadins was not altered when both were coexpressed in the

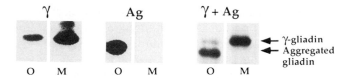

Fig. 2 Secretion of the wheat gliadins and aggregated gliadins from the oocytes. Oocytes were injected with SP6 mRNAs of a γ-gliadin (γ), aggregated gliadin (Ag), or both mRNAs together (γ + Ag), and left overnight to allow mRNA recruitment onto polysomes. The oocytes were incubated in medium containing [^{35}S]methionine for 72 h and harvested. Proteins soluble in 70% EtOH from homogenized oocytes (O) or the incubation medium (M) were analyzed by SDS–PAGE and fluorography (modified from Altschuler and Galili, 1993).

same group of oocytes, suggesting that each of the gliadins was transported independently of the other. The efficiency of secretion of the γ-gliadin varied between different experiments ranging between ~50 and 80%. Thus, care should be taken to account for such variations by repeating the experiments and comparing between treatments within a given experiment.

We have also analyzed whether the Golgi complex was involved in secretion of the γ-gliadin by treating the oocytes with monensin. This compound caused a significant retardation in the secretion of the γ-gliadin (Simon *et al.*, 1990). Similarly, treatment of the oocytes with tunicamycin significantly impeded the transport of the γ-gliadin (Simon *et al.*, 1990), suggesting that ER resident proteins are involved in the assembly and transport of this protein. To study the role of specific domains within the γ-gliadin in its transport via the ER, we have expressed various deletion and insertion mutants of this protein within the oocytes (Altschuler *et al.*, 1993). These studies showed that the N-terminal repetitive region signals for retention and assembly into protein bodies within the ER, whereas the C-terminal region makes the protein competent for export from the ER to the Golgi complex.

We have also used metrizamide density gradients to analyze the assembly of the gliadins into protein bodies within the ER of the oocytes. This study showed that gliadins expressed in the oocytes could be packaged into protein bodies with physical characteristics similar to those of native protein bodies from wheat endosperm cells (Altschuler *et al.*, 1993). The utilization of a similar technique, using sucrose density gradients, is described in detail below.

B. Assembly and Glycan Modifications of Bean Phaseolin Storage Proteins

Bean phaseolins (PHSL) are trimeric, globulin-type glycoproteins. These storage proteins assemble into trimers within the ER of the cotyledonary cells and are then transported via the Golgi complex to storage vacuoles where they accumulate. The native PHSL polypeptides can be divided into two size classes, α and ß. Both α and ß polypeptides appear in two different glycoforms (α′, α″, ß′, ß″), bearing either one or two oligosaccharide chains. This is due to the fact that only one of the two potential glycosylation sites present in PHSL is used with 100% efficiency (Bollini *et al.,* 1983).

Xenopus oocytes were evaluated as an expression system for studying the characteristics of PHSL that are important for the formation of the trimeric structure and for the acquisition of transport competence. To this end, ß-type PHSL was expressed in these cells by injecting a synthetic mRNA encoding this protein. Following overnight recovery, microinjected oocytes were labeled for 2 h with [^{3}H]leucine (1 mCi/ml). At the end of the labeling period, some oocytes were directly homogenized, whereas others were chased for 24 h in the presence of 10 mM cold leucine before homogenization. Oocyte homogenates and incubation media were frozen in liquid nitrogen and stored at −80°C before analysis. During the chase period, some of the newly synthesized, radiolabeled PHSL was secreted into the medium. Both the oocyte-associated and secreted PHSL were tested for their state of assembly and for the presence of Golgi-mediated processing of the oligosaccharide chains. The assembly state was determined by sedimentation velocity centrifugation on sucrose gradient (Fig. 3). After completion of the run, the PHSL polypeptides contained within the gradient fractions were immunoprecipitated using a rabbit polyclonal antiserum. After 2 h of pulse labeling, PHSL present inside the oocytes was mainly in a monomeric form (Fig. 3, 0-h oocytes; upper band represents ß″; lower band represents ß′ subunits). After 24 h of chase, the oocyte-associated PHSL was found to be distributed between monomers and trimers (Fig. 3, 24-h oocytes), whereas only trimeric PHSL was found in the incubation medium (Fig. 3, 24-h medium). This analysis demonstrated that PHSL oligomerization is a post-translational event and that only trimers are transported to the cell surface.

The polypeptide pattern of monomeric PHSL following 24 h of chase was different from that found following 0 h of chase, the first lacking single-

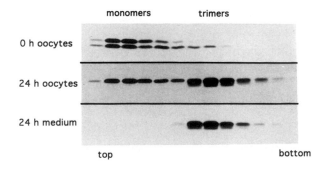

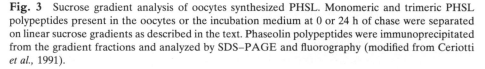

Fig. 3 Sucrose gradient analysis of oocytes synthesized PHSL. Monomeric and trimeric PHSL polypeptides present in the oocytes or the incubation medium at 0 or 24 h of chase were separated on linear sucrose gradients as described in the text. Phaseolin polypeptides were immunoprecipitated from the gradient fractions and analyzed by SDS–PAGE and fluorography (modified from Ceriotti *et al.,* 1991).

glycosylated polypeptides. This was due to a slow post-translational glycosylation of the monomeric single-glycosylated PHSL, which occurred during the chase period (Ceriotti *et al.*, 1991). Trimeric PHSL present in the oocytes and in the incubation medium appeared as two closely spaced bands as the result of a Golgi-mediated processing of single-glycosylated polypeptides. In bean cotyledonary cells, only single-glycosylated polypeptides are subjected to Golgi-mediated processing, indicating that the presence of the second oligosaccharide chain inhibits processing of the first (Sturm *et al.*, 1987). EndoH treatment was also used to establish that trimeric PHSL has been transported through the Golgi complex on its way to the cell surface. Fractions from the gradients shown in Fig. 3, corresponding to the monomeric and trimeric PHSL, were pooled and PHSL polypeptides were immunoprecipitated. Following recovery, they were subjected to endoH treatment as described under Materials and Methods. The results of this analysis are shown in Figure 4. Following 0 h of chase both monomeric and trimeric PHSL were found to be fully susceptible to endoH treatment. After 24 h of chase, trimeric PHSL that was present inside the oocytes or in the incubation medium was partially resistant to endoH cleavage, indicating that the oocyte-associated trimers have been exported from the endoplasmic reticulum and that secreted trimers have passed through the Golgi complex on their way to the cell surface. The finding that only a fraction of the oligosaccharide chains was processed could be ascribed to the fact that, as found in bean cotyledonous cells, only single-glycosylated PHSL is a substrate for Golgi-mediated processing. To conclude, the above data indicate that *Xenopus* oocytes can perform assembly and transport of bean PHSL, providing a convenient system for the study of the effects of introduced mutations on the efficiency of these processes (Ceriotti *et al.*, 1991).

When protein maturation requires interactions between different subunits, the level of expression of the interacting subunits can have a profound effect on the

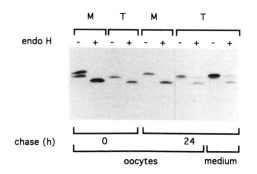

Fig. 4 EndoH treatment of monomeric and trimeric PHSL. Monomeric and trimeric PHSL polypeptides from the gradients shown in Fig. 3 were immunoprecipitated and treated with (+) or without (−) endo H before analysis by SDS–PAGE and fluorography. M, monomeric PHSL; T, trimeric PHSL (modified from Ceriotti *et al.*, 1991).

rate of assembly. For example, injection of as low as 1.3 ng/oocyte of total RNA from developing bean cotyledons still resulted in the synthesis of a discrete amount of the subunits of the bean storage protein PHSL. However, this expression level apparently did not lead to adequate levels of the protein needed to support efficient formation of PHSL trimers and thus the protein was retained as a monomer within the ER (Ceriotti *et al.*, 1991). In another study, (Altschuler *et al.*, 1993), packaging of a wheat gliadin into dense protein bodies was shown to be clearly dependent on its expression level in the oocytes. It is important to note in this regard that the amount of expressed protein needed to obtain efficient assembly may differ considerably between protein types. For example, in the case of immunoglobulins, complete assembly can be obtained even when the expression level is reduced to the minimum required to perform the analysis (Ceriotti and Colman, 1990).

The large size of the oocytes renders them a unique system to study diffusion of macromolecules within a given cell. After the mRNA is injected into one hemisphere, the oocytes are incubated for increasing time periods, and then dissected at the equator and each hemisphere is analyzed separately. This can provide information about the association of the protein with membranes or scaffolds within the oocyte. Similarly, in the case of cytosolic proteins, diffusion can be measured after direct microinjection of the protein. Soluble proteins diffuse rapidly (much more rapidly than mRNA) through the oocyte cytosol (Bonner, 1975; Drummond *et al.*, 1985). In the case of rabbit globin, an equilibrium of 85:15 (animal:vegetal) was found in its distribution within 6 h following mRNA injection into the vegetal pole. This is clearly different from the expected 60:40 distribution based on the estimated volume of solute-accessible cytoplasm in each hemisphere (Drummond *et al.*, 1985), suggesting that free diffusion by itself may not fully account for the final distribution of a given protein in the oocyte cytosol. Soluble proteins inserted into the ER have also been found to diffuse through the oocyte (Ceriotti and Colman, 1988; Drummond *et al.*, 1985) with a faster movement toward the animal pole than in the opposite direction. This diffusion has been interpreted as a consequence of the continuity of this organelle through the oocyte cytosol. Conversely, proteins that bind to the ER membrane or malfolded proteins that precipitate within the ER lumen have been found to remain localized to the site of injection (Altschuler and Galili, 1993; Ceriotti and Colman, 1988). It should be noted that if the interpretation of diffusion experiments is clear in the case of resident ER proteins, it becomes more problematic in the case of proteins that are rapidly secreted. For example, chicken lysozyme, which is small but rapidly secreted, remains mainly localized to the site of synthesis, whereas ovalbumin, a larger but slowly secreted molecule, diffuses efficiently into the opposite half. It has been speculated that this might be the consequence of a lack of continuity through the oocyte of organelles of the secretory pathway other than the ER (Drummond *et al.*, 1985). The use of *Xenopus* oocytes is therefore best suited for the study of proteins that, after insertion in the ER, are permanently retained in this compartment, or those that are secreted at a relatively slow rate, and are therefore mainly restricted to the ER during the period when diffusion is monitored.

In the case of wheat gliadins, the fact that a significant amount of the proteins are retained within the ER allows the study of their diffusion within the lumen of this organelle. It was found that wild-type γ and aggregated gliadins diffuse rather efficiently within the ER, suggesting that their retention within the organelle is not due to protein insolubility and precipitation (Altschuler and Galili, 1993). This was supported by the fact that mutant γ-gliadins, lacking specific conserved cysteine residues in the C-terminal region, diffused at a much reduced rate (Altschuler and Galili, 1993).

Acknowledgments

We thank Prof. Y. Oron and his group from Tel Aviv University for their critical comments on oocyte manipulations, particularly the collagenase treatments. We also acknowledge the technical assistance of Ms. H. Levanony in the experiments described in this chapter and Dr. A. Vitale and Mr. Y. Avivi for critical reading of the manuscript. Research in the laboratory of G.G. was supported in part by the Leo and Julia Forchheimer Center for Molecular Genetics and the Angel Faivovich Foundation for Ecological Research. Research in the laboratory of A.C. was supported in part by the Progetti Finalizzati CNR "Biotecnologie e Biostrumentazione" and "RAISA" Subproject 2, Paper No. 1954. G.G. is an incumbent of the Abraham and Jenny Fialkow Career Development Chair in Biology.

References

Altschuler, Y., and Galili, G. (1993). Role of conserved cysteines of a wheat gliadin in its transport and assembly into protein bodies in *Xenopus* oocytes. *J. Biol. Chem.* **269**, 6677–6682.

Altschuler, Y., Harel, R., and Galili, G. (1993). The N-terminal and C-terminal regions regulate the transport of wheat γ-gliadin throughout the endoplasmic reticulum in *Xenopus* oocytes. *Plant Cell* **5**, 443–450.

Ballas, N., Broido, S., Soreq, H., and Loyter, A. (1989). Efficient functioning of plant promoters and poly (A) sites in *Xenopus* oocytes. *Nucleic Acids Res.* **17**, 7891–7903.

Bassüner, R. A., Mmanteuffel, H. R., and Rapoport, T. (1983). Secretion of plant storage globulin polypeptides by *Xenopus laevis* oocytes. *Eur. J. Biochem.* **133**, 321–326.

Bollini, R., Vitale, A., and Chrispeels, M. J. (1983). *In vivo* and *in vitro* processing of seed reserve protein in the endoplasmic reticulum: Evidence for two glycosylation steps. *J. Cell Biol.* **96**, 999–1007.

Bonner, W. M. (1975). Protein migration into nuclei. I. Frog oocyte nuclei *in vivo* accumulate microinjected histones, allow entry to small proteins and exclude large proteins. *J. Cell Biol.* **64**, 421–430.

Braakman, I., Hoover-Litty, H., Wagner, K., and Helenius, A. (1991). Folding of influenza hemagglutinin in the endoplasmic reticulum. *J. Cell Biol.* **114**, 401–411.

Ceriotti, A., and Colman, A. (1988). Binding to membrane proteins within the endoplasmic reticulum cannot explain the retention of the glucose-regulated protein GRP78 in *Xenopus* oocytes. *EMBO J.* **7**, 633–638.

Ceriotti, A., and Colman, A. (1990). Trimer formation determines the rate of influenza hacmagglutinin transport in the early stages of secretion in *Xenopus* oocytes. *J. Cell Biol.* **111**, 409–420.

Ceriotti, A., Pedrazzini, E., Fabbrini, S., Zoppè, M., Bollini, R., and Vitale, A. (1991). Expression of the wild-type and mutated vacuolar storage protein phaseolin in *Xenopus* oocytes reveals relationships between assembly and intracellular transport. *Eur. J. Biochem.* **202**, 959–968.

Cestari, I. N., Albuquerque, E. X., and Burt, D. R. (1993). A PCR shortcut to oocyte expression. *Biotechniques* **14,** 404–407.

Colman, A. (1984). Translation of eukaryotic messenger RNA in *Xenopus* oocytes. *In* "Transcription and Translation" (B. D. Hames and H. J. Higgins, eds.), pp. 271–300. Oxford: IRL Press.

Colman, A., Besly, J., and Valle, G. (1982). Interactions of mouse immunoglobulin chains with *Xenopus* oocytes. *J. Mol. Biol.* **160,** 459–474.

Colman, A., Bhamra, S., and Vale, G. (1984). Post-translational modification of exogenous proteins in *Xenopus laevis* oocytes. *Biochem. Soc. Trans.* **12,** 932–937.

Drummond, D. R., Armstrong, J., and Colman, A. (1985). The effect of capping and polyadenylation on the stability, movement and translation of synthetic messenger RNAs in *Xenopus* oocytes. *Nucleic Acids Res.* **13,** 7375–7394.

Drummond, D. R., McCrae, M. A., and Colman, A. (1985). Stability and movement of mRNAs and their encoded proteins in *Xenopus* oocytes. *J. Cell Biol.* **100,** 1148–1156.

Dumont, J. N. (1972). Oogenesis in *Xenopus laevis* (*Daudin*). I. Stages of oocyte development in laboratory maintained animals. *J. Morphol.* **136,** 153–180.

Fabbrini, M. S., Zoppè, M., Bollini, R., and Vitale, A. (1988). 1-Deoxymannojirimycin inhibits Golgi-mediated processing of glycoprotein in *Xenopus* oocytes. *FEBS Lett.* **234,** 489–492.

Furuichi, Y., LaFaiandra, A., and Shatkin, A. J. (1977). 5′-Terminal structure and mRNA stability. *Nature* (*London*) **266,** 235–239.

Galili, G., Altschuler, Y., and Levanony, H. (1993). Assembly and transport of seed storage proteins. *Trends Cell Biol.* **3,** 437–442.

Galili, G., Kawata, E. E., Smith, L. D., and Larkins, B. A. (1988). Role of the 3′ poly (A) sequence in translational regulation of mRNAs in *Xenopus laevis* oocytes. *J. Biol. Chem.* **263,** 5764–5770.

Galili, G. K., Kawata, E. E., Cuellar, R. E., Smith, L. D., and Larkins, B. A. (1986). Synthetic oligonucleotide tails inhibit *in vitro* and *in vivo* translation of SP6 transcripts of maize zein cDNA clones. *Nucleic Acids Res.* **14,** 1511–1524.

Gallie, D. R., Sleat, D. E., Watts, J. W., Turner, P. C., Michael, T., and Wilson, A. (1987). The 5′-leader sequence of tobacco mosaic virus RNA enhances the expression of foreign gene transcripts *in vitro* and *in vivo*. *Nucleic Acids Res.* **15,** 3257–3273.

Gurdon, J. B., and Wickens, M. P. (1983). The use of *Xenopus* oocytes for the expression of clone genes. *Methods Enzymol.* **101,** 370–386.

Hoffman, L. M., and Donaldson, D. D. (1988). Vector for *in vitro* synthesis of poly (A)⁺ transcripts. *Gene* **67,** 137–140.

Hsich, P., Rosner, M. R., and Robbins, P. W. (1983). Selective cleavage by endo-B, N-acetylglucosaminidase H at individual glycosylation sites of *Sindbis* virion glycoproteins. *J. Biol. Chem.* **258,** 2555–2561.

Hurkman, W. J., Smith, L. D., Richter, J., and Larkins, B. A. (1981). Subcellular compartmentalization of maize storage proteins in *Xenopus* oocytes with zein mRNAs. *J. Cell Biol.* **89,** 292–299.

Kawata, E. E., Galili, G., Smith, L. D., and Larkins, B. A. (1988). Translation in *Xenopus laevis* oocytes of mRNAs transcribed *in vitro*. *In* "Plant Molecular Biology Manual" (S. B. Gelvin and R. A. Schilperoort, eds.), Vol. B7, pp 1–22. Dordrecht, The Netherlands: Kluwer.

Krieg, P. A., and Melton, D. A. (1984). Functional messenger RNAs are produced by SP6 *in vitro* transcription of cloned cDNAs. *Nucleic Acids Res.* **12,** 7057–7070.

Leaf, D. S., Roberts, S. J., Gerhart, J. C., and Moore, H. (1990). The secretory pathway is blocked between the trans-Golgi and the plasma membrane during meiotic maturation in *Xenopus* oocytes. *Dev. Biol.* **141,** 1–12.

Mager, S., Naeve, J., Quick, M., Labarca, C., Davidson, N., and Lester, H. A. (1993). Steady-states, charge movements and rates for a cloned gaba transporter expressed in *Xenopus* oocytes. *Neuron* **10,** 177–188.

Matthews, G., and Colman, A. (1991). A highly efficient, cell-free translation/translocation system prepared from *Xenopus* eggs. *Nucleic Acids Res.* **19,** 6405–6412.

Melton, D. A. (1987). Translation of messenger RNA in injected frog oocytes. *Methods Enzymol.* **152,** 288–296.

Pelletier, J., and Sonenberg, N. (1985). Insertion mutagenesis to increase secondary structure within the 5′ noncoding region of a eukaryotic mRNA reduces translational efficiency. *Cell* **40,** 515–526.

Pfaff, S. L., Tamkun, M. M., and Taylor, W. L. (1990). pOEV: A *Xenopus* oocyte protein expression vector. *Anal. Biochem.* **188,** 192–199.

Quick, M. W., Naeve, J., Davidson, N., and Lester, H. A. (1992). Incubation with horse serum increases viability and decreases background neurotransmitter uptake in *Xenopus* oocytes. *Biotechniques* **13,** 357–361.

Richter, J. D., Jones, N. C., and Smith, L. D. (1982). Stimulation of *Xenopus* oocyte protein synthesis by microinjected adenovirus RNA. *Proc. Natl. Acad. Sci. U.S.A.* **79,** 3789–3793.

Richter, J. D., and Smith, L. D. (1981). Differential capacity for translation and lack of competition between mRNAs that segregate to free and membrane bound polysomes. *Cell* **27,** 183–191.

Schenborn, E. T., and Mierendorf, R. C. (1985). A novel transcription property of SP6 and T7 polymerases: Dependence on template structure. *Nucleic Acids Res.* **13,** 6223–6236.

Sheares, B. T., and Robbins, P. W. (1986). Glycosylation of ovalbumin in a heterologous cell: Analysis of oligosaccharide chains of the cloned glycoprotein in mouse L cells. *Proc. Natl. Acad. Sci. U.S.A.* **83,** 1993–1997.

Simon, P., and Jones, R. L. (1988). *Xenopus* oocytes injected with barley mRNAs. *Eur. J. Cell Biol.* **47,** 213–221.

Simon, R., Altschuler, Y., Rubin, R., and Galili, G. (1990). Two closely related wheat storage proteins follow a markedly different subcellular route in *Xenopus laevis* oocytes. *Plant Cell* **2,** 941–950.

Smith, L. D., Weilong, X. U., and Varnold, R. L. (1991). Oogenesis and oocyte isolation. *Methods Cell Biol.* **36,** 45–60.

Soreq, H., and Seidman, S. (1992). *Xenopus* oocyte microinjection: From gene to protein. *Methods Enzymol.* **207,** 225–265.

Sturm, A., Kuik, J. A., Vliegenthart, J. F. G., and Chrispeels, M. J. (1987). Structure, position and biosynthesis of the high mannose and complex oligosaccharide side chains of the bean storage protein phaseolin. *J. Biol. Chem.* **262,** 13392–13403.

Swick, A. G., Janicot, M., Chenevel-Kastelic, T., McLenithan, J. C., and Lane, M. D. (1992). Identification of a transcriptional repressor down-regulated during preadipocyte differentiation. *Proc. Natl. Acad. Sci. U.S.A.* **89,** 1812–1816.

Tate, T. S., Urade, R., Micanovic, R., Gerber, L., and Underfriend, S. (1990). Secreted alkaline phosphatase: An internal standard for expression of injected mRNAs in the *Xenopus* oocyte. *FASEB J.* **4,** 227–231.

Vitale, A., Sturm, A., and Bollini, R. (1986). Regulation of processing of a plant glycoprotein in the Golgi complex: A comparative study using *Xenopus* oocytes. *Planta* **169,** 108–116.

Vitale, A., Zoppè, M., Fabrini, M. S., Genga, A., Rivas, L., and Bollini, R. (1989). Synthesis of lectin-like protein in developing cotyledons of normal and phytohemagglutinin-deficient *Phaseolus vulgaris. Plant Physiol.* **90,** 1015–1021.

Wallace, J. C., Galili, G., Kawata, E. E., Cuellar, R. E., Shotwell, M. A., and Larkins, B. A. (1988). Aggregation of lysine-containing zeins into protein bodies in *Xenopus* oocytes. *Science* **240,** 662–664.

Woodland, H. R. (1979). The modification of stored histones H3 and H4 during oogenesis and early development of *Xenopus* laevis. *Dev. Biol.* **68,** 360–370.

Wormington, M. (1991). Preparation of synthetic mRNAs and analysis of translational efficiency in microinjected *Xenopus* oocytes. *Methods Cell Biol.* **36,** 167–183.

Yet, M.-G., Shao, M.-C., and Wold, F. (1988). Effects of the protein matrix on glycan processing in glycoproteins. *FASEB J.* **2,** 22–23.

CHAPTER 37

Heterologous Expression of Higher Plant Transport Proteins and Repression of Endogenous Ion Currents in *Xenopus* Oocytes

Julian I. Schroeder

Department of Biology and Center for Molecular Genetics
University of California, San Diego
La Jolla, California 92093-0116

I. Introduction

Recent research has led to considerable progress toward an understanding of the molecular, structural, and biophysical mechanisms of transport and membrane signal transduction in higher plant cells. The cloning of cDNAs has been reported which encode plasma membrane transport proteins from higher plants, including ion pumps, proton-coupled carriers, and ion channels. Several types of these plant membrane transporters can be functionally expressed in *Xenopus* oocytes. These studies allow structure–function studies of the gene products and in some cases have for the first time enabled voltage-clamp studies revealing the biophysical mechanism of transport of these membrane proteins. This article focuses on methods for the use of *Xenopus* oocytes as a heterologous expression system for plant transporters and describes methods for the repression of endogenous inward-rectifying ion currents in oocytes.

The ability of higher plants to take up and shuttle nutrients throughout the plant as well as the ability to respond to various environmental signals via receptors plays a central role in the growth and development of plants. Knowledge of the mechanisms of transmembrane transport and of membrane-associated signal transduction has recently been extended to the molecular biophysical and structural level by use of novel experimental approaches, including patch clamp techniques, molecular biological and genetic approaches, membrane protein purification, cell biological methods, and microelectrode recordings. The difficulty in voltage-clamping higher plant membranes has left a void in information with respect to the mechanisms of membrane transport and signal transduction in higher plants. Problems hindering voltage clamp recordings have been the efficient cell–cell coupling between plant cells via plasmodesmata, the small size of most higher plant cells, the large size of intracellular vacuolar organelles, and the rigidity of plant cell walls. Giant algal cells circumvent several of these problems and have therefore been used for classical membrane excitability studies (Cole and Curtis, 1938; for rev. Tazawa *et al.,* 1987). Many of the processes and signals that regulate plant growth and development, however, need to be studied in specialized plant cell types. Patch clamp studies on higher plant cells and vacuoles have circumvented many of the problems associated with voltage clamping.

In parallel to these advances, recent research in plant molecular biology and molecular genetics is currently leading to the cloning of many membrane-associated proteins from higher plants. Transgenic plant studies (e.g., overexpression or antisense expression) or functional analysis in heterologous systems such as yeast, *Xenopus* oocytes, and insect or mammalian cell lines can be pursued to characterize the functions of these cDNAs. Previous studies have shown that higher plant cDNAs encoding soluble proteins can be functionally expressed in *Xenopus* oocytes (Wallace *et al.,* 1988; Aoyagi *et al.,* 1990; see also Galili *et al.,* this volume). Recently cDNAs encoding higher plant membrane transporters have been functionally expressed in oocytes. We will use these transport proteins as examples in the description of methods for studying higher plant membrane transporters. Potential problems, such as endogenous ion currents in oocytes, will also be discussed, and novel approaches for circumventing or accounting for such problems will be described. Several of the approaches outlined here are also applicable to heterologous expression studies in mammalian or insect cell lines. Detailed expression studies have shown that molecular dynamical properties of ion transport proteins are retained when expressed in oocytes, including submillisecond flickers of ion channels (Sakmann *et al.,* 1986; for rev. Soreq and Seidman, 1992). Furthermore, the hallmark functional properties of plant K^+ channels expressed in oocytes (Schachtman *et al.,* 1992) are similar to these properties identified in higher plant protoplasts (e.g., Schroeder *et al.,* 1987) showing that the molecular dynamics and structure of these membrane proteins are stable and retained.

II. Materials

Xenopus oocytes are isolated and injected as described elsewhere (Cao *et al.,* 1992; Soreq and Seidman, 1992). In brief, adult female *Xenopus laevis* frogs are placed on ice and anesthetized. A small incision is made on the side of the abdomen to remove several ovarian lobes. The lobes are separated and immersed in sterile Barth's solution supplemented with gentamycin [m*M:* 88 NaCl, 1 KCl, 2.4 NaHCO$_3$, 10 Hepes–NaOH, 0.33 Ca (NO$_3$)$_2$ · 4 H$_2$O, pH 7.4, 0.41 CaCl$_2$ · 2 H$_2$O, 0.82 Mg SO$_4$ plus 1 ml/liter gentamycin (Sigma G-1397)]. Before microinjection and voltage clamp recording, the follicular cell layer surrounding oocytes must be removed. Oocytes are enzymatically defolliculated by incubation in Barth's solution containing 5 mg/ml collagenase (type A, Boehringer) at 24°C for 1 to 2 h. Subsequently, oocytes are immersed in a potassium phosphate solution (100 m*M* K$_2$HPO$_4$, pH 6.5, 1 g/liter BSA, fraction V) for 1 h at 24°C on a circulatory shaker (60 RPM) to remove follicular layers. Subsequently oocytes are washed gently five times with Barth's solution containing BSA (1 g/liter) at the end of the incubation. Stage 5 to 6 oocytes (Dumont, 1972) are selected and stored at 18°C in the dark in Barth's solution containing gentamycin (1 ml/liter) for 24 h before microinjection.

Special vectors for *in vitro* transcription of mRNA are recommended for efficient expression in oocytes (see Schroeder, 1994; Chapter 36, this volume). Messenger RNA transcripts (1–50 ng per oocyte) are injected into oocytes in a final volume of 50 nl (see Soreq and Seidman, 1992). Oocytes are incubated in Barth's solution supplemented with 1 ml/liter gentamycin for 1–6 days at 18°C in the dark before recording. The sterile Barth's solution is exchanged every day and any slightly damaged oocytes are removed daily. Electrophysiological recordings of membrane transport proteins are usually performed within 1–6 days after mRNA injection (Cao *et al.,* 1992; Stühmer, 1992).

III. Critical Aspects: Endogenous Ion Currents in *Xenopus* Oocytes

For higher plants, most structure–function studies of ion transport proteins have focused on uptake mechanisms into cells. Uptake of ions or uncharged solutes into higher plant cells usually takes place at resting potentials ranging from −100 to −200 mV (Glass, 1988). This membrane potential range is exceedingly more negative than that used in most oocyte expression studies of animal transport systems, which are often activated by membrane depolarization. Contrary to possible expectations from animal voltage clamping, healthy oocytes do not show dielectric breakdown of the membrane when the membrane potential is held in the range -100 to -180 mV for up to several seconds.

When analyzing heterologously expressed membrane transporters in any cell type, it is important to ensure that endogenous activities similar to those of the studied transporter are either absent or minimized in magnitude. When recording membrane currents at hyperpolarized potentials from approximately -100 to -160 mV (Fig. 1A), endogenous currents appear in some batches of oocytes and need to be considered. Figure 1B shows an example of an oocyte from a batch that showed unusually large endogenous currents. These inward currents (downward deflections in Figs. 1B and 1C) are activated by hyperpolarizing the membrane to strongly negative potentials. These currents have been previously studied and described as Cl^- channels (Parker and Miledi, 1988). The activation and deactivation of these endogenous oocyte currents show a slow time dependence that is similar to time dependencies of several ion channel types in plant cells (Schroeder *et al.,* 1987; Terry *et al.,* 1991; Schroeder and Keller, 1992).

We have studied these endogenous oocyte currents in detail ($n > 200$) and found several approaches that can help to greatly reduce or abolish these currents (Fig. 1D). In the following are listed several of the main approaches that aid in reducing these currents to low absolute levels (<5 nA at -150 mV).

1. The magnitude of endogenous oocyte currents observed at hyperpolarizations (I_{hyp}), if measureable, often diminishes within 2–3 days after oocyte isolation. Therefore, on Day 3 after oocyte isolation this current becomes less prevalent in control experiments (e.g., Fig. 1D).

2. In most oocytes, I_{hyp} activates at potentials more negative than -150 mV. Therefore, a voltage range of -100 to -150 mV can be employed in analyses without significant interference by this background current.

3. A small fraction of oocyte batches (≈ 5 to 10%) show large endogenous I_{hyp} at potentials more positive than -150 mV (Figs. 1B and 1C). Therefore, the level of endogenous I_{hyp} in several control oocytes needs to be measured on each day. This procedure is appropriate since the within-batch variance of the magnitude of I_{hyp} is low (Parker and Miledi, 1988; Cao *et al.,* 1992).

4. Use of glutamate salts rather than chloride salts in bathing media leads to a reduction or disappearance of I_{hyp} currents ($n > 40$).

5. Oocytes are known to contain Ca^{2+}-activated Cl^- channels that have been effectively used as tools for expression cloning and funtional studies of receptors (Soreq and Seidman, 1992). It has been suggested that I_{hyp} differs from the typically studied Ca^{2+}-activated Cl^- channels in oocytes (Parker and Miledi, 1988). Nevertheless, Ca^{2+} ions are involved in the activation of I_{hyp} (Cao and Schroeder, unpublished), and approaches that avoid elevation in the cytosolic Ca^{2+} concentration (i.e., BAPTA injection) can efficiently abolish I_{hyp}.

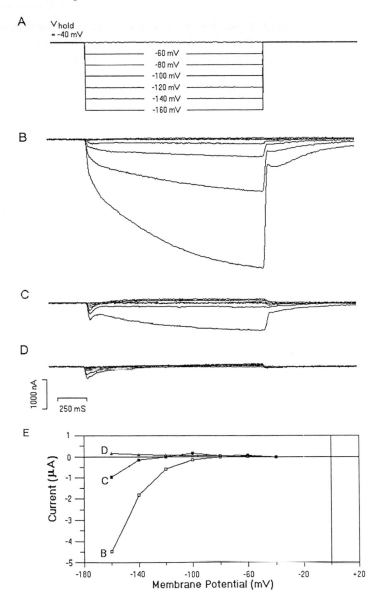

Fig. 1 Some batches of oocytes show a large endogenous current, which is activated by hyperpolarization. (A) Superposition of seven imposed voltage pulses recorded by the voltage electrode. (B, C, and D) Currents activated by hyperpolarizing voltage steps from three different oocyte batches that showed different levels of the endogenous current, I_{hyp} (see text). The large magnitude of I_{hyp} shown in (B) is not typical and occurs in <5% of oocyte batches prepared in the author's laboratory, (D) Typical I_{hyp} levels in most oocyte batches. (E) Steady-state current–voltage relationships of (B), (C), and (D). Currents were recorded in a bath solution containing 117.5 mM KCl, 1 mM CaCl$_2$, 2 mM MgCl$_2$, 10 mM Hepes, pH 7.2. Leak currents were subtracted in (B)–(E) by a P/6 procedure (see Cao *et al.*, 1992).

6. Despite the above precautions, it is important to rule out that a small fraction of endogenous I_{hyp} is prevalent when studying heterogously expressed ion transporters at strong hyperpolarizations. For example, inward-rectifying K⁺ channels show a more rapid but similar time dependence of activation to I_{hyp} (Schroeder *et al.*, 1987; Schachtmann *et al.*, 1992). One method that can be used to distinguish between these K⁺ channel currents and I_{hyp} lies in the slow deactivation of I_{hyp} when compared to K⁺ channels (see, for example, the slow relaxation at the end of the voltage pulse in Fig. 1B). Appearance of such a slowly relaxing current upon repolarization indicates that a fraction of the recorded current may be carried by I_{hyp}. Futhermore, repolarization to potentials more positive than ≈ 0 mV results in large transient outward currents, when I_{hyp} is activated (Fig. 2). These large currents serve as a potent indicator that I_{hyp} has been activated during the preceding hyperpolarization. Combining the procedures outlined above provides an approach to abolish endogenous ion currents, and also to monitor the magnitude of I_{hyp}.

IV. Methods and Approaches

A. Composition of Extracellular Solutions

Most transport processes studied in mammalian systems can be analyzed in modified frog NaCl ("Ringer") solution (in m*M:* 115 NaCl, 2.5 KCl, 1.8 $CaCl_2 \cdot 2H_2O$, 1 $NaHCO_3$, 10 Hepes, 1 $MgCl_2 \cdot 6\ H_2O$, pH 7.4). Nevertheless, in many voltage clamp studies the ionic composition of the bathing medium is greatly varied to study separately specific currents while abolishing other currents. For example, when Ca^{2+} channel currents are studied, extracellular Na^+ is typically replaced by isotonic extracellular Ba^{2+} or Ca^{2+} (Nowycky *et al.*, 1985). Most higher plant cells are not exposed to high millimolar extracellular concentrations of Na^+. It is therefore best to design and optimize extracellular solutions for the study of specific membrane transporters. Examples of extracellular solutions are presented in the following for the study of different ion transporters. These methods should be useful as a starting point for analyzing their specific functions.

1. Solutions for Studying K⁺Channels and K⁺ Transporters

Both depolarization-activated and hyperpolarization-activated K⁺ channels have been described in higher plant cells (Schroeder *et al.*, 1987; for rev. Schroeder *et al.*, 1994). Furthermore, both types of K⁺ currents have been induced in *Xenopus* oocytes after injection of plant mRNAs (Cao *et al.*, 1992; Schachtman *et al.*, 1992). It is possible that high Na^+ concentrations in the extracellular solution may block or alter properties of higher plant K⁺ channels (Kourie and Goldsmith, 1992; Bush *et al.*, 1988). Certain plant K⁺ channels may not interract with extracel-

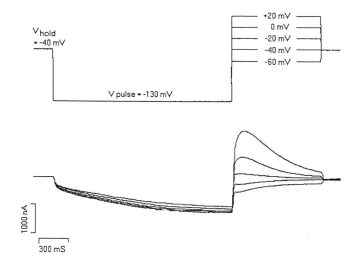

Fig. 2 Large endogenous transient outward currents observed in *Xenopus* oocytes. The top traces show imposed voltage pulses recorded by the voltage electrode. The lower traces show large transient endogenous currents that are observed when the membrane is hyperpolarized and subsequently depolarized only when endogenous I_{hyp} currents are activated (see text).

lular Na^+. For example, expression of the inward-rectifying K^+ channel cRNA *KAT1* in oocytes produces K^+ currents that are not significantly blocked by extracellular Na^+ ($n = 3$) (W. Gassmann and J. I. Schroeder, unpublished). Extracellular sodium in standard frog Ringer solution can be entirely replaced by K^+. If lower concentrations of K^+ are used, a standard procedure is to adjust the final osmolality of the solution by adding D-mannitol (final osmolality approximately 240 mosmol kg^{-1}). Continuous perfusion with extracellular solution is necessary to avoid a slow decline in inward-rectifying K^+ channel current (A. Ichida, unpublished). This may be due to extracellular depletion of K^+, because of the highly folded oocyte membrane.

High-affinity K^+ uptake transport into plant cells shows an apparent affinity for K^+ (K_m) in the range 10–20 μM (Epstein *et al.*, 1963). K^+ uptake by high-affinity K^+ uptake transporters would be best resolved in voltage clamp studies by exchanging K^+ concentrations (10 μM to 10 mM K^+) in the bath solution. Both extracellular Na^+ and Ca^+ may strongly interact with these high-affinity K^+ uptake transporters (Rains and Epstein, 1967; LaHaye and Epstein, 1969). It is therefore recommendable in initial studies to remove any Na^+ or other alkali metal ions from the bathing medium. A limited concentration of Ca^{2+} is required for oocyte stability (empirically 0.15 mM; see next section).

2. Nitrate and Anion Transporters

Uptake of nitrate as well as other anions into plant cells is of prime importance for plant nutrition (Glass, 1988). Because the resting potential of plant cells during uptake is usually more negative than -100 mV, the uptake of anions such as nitrate occurs via an active transport mechanism. The uptake of many substrates across the higher plant plasma membrane is coupled to proton influx (Slayman, 1987; Harper and Sussman, 1989). A detailed study of nitrate uptake in roots, using a combination of tracer flux measurements and extracellular and intracellular electrode recordings, has provided evidence that for each nitrate ion transported into cells, two protons are cotransported (McClure *et al.,* 1990). Thus, the nitrate carrier would transport a net positive charge into higher plant cells, which may be driven by hyperpolarization and by H^+ and substrate gradients.

A first nitrate transporter gene was isolated by cloning the genetic *CHL-1* locus in *Arabidopsis* (Tsay *et al.,* 1993). The *CHL-1* locus is associated with low-affinity nitrate uptake and also transports other anions such as chlorate. The CHL-1 protein may also transport Cl^- ions. For voltage clamp studies in *Xenopus* oocytes a small amount of Cl^- ions should be present in bathing solutions (e.g., 0.3 mM Cl^-) to stabilize electrical potentials of bath electrodes. Furthermore, we have found that a small amount of Ca^{2+} needs to be added to the bath solution to produce low-background voltage clamp recordings from oocytes (W. Gassman and J. I. Schroeder, unpublished). To meet these requirements a minimal solution was developed that suffices for obtaining stable voltage clamp recordings from oocytes. This solution contained 0.15 mM $CaCl_2$, 10 mM Hepes, pH 7.4, and D-mannitol (230 mM). To record large membrane currents (e.g., >100 nA) in oocytes, additional salts need to be added to the bath to reduce the electrical resistance of the bath solution (Sherman-Gold, 1993). It is important that the resistance of the bath solution remain significantly lower (e.g., >20 times lower) than the membrane resistance during full transporter activation. In order to assay CHL1-mediated transport, NO_3^- was added to the extracellular solution by perfusion. As expected from uptake studies (McClure *et al.,* 1990) no significant currents were found at neutral pH. However, when the extracellular pH was shifted to physiological values of 5.5, a net inward current was produced in the presence of NO_3^- (Tsay *et al.,* 1992). Parallel studies showing NO_3^- uptake into oocytes indicate that CHL-1 encodes a proton nitrate cotransporter (Tsay *et al.,* 1993). Biophysical considerations for analyzing the stoichometry of an H^+-coupled transporter are discussed in a later section (carrier stoichiometry).

When investigating the transport activity of any anion transporter, whether anion carrier or channel, bath electrode potentials need to be considered. Bath electrode potentials shift with changes in the bath chloride concentration. Therefore it is important to maintain the Cl^- concentration at a constant level. The standard approach to avoid or minimize such bath electrode potential shifts is to use agar bridges for bath electrodes containing high chloride salt concentrations (e.g., 1–3 M KCl). A pitfall when using high-salt agar bridges for bath electrodes is that diffusion of salts from the agar bridge may significantly change the ionic content of the bathing medium (see Schroeder and Fang, 1991; Neher,

1992). To avoid diffusion of salts from bath electrodes into the bathing medium, we have placed these electrodes in a small exit chamber for the perfusion tubing, and continuously perfuse the bath (Schroeder and Fang, 1991). Other electronic "bath clamp" methods can also be used to minimize shifts in bath electrode potentials (Sherman-Gold, 1993).

3. Glucose, Sucrose, and Amino Acid Transporters

Use of *Chlorella* mutants, as well as complementation of yeast mutants, has recently led to the cloning of first higher plant glucose, sucrose, and amino acid transporters (Sauer *et al.*, 1990; Frommer *et al.*, 1993; Hsu *et al.*, 1993). One of these transporters, an *Arabidopsis* glucose transporter, has been functionally expressed in oocytes (Boorer *et al.*, 1992). Addition of amino acids, glucose, or sucrose to frog Ringer may allow studies of electrogenic cotransport with protons. Higher plant carriers make use of the acidic extracellular pH and the negative membrane potential by using protons to create a net influx of positive charge into the cell when transporting uncharged solutes. Homologues or functional equivalents to these transporters in animal systems capitalize on the large Na^+ gradient from the extracellular space to the cytosol by coupling uptake to Na^+ influx. It is therefore possible that the large Na^+ concentration in frog Ringer solution (116 mM) may provide a driving force for heterologously expressed plant coupled carriers which should be analyzed.

4. Water-Transporting Aquaporins

TIP is a vacuolar membrane protein belonging to the MIP family of membrane proteins. When expressed in oocytes some TIP and MIP homologues allow a high degree of water permeability while limiting ion transport (Preston *et al.*, 1992; Maurel *et al.*, 1993). A combination of various salt solutions during voltage clamp recordings were used to determine whether ions are transported in parallel to the large fluxes of water across the oocyte plasma membrane (see Maurel *et al.*, 1993). It was estimated that approximately one ion may be transported across the membrane per 10^4 to 2×10^5 water molecules (Maurel *et al.*, 1993). Lack of permeation of small solutes suggests energetic constraints for H_2O selectivity (Maurel *et al.*, 1993). It should be noted that the MIP family does not exclusively encode aquaporins but that certain MIP homologues encode transporters for other solutes. Therefore each MIP homologue needs to be investigated with various approaches to determine their respective transport specificities (see Maurel *et al.*, 1993).

5. Carrier Stoichiometry

Detailed information can be extracted with regard to the selectivity and stoichiometry of transporters. We will outline an example for studying the reversal potential of such a transporter. A proton-coupled K$^+$ transporter has been suggested as a high-affinity mechanism for K$^+$ uptake into fungi (Rodriquez-Navarro *et al.*, 1986) and higher plants. Assuming that this transporter has a stoichiometry of one proton entering the oocyte per K$^+$ ion we would expect the reversal potential to be determined by the following equation following a chemi-osmotic model (at $T = 25°C$):

$$V_{rev} = \frac{59}{2} \, mV \, (pH_{cyt} - pH_{ext}) - \frac{59}{2} \log \frac{[K^+_{cyt}]}{[K^+_{ext}]} \tag{1}$$

where pH$_{ext}$ and pH$_{cyt}$ are the pH values of the bath solution and the cytosol, respectively, and [K$^+$cyt] and [K$^+$ext] represent the cytosolic and extracellular K$^+$ concentrations, respectively. Using this equation a proton-coupled K$^+$ transporter can be analyzed by measuring several voltage ramps while modifying one parameter (e.g., pH) and maintaining other parameters constant. For all experiments it is important to ensure that control (uninjected or water injected) oocytes do not respond to these treatments (Schachtman and Schroeder, 1994).

The theoretical shift in reversal potentials assuming a stoichiometry of one proton entering the cell per potassium taken up can be estimated using Eq. (1). For external acidification by one pH unit, and assuming constant extracellular K$^+$, a cytosolic K$^+$ concentration of 100 mM, and a cytosolic pH of 7.5, a shift in reversal potentials with respect to the background membrane resistance by approximately +29 mV would indicate that the transporter has a stoichiometry of one proton entering the cell per potassium taken up. A shift in reversal potentials by $\approx$ 20 mV would indicate one proton being taken up per two K$^+$ ions. In addition to reversal potential shift measurements, the high time resolution of the cut-open oocyte technique (Taglialatela *et al.*, 1992) enables detailed studies of molecular biophysical properties, such as transport rate constants and gating charge movements, of carriers and pumps by studying non-steady-state relaxation currents. These approaches are described in detail in a recent article (Lester, 1994).

B. Expression Cloning in Oocytes

The *Xenopus* oocyte system has been particularly successful for expression cloning of membrane receptors and transporters from animal systems (Soreq and Seidman, 1992). Many reports and detailed accounts of using *Xenopus* oocytes for expression cloning of membrane receptors and ion channels have appeared elsewhere (e.g., Masu *et al.*, 1987; Frech *et al.*, 1989; for rev. Soreq and Seidman, 1992). Because higher plant membrane transporters are functionally expressed in oocytes, this system may lend itself to expression cloning of transporter encoding cDNAs. In general this approach involves several major steps, including polyadenylated [poly(A$^+$)]mRNA isolation, size-fractionation of total poly(A$^+$)mRNA, and injection into oocytes. The size-selected poly(A$^+$)mRNA fraction showing maximal functional activity is used to construct a directional phage cDNA library

(Cao *et al.*, 1992). Pools of recombinant phages from the cDNA library are amplified and used to prepare DNA templates for RNA synthesis. RNA synthesized from the cDNA library (cRNA) is injected into oocytes to assay for functional expression of the previously identified activity. Several methods for subdividing the cDNA library have been described (Lübbert *et al.*, 1987; Masu *et al.*, 1987; Hollmann *et al.*, 1989). The difficulty in expression cloning is greatly potentiated, but not rendered impossible, when more than one cDNA is required for functional expression. A technique for cloning of multiple cDNAs has been described (Lübbert *et al.*, 1987), but this has been seldomly applied due to associated technical complexities.

One of the most powerful uses of the *Xenopus* oocyte system for studying membrane signaling events has been expression cloning and functional characterization of many mammalian membrane-bound receptors from animal systems (Masu *et al.*, 1987). Over 20 receptor cDNAs have been cloned and functionally analyzed using oocytes. Most of these receptors belong to the seven transmembrane domain class of receptors that couple to heterotrimeric GTP-binding proteins (G proteins). It remains unknown whether G-protein-coupled receptors exist in higher plants, although there are several lines of circumstantial evidence supporting this hypothesis (e.g. Ma *et al.*, 1990; M. Bennett, U. Warwick, personal communication).

V. Limitations of the Procedure

Although the methods presented here are applicable to the characterization of a wide variety of plant membrane transporters, several limitations, in addition to those referred to throughout this article, should be considered. Analysis of expressed ion transporters that do not transport a net charge (e.g., cation-coupled anion transporters with a stoichiometry of $1:1$) is not possible using voltage clamp recording. However, other assay methods, including the use of radioactive tracers or fluorescent probes, may be applicable for the detection of nonelectrogenic transport activity (e.g., for glycerol see Maurel *et al.*, 1993). Ion transporters that require more than one type of plant polypeptide, either directly or indirectly, pose additional difficulties. A related concern is that native regulation of many transporters is expected to require the coexpression of additional subunits, or proteins responsible for post-translational modification, such as protein kinases. The coexpression of ion transporters and putative regulatory proteins represents a powerful approach for future research. An example of a plant ion transporter that has not been amenable thus far to functional expression in *Xenopus* oocytes is the K^+ channel AKT1 (Cao and Schroeder, unpublished). Although AKT1 is functional when expressed in yeast (Sentenac *et al.*, 1992), the requirement for additional plant polypeptides for expression in oocytes cannot be excluded. Other possible explanations include a lack of targeting of AKT1 to the oocyte plasma membrane or instability of the protein in oocytes. Use of different heterologous expression systems, such as mammalian cells or insect cells, may circumvent expression problems in certain cases.

VI. Conclusions and Perspectives

The use of *Xenopus* oocytes for structure–function studies and for expression cloning of membrane transport proteins and receptors has exploded since the early demonstrations by Miledi and colleagues that these proteins can be functionally expressed in oocytes (e.g., Gunderson *et al.,* 1984). This heterologous system is therefore useful for expression cloning and structure–function studies of some membrane-associated proteins from higher plants. An application of the *Xenopus* oocyte system has become the identification of the precise function of cloned plant cDNAs. Because of efficient homology cloning techniques and ongoing genome projects, many cDNAs of unknown function are being cloned that show homology to genes encoding known functions. For membrane proteins, *Xenopus* oocytes have already proven useful in determining the molecular function of higher plant membrane transport processes (Schachtman *et al.,* 1992; Maurel *et al.,* 1993; Tsay *et al.,* 1993; Schachtman and Schroeder, 1994).

Transformation of plants by overexpression of either wild-type or structurally altered genes has proven powerful for identifying physiological functions of various genes. *Xenopus* oocytes provide a system in which the effect of point mutations on the biophysical function of individual membrane proteins can be analyzed at high resolution. Transformation of plants with such well-characterized and defined mutants should allow the correlation of function and molecular dynamics of individual proteins to physiological effects at the level of whole transgenic plants. For example, disruption of expression of a protein for nitrate uptake has significant effects on herbicide resistance in the associated mutant plants (Tsay *et al.,* 1993).

Acknowledgments

Ms. Judie Murray is acknowledged for excellent assistance, and John Ward and Walter Gassmann for helpful discussions and critical comments, during the preparation of the manuscript. I thank Meyer B. Jackson (UCLA) for allowing me to use oocyte research facilities for the development of initial functional expression studies of plant ion channel mRNAs in *Xenopus* oocytes from 1988 to 1990. I thank Mirka Anderova, Yongwei Cao, Walter Gassmann, Audrey Ichida, Christophe Maurel, and Daniel Schachtman, who have carried out at UCSD the recent research reviewed in this article. An article with similar but extended contents will appear in a volume on Methods for Ion Channel Reconstitution (Schroeder, 1994) with the permission of, and upon request by editors, Academic Press. Research in the author's laboratory was supported by the Powell Foundation, the National Science Foundation (Grant MCB9004977), an NSF Presidential Young Investigator Award, the U.S. Departments of Energy (Grant DE-FG03-94-ER20148) and Agriculture (Grant 92-37304-7757).

References

Aoyagi, K., Sticher, L., Wu, M., and Jones, R. L. (1990). The expression of barley α-amylase genes in *Xenopus laevis* oocytes. *Planta* **180,** 333–340.

Boorer, K. J., Forde, B. G., Leigh, R. A., and Miller, A. J. (1992). Functional expression of a plant plasma membrane transporter in *Xenopus* oocytes. *FEBS* **302,** 166–168.

Bush, D. S., Hedrich, R., Schroeder, J. I., and Jones R. L. (1988). Channel-mediated K^+ flux in barley aleurone protoplasts. *Planta* **176,** 368–377.

Cao, Y., Anderova, M., Crawford, N. M., and Schroeder, J. I. (1992). Expression of an outward rectifying potassium channel from maize mRNA and cRNA in *Xenopus* oocytes. *Plant Cell* **4,** 961–969.

Cole, K. S., and Curtis, H. J. (1938). Electrical impedance of *Nitella* during activity. *J. Gen. Physiol.* **22,** 37–64.

Dumont, J. N. (1972). Oogenesis in *Xenopus laevis* (Daudin). I. Stages of oocytes development in laboratory maintained animals. *J. Morphol.* **136,** 153–180.

Epstein, E., Rains, D. W., and Elzan, O. E. (1963). Resolution of dual mechanisms of potassium absorption by barley roots. *Proc. Natl. Acad. Sci. U.S.A.* **49,** 684–692.

Frech, G. C., VanDongen, A. M. J., Schuster, G., Brown, A. M., and Joho, R. H. (1989). A novel potassium channel with delayed rectifier properties isolated from rat brain by expression cloning. *Nature* **340,** 642–645.

Frommer, W. B., Hummel, S., and Riesmeier, J. W. (1993). Expression cloning in yeast of a cDNA encoding a broad specificity amino acid permease from *Arabidopsis thaliana. Proc. Natl. Acad. Sci. U.S.A.* **90,** 5944–5948.

Glass, A. D. M. (1988). "Plant Nutrition: An Introduction to Concepts." Boston, MA: Jones and Bartlett.

Gunderson, C., Miledi, R., and Parker, I. (1984). Slowly inactivating potassium channels induced in *Xenopus* oocytes by messenger ribonucleic acid from torpedo brains. *J. Physiol. (London)* **353,** 231–248.

Hollmann, M., O'Shea-Greenfield, A., Rogers, S., and Heinemann, S. (1989). Cloning by functional expression of a member of the glutamate receptor family. *Nature* **342,** 643–648.

Hsu, L.-C., Chiou, T.-J., Chen, L., and Bush, D. R. (1993). Cloning a plant amino acid transporter by functional complementation of a yeast amino acid transport mutant. *Proc. Natl. Acad. Sci. U.S.A.* **90,** 7441–7445.

Kourie, J., and Goldsmith, M. H. M. (1992). K^+ channels are responsible for an inwardly rectifying current in the plasma membrane of mesophyll protoplasts of *Avena sativa. Plant Physiol.* **98,** 1087–1097.

LaHaye, P. A., and Epstein, E. (1969). Salt toleration by plants: enhancement with calcium. *Science* **166,** 395–396.

Lester, H. A., Mager, S., Quick, M. W., and Corey, J. L. (1994). Permeation properties of neurotransmitter transporters. *Annu. Rev. Pharmacol. Toxicol.* **34,** 219–249.

Lübbert, H., Hoffman, B. J., Snutch, T. P., van Dyke, T., Levin, A. J., Hartig, P. R., Lester, H. A., and Davidson, N. (1987). cDNA cloning of a serotonin 5-HT1C receptor by electrophysiological assays of mRNA-injected *Xenopus* oocytes. *Proc. Natl. Acad. Sci. U.S.A.* **84,** 4332–4336.

Ma, H., Yanofsky, M. F., and Meyerowitz, E. M. (1990). Molecular cloning and characterization of GPA1, a G protein α subunit gene from *Arabidopsis thaliana. Proc. Natl. Acad. Sci. U.S.A.* **87,** 3821–3825.

Masu, Y., Nakayama, K., Tamaki, N., Harada, Y., Kuno, M., and Nakanishi, S. (1987). cDNA cloning of bovine substance-K receptor through oocyte expression system. *Nature* **329,** 836–838.

Maurel, C., Reizer, J., Schroeder, J. I., and Chrispeels, M. J. (1993). The vacuolar membrane protein γ-TIP creates water specific channels in *Xenopus* oocytes. *EMBO J.* **12,** 2241–2247.

McClure, P. R., Kochian, L. V., Spanswick, R. M., and Shaff, J. E. (1990). Evidence for cotransport of nitrate and protons in maize roots. *Plant Physiol.* **93,** 281–289.

Neher, E. (1992). Correction for liquid junction potentials in patch clamp experiments. *Methods Enzymol.* **207,** 123–131.

Nowycky, M. C., Fox, A. P., and Tsien, R. W. (1985). Three types of neuronal calcium channel with different calcium agonist sensitivity. *Nature* **316,** 440–443.

Parker, I., and Miledi, R. (1988). A calcium-independent chloride current activated by hyper-polarization in *Xenopus* oocytes. *Proc. R. Soc. London B* **233,** 191–199.

Preston, G. M., Carroll, T. P., Guggino, W. B., and Agre, P. (1992). Appearance of water channels in *Xenopus* oocytes expressing red cell CHIP28 protein. *Science* **256,** 385–387.

Rains, D. W., and Epstein, E. (1967). Sodium absorption by barley roots: Its mediation by mechanisms 2 of alkali cation transport. *Plant Physiol.* **42,** 319–323.

Rodriquez-Navarro, A., Blatt, M. R., and Slayman, C. L. (1986). A potassium-proton symport in *Neurospora crassa. J. Gen. Physiol.* **87,** 649–674.

Sakmann, B., Methfessel, C., Mishina, M., Takahashi, T., Takai, T., Kurasaki, M., Fukuda, K., and Numa, S. (1985). Role of acetylcholine receptor subunits in gating of the channel. *Nature* **318,** 532–543.

Sauer, N., Friedländer, K., and Graml-Wicke, U. (1990). Primary structure, genomic organization and heterologous expression of a glucose transporter from *Arabidopsis thaliana. EMBO J.* **9(10),** 3045–3050.

Schachtman, D. P., and Schroeder, J. I. (1994). Structure and transport mechanism of a high-affinity K⁺ uptake transporter from higher plants. *Nature* **370,** 655–658.

Schachtman, D. P., Schroeder, J. I., Lucas, W. J., Anderson, J. A., and Gaber, R. F. (1992). Expression of an inward-rectifying potassium channel by the *Arabidopsis* KAT-1 cDNA. *Science* **258,** 1654–1658.

Schroeder, J. I., Raschke, K., and Neher, E. (1987). Voltage-dependence of K⁺ channels in guard cell protoplasts. *Proc. Natl. Acad. Sci. U.S.A.* **84,** 4108–4112.

Schroeder, J. I., and Fang, H. H. (1991). Inward-rectifying K⁺ channels in guard cells provide a mechanism for low affinity K⁺ uptake. *Proc. Natl. Acad. Sci. U.S.A.* **88,** 11583–11587.

Schroeder, J. I., and Keller, B. U. (1992). Two types of anion channel currents in guard cells with distinct voltage regulation. *Proc. Natl. Acad. Sci. U.S.A.* **89,** 5025–5029.

Schroeder, J. L., Ward, J. M., and Gassmann, W. (1994). Perspectives on the physiology and structure of inward-rectifying K⁺ channels in higher plants: Biophysical implications for K⁺ uptake. *Annu. Rev. Biophys. Biomol. Struct.* **23,** 441–471.

Schroeder, J. I. (1994). Heterologous expression and functional analysis of higher plant transport proteins in *Xenopus* oocytes. *Methods: A companion to Methods in Enzymol.* **6,** 70–81.

Sentenac, H., Bonneaud, N., Minet, M., Lacroute, F., Salmon, J.-M., Gaymard, F., and Grignon, C. (1992). Cloning and expression in yeast of a plant potassium ion transport system. *Science* **256,** 663–665.

Sherman-Gold, R. (ed.) (1993). "The Axon Guide for Electrophysiology and Biophysics." p. 282. Laboratory Techniques, Axon Instruments, Inc.

Slayman, C. L. (1987). The plasma membrane ATPase of *Neurospora* a proton pumping electroenzyme. *J. Bioenerg. Biomembr.* **19,** 1–20.

Soreq, H., and Seidman, S. (1992). *Xenopus* oocyte microinjection: From gene to protein. *Methods Enzymol.* **207,** 225–265.

Stühmer, W. (1992). Electrophysiological recording from *Xenopus* oocytes. *Methods Enzymol.* **207,** 319–339.

Sussman, M. R., and Harper, J. F. (1989). Molecular biology of the plasma membrane of higher plants. *Plant Cell* **1,** 953–960.

Taglialatela, M., Toro, L., and Stefani, E. (1992). Novel voltage clamp to record small, fast currents from ion channels expressed in *Xenopus* oocytes. *Biophys. J.* **61,** 78–82.

Tazawa, M., Shimmen, T., and Mimura, T. (1987). Membrane control in the Characea. *Annu. Rev. Plant Physiol.* **38,** 95–117.

Terry, B. R., Tyermann, S. D., and Findlay, G. P. (1991). Ion channels in the plasma membrane of *Amaranthus* protoplasts: One cation and one anion channel dominate the conductance. *J. Membr. Biol.* **121,** 223–236.

Tsay, Y. F., Schroeder, J. I., Feldmann, K. A., and Crawford, N. M. (1993). The herbicide sensitivity gene CHL1 of *Arabidopsis* encodes a nitrate-inducible nitrate transporter. *Cell* **72,** 705–713.

Wallace, J. C., Galili, G., Kawata, E. E., Cuellar, R. E., Shotwell, M. A., and Larkins, B. A. (1988). Aggregation of lysine-containing zeins into protein bodies in *Xenopus* oocytes. *Science* **240,** 662–664.

INDEX

Extracellular solutions, composition, 524–528
Extraction
  and hydrolysis, cell protein, 43–44
  rapid, proteins, 132–133
  sequential, polysomes, 216
Extracts
  nuclear and cytoplasmic protein, 475–476
  standardization, 430–431

F

Fab, recombinant
  high-diversity library, 87–88
  soluble subcloning, 92–93
Farnesyl, incorporated into proteins, 36–38
Flow cytometry
  chromosome suspensions, 68–72
  with FALS discriminator, 8–9
Flow-karyotyping
  chromosomes in suspension, 78–79
  high-resolution, 74
  isolated chromosomes, 63–64
Flow-sorting
  chromosomes in suspension, 69–72, 79
  grounding of sorting tubes during, 74
  isolated chromosomes, 63–64
Fluorescamine, α-amine-reacting reagent, 42–43
Fluorescence
  chlorophyll, kinetics, 15–29
  induction and decay, 21–23
  as probe for PS II, 24–28
  variable
    caveat, 19
    monitoring donor-side modifications in PS II, 27–29
    for probing PS II acceptor side, 25–27
    and Q_A^-, 24–25
    in stacked thylakoids, 23
Fluorescence emission profiles, altered, 12
Fluorescence microscopy, in cytoskeleton fraction analysis, 230–232
Fluorescence pulse height, and DNA, histogram, 69
Fluorogenic GUS assay, 431, 434
Fluorogenic substrates, access problems, 4
Fluorometer, Waltz, computer interfaced, 19, 22
Formaldehyde, isolated nuclei and chromosomes, 73
$F_1\beta$ precursor, *Neurospora,* import, 277–278
Fractionation, *see also* Subfractionation
  in analysis of nuclear transport, 288
  chloroplasts, 172–173
  mitochondria, 167

nuclear and cytosolic, 291–292
  subcellular, *see* Subcellular fractionation
Fruit, *Capsicum annum,* chromoplast isolation, 198
Frustules, diatom, removal from filtrate, 180–184
Fusicoccin, binding to membranes, 141
Fusion proteins
  GUS, 285–293
  *mal*E–ALS, 95

G

Gel electrophoresis
  band resolution, 233
  in cytoskeleton isolation, 226–227
  SDS–PAGE
    analysis of translation products, 302–305
    isoprenylated proteins, 34–35
Gel filtration, denaturing, recombinant protein, 464–465
Gene fusion, with reporter system based on GUS, 284–285
Gene fusion constructs, toxic, on appropriate vector, 443–444
Genes, *see also* Reporter genes
  chimeric constructs, 393
  chromoplast-specific, 194–195
  endogenous, expression control, 421–423
  gus, 418
  GUS and BASTA, 371–372
  human V, repertoire amplification and assembly, 87
  as internal controls, 126
  MAS, promoter activity, 394–395
  newly isolated, functional analysis, 401–409
  PGK1, 484
  precursor protein, transcription–translation, 256–258
  SLG, fusion with DT-A, 444
  transient expression
    analysis, 390–391, 393–395
    in plant protoplasts, 336–340
Gene systems, inducible/repressible, 411–423
Genetic ablation
  methods, 440–441
  in plants, 442–446
Gene transfer, *see also* Direct gene transfer
  Agrobacterium tumefaciens-mediated, 418
  PEG-mediated, 387–390
Geranylgeranyl, incorporated into proteins, 36–38
Germ, wheat or rice, nuclei isolation, 104–108
GFP, *see* Green-fluorescent protein
GGPP synthase, encoding gene, 194

VOLUMES IN SERIES

Founding Series Editor
DAVID M. PRESCOTT

Volume 1 (1964)
Methods in Cell Physiology
Edited by David M. Prescott

Volume 2 (1966)
Methods in Cell Physiology
Edited by David M. Prescott

Volume 3 (1968)
Methods in Cell Physiology
Edited by David M. Prescott

Volume 4 (1970)
Methods in Cell Physiology
Edited by David M. Prescott

Volume 5 (1972)
Methods in Cell Physiology
Edited by David M. Prescott

Volume 6 (1973)
Methods in Cell Physiology
Edited by David M. Prescott

Volume 7 (1973)
Methods in Cell Biology
Edited by David M. Prescott

Volume 8 (1974)
Methods in Cell Biology
Edited by David M. Prescott

Volume 9 (1975)
Methods in Cell Biology
Edited by David M. Prescott

Volume 10 (1975)
Methods in Cell Biology
Edited by David M. Prescott

Volume 11 (1975)
Yeast Cells
Edited by David M. Prescott

Volume 12 (1975)
Yeast Cells
Edited by David M. Prescott

Volume 13 (1976)
Methods in Cell Biology
Edited by David M. Prescott

Volume 14 (1976)
Methods in Cell Biology
Edited by David M. Prescott

Volume 15 (1977)
Methods in Cell Biology
Edited by David M. Prescott

Volume 16 (1977)
Chromatin and Chromosomal Protein Research I
Edited by Gary Stein, Janet Stein, and Lewis J. Kleinsmith

Volume 17 (1978)
Chromatin and Chromosomal Protein Research II
Edited by Gary Stein, Janet Stein, and Lewis J. Kleinsmith

Volume 18 (1978)
Chromatin and Chromosomal Protein Research III
Edited by Gary Stein, Janet Stein, and Lewis J. Kleinsmith

Volume 19 (1978)
Chromatin and Chromosomal Protein Research IV
Edited by Gary Stein, Janet Stein, and Lewis J. Kleinsmith

Volume 20 (1978)
Methods in Cell Biology
Edited by David M. Prescott

Advisory Board Chairman
KEITH R. PORTER

Volume 21A (1980)
Normal Human Tissue and Cell Culture, Part A: Respiratory, Cardiovascular, and Integumentary Systems
Edited by Curtis C. Harris, Benjamin F. Trump, and Gary D. Stoner

Volume 21B (1980)
Normal Human Tissue and Cell Culture, Part B: Endocrine, Urogenital, and Gastrointestinal Systems
Edited by Curtis C. Harris, Benjamin F. Trump, and Gary D. Stoner

Volume 22 (1981)
Three-Dimensional Ultrastructure in Biology
Edited by James N. Turner

Volume 23 (1981)
Basic Mechanisms of Cellular Secretion
Edited by Arthur R. Hand and Constance Oliver

Volume 24 (1982)
The Cytoskeleton, Part A: Cytoskeletal Proteins, Isolation and Characterization
Edited by Leslie Wilson

Volume 25 (1982)
The Cytoskeleton, Part B: Biological Systems and *in Vitro* Models
Edited by Leslie Wilson

Volume 26 (1982)
Prenatal Diagnosis: Cell Biological Approaches
Edited by Samuel A. Latt and Gretchen J. Darlington

Series Editor
LESLIE WILSON

Volume 27 (1986)
Echinoderm Gametes and Embryos
Edited by Thomas E. Schroeder

Volume 28 (1987)
***Dictyostelium discoideum:* Molecular Approaches to Cell Biology**
Edited by James A. Spudich

Volume 29 (1989)
Fluorescence Microscopy of Living Cells in Culture, Part A: Fluorescent Analogs, Labeling Cells, and Basic Microscopy
Edited by Yu-Li Wang and D. Lansing Taylor

Volume 30 (1989)
Fluorescence Microscopy of Living Cells in Culture, Part B: Quantitative Fluorescence Microscopy—Imaging and Spectroscopy
Edited by D. Lansing Taylor and Yu-Li Wang

Volume 31 (1989)
Vesicular Transport, Part A
Edited by Alan M. Tartakoff

Volume 32 (1989)
Vesicular Transport, Part B
Edited by Alan M. Tartakoff

Volume 33 (1990)
Flow Cytometry
Edited by Zbigniew Darzynkiewicz and Harry A. Crissman

Volume 34 (1991)
Vectorial Transport of Proteins into and across Membranes
Edited by Alan M. Tartakoff

Selected from Volumes 31, 32, and 34 (1991)
Laboratory Methods for Vesicular and Vectorial Transport
Edited by Alan M. Tartakoff

Volume 35 (1991)
Functional Organization of the Nucleus: A Laboratory Guide
Edited by Barbara A. Hamkalo and Sarah C. R. Elgin

Volume 36 (1991)
***Xenopus laevis:* Practical Uses in Cell and Molecular Biology**
Edited by Brian K. Kay and H. Benjamin Peng

Series Editors
LESLIE WILSON AND PAUL MATSUDAIRA

Volume 37 (1993)
Antibodies in Cell Biology
Edited by David J. Asai

Volume 38 (1993)
Cell Biological Applications of Confocal Microscopy
Edited by Brian Matsumoto

Volume 39 (1993)
Motility Assays for Motor Proteins
Edited by Jonathan M. Scholey

Volume 40 (1994)
A Practical Guide to the Study of Calcium in Living Cells
Edited by Richard Nuccitelli

Volume 41 (1994)
Flow Cytometry, Second Edition, Part A
Edited by Zbigniew Darzynkiewicz, J. Paul Robinson,
and Harry A. Crissman

Volume 42 (1994)
Flow Cytometry, Second Edition, Part B
Edited by Zbigniew Darzynkiewicz, J. Paul Robinson,
and Harry A. Crissman

Volume 43 (1994)
Protein Expression in Animal Cells
Edited by Michael G. Roth

Volume 44 (1994)
***Drosophila melanogaster:* Practical Uses in Cell and Molecular Biology**
Edited by Lawrence S. B. Goldstein and Eric A. Fyrberg

Volume 45 (1994)
Microbes as Tools for Cell Biology
Edited by David G. Russell

Volume 46 (1995)
Cell Death
Edited by Lawrence M. Schwartz and Barbara A. Osborne

Volume 47 (1995)
Cilia and Flagella
Edited by William Dentler and George Witman

Volume 48 (1995)
***Caenorhabditis elegans:* Modern Biological Analysis of an Organism**
Edited by Henry F. Epstein and Diane C. Shakes

ISBN 0-12-564152-4

90018

9 780125 641524